Springer Geophysics

The Springer Geophysics series seeks to publish a broad portfolio of scientific books, aiming at researchers, students, and everyone interested in geophysics. The series includes peer-reviewed monographs, edited volumes, textbooks, and conference proceedings. It covers the entire research area including, but not limited to, applied geophysics, computational geophysics, electrical and electromagnetic geophysics, geodesy, geodynamics, geomagnetism, gravity, lithosphere research, paleomagnetism, planetology, tectonophysics, thermal geophysics, and seismology.

More information about this series at http://www.springer.com/series/10173

Danis Nurgaliev · Valery Shcherbakov
Andrei Kosterov · Simo Spassov
Editors

Recent Advances in Rock Magnetism, Environmental Magnetism and Paleomagnetism

International Conference on Geomagnetism, Paleomagnetism and Rock Magnetism (Kazan, Russia)

Springer

Editors
Danis Nurgaliev
Kazan Federal University
Kazan, Russia

Valery Shcherbakov
Borok Geophysical Observatory branch
Institute of Earth Physics of Russian
Academy of Science
Borok, Russia

Andrei Kosterov
Saint Petersburg State University
St. Petersburg, Russia

Simo Spassov
The Geophysical Center of Dourbes
Royal Meteorological Institute of Belgium
Dourbes (Viroinval), Belgium

ISSN 2364-9127 ISSN 2364-9119 (electronic)
Springer Geophysics
ISBN 978-3-030-08014-3 ISBN 978-3-319-90437-5 (eBook)
https://doi.org/10.1007/978-3-319-90437-5

This Springer imprint is published by the registered company Springer Nature Switzerland AG
The registered company address is: Gewerbestrasse 11, 6330 Cham, Switzerland

Contents

Part I Paleomagnetism

1 **Results of Paleomagnetic and Geochronological Studies of Sedimentary Rocks from Kema and Silasa Formations of the Sikhote-Alin Orogen** 3
M. V. Arkhipov, A. Yu. Peskov, A. N. Didenko, S. Otoh, A. V. Kudymov, M. Nagata, Y. Kouchi and K. Yamamoto

2 **Late Paleozoic Remagnetization: Evaluation of the Sequence of Folding in the South Urals** 13
Inessa Golovanova, Konstantin Danukalov and Raushaniya Sal'manova

3 **Paleomagnetic Directions Distortion Caused by Viscous-Plastic Deformations Estimated from Anisotropy of Magnetic Susceptibility (Case Study of Berriasian Clays from East Crimea)** 25
V. A. Grishchenko and A. Yu. Guzhikov

4 **Carboniferous of the Russian Platform: Paleomagnetic Data** 37
A. G. Iosifidi, V. A. Mikhailova, V. V. Popov, E. S. Sergienko, A. V. Danilova, N. M. Otmas and A. V. Zhuravlev

5 **Evidence for the Existence of the Gothenburg and Mono Lake Excursions Based on Paleomagnetic Data from Baunt Lake Sediments (Northern Transbaikalia)** 55
M. A. Krainov, E. V. Bezrukova, A. A. Shchetnikov and A. Yu. Peskov

6 **Intrusions of the Kulumbe River Valley, NW Siberian Traps Province: Paleomagnetism, Magnetic Fabric and Geochemistry** . . . 67
A. V. Latyshev, N. A. Krivolutskaya, P. S. Ulyahina, Ya. V. Bychkova and B. I. Gongalsky

7 Paleomagnetism of Basalts from the Mid-Atlantic Ridge and the Bouvet Triple Junction 83
V. I. Maksimochkin and A. N. Tselebrovskiy

8 Archaeomagnetic Studies of the Material of the Archaeological Monument Dmitrievskaya Sloboda II of the Second Millennium B.C 97
O. V. Pilipenko, I. E. Nachasova, S. K. Gribov and O. V. Zelentsova

Part II Rock and Environmental Magnetism

9 Influence of Magnetostatic Interaction on Magnetic Characteristics of Decay Products of Nanodisperse Titanomagnetites 111
S. V. Anisimov, L. L. Afremov and I. G. Iliushin

10 An Estimate of the Remanent Magnetization in the Case of a Ferromagnet Transformation Accompanied by a Change in the Curie Temperature 123
V. I. Belokon, E. V. Chibiriak and O. I. Dyachenko

11 A 13,000-Yr Record of Environmental Change from Tschuchye Lake in Northeast Yakutia 133
S. S. Burnatny, A. N. Naumov and Yu. A. Korzun

12 Comparison of the Porosity Determination on the Whole Core and Petrophysical Samples 151
A. V. Fattakhov, V. E. Kosarev, D. L. Melnikova, V. D. Skirda and A. V. Starovoytov

13 Magnetic Properties of Soils from the Volga-Kama Forest-Steppe 161
L. A. Fattakhova, L. R. Kosareva and A. A. Shinkarev

14 Magnetic Properties of Artificial CRM Created on Titanomagnetite-Bearing Oceanic Basalts 173
S. K. Gribov, V. P. Shcherbakov and N. A. Aphinogenova

15 Blocking Temperature and Hysteresis Characteristics of Nanoparticles of Oxidated Magnetite 195
Ilia Iliushin and Leonid Afremov

16 Low-Temperature Magnetic Properties and Magnetic Mineralogy of the Ropruchey Sill (Russian Karelia) 205
Andrei Kosterov, Elena S. Sergienko, Petr V. Kharitonskii, Svetlana Yu. Yanson and Irina A. Vasilieva

17 **Preliminary Magnetic Investigation of Samples from Reference Permian-Triassic Sequence, Kemerovo Region, Russia** 225
Dilyara Kuzina, Lina Kosareva, Ilmir Gilmetdinov, Radmir Aupov, Vladimir Silantiev, Vladimir Davydov, Irina Dogadina and Natalia Kuzmina

18 **Magnetic Properties and Concentration of Heavy Metals in Soils of the Krasnyi Kut Town (Saratov, Russia)** 235
M. V. Reshetnikov, A. S. Sheshnev, E. S. Sokolov and S. D. Shkodin

19 **Rock-Magnetism and Granulometry of Late Palaeolithic Sites in the Tunka Rift Valley (Pribaikalie) as a Tool for Reconstruction of Sedimentary Environment** 245
G. G. Matasova, A. A. Shchetnikov, I. A. Filinov, A. Yu. Kazansky, G. A. Vorobyeva, N. E. Berdnikova, E. O. Rogovskoy, E. A. Lipnina, I. M. Berdnikov and L. V. Lbova

20 **Application of Methods of Rocks Magnetism for Determination of Criteria of Perspectiveness of Ore-Formation** 265
J. H. Minasyan

21 **Rock Magnetic Properties of Pleistocene Tephras from the Polovinka Section of the Central Kamchatka Depression** 273
A. N. Naumov, S. S. Burnatny, P. S. Minyuk and A. G. Zubov

22 **Basalts of the Zhom-Bolok Lava River as a Possible Sources of Metallic Iron in Sediments of Local Lakes: Thermomagnetic and Microprobe Justification** 297
D. M. Pechersky, A. Yu. Kazansky, A. Kozlovsky, G. P. Markov, A. A. Shchetnikov and V. A. Tselmovich

Part III Magnetostratigraphy

23 **Magnetostratigraphy of the Reference Sections of the Cisuralian Series (Permian System)** 317
Yu. P. Balabanov, R. Kh. Sungatullin, G. M. Sungatullina, L. R. Kosareva, M. S. Glukhov, P. G. Yakunina, A. O. Zhernenkov, V. V. Antonenko and A. A. Churbanov

24 **General Magnetostratigraphic Scale: Present Status and Outlook of Development** 343
A. Yu. Guzhikov

25 **Existence of the Reversal Polarity Zones in Turonian-Coniacian from the Lower Volga (Russia): New Data** 353
A. A. Guzhikova, A. Yu. Guzhikov, E. M. Pervushov, I. P. Ryabov and A. M. Surinskiy

26 Magnetostratigraphy of the Key Loess-Palaesol Sequence at Roxolany (Western Black Sea Region) ... 371
D. V. Hlavatskyi and V. G. Bakhmutov

27 Magnetic Polarity Stratigraphy of the Upper Cenozoic Deposits of Near-Sea Dagestan (Russia) and the Age of the Early Paleolithic Site Rubas-1 ... 383
A. Yu. Kazansky, A. A. Anoikin, A. P. Derevianko, G. G. Matasova and V. Yu. Bragin

28 Magnetic Stratigraphy of the Bazhenov Suite of Western Siberia and the Surrounding Deposits ... 411
A. G. Manikin, V. A. Grishchenko, A. Yu. Guzhikov and V. V. Kolpakov

Part IV Extraterrestrial Magnetism

29 Metallic Iron in the Planets Interior: Generalization of Thermomagnetic and Microprobe Results ... 427
D. M. Kuzina and D. M. Pechersky

30 Magnetic Properties of Tektite-like Impact Glasses from Zhamanshin Astrobleme, Kazakhstan ... 445
Vyacheslav A. Starunov, Andrei Kosterov, Elena S. Sergienko, Svetlana Yu. Yanson, Gennady P. Markov, Petr V. Kharitonskii, Aleksandr S. Sakhatskii, Irina E. Lezova and Evgenii V. Shevchenko

31 Background Magnetic Component of Holocene Cosmic Dust Allocated from Peat ... 467
V. A. Tselmovich and A. Yu. Kurazhkovskii

Part V Geomagnetic Field and Magnetic Surveys

32 Preliminary Results of the Geohistorical and Paleomagnetic Analysis of Marine Magnetic Anomalies in the Northwestern Indian Ocean ... 479
S. A. Ivanov and S. A. Merkuriev

33 Geomagnetic Variations for Axial Dipole Hypothesis and Dynamo Parameters ... 491
S. V. Starchenko

34 Simple Estimations for Planetary Convection Turbulence and Dynamo Magnetism from Optimized Scaling and Observations ... 501
S. V. Starchenko

35 MHD Sources, 1600-2005 Evolution and 1900-2005 Probabilistic Time Analysis for Logarithmic Time-Derivatives of Geomagnetic Spherical Harmonics 513
S. V. Starchenko and S. V. Yakovleva

36 Unmanned Airborne Magnetic Survey Technologies: Present and Future 523
Tsirel Vadim, Parshin Alexander, Ancev Vasily and Kapshtan Dmitry

Part I
Paleomagnetism

Chapter 1
Results of Paleomagnetic and Geochronological Studies of Sedimentary Rocks from Kema and Silasa Formations of the Sikhote-Alin Orogen

M. V. Arkhipov, A. Yu. Peskov, A. N. Didenko, S. Otoh, A. V. Kudymov, M. Nagata, Y. Kouchi and K. Yamamoto

Abstract The paper presents the results of paleomagnetic and geochronological studies on the Lower Cretaceous sandstones of the Silasa and Kema Formations (Kiselevka-Manoma and Kema terranes, respectively). Source areas of rocks are determined and paleolatitudes at which they formed are constrained. The fact that ages of zircons from sandstones of the studied formations differ essentially suggests that the material from a source area in continental Eurasia was not transported to the deposition area of the Silasa Formation. The reason is that sediments of the Silasa Formation accumulated in the outer island arc zone separated by the fore-arc basin from the inner island arc zone where the material from the continent was transported to the deposition area of the Kema Formation.

Keywords Geochronology · Paleomagnetism · Sedimentary rocks
Sikhote-Alin

M. V. Arkhipov (✉) · A. Yu. Peskov · A. N. Didenko · A. V. Kudymov
Kosygin Institute of Tectonics and Geophysics, Far Eastern Branch,
Russian Academy of Science (ITiG FEB RAS), Khabarovsk, Russia
e-mail: arkhipov_mv@mail.ru

S. Otoh · M. Nagata · Y. Kouchi
Graduate School of Science and Engineering, University of Toyama,
Toyama, Japan

A. N. Didenko
Pacific National University, Khabarovsk, Russia

K. Yamamoto
Graduate School of Environmental Studies, Nagoya University,
Nagoya, Japan

D. Nurgaliev et al. (eds.), *Recent Advances in Rock Magnetism, Environmental Magnetism and Paleomagnetism*, Springer Geophysics,
https://doi.org/10.1007/978-3-319-90437-5_1

Introduction

The Sikhote-Alin orogenic belt is one of the most important structures of the Eurasia eastern margin, which is composed of terranes differing in age and formation conditions (Fig. 1.1) (Golozubov 2006; Khanchuk and Kemkin 2015). Many researchers addressed the issue of its evolution, but the problem has not yet been fully resolved. The task of reconstructing the sequence of events related to the tectonic incorporation of terranes into the Sikhote-Alin orogen, the kinematics and the paleolatitude of their formation is a challenge calling for the use of a range of methods. Towards this end, we present the results of geochronological and paleomagnetic studies on Cretaceous sedimentary rocks from the Silasa Formation of the Kiselevka-Manoma terrane and from the Kema Formation of the Kema terrane.

The Kiselevka-Manoma terrane located in the Lower Amur region can be discontinuously traced in the NW direction over a distance of 700 km along the Amur River and is 20–40 km wide. The terrane is made up of packages of tectonic slices mainly composed of Early Cretaceous siltstones and turbidites, as well as Jurassic and Early Cretaceous siliceous and siliceous-clayey rocks with bodies of basalt and limestone (Malinovsky 2010).

The Kema terrane, which is located in the eastern part of the Sikhote-Alin orogen and extends along the coast of the Sea of Japan, is 80 km wide and 850 km long. The terrane consists of horizons of mixtites and siltstones, turbidites and layers of basic volcanics and their pyroclasts. They are interpreted as deposits of the back-arc basin of the Early Cretaceous Moneron-Samarga island-arc system (Malinovsky 2010).

Rock samples of the greywacke suite of the Lower Silasa Subformation of the Silasa Formation collected from outcrops along the shore of Lake Udyl' constituted the first target of our study (Cape Skalnyi). Deposits of the Kiselevka-Manoma terrane under consideration are assigned to the Silasa Formation as a result of the revision mapping survey carried out by Anoikin and Kaidalov (2000). The greywacke suites represented by Albian-Lower Cenomanian sandstone and siltstone that are products of destruction of epioceanic island-arc system fragments. Rocks of the suite belong to the Kiselevka-Manoma terrane of the Sikhote-Alin orogen which is a fragment of the middle Cretaceous prism.

Another focus of research was sedimentary rocks of the Upper Kema Subformation of the Kema Formation in the Kema terrane which is a fragment of the back-arc basin of a Hauterivian-Albian island-arc system. The Kema Formation contains Late Albian beds of siltstone and sandstone rhythmically alternating with distinctive beds of subaqueous slump deposits, sandstone and siltstone. In the lower and middle parts of the subformation, organic remains of ammonites and aucelline bivalves were found indicating an early-middle Albian age of host deposits (Golozubov 2006; Markevich et al. 2000). In addition, the palynological content of siltstone from the upper part of the section suggests that host deposits are of late Albian age (Markevich et al. 2000).

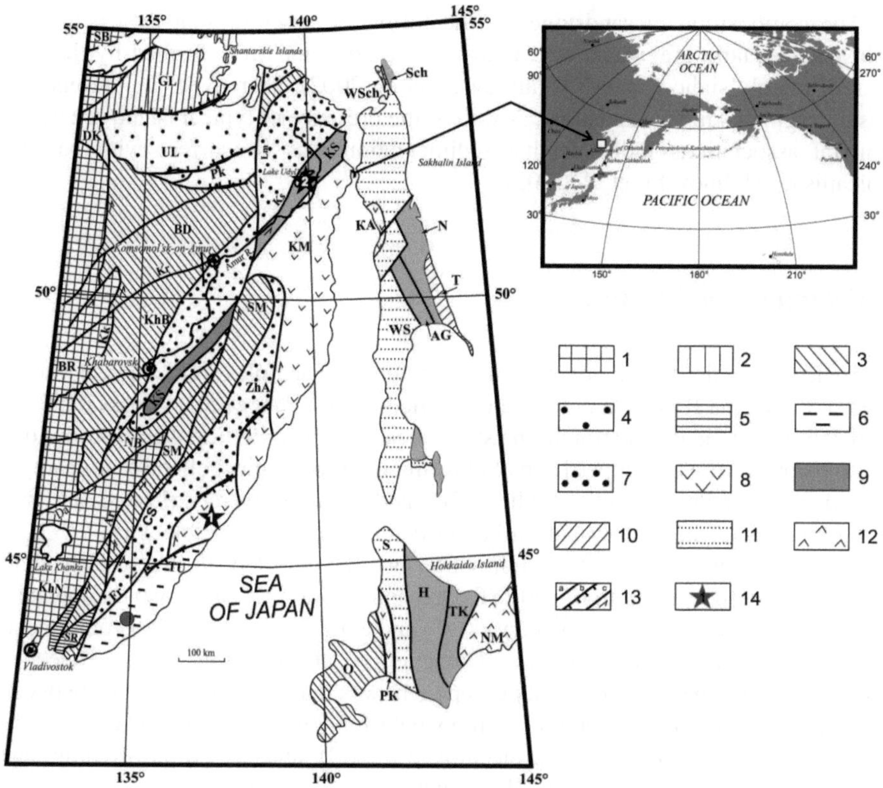

Fig. 1.1 Schematic map of terranes of the Sikhote-Alin-Northern Sakhalin orogenic belt according to Khanchuk and Kemkin (2015) with additions. 1—Precambrian and Early Paleozoic cratons and superterranes (BR, Bureya; SB, Siberian; KhN, Khankai); 2—Paleozoic terranes (GL, Galam; DzhK, Dzhagda-Kerbi); 3, 4—Jurassic terranes: 3—fragments of accretionary prisms (BD, Badzhal; NB, Nadanhada-Bikin; O, Oshima; SM, Samarka; KhB, Khabarovsk), 4—near-continental turbidite basin (UL, Ulban); 5—Precambrian-Early Paleozoic fragments in the structure of the Jurassic accretionary prism (SR, Sergeevsky complex of the Samarka terrane); 6–9—Cretaceous, essentially Early Cretaceous, terranes: 6—Neocomian accretionary prism (TU, Taukha), 7—near-continental pull-apart turbidite basin (ZhR, Zhuravlevsk-Amur), 8—Barremian-Albian island-arc system (WSch, western Schmidt; KA, Kamyshovyi; KM, Kema; RK, Rebun-Kobato), 9—Albian Accretionary prism (AG, Aniva-Gomon; KS, Kuselevka-Manoma; N, Nabilsky; TK, Tokoro; H, Hidaka; Sch, Schmidt); 10–11—Early-Late Cretaceous: 10—island-arc (T, Terpeniya); 11—turbidite (WS, Western Sakhalin; S, Sorachi-Yezo); 12—Late Cretaceous-Paleogene flysch-island-arc (Nm, Nemuro); 13—main faults (a—with an unknown strike-slip component, b—thrusts, c—strike-slip faults including: Ar, Arsen'evsky; Da, Dunmi-Alchansky; Kk, Kukansky, Kr, Kursky; Ks, Kiselevsky; Lm, Limurchansky; Pk, Paukansky; Fr, Furmanovsky; CS, Central Sikhote-Alinsky; 14—regions of detailed paleomagnetic studies (1—Kema Formation of the Kema terrane, 2—Silasa Formation of the Kiselevka-Manoma terrane)

The composition of sandstones of the formation suggests that they belong to feldspathic-quartz greywacke and feldspathic arkose and were formed by the erosion of mainly sialic material (Malinovsky et al. 2002). Moreover, the presence of basalt fragments in the sandstone, as well as the presence of typical island-arc heavy mineral associations indicates that sedimentation was concurrent with basaltic volcanism (Malinovsky et al. 2002).

Paleomagnetic Studies

For paleomagnetic studies of the Kema Formation, 47 samples were collected in the area of the Kema River. Samples were oriented using a magnetic compass (45.7°N, 136.7°E; magnetic declination in the study area was 11°). Particle size distributions of sandstones from the Upper Kema Subformation proved most effective for sample selection and paleomagnetic studies. Oriented samples were collected at three different sites of the section with different bedding attitudes (Table 1.1).

Specimens from the Kema Formation were subjected to alternating field (AF) demagnetization up to 65–90 mT. Demagnetization revealed two components of magnetization, i.e., a low-coercivity (LC) and a high-coercivity (HC) component. The low-coercivity component was demagnetized at alternating fields between 0 and 30 mT, and the high-coercivity component was demagnetized at fields between 30 and 90 mT. The high-coercivity component of magnetization is directed towards the origin of the orthogonal projection (Fig. 1.2) and has a pre-folding age (Table 1.1). Examples of thermal demagnetization of specimens are in (Fig. 1.2).

Two components of magnetization were isolated for sample K15/6-4 (Fig. 1.2a). The low-coercivity component direction is: Dec_s = 321.2°, Inc_s = 66.7°, a_{95} = 5.0°. This component is isolated in the range from 0 to 25 mT. The high-coercivity component of this sample is isolated in the range 25–80 mT and is directed towards the origin of coordinates (Dec_s = 350.8°, Inc_s = 78.6°, a_{95} = 5.0°).

Figure 1.2b illustrates demagnetization of sample K15/7-2. As in the previous case, two components of magnetization were isolated for this sample. The high-coercivity component of this sample is isolated in the range of 30–65 mT and is directed towards the origin of coordinates.

Figure 1.3a–c show projections of the high-coercivity magnetization component directions in three sites. The site-mean directions in geographic and stratigraphic coordinates are in Fig. 1.3d, e. Analysis of distributions of the site-mean directions in stratigraphic and geographic coordinates (Table 1.1) shows a significant increase in grouping in the stratigraphic system: Ks/Kg = 3.3. The stepwise application of the fold test (Fig. 1.3f) clearly demonstrates dominance of a "pre-fold" component in the NRM of rocks. The best grouping of the high-coercivity component (Kmax) is reached at 117% unfolding. In the course of the study, coordinates of the paleomagnetic pole for Kema Formation (Plat = 77.1°, Plong = 354.5°, dp = 17.9°, dm = 12.6°) and the paleolatitude at which Kema rocks formed (35°N) were calculated.

Table 1.1 Results of paleomagnetic studies

Sites	φ (°)	λ (°)	Rock type	N_1	N_2	Dip azim. (°)	Dip angle (°)	Dec_g (°)	Inc_g (°)	K_g	a_{95} (°)	Dec_s (°)	Inc_s (°)	K_s	a_{95} (°)
K15/5	45.7	136.7	Sandstone	30	15	304	50	81.5	57.8	32.6	6.8	346.3	57.7	32.6	6.8
K15/6	45.7	136.7	Sandstone	9	7	310	35	55.5	58.2	17.6	14.8	3.7	51.0	17.6	14.8
K15/7	45.7	136.7	Sandstone	8	7	305	60	81.4	36.8	11.2	18.9	12.4	53.3	11.2	18.9
Site-mean K15/5, K15/6, K15/7								74.2	51.6	28.8	23.4	1.4 (**350.4**)	54.5	95.1	12.7
Paleomagnetic pole coordinates								Plat = 77.1°; Plong = 354.4°; d_p = 17.9; d_m = 12.6							

Note φ, λ—geographical latitude and longitude of sampling site, respectively; N_1—number of collected samples; N_2—number of samples used in calculating the mean; Dec, Inc—paleomagnetic declination and inclination, respectively, in geographic (g) and stratigraphic (s) coordinates; K—grouping; a_{95}—radius of the confidence ellipse around the mean; Plat, Plong—latitude and longitude of the paleomagnetic pole, respectively; d_p; d_m—semi-axes of the confidence ellipse around the pole. The site mean direction in stratigraphic coordinates is given in parentheses considering the magnetic declination at the sampling site

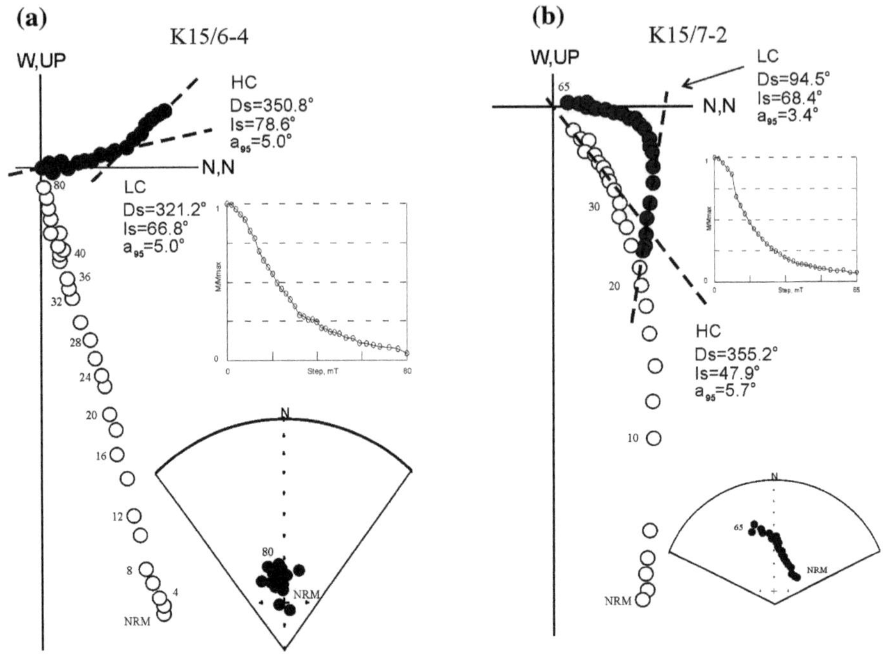

Fig. 1.2 Zijderveld plot for samples K15/6-4 (**a**) and K15/7-2 (**b**) of Kema Formation

Previously, paleomagnetic studies were carried out on over 200 samples from the Silasa Formation collected near Lake Udyl' (52.2°N, 140.1°E) (Didenko et al. 2017). Coordinates of the paleomagnetic pole (Plat = 26.3°, Plong = 70.5°, dp = 4.8°, dm = 7.0°) and the paleolatitude at which Silasa rocks formed (32.8°N) were calculated in the course of paleomagnetic studies.

Geochronological Studies

The sample preparation and isolation of zircons from rock chips was carried out at the Graduate School of Science and Engineering of Toyama University and U-Pb dating of zircons (LA-ICP-MS) at the Graduate School of Applied Research of Nagoya University using the mass spectrometer with inductively coupled plasma Agilent 7700x following the procedure developed in (Kouchi 2015). Measured $^{206}Pb/^{238}U$ and $^{207}Pb/^{235}U$ ratios for each grain (http://itig.as.khb.ru/sved_obraz/pdf/geochron_data.pdf) were recorded with a measurement error at $\pm 2\sigma$ level. The final age of concordant detrital zircons was calculated based on the $^{206}Pb/^{238}U$ ratio; measured zircon age distribution and age probability density distribution were calculated using the Isoplot 3.7 toolkit (Ludwig 2012).

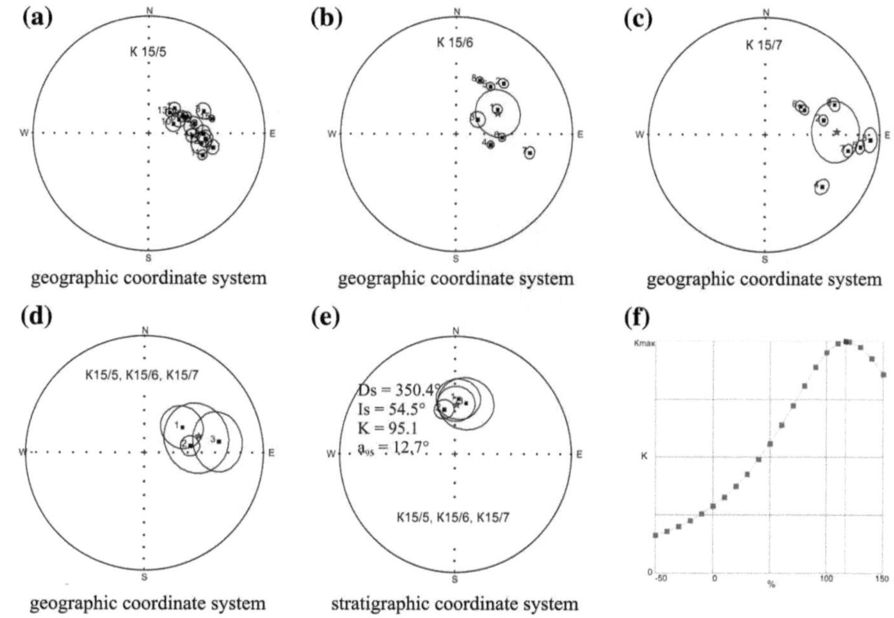

Fig. 1.3 Stereograms of distribution of ancient magnetization component for samples of Kema Formation of sites K15/5 (**a**), K15/6 (**b**), K15/7 (**c**); The averaged value of the distribution of ancient components for these sites in the geographical (**d**) and stratigraphic (**e**) coordinate systems. **f** Fold test

For the purposes of geochronological studies a rock chip sample (10 kg) of silty sandstone from the Lower Silasa Subformation (A-15/5) was collected near the Cape Skalnyi (Lake Udyl' east shore; 52.13°N, 140.01°E) and a rock chip sample (10 kg) of sandstone from the Upper Kema Subformation, the most recent part of the section, was collected from the Kholmogorka river-bank outcrop in the Kema River basin. Geochronological studies on detrital zircons from the sandstone of these formations revealed several similarities. Both have a group of "young" Albian zircons (Fig. 1.4) whose age correlates well with the biostratigraphic age of these formations (Kaidalov et al. 2011; Malinovsky et al. 2002; Markevich et al. 1997). Another similarity recognized at the 168–320 Ma interval is the presence of four zircon groups: 168–178, 196–247, 261–286, 304–320 Ma (Fig. 1.4). Conceivably these age intervals might correspond to the initiation of the volcanic island arc formation in the Middle Permian and to the following episodes of intensification of magmatic processes, respectively.

Significant differences in the distribution of concordant ages are as follows: (1) in the sandstone of the Kema Formation there are zircons with a wide age range from the early Proterozoic (2.55 Ga) to the end of the Early Cretaceous (107.5 ± 2.3 Ma), whereas the oldest zircon from the sandstone of the Silasa Formation is Devonian in age (382 Ma); (2) the source area of Early Paleozoic and

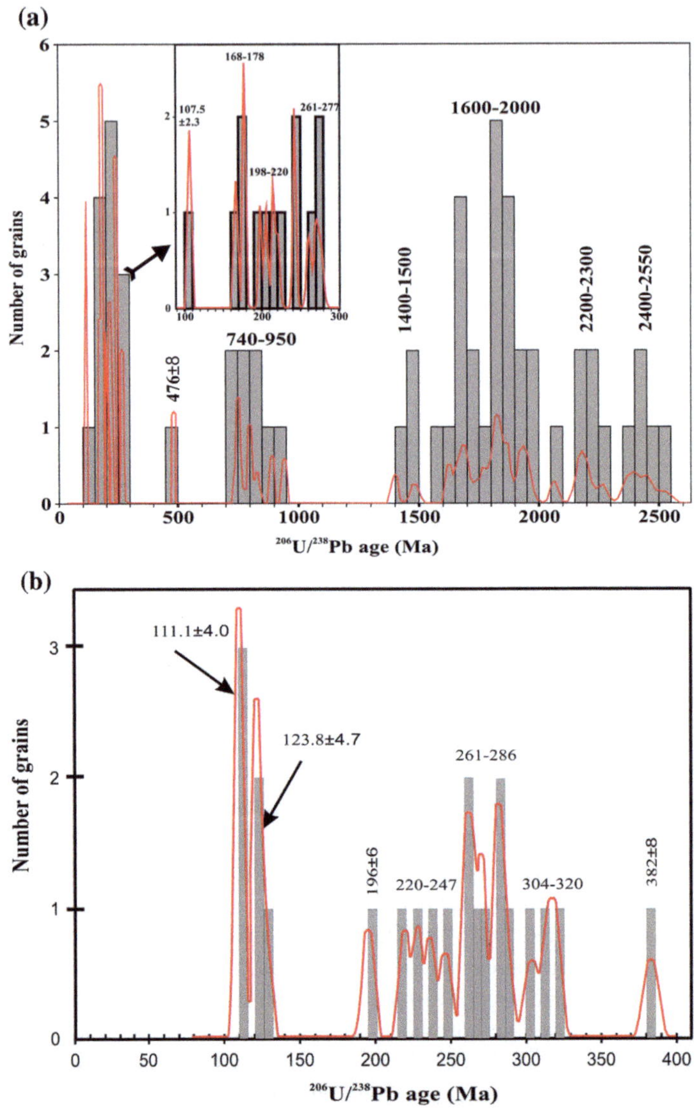

Fig. 1.4 Diagram of U-Pb dating of detrital zircons from sandstones of Kema (**a**) and Silasa (**b**) formations

Precambrian zircons in the sandstone of the Kema Formation was in continental Eurasia, whereas any kind of loose rock material of continental origin is entirely absent in sediments of the Silasa Formation. The source of the Early Paleozoic zircon (476 ± 8 Ma) was most probably collision granites of the Khankai (Bureya-Khankai) terrane (Fig. 1.1) aged between 490 and 460 Ma (Sorokin et al. 2011). Precambrian zircons form three well-defined age groups: 0.74–0.79, 1.5–2 and 2–2.2 Ga. The age of these zircon groups correlates well with the timing of supercontinent cycles identified in the known global age distribution of detrital zircons (Hawkesworth et al. 2010). The absence of continental zircons in sediments of the Silasa Formation, which accumulated in the outer island arc setting, is attributed to the width and depth of the fore-arc basin, which separated the outer island arc from the inner island arc where sediments of the Kema Formation accumulated.

Conclusions

Based on the paleomagnetic and geochronological data on sedimentary rocks of the Kema and Kiselevka-Manoma terranes of the Sikhote-Alin orogenic belt the following may be concluded. Rocks of the Kema and Silasa Formations formed at approximately the same paleolatitude (32–35°N). The studied rocks are close in age and assigned to the Albian stage, however, provenance areas of the rocks from the Silasa and Kema Formations differ considerably. In the rocks of the Kema Formation, there is evidence for rock particles which were transported from the continent and yielded zircon ages varying from 107 Ma to 2.55 Ga. No continental material is found in the rocks of the Silasa Formation, the oldest zircon having the Devonian age (382 Ma).

Acknowledgements The study was supported by the Russian Foundation for Basic Research (project no. 18-05-00117) and as part of the State Assignment to the ITiG FEB RAS.

References

Didenko, A.N., Peskov, A.Yu., Kudymov, A.V., Voinova, I.P., Tikhomirova,A.I., Arkhipov M.V. Paleomagnetism and accretionary tectonics of the northern Sikhote-Alin // *Izvestiya, Physics of the Solid Earth*. 2017. V. 53. N. 5. P. 733–749.

Golozubov, V.V. Tectonics of the Jurassic and Lower Cretaceous Complexes of the Northwestern Pacific Framing. Vladivostok: Dal'nauka. 2006. 239 pp. (in Russian).

Hawkesworth C., Dhuime B., Pietranik A., Cawood P., Kemp T., Storey C. The generation and evolution of the continental crust: Journal of the Geological Society. 2010. V. 167. P. 229–248.

Kaidalov, V.A., Anoikin, V.I., Belomestnova T.D. State Geological Map of the Russian Federation. Scale: 1:200 000. Second Edition. Nikolayevsk Series. Sheet N-54-XXXII. St. Petersburg: VSEGEI, 2011.

Khanchuk A.I., Kemkin I.V. Jurassic geodynamic history of the Sikhote-Alin-Priamurye region / In Anderson, T.H., Didenko, A.N., Johnson, C.L., Khanchuk, A.I., and MacDonald, J.H., Jr., eds., Late Jurassic Margin of Laurasia–A Record of Faulting Accommodating Plate Rotation: Geological Society of America Special Paper 513. 2015. P. 509–526.

Kouchi, Y., Y. Orihashi, H. Obara, T. Fujimoto, Y. Haruta and K. Yamamoto. 2015. Zircon U-Pb dating by 213 nm Nd: YAG laser ablation inductively coupled plasma mass spectrometry: Optimization of the analytical condition to use NIST SRM 610 for Pb/U fractionation correction. Chikyukagaku (Geochemistry) 49: 19–35. (in Japanese with English abstract).

Ludwig, K. R. 2012. User's manual for Isoplot 3.75: A Geochronological Toolkit for Microsoft Excel. Berkeley Geochronology Centre Special Publication 5, 75 pp.

Malinovsky, A. I. Lithological composition of island-arc complexes of the Russian Far East // *Lithology and Mineral Resources.* 2010. V. 45. N 1. P. 24–38.

Malinovsky, A.I., Filippov, A. N., Golozubov, V.V., Simanenko, V.P., Markevich, V.S. Lower Mesozoic deposits of the Kema River basin, Eastern Sikhote-Alin: sedimentary filling of the back-arc basin // *Tikhookeanskaya geologiya.* 2002. V. 21. N 1.P. 52–66. (in Russian with English abstract).

Markevich, P.V., Filippov, A.N., Malynovskyi, A.I., Zyabrev, S.V., Nechaev, V.P., Vysotsky, S.V. Cretaceous volcanogenic-sedimentary rocks of Lower Priamurye (structure, composition and sedimentation environment). Vladivostok: Dal'nedra. 1997. 300 pp. (in Russian).

Markevich, P.V., Konovalov, V.P., Malinovsky, A.I., etc. Lower Cretaceous Deposits of the Sikhote-Alin. Vladivostok: Dal'nedra. 2000. 283 pp. (in Russian).

Sorokin, A.A., Kotov A.B., Salnikova, E. B., Sorokin, A.P., Yakovleva, S.Z., Plotkina, Yu.V., Gorokhovskii, B.M. Early Paleozoic age of granitoids of the Kiviliisky complex of the Bureya terrane (Eastern of flank the Central Asian fold belt) // *Doklady Earth Sciences.* 2011. 440. No. 1. P. 1253–1257.

State Geological Map of the Russian Federation. Scale: 1:200000. Second Edition. Nikolayevsk Series. Sheet N-54-XXXII: Explanatory Notes / Compiled by V.A. Kaidalov and V.I. Anoikin / Ed. by G.V. Roganov. St. Petersburg: VSEGEI, 2011.

Chapter 2
Late Paleozoic Remagnetization: Evaluation of the Sequence of Folding in the South Urals

Inessa Golovanova, Konstantin Danukalov and Raushaniya Sal'manova

Abstract Analysis of secondary magnetization components (Late Paleozoic remagnetization) makes it possible to trace the formation of the South Ural folded system. This paper presents new data on Late Paleozoic remagnetization in the rocks of the eastern Bashkir Meganticlinorium. Combined analysis of the newly obtained results and previously published data on Late Paleozoic remagnetization along the western slope of the South Urals shows that in the western segment the intermediate-temperature magnetization component (ITC) was acquired prior to or at the initial stages of deformation. In contrast, the ITC observed in the eastern segment postdates folding. Collision processes which occurred in the South Urals from the Late Devonian until the Late Permian time had undoubtedly left their trace in the western structures (in present-day coordinates). Within the passive margin of Baltica the collision resulted in the formation of a lateral series of fold-and-thrust structures. They started from the Main Uralian Fault westward, with possible gradual termination of this process towards the Pre-Uralian Foredeep. Paleomagnetic results of this research work support these assumptions.

Keywords Paleomagnetism · South Urals · Late Paleozoic · Remagnetization

Paleomagnetic data can help in studying the history of mountain formation, including the determination of the nature and scales of deformations at its final stages. Paleomagnetic research in the South Urals shows that a Late Paleozoic remagnetization component occurs almost in all studied sections of both Late Paleozoic and Precambrian rocks (Danukalov et al. 1983; Khramov et al. 1982; Svyazhina et al. 2003; Stratotip… 1982; Levashova et al. 2013). Only a few papers are known to summarize and analyze the data on secondary rock magnetization for

I. Golovanova (✉) · K. Danukalov · R. Sal'manova
Institute of Geology, Ufa Scientific Centre, Russian Academy of Sciences, Ufa, Russia
e-mail: golovanova@ufaras.ru

R. Sal'manova
e-mail: danukalov@mail.ru

D. Nurgaliev et al. (eds.), *Recent Advances in Rock Magnetism, Environmental Magnetism and Paleomagnetism*, Springer Geophysics,
https://doi.org/10.1007/978-3-319-90437-5_2

some individual sites (Levashova et al. 2013; Golovanova et al. 2017), however, no systematic analysis was performed for the whole area. The authors know only the work of Shipunov (1996), who analyzed the sequence of folding in this region using the example of the Katav formation, considering its magnetization to be metachronous. However, the age of the natural remanent magnetization of the rocks of the Katav formation, whose direction is close to the directions of the Late Paleozoic age, is questioned (Pavlov and Gallet 2009). We think that more thorough analysis of remagnetization components will enable us to trace the origin of folded structures at the final stages of the formation of the South Urals. This paper presents new results of paleomagnetic investigations in the South Urals and combined analysis of the newly obtained results and previously published data on Late Paleozoic remagnetization components. An attempt was made to use Late Paleozoic remagnetization in studying the timing and succession of deformations occurred within the passive continental margin of the East European Platform (Baltica paleocontinent) and the western Magnirogorsk Zone during the formation of the Ural orogen.

Very generally, the Urals are the result of collision between the Precambrian East European and Paleozoic Kazakhstan continents and are distinctly divided into two parts separated by the Main Uralian Fault (Puchkov 2003). The western Uralian margin of the East European continent began to form during the Late Cambrian or Early Ordovician, remained passive until the Late Devonian and underwent deformations in the Carboniferous and Permian periods associated with the formation of the Uralian fold and thrust belt. The eastern Urals represent an intricate combination of oceanic, island-arc and microcontinental blocks formed since the Late Devonian. It was shown earlier that as the western tectonic units of the Urals are a part of the cratonic Baltica margin, deformed in Permian time (Puchkov 2003; Levashova et al. 2013, 2015) we can extrapolate the paleomagnetic data from the western Urals to Baltica.

We examined Upper Proterozoic and Paleozoic sections of different rock types along the western part of the Zilair Synclynorium in the latitudinal and upper meridional course of the Belaya River (Fig. 2.1). Consideration was given, where possible, to tectonically undisturbed sections that form complete sequences and are comprehensively characterized by different fauna groups. The Apshak and Krivaya Luka sections (Table 2.1) were studied including Upper Riphean and Vendian sandstones, aleurolites and argillites and also the diabase dike. Ordovician sedimentary rocks were studied in the sections located near the villages of Nabievo, Kurgashly, Mindigulovo and Kiekbaevo. The rocks were sampled near Nabievo (thin-layered dolomitized quartz sandstones), Kurgashly (dolomitized sandstones), Mindigulovo (dolomites and argillites) and Kiekbaevo (dolomites and dolomitized sandstones). Silurian sedimentary rocks were studied in the sections near Mindigulovo (dolomites and argillites) and Kiekbaevo (silicified dolomites). Devonian sedimentary rocks were studied in the Kaga and Mindigulovo sections. In the Belaya River valley near the village of Kaga we collected grey limestones of the Irgizly Formation, quartz sandstones of the Vanyashkino Formation, clay interlayered black bituminous limestones of the Vyazovskaya Formation and limestones of the Koyva Formation. However, the sandstones of the Vanyashkino Formation turned out to be unsuitable for laboratory testing, i.e. they did not provide any paleomagnetic results. A small group of Lower Devonian dolomite samples were collected in the Mindigulovo section.

Paleomagnetic samples were collected as hand-sized blocks oriented with a magnetic compass, one sample being taken at a stratigraphic level. Eight to twelve or even more samples collected from 3 to 10, sometimes up to 30 m thick section

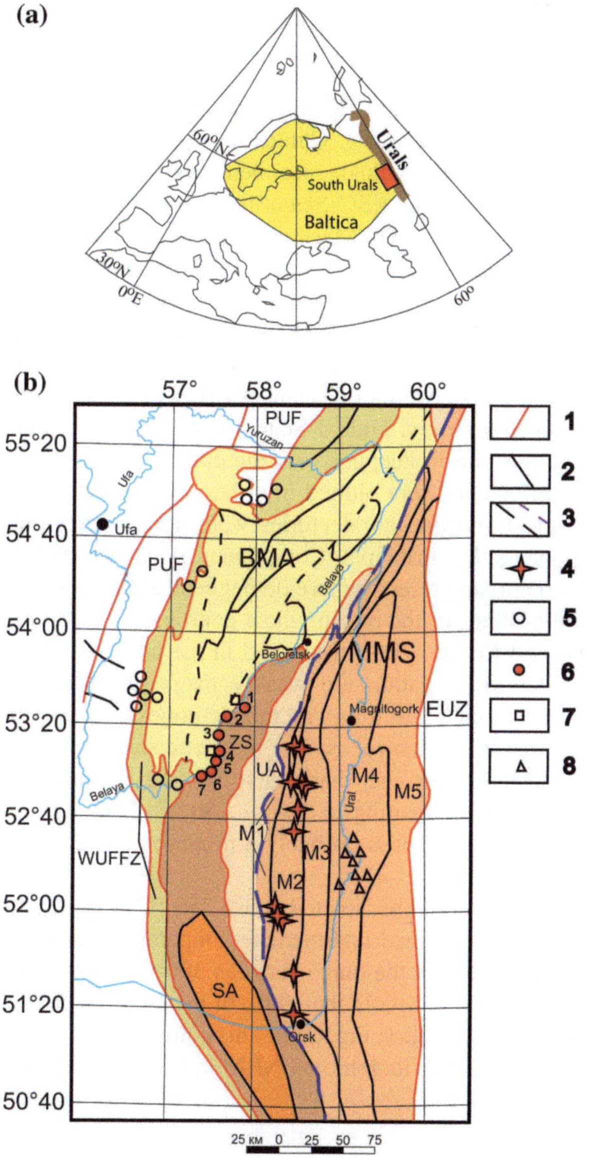

◂**Fig. 2.1** **a** Location of the Baltica block with Precambrian basement (shaded), the Urals (brown band), and the area investigated in the South Urals (SU, red rectangle). **b** Schematic map of the South Urals, simplified after Kozlov (2002). Legend: (1) Boundary between the 1-order structures; (2) Boundary between the 2-order structures; (3) The largest tectonic faults, thick blue line is the Main Uralian Fault. Tectonic structures: PUF—Pre-Uralian foredeep; WUFFZ—West Uralian frontal folding zone; BMA—Bashkir Meganticlinorium; ZS—Zilair Synclinorium; UA—Uraltau Anticlinorium; MUF—Main Uralian Fault; MMS—Magnitogorsk Megasynclinorium; M1—Voznesensk-Prisakmarskaya Subzone; M2—Irendyk Subzone; M3—Kizil-Utrazym Subzone; M4—Central Magnitogorsk Zone; M5—East Magnitogorsk Zone; EUZ—East Uralian Zone. Sampling sites: (4) according to Golovanova et al. (2016); (5) according to Golovanova et al. (2017); (6) filled red circles indicate the sections under investigation numbered in accordance with the Table 2.1; (7) according to Vinogradov (2016); (8) according to Khidiyatov et al. (2015)

parts, depending on the exposure quality and the availability of suitable rock varieties, are considered to be a site. Cubic specimens of 8-cm^3 volume were cut from each hand sample. Individual specimens were stepwise heated in 12–20 increments up to 700 °C in an ASC Scientific TD-48 thermal demagnetizer with internal residual fields of <10 nT. Measurements of the natural remanent magnetization, NRM, were done with a JR-4 or JR-6 spinner magnetometer with a noise level of 0.05 mA/m. For some part of the collection (Kaga, Nabievo, Kurgashly and Mindigulovo sections) NRM measurements were done in the Laboratory of the main geomagnetic field and petromagnetism of the Schmidt Institute of Physics of the Earth of the Russian Academy of Sciences (Moscow) with a cryogenic SQUID magnetometer (2G Enterprises, United States) installed in the nonmagnetic room. The laboratory studies are in keeping with the currently accepted procedures.

Demagnetization results were plotted in orthogonal vector diagrams (Zijderveld 1967). Visually identified linear trajectories were used to determine directions of magnetic components by Principal Component Analysis (PCA), employing a least-squares fit comprising three or more demagnetization steps (Kirschvink 1980), anchoring the fitting lines to the origin where appropriate. Paleomagnetic software by Enkin (1994) and Cogné (2003) was used in the analysis.

The rocks under investigation are weakly magnetic. Their natural remanent magnetization varies from 0.1 to 8 mA/m, mostly not exceeding 5 mA/m and being less than 1.5 mA/m for Silurian rocks. The magnetic susceptibility varies from 0.1 to 20×10^{-5} SI, usually no higher than 2.5×10^{-5} SI. An unstable magnetization component is destroyed by heating to 200 or 250 °C. Directions of this magnetization are nearly chaotic, and it is probably a mixture of viscous and the so-called transportation components.

No other components could be isolated above 250 °C in some part of the samples studied. However, a well-defined intermediate-temperature component, ITC, that is isolated from other samples in the range between 300–350 and 510–540 °C or even 600 °C and often accounts for more than 90% of the total NRM intensity is reliably grouped at many sites (Table 2.1; Fig. 2.2). This component is of reverse polarity in all the samples studied. After its removal NRM values reduce to zero.

Table 2.1 Intermediate-temperature component directions in the rocks of the eastern Bashkir Meganticlinorium

Id	Age	n	S/D	In situ				Tilt-corrected			
				D	I	k	α_{95}	D	I	k	α_{95}
1. Kaga (N: 53.52, E: 57.70)											
1-1	D	7/7	10/38	223.1	−46.1	100.0	6.1	243.1	−21.2	64.6	7.6
1-2	D	8/7	19/28	226.0	−39.8	36.1	10.2	240.7	−23.2	40.2	9.6
1-3	D	6/6	27/38	231.7	−49.5	85.0	7.3	256.6	−26.0	81.2	7.5
1-4	D	9/5	22/43	229.1	−43.1	62.7	9.7	249.2	−15.4	58.0	10.1
1-5	D	8/8	26/52	226.5	−50.8	85.0	6.0	257.6	−18.4	47.3	8.1
2. Krivaya Luka (N: 53.40, E: 57.53)											
2-1	R	10/5	200/46	227.2	−38.3	169.1	6.1	180.8	−41.6	35.3	13.4
2-2	dike	9/5	197/55	239.8	−30.9	29.9	14.2	189.2	−50.5	29.9	14.2
2-3	V	15/6	197/49	215.4	−43.6	26.4	13.3	191.9	−30.7	3.4	42.8
2-4	V	14/7	180/50	262.2	−28.5	15.1	16.2	221.3	−70.1	15.1	16.2
2-5	V	12/6	212/76	228.5	−34.4	14.9	17.9	178.1	−19.6	18.2	16.1
3. Apshak (N: 53.32, E: 57.53)											
3-1	R	12/5	85/25	222.5	−44.8	28.4	14.6	249.2	−62.0	22.9	16.3
3-2	V	13/6	50/35	244.5	−35.3	12.5	19.7	265.3	−27.9	14.4	18.2
4. Kurgashly (N: 53.28, E: 57.53)											
4-1	O	8/8	189/55	222.1	−16.3	17.1	13.9	194.1	−36.2	16.2	14.3
4-2	O	6/5	342/20	250.5	−45.3	23.9	16.0	255.7	−25.5	28.7	14.5
5. Nabievo (N: 53.27, E: 57.52)											
5-1	O	7/5	170/38	235.4	−31.9	25.3	15.5	207.8	−60.1	29.0	14.4
5-2	O	7/6	173/21	229.7	−42.4	26.4	13.3	219.2	−54.7	35.2	11.4
5-3	O	7/5	333/24	221.5	−54.5	16.8	19.2	229.6	−28.4	20.9	17.1
5-4	O	7/5	341/31	234.9	−46.8	50.6	10.9	239.3	−24.0	28.3	14.6
6. Mindigulovo (N: 53.07, E: 57.35)											
6-1	O	15/8	47/22	257.5	−25.9	32.2	9.9	264.4	−16.7	34.6	9.5
6-2	S	11/6	41/23	247.9	−28.8	83.7	7.4	257.4	−20.7	83.7	7.4
6-3	S	13/11	22/29	249.4	−36.0	38.5	7.5	258.1	−13.0	30.3	8.4
6-4	D	8/5	19/42	241.0	−36.0	114.8	7.2	253.0	−7.2	36.8	12.8
7. Kiekbaevo (N: 53.05, E: 57.27)											
7-1	O	18/13	36/11	219.8	−42.7	38	6.8	229.6	−41.1	36	7.0
7-2	S	9/5	344/25	235.7	−21.9	59.7	10.0	237.1	1.9	82.0	8.5
ALL		**24**		**232.0**	**−38.8**	**34.0**	**5.1**	**233.1**	**−34.0**	**6.9**	**12.2**
	3.8%			**232.2**	**−38.8**	**34.2**	**5.1**				
ALL*		**30**		**229.7**	**−39.1**	**36.9**	**4.4**	**230.7**	**−34.2**	**6.8**	**10.9**
	4.6%			**229.9**	**−39.1**	**37.4**	**4.4**				

Notes S/D, strike and dip angle. n, number of samples (sites) studied/used. D, Declination, I, Inclination, k, Concentration parameter, α_{95}, radius of 95% confidence circle. Localities are labeled as in Fig. 2.1b; ALL is the overall mean for all sites listed in the table; ALL* are values calculated with account for two sections (6 sites) in the same area as in Vinogradov (2016)

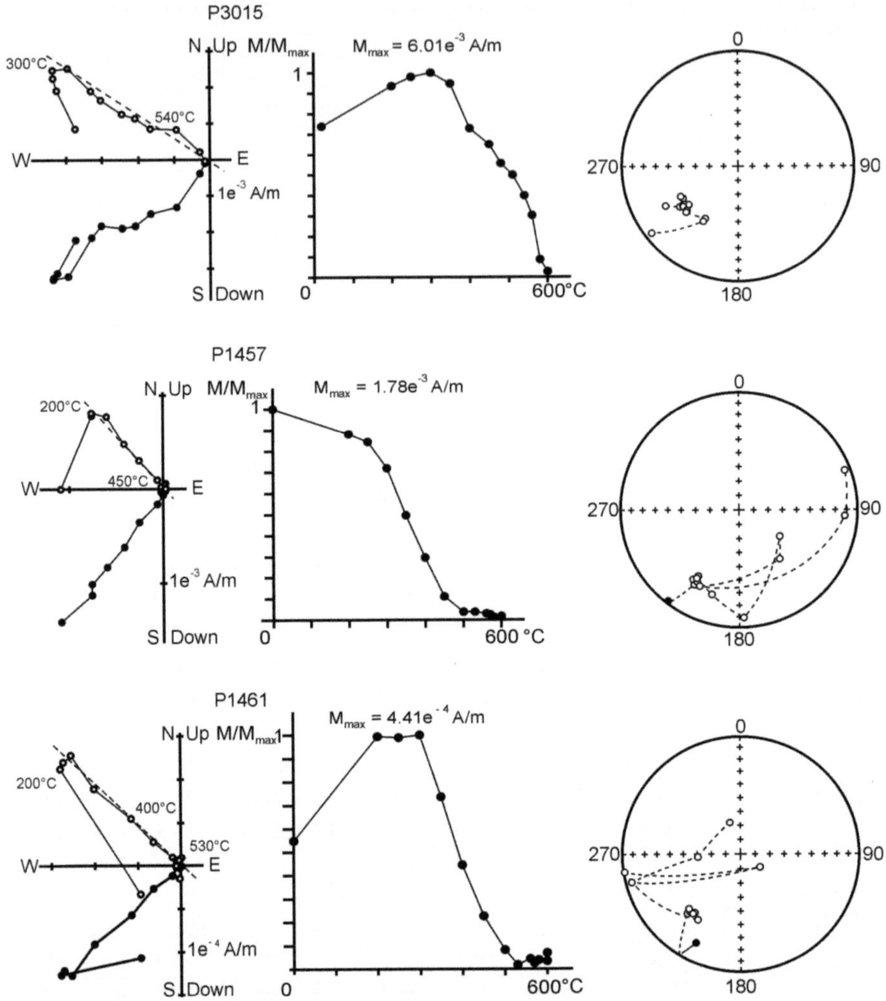

Fig. 2.2 Examples of thermal demagnetization of specimens from the eastern Bashkirian Meganticlinorium in the South Urals (Nabievo—P3015 and Kiekbaevo—P1457, P1461), in stratigraphic coordinates. Solid (open) circles represent vector endpoints projected onto the horizontal (vertical) plane. Isolated intermediate-temperature components are denoted by thick dashed lines

Judging from the previously published results of thermal demagnetization and petromagnetic data (Danukalov et al. 1983; Svyazhina et al. 2003; Vinogradov 2016), magnetite is an ITC carrier in all the samples (Fig. 2.3).

It should be noted that our results do not always coincide with previously published data. I. A. Svyazhina who studied Ordovician rock paleomagnetism in the vicinity of the village of Baynazarovo (near the villages of Nabievo and Kurgashly) (Svyazhina et al. 2003) also isolated a higher-temperature component

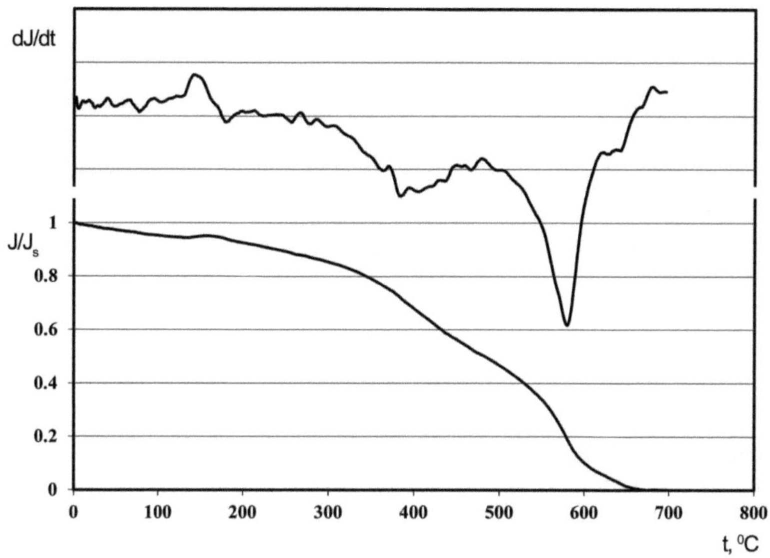

Fig. 2.3 Example of thermomagnetic behavior of a sample of thin-layered dolomitized quartz sandstones from the section near the village of Nabievo

on heating above 600 °C. She associated this component with primary magnetization formed during the Ordovician. As already noted, however, any attempts to isolate the high-temperature component in the collection in question were in vain. Besides, the results of thermomagnetic analysis revealed no hematite in the samples studied.

There was no convergence of views among the researchers who worked previously in the vicinity of Kaga. Bachtadse (Bachtadse et al. 1998) who studied there Ordovician and Silurian rocks along with Devonian ones believed that paleomagnetic directions isolated as a result of heating or alternating magnetic field affecting the samples from rocks of different age corresponded to the Late Paleozoic geomagnetic field. Burtman et al. (2000) obtained another direction with a quite different inclination. Relying on test results, these authors inferred the primary nature of the high-temperature component they isolated. Petrov and Svyazhina supposed the pre-folding nature of the intermediate and high-temperature magnetization components (Petrov and Svyazhina 2006).

According to our data, ITC site-means are better grouped in geographic coordinates than in stratigraphic coordinates (Fig. 2.4). Upon stepwise unfolding, the best data grouping (= dispersion minimum) is reached at 4.3% unfolding. The difference between the maximum and values of concentration parameters in situ is not "convincingly" significant. This attests to the fact that the corresponding magnetization component postdates folding. The calculated mean paleomagnetic pole has the latitude Plat = −39.5° and longitude Plong = 346.2°, the radius of the 95% confidence circle is A_{95} = 4.7°, N = 24.

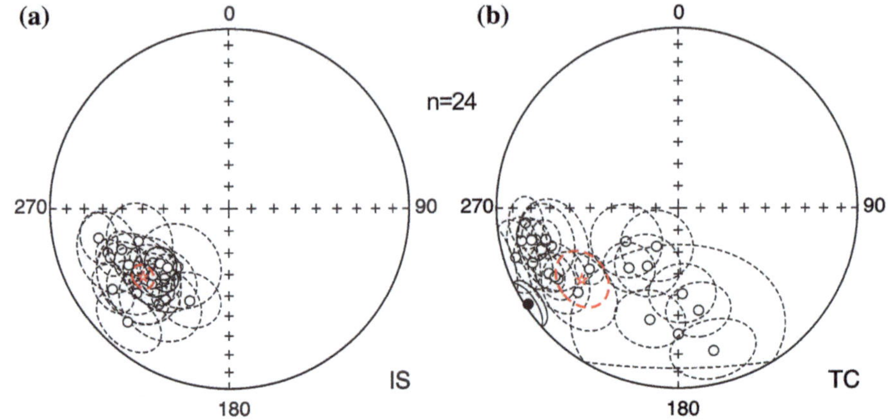

Fig. 2.4 Stereoplots of intermediate-temperature component, ITC, site-mean directions in situ (**a**) and after tilt correction (**b**) in sediment rocks studied from the eastern Bashkir Meganticlinorium in the South Urals. Stars are the overall mean directions with associated confidence circles. Solid (open) symbols are projected onto the lower (upper) hemisphere

A comparison with reference directions recalculated from the apparent wander path of a paleomagnetic pole for Baltica (Torsvik et al. 2012) shows that poles calculated from ITC average directions best agree with the reference result corresponding to the age 290–310 Ma (Fig. 2.5). Such ITC age also agrees with predominantly reverse polarity of this component, which is indicative of its origin in the second half of the reverse polarity Kiaman Superchron, and also with geological data on the Kungurian (272–279 Ma) age of the final stage of deformations in the Urals (Puchkov 2003).

Neoproterozoic and Lower Paleozoic rocks under investigation show no primary magnetization component preserved.

The remagnetization process has encompassed almost the whole South Urals. Analysis of our previous results and data published by other authors shows that rocks in almost all Paleozoic sections were remagnetized during the Late Paleozoic. The same magnetization component is also isolated in Upper Riphean and Vendian rocks. Some sections show a pre-folding remagnetization, while in others it postdates folding. Thus, Late Paleozoic remagnetization registered in the paleomagnetic record is of a regional nature and probably reflects a collision event at the final stage of the evolution of the Uralian Paleoocean.

Analysis of available data on Late Paleozoic remagnetization makes it possible to trace, to some extent, the formation of folded structures in the Uralian Fold Region at the final stages of the closure of the Uralian Paleoocean.

Using the results on many rock sections of different age it was concluded for the western part of the West Uralian Megazone in the South Urals that the intermediate-temperature magnetization component is pre-folding or has originated at the initial stages of deformations during the Kungurian time (279–272 Ma) (Puchkov 2003; Levashova et al. 2013). In contrast, in its eastern part, in meridional

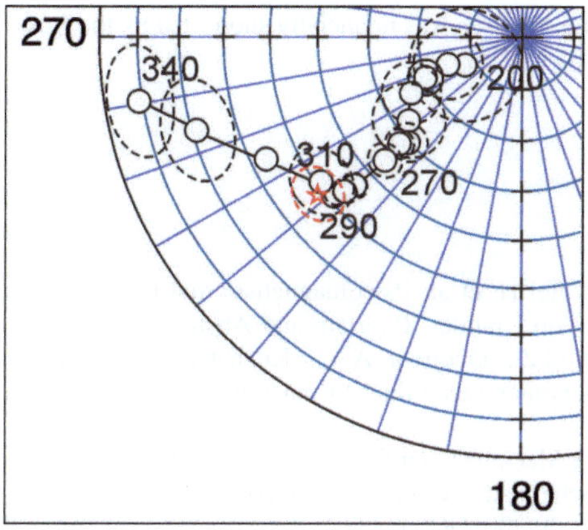

Fig. 2.5 Stereoplot of reverse reference directions [circles connected with thin line for 10 Ma intervals recalculated from 200 to 340 Ma segment of the APWP for Baltica (Torsvik et al. 2012)] to a point at 54°N, 58°E. Thin dashed lines are confidence circles for reference directions. Red star and thick dashed line are overall mean direction with associated confidence circle. Open symbols and dashed lines are projected onto the upper hemisphere

course of the Belaya River, the isolated magnetization component is post-folding. This is evidenced by the results of this research work and also the data given in (Vinogradov 2016). We think that analysis of the natural remanent magnetization components makes it possible to trace the formation of folded structures at the final stages of the formation of the western South Urals as the deformed margin of the East European Platform. Collision processes occurring in the South Urals from the Late Devonian to the Late Permian have undoubtedly found their reflection in the western flank (in present coordinates) structures. Within the passive margin of Baltica the collision resulted in the formation of a lateral series of fold-and-thrust structures. They started from the Main Uralian Fault westward, with possible gradual termination of this process towards the Pre-Uralian Foredeep. Paleomagnetic results of this research work support this hypothesis.

In the Magnitogorsk Zone and in most of the sites under investigation we isolate an intermediate-temperature magnetization component corresponding to Late Paleozoic remagnetization. In Devonian rocks of the western Magnitogorsk Zone this component is post-folding (Golovanova et al. 2016). This result enables us to propose that deformations there took place long before the final stages of the formation of the Urals, most probably at the end of the Devonian. Paleomagnetic data on Lower Carboniferous rocks from the Magnitogorsk-Bogdanovka Graben attest to the fact that this Late Paleozoic magnetization was formed either in the process of deformations or only shortly prior to the fold formation and at the initial stages of deformations (Khidiyatov et al. 2015). Thus, it can be supposed that the fold formation across the area under investigation in the central part of the Magnitogorsk Zone took place at the final stage of the formation of the Urals during the Middle Permian (Kungurian time).

Acknowledgements This research was financially supported by the Russian Foundation for Basic Research, grant No. 15-05-03036.

References

Bachtadse V., Tait J., Soffel H. et al. Paleomagnetism and plate kinematics of the South Urals. Europrobe Workshop on Uralides. Program and Abstracts. Moscow, 1998, p. 23.

Burtman, V.S., Gurarii G.Z., Dvorova A.V., Kuznetsov N.B., Shipunov S.V. The Uralian Paleoocean in the Devonian (according to paleomagnetic data). Geotektonics, 2000, vol. 34, no. 5, pp. 397–406.

Cogné, J.P. PaleoMac: A Macintosh application for treating paleomagnetic data and making plate reconstructions. Geochemistry, Geophysics, Geosystems, 2003, vol. 4 (1), article no. 1007. http://dx.doi.org/10.1029/2001GC000227.

Danukalov, N.F., Kondruchina, L.S., Chernikov, A.P. Paleomagnetizm paleozoya Yuzhnogo i Srednego Urala [Paleozoic paleomagnetism of the South and Middle Urals]. Ufa: BFAN SSSR, 1983. 120 p. (In Russian).

Enkin, R.I. A Computer Program Package for Analysis and Presentation of Paleomagnetic Data, Sidney, Canada: Pacific Geoscience Centre, Geological Survey of Canada, 1994.

Golovanova, I.V., Danukalov, K.N., Kadyrov A.F., Khidiyatov, M.M., Sal'manova, R.Yu., Shakurov, R.K., Levashova, N.M., Bazhenov, M.L. Paleomagnetism of sedimentary strata and the origin of the structures in the western slope of South Urals. Izvestiya, Physics of the Solid Earth, 2017, vol. 53, no. 2, pp. 311–319.

Golovanova, I.V., Danukalov, K.N., Kosarev, A.M., Khidiyatov, M.M., Bazhenov, M.L Paleomagnetizm Magnitogorskoy zony Yuzhnogo Urala. Predvaritelnye rezultaty [Paleomagnetism of the Magnitogorsk Zone in the South Urals. Preliminary results]. In: Materialy mezhdunarodnoy shkoly-seminara "Problemy paleomagnetizma i magnetizma gornykh porod" [Proceedings of the International Workshop-School "Problems of Rock Paleomagnetism and Magnetism"]. Shmidt Institute of Physics of the Earth, St. Petersburg State University. Yaroslavl: Filigran, 2016, pp. 29–35 (In Russian).

Kirschvink, J.L. The least-square line and plane and the analysis of palaeomagnetic data. Geophysical Journal of the Royal Astronomical Society, 1980, vol. 62, pp. 699–718.

Khidiyatov, M.M., Golovanova, I.V., Danukalov, K.N., Kadyrov A.F., Levashova, N.M., Bazhenov, M.L. Paleomagnetizm nizhnekamennougolnykh vulkanogennykh i osadochnykh tolshch Magnitogorsko-Bogdanovskogo grabena na Yuzhnom Urale. Predvaritelnye rezultaty [Paleomagnetism of Lower Carboniferous volcanogenic and sedimentary rocks in the Magnitogorsk-Bogdanovka Graben of the South Urals]. In: Paleomagnetizm i magnetism gornykh porod: teoriya, praktika, eksperiment [Paleomagnetism and rock magnetism: Theory, [practice, experiment]. "Borok" Geophysical Laboratory, Division of the Shmidt Institute of Physics of the Earth. Yaroslavl: Filigran, 2015, pp. 229–234 (In Russian).

Khramov, A.N., Goncharov, G.I., Komissarova, R.A., Pisarevskiy, S.A., Pogarskaya, I.A., Rzhevskiy, Yu.S., Rodionov, V.P., Slautsetays, I.P. Paleomagnitologiya [Paleomagnetology]. Leningrad: Nedra, 1982. 312 p. (In Russian).

Kozlov, V.I., (Ed.), 2002. Geological Map of the Russian Federation and Adjacent Territory of the Republic of Kazakhstan. Scale: 1:1 000000 (a New Series). Sheet N-40(41)-Ufa. St. Petersburg: VSEGEI. (In Russian).

Levashova, N.M., Bazhenov, M.L., Meert, J.G., Kuznetsov, N.B., Golovanova, I.V., Danukalov, K.N., Fedorova, N.M. Paleogeography of Baltica in the Ediacaran: Paleomagnetic and geochronological data from the clastic Zigan Formation, South Urals. Precambrian Res., 2013, vol. 236, pp. 16–30.

Levashova, N.M., Bazhenov, M.L., Meert, J.G., Danukalov, K.N., Golovanova, I.V., Kuznetsov, N.B., Fedorova, N.M. Paleomagnetism of upper Ediacaran clastics from the South Urals: Implications to paleogeography of Baltica and the opening of the Iapetus Ocean. Gondwana Research, 2015, vol. 28, pp. 191–208.

Pavlov V.E., Gallet Y. Katav limestones: a unique example of remagnetization or an ideal recorder of the neoproterozoic geomagnetic field. Izvestiya. Physics of the Solid Earth, 2009, vol. 45, no. 1, pp. 31–40.

Petrov, G.A., Svyazhina, I.A. Korrelyatsiya ordoviksko-devonskikh sobytiy na Uralskoy i Skandinavskoy okrainakh Baltiki: geologicheskie i paleomagnitnye dannye [Correlation of the Ordovician-Silurian events on the Uralian and Scandinavian margins of Baltica: Geological and paleomagnetic data]. Lithosphere, 2006, no. 4, pp. 23–39 (In Russian).

Puchkov V.N. The Uralides and Timanides: Their structural relationship and position in the geologic history of the Ural-Mongolian fold belt. Russian Geology and Geophysics, 2003, vol. 44 (1–2), pp. 28–39.

Shipunov S.V. Synfold magnetization: estimation of its direction and geological implications. Izvestiya. Physics of the Solid Earth, 1996, vol. 31, no. 11, pp. 948–955.

Stratotip rifeya. Paleontologiya. Paleomagnetizm [The stratotype of the Riphean. Paleontology. Paleomagnetism]. B.M. Keller (ed.). Moscow: Nauka, 1982. 183 p. (In Russian).

Svyazhina, I.A., Puchkov, V.N., Ivanov, K.S., Petrov, G.A., Paleomagnetizm ordovika Urala [Ordovician paleomagnetism in the Urals]. Ekaterinburg: UrO RAN, 2003. 136 p. (In Russian).

Torsvik, T.H., Van der Voo, R., Preeden, U., MacNiocaill, C., Steinberger, B., Doubrovine, P.V., van Hinsbergen, D.J.J., Domeir, M., Gaina, C., Tohver, E., Meert, J.G., McCausland, P.J.A., Cocks, R.M., Phanerozoic polar wander, palaeogeography and dynamics. Earth Science Reviews, 2012, vol. 114, pp. 325–368.

Vinogradov E.V. Paleomagnetizm ordoviksko-siluriyskikh otlozheniy Zapadno-Zilairskoy zony Yuzhnogo Urala [Paleomagnetism of Ordovician-Silurian deposits in the western Zilair Zone of the South Urals]. Materialy mezhdunarodnoy shkoly-seminara "Problemy paleomagnetizma i magnetizma gornykh porod" [Proceedings of the International Workshop-School "Problems of Rock Paleomagnetism and Magnetism"]. Shmidt Institute of Physics of the Earth, St. Petersburg State University. Yaroslavl: Filigran, 2016. pp. 18–23 (In Russian).

Zijderveld, J.D.A. AC demagnetization of rocks: Analysis of results. In: Methods in Paleomagnetism. D.W. Collinson, K.M. Creer (eds.). Elsevier, Amsterdam, 1967, pp. 254–286.

Chapter 3
Paleomagnetic Directions Distortion Caused by Viscous-Plastic Deformations Estimated from Anisotropy of Magnetic Susceptibility (Case Study of Berriasian Clays from East Crimea)

V. A. Grishchenko and A. Yu. Guzhikov

Abstract Deformations inside clayey strata may result in a mechanical rotation of ferromagnetic particles and of associated paleomagnetic vectors. Accordingly, restoring the orientation of magnetic susceptibility (AMS) ellipsoids to the position corresponding to the moment of magnetization fixation would be expected to improve the paleomagnetic statistics. The Upper Berriasian clays from Zavodskaya Balka section in the vicinity of Feodosiya was chosen to test this hypothesis. Paleomagnetic quality parameters in the examined rocks improved after the restoring the AMS ellipsoids into their positions at the moment of magnetization acquisition. The reversal test, negative when the original paleomagnetic directions were considered, became positive at the B level (McFadden and McElhinny 1990) after this procedure. Fold test also used to be negative, after the correction revealed a synfolding component throughout the entire section. Thus, our results validate the hypothesis on the possibility of correcting paleomagnetic data by means of the data on anisotropy of magnetic susceptibility. Nevertheless, the technique we have tested requires further elaboration and verification at other objects composed of argillaceous deposits.

Keywords Magnetic susceptibility · Anisotropy of magnetic susceptibility
Paleomagnetism · Berriasian · Tectonic deformations · East Crimea

Deformations inside clayey strata may result in a mechanical rotation of ferromagnetic particles and of associated paleomagnetic vectors. Accordingly, restoring the orientation of magnetic susceptibility (AMS) ellipsoids to the position corre-

V. A. Grishchenko (✉) · A. Yu. Guzhikov
Geological Faculty, Saratov State University, Saratov, Russia
e-mail: grishenko-vladimir@bk.ru

A. Yu. Guzhikov
e-mail: aguzhikov@yandex.ru

D. Nurgaliev et al. (eds.), *Recent Advances in Rock Magnetism, Environmental Magnetism and Paleomagnetism*, Springer Geophysics,
https://doi.org/10.1007/978-3-319-90437-5_3

sponding to the moment of magnetization fixation would be expected to improve the paleomagnetic statistics. The Upper Berriasian clays from Zavodskaya Balka section in the vicinity of Feodosiya was chosen to test this hypothesis. Paleomagnetic quality parameters in the examined rocks improved after the restoring the AMS ellipsoids into their positions at the moment of magnetization acquisition. The reversal test, negative when the original paleomagnetic directions were considered, became positive at the B level (McFadden and McElhinny 1990) after this procedure. Fold test also used to be negative, after the correction revealed a synfolding component throughout the entire section. Thus, our results validate the hypothesis on the possibility of correcting paleomagnetic data by means of the data on anisotropy of magnetic susceptibility. Nevertheless, the technique we have tested requires further elaboration and verification at other objects composed of argillaceous deposits.

Introduction

In thick, lithologically uniform clayey strata, visual identification of bedding elements at every section level is difficult or impossible. As a rule, one is forced to regard the levels within clay members by means of interpolating or extrapolating the bedding elements measured from the nearest hard-rock interlayers with clearly defined bedding planes. Meanwhile, the true azimuths and layer dip angles may differ substantially due to plastic deformations of diverse nature.

Errors in identifying the strata bedding elements may be important or occasionally decisive for acquiring paleomagnetic information. Disregarding the errors leads to distortions in paleomagnetic directions and, consequently, to negative results of field tests (fold test, reversal test), providing reasons for data rejection (Opdyke and Channell 1996; Supplement … 2000; Van der Voo 1993). Therefore, clays do not make favorable study objects for solving magnetotectonic problems, but are widely used in magnetostratigraphy (e.g. Arkadiev et al. 2010, 2015a, b; Bragin et al. 2013; Dzyuba et al. 2017; Grishchenko et al. 2016; Guzhikov et al. 2003, 2014).

Submicron ferromagnetic particles are as a rule aggregated on the clayey mineral scales, producing a planar anisotropy of magnetic susceptibility (AMS) with (sub) vertical arrangement of short axes of magnetic ellipsoids (K3) in the case of sediment deposition in quiet hydrodynamic settings. This fact allows to interpret deviations of the average K3 direction ($K3_{av}$) from the stereogram center (in the geographic coordinate system) as the elements of clay layer bedding that cannot be measured with a geological compass in the case of subhorizontal bedding within the platform sedimentary cover (Dzyuba et al. 2017).

Paleomagnetic data on the Tithonian-Valanginian clays from Mountainous Crimea (Arkadiev et al. 2010, 2015a, b; Grishchenko et al. 2016; Guzhikov et al. 2012, 2014) are generally peculiar for low precision parameters and negative results of the fold test. The reversal test results are positive only at low significance levels

["C" grade according to McFadden and McElhinny (1990)]; otherwise, they are negative. AMS data from these clays show numerous deviations from the ideal pattern of the sediment magnetic texture, the one with the short axes of magnetic ellipsoids tending towards vertical positions and the long (K1) and medium (K2) axes towards the stereogram equator. In Mountainous Crimea, such deviations may result from tectonic and non-tectonic deformations of both non-lithified clayey sediments and mature plastic clays.

In the course of deformations, a mechanical rotation of ferromagnetic particles and associated paleomagnetic vectors takes place. Consequently, recovering the positions acquired by ellipsoids of magnetic susceptibility at the moment of magnetization fixation would contribute to improving paleomagnetic statistics. The attempts to apply this effect while analyzing paleomagnetic data are not known to us, but there exists a significant literature on the problem of synfolding remagnetization and the paleomagnetic fold test (e.g., Borradaile 1997; Cogné and Perroud 1987; Facer 1983; Kodama 1988; Stamatakos and Kodama 1991; Van der Pluijm 1987). In each of these previous studies, reorientation of particles and their remanent vectors is considered to take place in response to tectonic stresses, and the vectors rotate away from the compression axis. In our model, the magnetic fabric forms in an incompletely lithified layer which is nevertheless tilted by tectonic compression. Within this layer particles (and hence remanence vectors and anisotropy ellipsoids) rotate in response to gravitational forces, so that vectors and long axes rotate away from the vertical and towards the tectonic compression axis. The aim of conducted study was to make an experimental examination of theoretical supposition that the restitution of ellipsoids of magnetic susceptibility to their position at the moment of acquisition of magnetization promotes the improvement of paleomagnetic statistics parameters.

Object and Methods of Investigations

The Upper Beriasian Zavodskaya Balka section in the vicinity of Feodosiya was chosen to test this experimental statement. The section is represented by gray, massive, unstratified, unctuous, plastic, slightly silty (below 5%), slightly micaceous (less than 5% of mica scales) carbonate clays of the Sultanovka suite, with the apparent thickness over 70 m (Arkadiev et al. 2010).

Results paleomagnetic and petromagnetic examination including the data on anisotropy of magnetic susceptibility of this section have been published previously (Arkadiev et al. 2010, 2015a, b; Guzhikov et al. 2014).

Despite the good quality of Zijderveld diagrams and the presence of characteristic components (**ChRM**) of normal as well as reverse geomagnetic field polarities in the section, paleomagnetic interbed precision parameters, calculated for the directions of each polarity particularly, are not high (11.2—for N and 6.5—for R-polarity), the reversal test is negative, and the fold test is also negative. Positive reversal test at the C level (McFadden and McElhinny 1990) was obtained only with the samples composed of specimens with minimum-sized ferromagnetic grains as evaluated from the K/J_{rs} ratio (*K*—magnetic susceptibility, J_{rs}—remanent saturation magnetization) (Guzhikov et al. 2014).

The studied clays are characterized by the orientation of long axes of magnetic ellipsoids in the NWW-SEE direction (Fig. 3.1a–c) and a planar magnetic anisotropy (Fig. 3.2a). A hypothesis that this is due to a flattened shape of clay flakes, on which the submicron ferromagnetic particles are aggregated, is confirmed by microscopic observations (Fig. 3.2b).

The method to "restore" the magnetic particles into their primary orientation corresponding to the moment of remanent magnetization fixation is based on the model of the magnetic fabric formation in the Zavodskaya balka sediments, which is illustrated in Fig. 3.3 and boils down to the following:

1. We suppose that deviations of the axes of magnetic ellipsoids from their primary positions are caused mostly by tectonic deformations of sediments during late Berriasian orogeny (Nikishin et al. 1997). These processes are reflected in the orientation of long axes of AMS ellipsoids (Fig. 3.1), perpendicular to the main direction of collision compression in the Mountain Crimea (Bagaeva and Guzhikov 2014; Guzhikov et al. 2014).
2. In the hard bed there is no rotation of ferromagnetic particles during tectonic deformations (Fig. 3.3a), and therefore there is no need to reconstruct the primary positions of magnetic ellipsoids in it. In this case the main position of short axes ($K3_{av}$) in the geographic coordinate system deviate from the vertical by the angle equal to the bedding dip ($\angle$ **1**), in the paleogeographic coordinate system the projection of ($K3_{av}$) coincide with the center of stereogram (Fig. 3.3a). The age of magnetization is prefolding. The angle between the horizontal plane and the magnetic foliation plane, where K1 and K2 are located, coincide with the dip angle of bedding ($\angle$ **1**).
3. During the synsedimentary folding deformations occur in a not completely lithified layer, within which flattened particles strive to take a horizontal position again, but because of the high viscosity of the sediment they cannot reach it. Anyway, the position of K3 deviates from the perpendicular to the underlying hard surface of bedding and on the AMS stereograms in geographic coordinate system $K3_{av}$ deviates from the vertical by the angle $\angle$ **1**, as in the case of hard layer. In the paleogeographic coordinate system $K3_{av}$ does not coincide with the center of stereogram (Fig. 3.3b).
   The direction of $K3_{av}$ corresponds to the average orientation of magnetic ellipsoids during the fixation of remanent magnetization in the rocks. For this reason to improve statistical paleomagnetic parameters it is desirable to rotate each paleomagnetic vector in such a way, that the related K3 coincide with $K3_{av}$.
4. If the final lithification of a sediment and fixation of magnetization takes place after the termination of folding, the flattened magnetic particles will strive to take a horizontal position. In this case K3 will tend to the vertical position in the geographic coordinate system, while in the paleogeographic coordinate system,

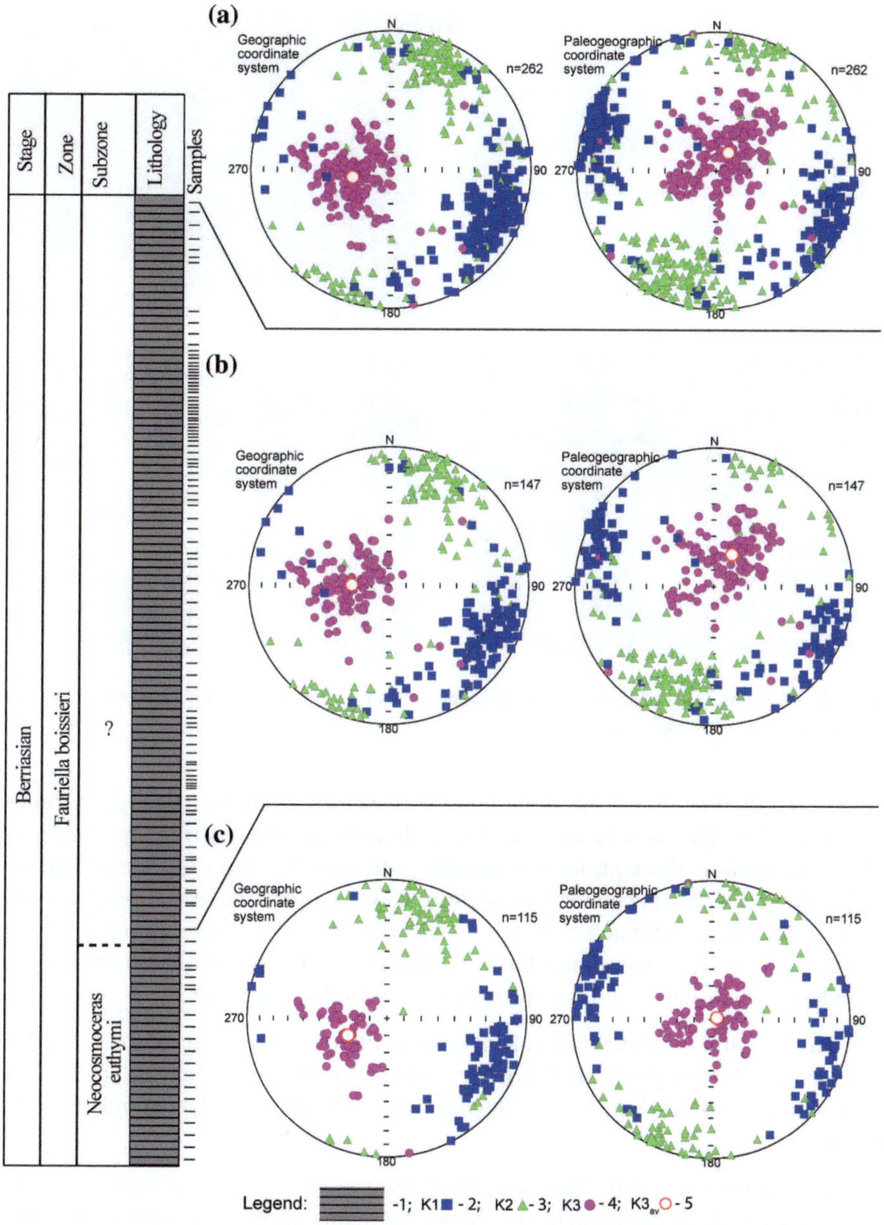

Fig. 3.1 AMS data on the "Zavodskaya balka" section: **a** full section, **b** upper part of the section, **c** lower part of the section. Legend: 1—clay; 2, 3, 4—projections of long (K1), medium (K2), and short (K3) axes of magnetic ellipsoid, respectively; 5—projections of mean direction of short axes ($K3_{av}$) of magnetic ellipsoid

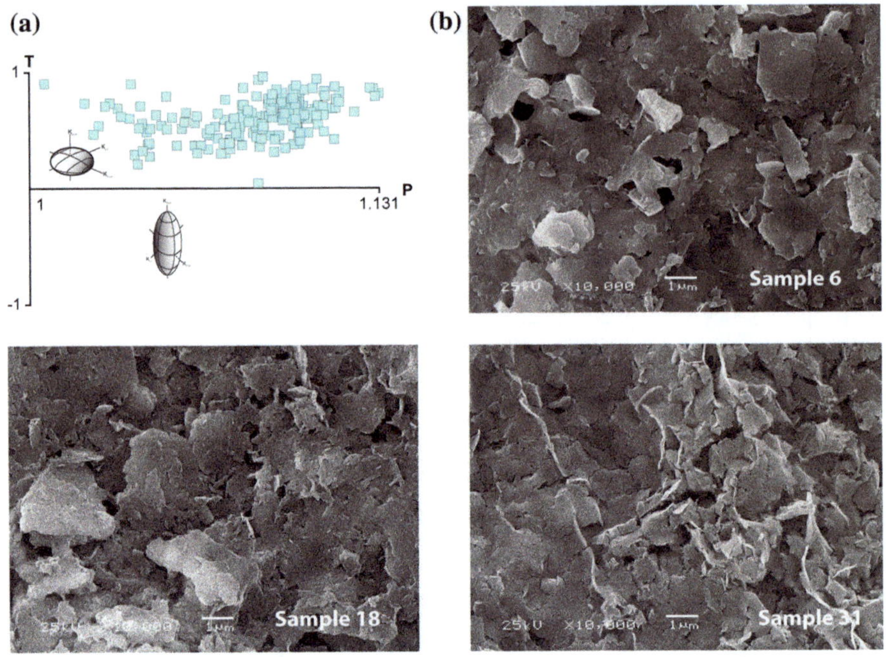

Fig. 3.2 Diagram T-P (AMS parameters) as an indicator of magnetic ellipsoids shape (**a**) and SEM images (**b**) of studied clay

on the contrary, K3 projections will be displaced from the stereogram center (Fig. 3.3c). The results of fold test will indicate the postfolding nature of magnetization, although its age actually will not differ from the age of folding. In this case the paleomagnetic determinations reflecting most precisely the direction of geomagnetic field in which the sediment were formed will be obtained in the modern coordinate system, while the extrapolation of bed orientation from the underlying levels would introduce an error.

Therefore, the correction of paleomagnetic directions by returning AMS ellipsoids to their presumptive initial orientation is required only in the case, mentioned in the clause 3 above (Fig. 3.3b). The algorithm of **ChRM** recalculation is then the following:

1. The totality of magnetic ellipsoid axes is reduced to be arranged so that the $K3_{av}$ assumes vertical position in the paleogeographic coordinate system. For this purpose, in the Anisoft 4.2 program, designed to analyze the AMS data, in the "Edit data" regime, values $\mathbf{180° + D_{K3av}}$ and $\mathbf{90° - I_{K3av}}$ are inserted for all the samples in columns F1d and F1i, respectively. (In this case $\mathbf{D_{K3av}}$ and $\mathbf{I_{K3av}}$ are, respectively, the $K3_{av}$ declination and inclination in the geographical coordinate system.)

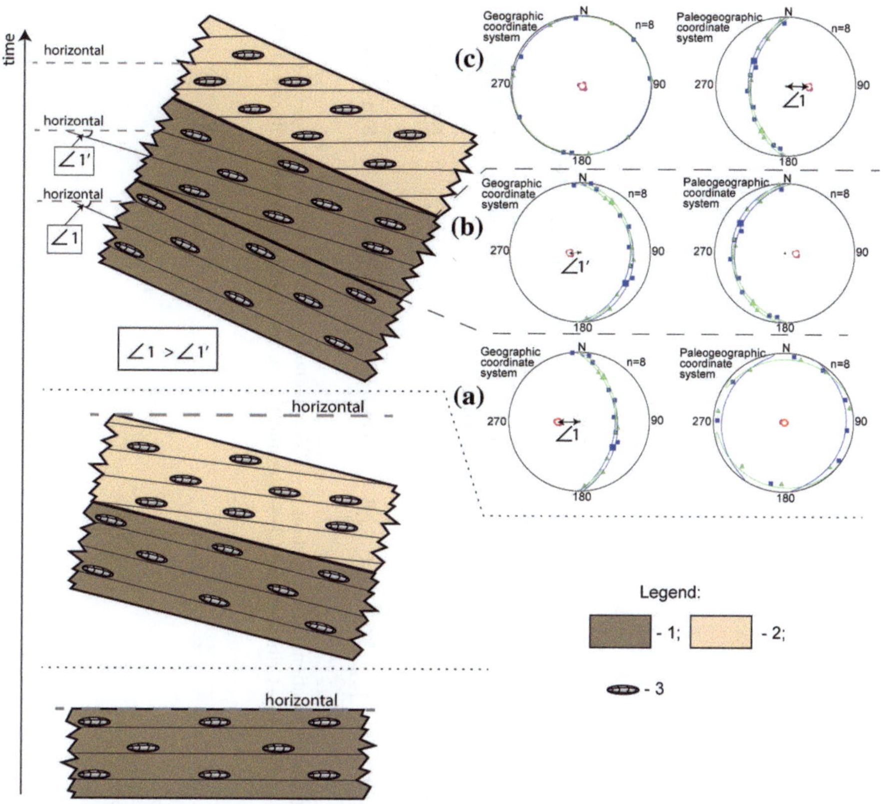

Fig. 3.3 Models of magnetic texture (AMS) forming in rocks, affected by tectonic deformations in the hard (**a**) and viscous-plastic (**b**) state. Legend: 1, 2—Hard and viscous-plastic sediment respectively; 3—magnetic particles. Other legend symbols: see Fig. 3.1

2. Parameters $\mathbf{A_i = 180° + D_{K3i}}$ and $\mathbf{B_i = 90° - I_{K3i}}$ are used as the strata dip azimuth and the strata dip angle, respectively, while converting the **ChRM** from the geographic to the stratigraphic coordinate system. (In this case, $\mathbf{D_{K3i}}$ and $\mathbf{I_{K3i}}$ are, respectively, $K3_i$ declination and inclination in the paleogeographic coordinate system, **i**–the sample number varying from 1 to n.)

To make the short axis of each magnetic ellipsoid in the set of n samples coincide with $K3_{av}$, values $\mathbf{A_i}$ and $\mathbf{B_i}$, should be inserted into the F1d and F1i columns, respectively, in the Anisoft 4.2 program in the "Edit data" regime. Therefore, use of $\mathbf{A_i}$ and $\mathbf{B_i}$ as the strata bedding elements results in the **ChRM** rotating by the angle equal to the angular distance between the short axis of the corresponding AMS ellipsoid and $K3_{av}$.

The paleomagnetic directions were analyzed by means of the Remasoft 3.0. software.

Table 3.1 Statistic paleomagnetic characteristics of the from the Zavodskaya Balka section

	Polarity	n	D°	I°	k	α_{95}°	Reversal test (McFadden and McElhinny 1990)			Fold test (McFadden and McElhinny 1990)		
							A (°)	A_c (°)	Cl.	Whole section	Lower part of section	Upper part of section
without regard for the AMS data	N	94	344	40.3	11.25	4.6	22.4	8.7	–	Not correct	Pre-folding	Not correct
	R	53	134.1	−42.2	6.50	8.3						
with regard for the AMS data	N	94	312.3	45.3	12.50	4.3	5.5	8.1	B	Syn-folding	Pre-folding	Syn-folding
	R	53	129.1	−40.3	7.52	7.8						

n—Number of samples in the selection, D and I—Average paleomagnetic declination and inclination, κ—Paleomagnetic precision parameter, α_{95}—Radius of the vector confidence circle, A—Angle between average N and R vectors, Ac—Critical angle, Cl.—Classification after McFadden and McElhinny (1990)

Research Results and Conclusions

In the whole Zavodskaya Balka section, $K3_{av}$ deviates from the vertical by 13°, with a substantial scatter of K3 directions relative to the mean (Fig. 3.1a). More thorough analysis of the AMS data has revealed that the distance between the $K3_{av}$ and the stereogram center is substantially larger in the upper part of the section (Fig. 3.1b) than in the lower one (Fig. 3.1c): 21.8° and 3.3°, respectively.

Thus, the AMS data from the lower half of the section (Fig. 3.1c) comply with the magnetic fabric of a hard stratum (Fig. 3.3a), while the distribution of magnetic ellipsoid axes in the upper half of the section (Fig. 3.1b) agrees with the theoretical model of synsedimentation folding (Fig. 3.3b). Therefore, the **ChRM** directions in the upper section have been recalculated with regard for the AMS data according to the procedure proposed above.

Results of statistical analyses of paleomagnetic determinations made with and without regard for the AMS data are compared in Table 3.1 and Fig. 3.4. The comparison reveals an improvement of paleomagnetic quality indicators in the examined rocks after AMS ellipsoid restoring to the position corresponding to purely depositional magnetic fabric.

Values of the interbed precision parameters have not changed appreciably within the section. But the reversal test that was negative when all samples were considered, and produced positive results of the C class only when analyzing the sample subset with minimal values of the K/J_{rs} ratio (Guzhikov et al. 2014), after correcting for AMS became positive for the entire section at the B level (McFadden

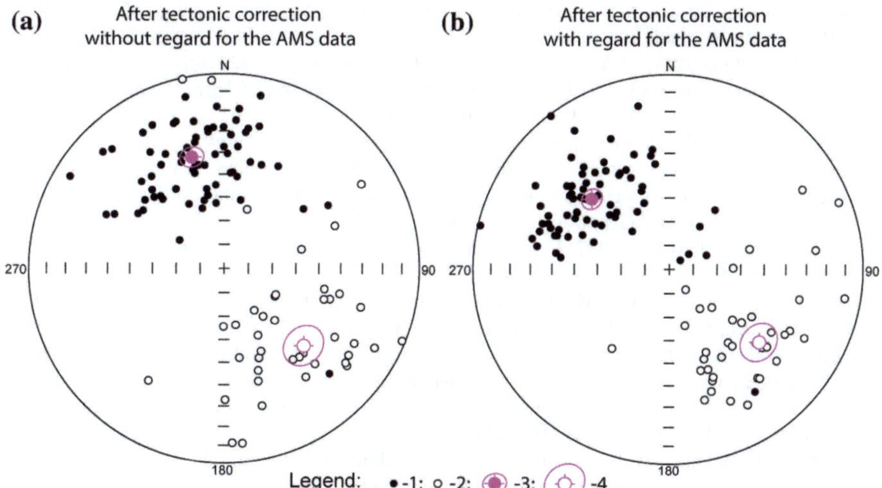

Fig. 3.4 Summary stereograms of ChRM on the "Zavodskaya Balka" section after the tectonic correction based on the bedding elements of layers, measured with geological compass (**a**) and on the AMS data (**b**). Legend: 1, 2—stereographic projections of $\mathbf{J_n}$ directions on the lower semi-sphere and upper semi-sphere respectively; 3, 4—mean direction of $\mathbf{J_n}$ vectors

and McElhinny 1990). Furthermore, after correction of paleomagnetic directions the fold test (McFadden 1990) reveals a synfolding component both in the upper part of the section and throughout the entire section.

Thus, results of our research validate the hypothesis on the possibility of correcting paleomagnetic data by means of the data on anisotropy of magnetic susceptibility. Nevertheless, this technique requires further elaboration and verification at other objects composed of argillaceous deposits.

Acknowledgements We are very grateful to Mike Jackson for reviewing our article.

The research was financially supported by the RFBR (projects № 16-35-00112 mol_a and № 17-05-00716 A).

References

Arkadiev V.V., Baraboshkin E.Yu., Bagaeva M.I., Bogdanova T.N., Guzhikov A.Yu., Manikin A. G., Piskunov V. K., Platonov E. S., Savel'eva Yu.N., Feodorova A.A., Shurekova O.V. (2015 b). New Data on Berriasian Biostratigraphy, Magnetostratigraphy, and Sedimentology in the Belogorsk Area (Central Crimea) // *Stratigraphy and Geological Correlation.* 23, 2. P. 155–191.

Arkadiev V.V., Guzhikov A.Yu., Savelieva J.N., Feodorova A.A., Shurekova O.V., Bagaeva M.I., Grishchenko V.A., Manikin A.G. (2015a). New data on bio- and magnetostratigraphy of Upper Berriasian section "Zavodskaya balka" (Eastern Crimea, Feodosiya) // *Bull. Saint Petersburg St. Univ., Geology, Geography*, 7, 4. P. 4–36 (in Russian).

Arkadiev V.V., Bagaeva M.I., Guzhikov A.Yu., Manikin A.G., Yampolskaya O.B (2010). Bio- and magnetostratigraphy characteristic of the Upper Berriasian section "Zavodskaya balka" (Eastern Crimea, Feodosia) // *Bull. Saint Petersburg St. Univ., Geology, Geography,* 7, 2. P. 3–16 (in Russian).

Bagaeva M.I., Guzhikov A.Yu. (2014). Magnetic textures as indicators of formation of Tithonian-Berriasian rocks of the Mountain Crimea // *Izvestiya of Saratov University. New. ser. Ser. Earth science.* 14, 1. P. 41–47 (in Russian).

Borradaile, G. J. (1997) Deformation and paleomagnetism //, *Surv. Geophys.*, *18*, P. 405–435.

Bragin V., Dzyuba O.S., Kazansky A., Shurygin B.N. (2013) New data on the magnetostratigraphy of the Jurassic-Cretaceous boundary interval, Nordvik Peninsula (northern East Siberia) // *Russian Geology and Geophysics.* 54, 3. P. 335–348.

Cogné, J.-P., and H. Perroud (1987) Unstraining paleomagnetic vectors: the current state of debate, // *Eos Trans. Am. Geophys. Union.*, *68*, P. 705–712.

Dzyuba O.S., Guzhikov A.Yu., Manikin A.G., Shurygin B.N., Grishchenko V.A., Kosenko I.N., Surinskii A.M., Seltzer V.B., Urman O.S. (2017) Magneto- and carbon-isotope stratigraphy of the Lower–Middle Bathonian in the Sokur section (Saratov, Central Russia): implications for global correlation // *Russian Geology and Geophysics*, 58, 2. P. 206–224.

Facer, R. A. (1983) Folding, strain, and Graham's fold test in paleomagnetic investigations, // *Geophys. J. R. Astr. Soc.*, *72*, P. 165–171.

Grishenko V.A., Arkadiev V.V., Guzhikov A.Yu. Manikin A.G., Platonov E. S., Savel'eva Yu.N., Feodorova A.A., Shurekova O.V. (2016). Bio-, magneto - and cyclostratigraphy of the Upper Berriasian section near Alekseevka village (Belogorsk district, Republic of Crimea). Article 1. Ammonites. Magnetostratigraphy. Cyclostratigraphy (in Russian) // *Izvestiya of Saratov University. New. ser. Ser. Earth science.* 16, 3. P. 162–172 (in Russian).

Guzhikov A., Bagayeva M., Arkadiev V. (2014) Magnetostratigraphy of the Upper Berriasian "Zavodskaya Balka" section (East Crimea, Feodosiya) // *Volumina Jurassica*, XII, 1. P. 175–184.

Guzhikov A.Yu., Arkad'ev V.V., Baraboshkin E.Yu., Bagaeva M.I., Piskunov V.K., Rud'ko S.V., Perminov V.A., Manikin A.G. (2012) New Sedimentological, Bio-, and Magnetostratigraphic Data on the Jurassic–Cretaceous Boundary Interval of Eastern Crimea (Feodosiya) // *Stratigraphy and Geological Correlation*. 20, 3. P. 261–294.

Guzhikov A.Yu., Baraboshkin E.Yu., Birbina A.V. (2003) New paleomagnetic data for the Hauterivian–Aptian deposits of the Middle Volga region: A possibility of global correlation and dating of time-shifting of stratigraphic boundaries // *Russian Journal of Earth Sciences*. 5, 6. P. 1–30.

Kodama, K. P. (1988) Remanence rotation due to rock strain during folding and the stepwise application of the fold test, // *J. Geophys. Res. B*,. *93*, P. 3357–3371.

McFadden P.L. (1990) A new fold test for palaeomagnetic studies // *Geophysical Journal International*, 103. P. 163–169.

McFadden P.L., McElhinny M.W. (1990) Classification of the reversal test in paleomagnetism // *Geophysical Journal International*, 103. P. 725–729.

Nikishin A.M., Bolotov S.N., Baraboshkin E.Yu., Brune F.M., Ershov A.V., Clouting S., Kopaevich L.F., Nazarevich B.P. & Panov D.I. (1997) Mesozoic–Cenozoic history and geodynamics of the Crimean–Caucasus–Black Sea region // *Bull. of Moscow State University* 4, 3. P. 6–16 (in Russian).

Opdyke N.D., Channell J.E.T. (1996) Magnetic Stratigraphy. // N.Y.: Academic press. 344 p.

Stamatakos, J., and K. P. Kodama (1991) Flexural folding and the paleomagnetic fold test: An example of strain reorientation of remanence in the Mauch Chunk Formation, // *Tectonics*,. *10*, P. 807–819.

Supplements to the Stratigraphic Code of Russia (2000) / A.I. Zhamoida (ch. Ed.) (2000). // St. Petersburg: Izd. VSEGEI. 112 p (in Russian).

Van der Pluijm, B. A. (1987) Grain scale deformation and the fold test—evaluation of synfolding remagnetization, // *Geophys. Res. Lett.*, *14*, P. 155–157.

Van der Voo R. (1993) Palaeomagnetism of the Atlantic, Tethys, and Iapetus oceans. // *Cambridge: Cambridge Univ. Press.* 412 p.

Chapter 4
Carboniferous of the Russian Platform: Paleomagnetic Data

A. G. Iosifidi, V. A. Mikhailova, V. V. Popov, E. S. Sergienko, A. V. Danilova, N. M. Otmas and A. V. Zhuravlev

Abstract New results for the Carboniferous rocks (Moscovian, Bashkirian and Visean stages) from four sections of the Russian Platform are presented. Primary magnetite and secondary hematite are the main carriers of magnetization in the studied rocks. All isolated NRM components generally correspond to the geomagnetic field of reversed polarity. Only in the rocks of the Visean stage a small zone of normal polarity is found. Widespread late Paleozoic remagnetization of the Carboniferous rocks of the Russian Platform is detected.

Keywords Carboniferous deposits · Paleomagnetic poles · Russian platform

Introduction

Paleomagnetic studies of the Carboniferous deposits of the Russian Platform were mainly carried out in the 60s years of the last century and have low parameters of reliability criteria for definitions. The reinvestigation we undertook recently of some of the old collections allowed to assess the reliability of certain determinations made in the 60s years (Iosifidi and Khramov 2002; Iosifidi et al. 2010; Iosifidi and Mikhailova 2017). In none of the studied (reinvestigated) sections, it was possible

A. G. Iosifidi (✉) · V. A. Mikhailova · V. V. Popov · A. V. Danilova · N. M. Otmas
All-Russia Petroleum Research Exploration Institute (VNIGRI), St. Petersburg, Russia
e-mail: iosag@mail.ru

V. V. Popov · E. S. Sergienko
Saint-Petersburg State University, St. Petersburg, Russia

A. V. Zhuravlev
Institut of Geology, Komi Science Center, Ural Branch of the Russian Academy of Sciences, Syktyvkar, Russia

A. G. Iosifidi
Institute of Terrestrial Magnetism, Ionosphere and Radiowave Propogation, St. Petersburg Branch, St. Petersburg, Russia

D. Nurgaliev et al. (eds.), *Recent Advances in Rock Magnetism, Environmental Magnetism and Paleomagnetism*, Springer Geophysics,
https://doi.org/10.1007/978-3-319-90437-5_4

to identify the directions corresponding to the normal direction of the geomagnetic field in the Carboniferous time. Meanwhile, according to the current general magnetostratigraphic scales (Dopolneniya … 2000; A Geological Time … 2004; The Geologic Time … 2012), there are numerous zones of normal polarity in the interval 356–299 million years in the Carboniferous period. One of the tasks of our research is to identify normal and reverse magnetozons in the Carboniferous period. We present new palaeomagnetic data for Carboniferous rocks (Moscovian, Bashkirian and Visean stages) from four sections of the Russian Platform.

Objects of Study and Experimental Methods

Ryazan region. The collection of rocks from Shatskaya series of the Vereiskiy horizon, Moscovian stage were sampled on the right bank of the river Aza, near the village Kuchasevo by R. A. Komissarova in 1965, Fig. 4.1. The rocks are represented by red-brown clays, sandy clays (blue and violet-red) and red-violet sandstones (7 m of the section, 12 stratigraphic levels) (Geologiya SSSR. T. IV 1971). The bedding is close to horizontal, the mean coordinates of the sampling area are 54.09°N, 41.97°E.

Donbass I. The rocks of the Amvrosievskaya Formation of the Bashkirian Stage of the Middle Carboniferous were sampled 1 km east of village Grigorievka (Ambrosievskiy district, Donetsk region) (Opdyke et al. 1992), Fig. 4.1. The total thickness of the sampled sequence is 35 m, 20 stratigraphic levels (Stratigrafiya URSR 1969). The bedding varies within 270°–344° in the dip azimuth and within 32°–50° in the dip angle; the mean coordinates of the sampling area are 48.85°N, 38.80°E.

Novgorod region. The rocks from the Aleksinskiy horizon of the Visean stage were sampled at the road excavation 300 m past village Novinka (road from the district center of Lubytino), Fig. 4.1. The rocks are represented by clays, sandstones, siltstones, argillites and limestones (Geologiya SSSR. T. I. 1971; Geologicheskaya karta SSSR 1968), thickness is 17 m with 44 stratigraphic levels. The bedding is close to horizontal, the mean coordinates of the sampling area are 58.3°N, 33.6°E.

Donbass II. At the left bank of the river Mokraya Volnovakha, 3 km upstream from the village Styla, limestones of the Styl'skaya suite representing the upper part of the Mokrovolnovakhskaya series of the Visean stage of the Lower Carboniferous were sampled by R. A. Komissarova in 1961, Fig. 4.1. The total thickness of the sampled sequence is 25 m, with 11 stratigraphic levels (Stratigrafiya URSR 1969). The bedding varies within 358°–33° in the dip azimuth and within 7°–18° in the dip angle; the mean sampling coordinates are 47.68°N, 37.82°E.

Paleomagnetic studies and processing of the obtained results were carried out according to the standard procedure (Khramov 1987). Analysis of the obtained data

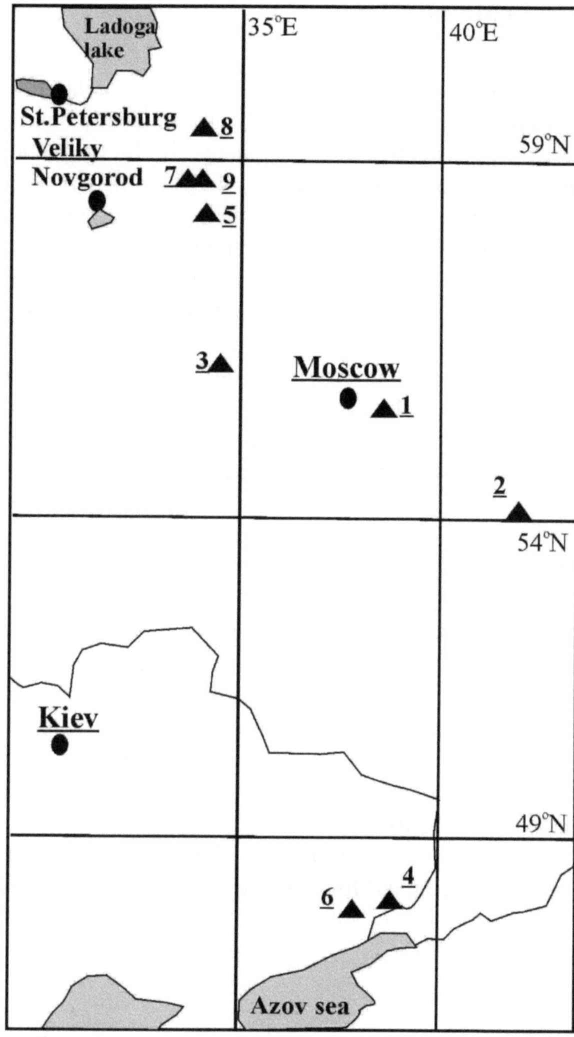

Fig. 4.1 Schematic map of natural Carboniferous outcrops of Russian platform: *1* Gzhel station; *2* r. Aza; *3* Rzev; *4* v. Grigorievka; *5* r. Msta; *6* M. Volnovakha; *7* Lubytino; *8* Boksitogorsk; *9* r. Priksha

and presentation of the results in a graphical form was conducted using software packages (Enkin 1994; Popov 2007, 2007). The fold test was applied in the modification of Watson and Enkin (1993), and the polarity reversal test according to McFadden and McElhinny (1990). The statistics were estimated by averaging over the hand samples representing stratigraphic levels.

The study was carried out at VNIGRI and the Research Park of St. Petersburg State University: Center for Geo-Environmental Research and Modeling (GEOMODEL); Center for Microscopy and Microanalysis; Center for X-ray Diffraction Studies.

Experimental Data

Magnetic properties. The values of the scalar magnetic characteristics: natural remanent magnetization (NRM) and magnetic susceptibility (Km) for the studied rocks vary widely. Curves of change in the magnetic susceptibility (Km-T curves) during heating are shown in Fig. 4.2 (cooling curves not shown). In all cases during the heating process significant chemical transformations take place, and upon cooling a new magnetic phase appears, most often magnetite.

For samples from the Shatskaya series of the Vereiskiy horizon of the Moscovian stage of the Carboniferous, as well as for the Styl'skaya suite, two magnetic phases with Curie points of 560–580 °C and 640–680 °C, respectively, are observed. Some samples from the Shatskaya series show an infection in the Km-T curves in the 320–340 °C range. Similar inflection can be seen for samples of the Styl'skaya suite in the 380–420 °C range.

Results of microscopic studies. SEM images of grains of ore minerals are shown in Figs. 4.3 and 4.4. The morphology of the observed grains and aggregates of ore minerals, as well as the results of microprobe analysis, allow us to conclude the following.

The ore fraction of the Shatskaya series samples is represented by relic magnetite (Fig. 4.3a, b). Often it contains impurities of up to weight 3% Al, and Ti. The grains are typically 5–20 μm in size, of different degree of roundness and destruction, but without signs of oxidation. Also, newly formed hematite is detected, sometimes with Ti impurities, with grain sizes from <1 to 10 μm. Hematite of different morphology is represented by individual oolite-like grains with a scaly structure (Fig. 4.4c), mineral aggregates of a nonmagnetic matrix in the form of skeletal-like forms or in the form of "shells", hematite roses (Fig. 4.3d). Also, native nickel and iron with individual grain sizes of up to 100 μm are reliably identified. In addition, there are individual finds in the form of copper oxides in chalcopyrite (apparently

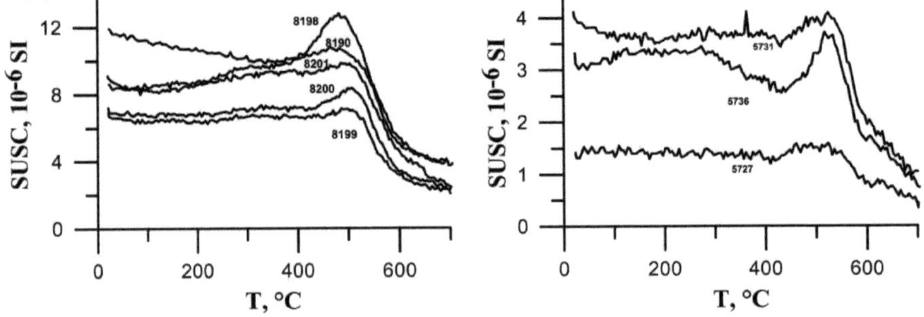

Fig. 4.2 Examples of thermomagnetic analysis of the Carboniferous rocks: the curves of the behavior of magnetic susceptibility at heating. **a** Samples of the Shatskaya series of the Vereyan horizon, Moscovian stage; **b** samples of the upper part of the Mokrovolnovakhskaya series of the Visean stage

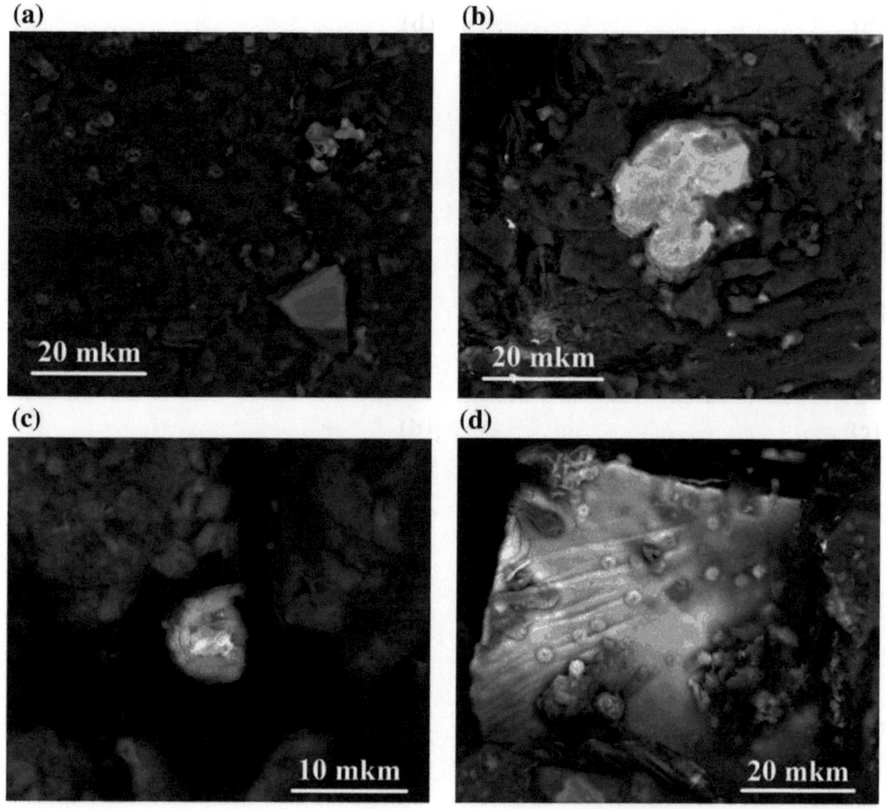

Fig. 4.3 SEM-images of microprobe samples of rocks of the Shatskaya series of the Vereyan horizon, Moscovian stage. Relic magnetite (**a** sample 8200, **b** sample 8199), oolite-like grain of hematite (**c** sample 8200), hematite release by leikoxene (**d** sample 8194)

relic). Thus, electron microscopic studies have shown that the rocks under investigation carry signs of deep reworking of the primary material, mainly under reducing conditions. This is confirmed by the results of powder X-ray diffraction (significant, up to 50% and more, content of quartz).

Samples from the Styl'skaya suite are clayey limestones. In the ore fraction, crystalline magnetite with grain sizes of 5–40 μm, weakly loosened, possibly primary (Fig. 4.4a, b), is found in the form of both individual grains and mineral aggregates. A single find of a grain in the form of spheroid is shown in Fig. 4.4. The grain morphology, lack of surface roughness and the results of the microsonde analysis suggest that it could be a two-phase grain consisting of a hematite shell and a magnetite core. Secondary minerals are represented by hematite and iron hydroxides. In Fig. 4.4d, the separation of the ore phase on smectite is shown. It may be lepidocrocite and hematite. In the figure, these are the light gray, almost white grains, with sizes less than 10 microns.

Fig. 4.4 SEM images of microprobe samples of the rocks of the upper part of the Mokrovolnovakhskaya series of the Visean stage. Crystalline magnetite (**a** sample 5727, **b** sample 5736), (spheroid—sample 5736), separation of the ore phase on smectite (**d** sample 5727)

Powder X-ray diffraction. In the samples of the Shatskaya, a considerable amount of clay minerals is present, which do not allow to perform an accurate quantitative analysis. Presumably, mixed-type minerals such as illite-smectite are present in the samples. The semiquantitative phase analysis of the samples showed the following composition (weight %): quartz 20-54; K-Na feldspars 15-30; mica 1-50; kaolinite 7-28; chlorite 0-11; ore minerals 1-5.

Component Analysis of NRM

The NRM components were isolated by stepwise thermal demagnetization and demagnetization by an alternating magnetic field. The single component NRM dominates in the rocks of the Shatskaya series in the high-temperature range of thermal demagnetization, as seen from Zijderveld diagrams (Fig. 4.5, samples

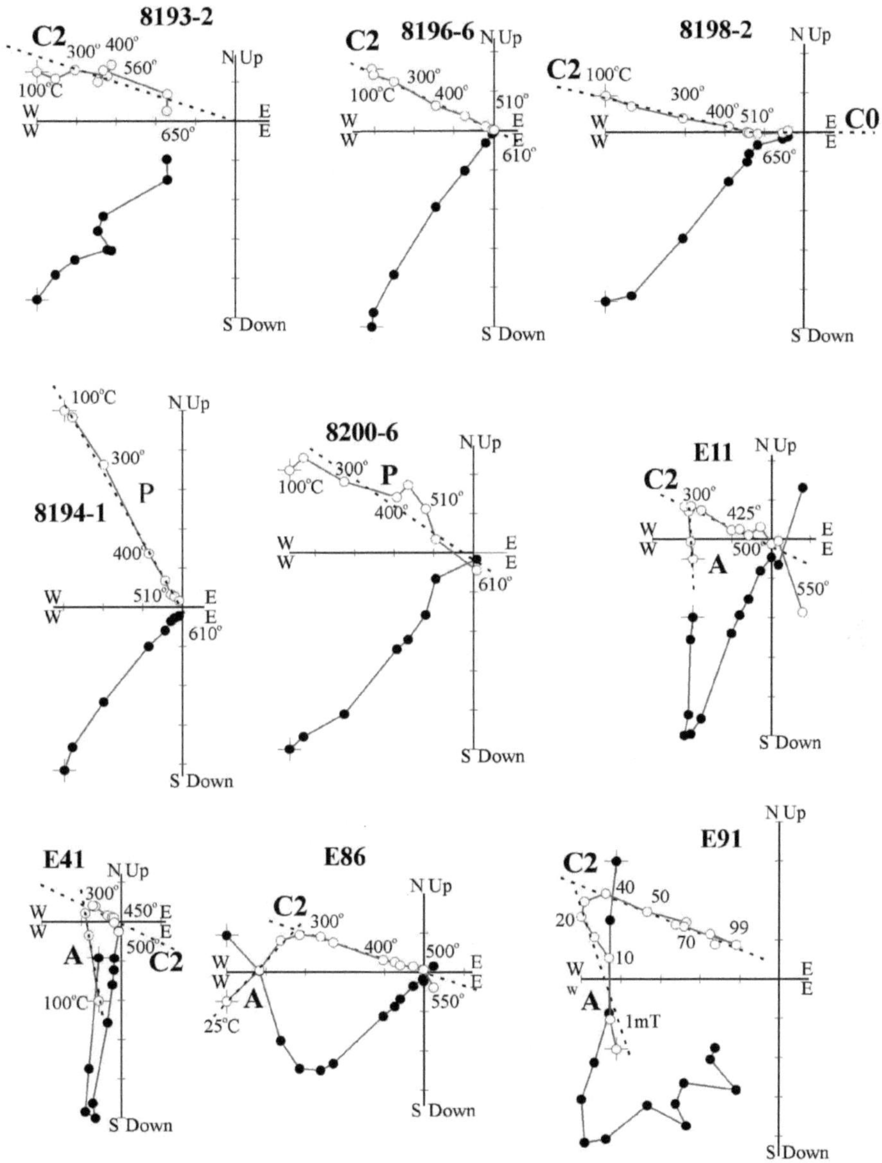

Fig. 4.5 Examples of stepwise thermodemagnetization of the Carboniferous samples. The Zijderveld diagram is shown in the stratigraphic coordinates; the open and filled circles are vector projections on the horizontal and vertical plane, respectively. The dashed line shows the intervals of isolation of the NRM components (their names are indicated in accordance with Table 4.1). The Zijderveld diagrams are constructed with the use of Remasoft 3.0 program (Chadima and Hrouda 2006)

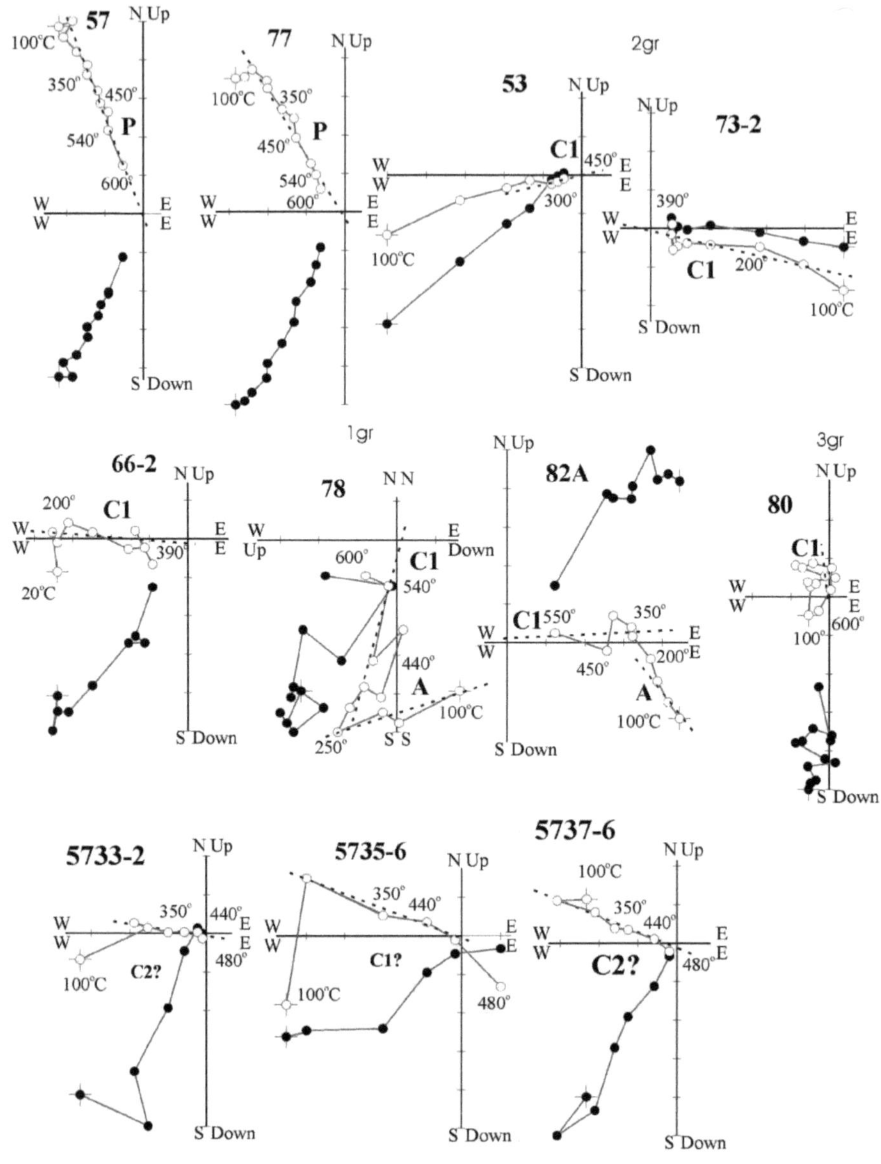

Fig. 4.5 (continued)

8193-2, 8196-6, 8194-1, 8200-6). In some samples, a second high-temperature component C0 is isolated, Fig. 4.5, sample 8198-2. The high temperature components P and C2 are detected in the temperature interval from 200 °C up to 610 °C and the components of C0 in the interval 510–654 °C. Statistical characteristics of isolated components are listed in Table 4.1, and Fig. 4.6 shows their distributions.

Table 4.1 Paleomagnetic directions and pole positions for the studied Carboniferous rocks

Object of study	Rock age	NRM component	Identification intervals, °C	N/n	s.c.	D°	I°	K	α_{95}°	Test	Φ °N	Λ °E	dp°	dm°	φ_m°
r. Aza (54.09°N, 41.97°E). Shatskay series, Vereyan horizon, Moscovian stage															
Redbeds (clay and sandstones)	C_2m	C2	200–650	15/24	g	224	−15	39	6		32	168	3	6	8
	C_2m	P	200–650	4/5	g	224	−41	67	11		45	157	8	13	24
		CO	500–650	4/5	g	258	3	29	17		6	143	9	17	2
v. Grigorievka (47.85°N, 38.80°E), amvrosievskay suite, Bashkirian stage															
Sandstones, siltstones, mudstones and limestones	C_2b	A	100–250	22/22	g	357	63	40	5	F−, 0.2±0.4	86	251	6	8	45
	C_2b	C2		17/17		203	−42	23	8						
			250–500		s	200	−7	35	6	F+, 0.7±0.4	42	191	3	6	4
		C2-1	250–500	9/9	s	209	−9	81	6	F~, 0.6±1.8	40	180	3	6	5
		C2-2	250–500	8/8	s	190	−6	53	8	F+, 0.7±0.5	44	205	4	8	3
				17		20	40		14						
		MSN			s	22	3		7		40	190	4	7	2
Lubytino (58.8°N, 33.6°E), Aleksinskiy horizon, Visean stage															
Clays, sandstones, siltstones, mudstones and limestones	C_1v	A	100–250	25/25	g	31	69	10	9		72	130	13	15	53
		P	100–600	13/15	g	222	−47	26	8		48	152	7	10	28
		C1-1grR	260–450	5/6	g	253	−3	40	12						
		C1-1grN	450–600	4/6	g	81	−3	46	14						
		C1-1grNR		9/12	g	257	−1	40	8	Rc, $\gamma/\gamma_{cr.}$=10/17	7	135	4	8	1
		C1-2grR	200–400	5/7	g	217	6	29	14						
		C1-2grN	300–600	2/3	g	29	7	253	16						
		C1-2grNR		7/10	g	215	2	32	11	Rc, $\gamma/\gamma_{cr.}$=15/16	24	175	6	11	1
		C1		16/22	g	238	1	11	12		16	152	6	12	1

(continued)

Table 4.1 (continued)

Object of study	Rock age	NRM component	Identification intervals, °C	N/n	s.c.	D°	I°	K	α_{95}°	Test	Φ °N	Λ °E	dp°	dm°	φ_m°
r. M. Volnovakha (47.68°N, 37.82°E), mokrovolnovakhskay series styl'skay' suite, Visean Stage															
Limestones	C_1v	C1?	280–480	19/29	g	211	−22	23	7						
					s	210	−8	26	7	F∼	39	178	4	7	4
		C1-1grN	280–480	6/10	g	230	−24	62	9						
					s	227	−10	69	8	F∼	32	159	4	8	5
		C1-2grN	280–480	9/14	g	206	−17	60	7						
					s	206	−3	76	6	F∼	39	184	3	6	2
		C1-3grN	280–480	4/5	g	192	−27	85	10						
					s	193	−15	115	9	F∼	49	198	5	9	8

Note N/n refers to number of samples and specimens; C.S. is the coordinate system: g-geographic, s-stratigraphic; D and I are the declination and inclination (both in degrees) of the mean directions; K is the concentration parameter; $\acute{\alpha}_{95}$ is the radius of the confidence circle for the average direction at 95% probability; Under 'Test' F+, F−, F∼ indicate that the fold test according to Watson and Enkin (1993) is positive, negative, indeterminate; Rc, γ/γ_{cr} indicate that the reversal test according to McFadden and McElhinny (1990) is positive, class C; Φ and Λ are the Northern latitude and Eastern longitude of the paleomagnetic pole; φ_m is the paleolatitude. MSN—New modification of the method of remagnetization planes proposed by Khramov, a method of "sliding normal" to the remagnetization circles (Khramov et al. 2011)

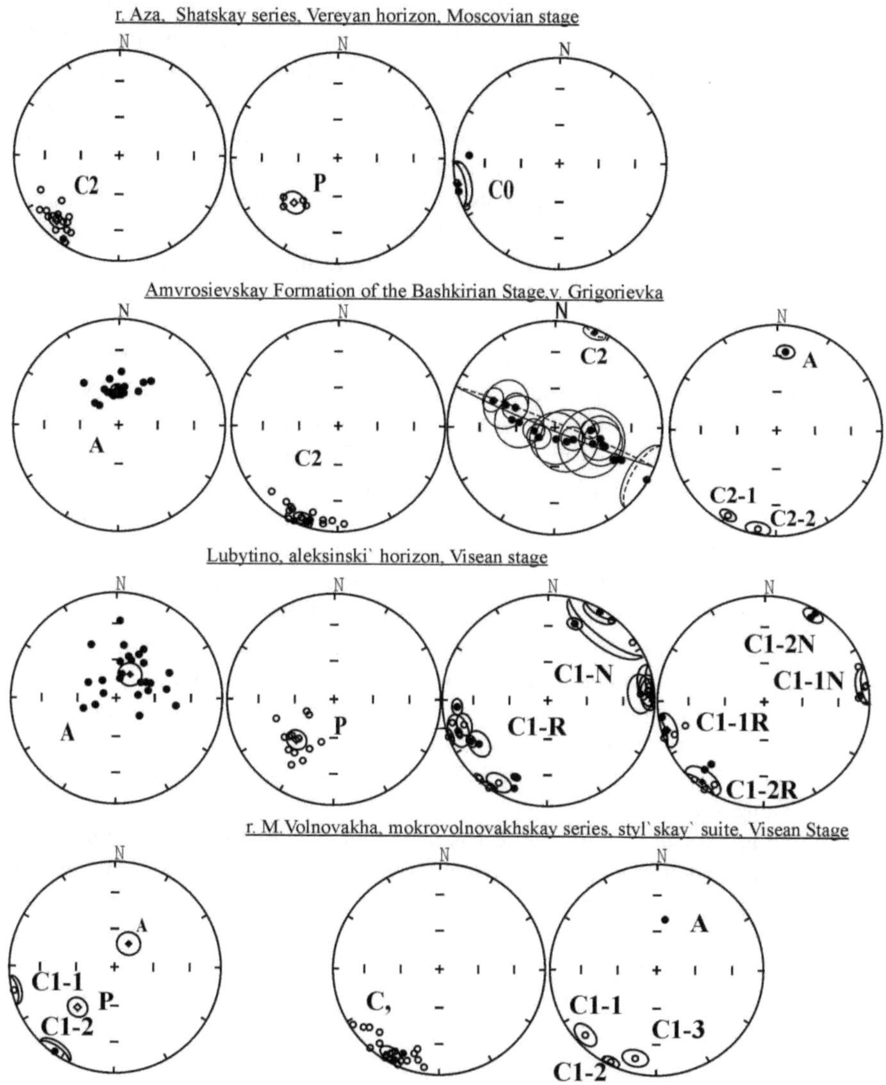

Fig. 4.6 The distributions of the directions of NRM components after demagnetizations for the Carboniferous rocks. Closed (open) symbols are in the lower (upper) hemisphere. Average directions and 95% confidence circles are indicated. The distributions are shown in the stratigraphic coordinate system (for component A in geographic coordinate system); full explanation in text

Two components of the NRM are found in the rocks of the Amvrosievskaya Formation of the Bashkirian Stage of the Middle Carboniferous. The first component A is isolated in the temperature range 100–250 °C and in alternating fields 1–15 mT. The high-temperature component C2 is isolated in the temperature range

250–500 °C and alternating fields 30–100 mT, (Fig. 4.5, samples E11, E41, E86, E91). The distributions of components A and C2 are shown in Fig. 4.6, and their statistical characteristics listed in Table 4.1. The fold test for component A is negative, and for component C2 is positive (Table 4.1). To estimate the direction of the component C2 we applied a new modification of the method of remagnetization planes, proposed by Khramov, a method of "sliding normal" to the remagnetization circles (Khramov et al. 2011). All obtained directions of the C2 component averaged over the stratigraphic levels are shown in Fig. 4.6. This figure shows examples of determining the position of sliding normal (circles through internal groups were drawn in a stratigraphic sequence at three points). The final result is shown in Table 4.1 and in Fig. 4.6.

The rocks of the Aleksinskiy horizon of the Visean stage have three NRM components. The first component A is isolated (Fig. 4.5, samples 78, 82A) in the temperature range 100–300 °C. The second component P is isolated in the temperature range 200–600 °C, Fig. 4.5, samples 57, 77. The bipolar component C1 is isolated in the medium and high-temperature intervals 200–450 and 200–600 °C, respectively, Fig. 4.5, samples 53, 73-2, 80, 66-2, 78, 82A. The bipolar component of C1 has two groups C1-1 and C1-2. The reversal test is positive for each of the groups, Table 4.1. The distributions of the components A, P, and C1 are shown in Fig. 4.6, and their statistical characteristics listed in Table 4.1.

The NRM of the Mokrovolnovakhskaya series of the Visean stage of the Lower Carboniferous has a single component (Fig. 4.5, samples 5732-2, 5735-6 и 5737-6). The high-temperature component C1 is isolated in the temperature interval 200–480°. The distribution of component C1 for the rocks from the outcrop on the river Mokraya. Volnovakha is shown in Fig. 4.6 and its statistical characteristics are listed in Table 4.1.

The fold test for the component C1 is uncertain. The distribution of component C1 is elongated (non-Fisherian). Declination varies from 230 to 195°. We divided all the directions into three parts (at this stage of the work, visually), and the average directions of the resulting groups of component C1 lie on a large remagnetization circle, Fig. 4.6.

Discussion

The obtained paleomagnetic data on the different age deposits of Carboniferous indicate the presence of both primary and secondary NRM components. According to the palaeogeographic data in the Carboniferous time on the Russian platform, a number of stages of transgressions and regressions of the ancient sea occurred. The conditions of both arid and humid climate were formed, which led to demagnetization and remagnetization of rocks (Alekseev et al. 1996; Osipova and Vel'skaya 1967; Makhlina et al. 1993, 2001a, b).

Microprobe analysis shows the presence of relic ore minerals in rocks of Shatskaya series and Styl'skaya suite. It can be assumed that these minerals carry the oldest component of natural remanent magnetization. Samples remagnetization resulting in acquisition of metachronous magnetization components was apparently due to the growth of secondary hematite and iron hydroxides. The presence of different generations of hematite implies that this mineral has been formed intermittingly, following the changing hydrothermal conditions. This is supported by paleomagnetic data as well, most of isolated NRM components showing elongated distributions, Fig. 4.6. It is noteworthy that in the results obtained, the NRM components of the reverse polarity dominate (with the exception of the Lyubytino section, where the bipolar component C1 is isolated). After averaging over stratigraphic levels, the bipolar component C1 can be divided into two groups (C1-1 and C1-2), in which both normal and reversed polarity components are present, Fig. 4.6. The details of the selective remagnetization in this section (the presence of component C1-2) are however not entirely clear at present. For the intervals where bipolar components C1-1 and C1-2 are present, a more detailed study on sister samples will be carried out aiming to clarify the reasons for the difference in the declination of these components.

Distributions for the components C2 (v. Grigorievka) and C1 (r. M. Volnovakha) are elongated from the west to the southeast (non-Fisherian). These distributions are divided into groups, indicating the presence of remagnetization process (all groups lie on a large circle with component A, Fig. 4.6). In accordance with the existing magnetostratigraphic scales (Dopolneniya 2000; The Geologic Time … 2012) in the Carboniferous period, the geomagnetic field with a frequent polarity change prevailed, intervals between reversals being about 1 My. In this case, it is difficult to admit the remagnetization of the entire section by a reversed-polarity field in the Carboniferous time. The component P corresponds to remagnetization in the middle Permian time. The poles position for components of C2 and C1 are close to the results of the first researchers, 31°N, 168°E (Khramov 1967) and 42°N, 181°E (Khramov and Komissarova 1963) respectively.

Positions of new paleomagnetic poles obtained in this study for the Carboniferous deposits of the Russian Platform are listed in Table 4.2. APWP for the Russian platform in the range 269–338 Ma is shown in Fig. 4.7. Paleomagnetic data from magmatic rocks of the early Carboniferous from the British Isles which pass the polarity reversal test according to (McFadden and McElhinny 1990) are listed in Table 4.2. The average age of these rocks is 338 Ma. The mean pole from the British Isles (Table 4.2) is also in agreement with the spline poles, which confirms the reality of such APWP for Russian platform in Carboniferous time, Fig. 4.7.

Thus, our studies show the direction of further work on the investigation (re-investigation of old collections) of Carboniferous deposits of the Russian Platform. This is primarily the study of collections (outcrops), in which the first researchers identified bipolar components of NRM and the rocks are represented not only by red-colored deposits.

Table 4.2 New paleomagnetic directions and pole positions for the Carboniferous rocks of the Russian platform

No	Index in the World paleomagnetic database (Result no)/reference	Age of NRM component	N/n	D°	I°	K	α^o_{95}	Φ, °N	Λ, °E	dp°	dm°	φ^o_m
1	2	3	4	5	6	7	8	9	10	11	12	13
Gzhel station, Shchelkovo clays, Gzhelian Stage												
1	(Iosifidi and Mikhailova 2017)	C_3gr (302 Ma)	9/9	217	−33	182	4	43	167	3	5	18
r. Aza, Vereyan horizon, Moscovian stage												
2	t.w.	C_2m (315 Ma)	15/24	224	−15	39	6	32	168	3	6	8
Rzhev, Rzhev layers, Kashirian horizon, Moscovian Stage												
3	(Iosifidi and Mikhailova 2017)	C_2m	8/14	216	−29	84	6	41	166	4	7	16
v. Grigorievka, amvrosievskay suite, Bashkirian stage												
4	t.w.	C_2b (318 Ma)	9/9	209	−9	81	6	40	180	3	6	5
r. Msta, Serpukhovian and Visean stages												
5	(Popov et al. 2016)	C_1s (330 Ma)	4/4	223	−22	36	16	33	161	9	17	11
		C_1v (333 Ma)	10/13	220	−21	32	9	34	168	5	10	11
r. M.Volnovakha, mokrovolnovakhskay series, styl'skay' suite, Visean Stage												
6	t.w.	C_1v (302 Ma)	19/29	210	−8	26	7	39	178	4	7	4
Lubytino, Aleksinskiy horizon, Visean stage												
		C_1v-1NR	9/12	257	−1	40	8	7	135	4	8	1
		C_1v-2NR	7/10	215	2	32	11	24	175	6	11	1
7	t.w.	C_1v-NR (337 Ma)	16/22	238	1	11	12	16	152	6	12	1
Leningrad region, Boksitogorsk, Tula horizon, Visean stage												
8	(Iosifidi and Mikhailova 2017)	C_1v (335 Ma)	12/23	227	−15	25	9	27	159	5	9	8
9		British Isles										
	6299/(53.2°N, 358.2°W)	C_1v (336 Ma), Rc	10	203	35.8	36	8.2	14.3	155.9	5.5	9.5	20

(continued)

Table 4.2 (continued)

No	Index in the World paleomagnetic database (Result no)/reference	Age of NRM component	N/n	D°	I°	K	α^{o}_{95}	Φ, °N	Λ, °E	dp°	dm°	φ^{o}_{m}
	3430/(56.0°N, 357.0°W)	C_1v (339 Ma), Rc	17	15.9	−28.5	23	7.6	18	161	5	9	15
	6315/(56.2°N, 356.9°W)	C_1v (339 Ma), Rc	13	205	31	31	7	14.1	152.2	4	8	17
	Mean	C_1v (338 Ma), Rc					N = 3	15	156	A_{95} = 5°		

Note The indices of the results correspond to the Resultno of the Worlds Paleomagnetic Database (WPDB) (McElhinny and Lock 1996; Pisarevsky 2005); t.w. stands for this work; the A^{o}_{95} confidence circle radius at 95% probability for direction of mean pole; other notations see note to Table 4.1

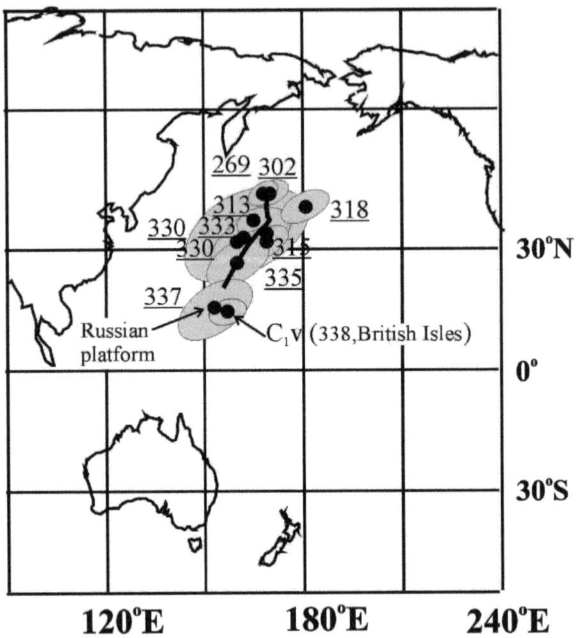

Fig. 4.7 The positions of the obtained new paleomagnetic poles for Russian platform and mean pole for British Isles (338 Ma) (according to the data of Table 4.2). APWP show's by black line. The plot is drawn with the use of GMAP program (Torsvik et al. 1990)

Acknowledgements We are grateful to A. Yu. Kazansky for criticism and constructive comments, A. A. Kosterov for help in translating the text into English.

Scientific research was carried out at VNIGRI and the Research Park of St. Petersburg State University: Center for Geo-Environmental Research and Modeling (GEOMODEL); Center for Microscopy and Microanalysis; Center for X-ray Diffraction Studies.

The work was supported by the Russian Foundation for Basic Research under project no. 16-05-00603a.

References

Alekseev A, Konovalova L., Nikishin A. The Devonian and Carboniferous of the Moscow Syneclise (Russian Platform): Stratigraphy and sea-level changes // Tectonophysics. 1996. V. 268. p. 149–168.

A Geological Time Scale 2004// Editors Gradstein F.M., Ogg J.G., Schmitz M.D. Cambridge University Press, Cambridge, p. 589.

Chadima, M. and Hrouda, F., Remasoft 3.0: a user-friendly paleomagnetic data browser and analyzer, *Travaux Geophysiques*, 2006, vol. XXVII, pp. 20–21.

Dopolneniya k Stratigraficheskomu kodeksu Rossii(Supplements to Stratigraphic Code of Russia), Zhamoid, A.I., Ed., St. Petersburg: VSEGEI, 2000.112 p.

Enkin, R.J., A Computer Program Package for Analysis and Presentation of Palaeomagnetic Data, Sidney: Pacific Geoscience Centre, Geol. Survey Canada. Sidney. 1994. 16 p.

Geologicheskaya karta SSSR. Masshtab 1: 200000. Seriya Tikhvinsko-Onezhskaya, O-36-IV (1: 200000 Geological Map of the USSR, Tikhvinsko-Onezhskaya Series, O-36-IV), Moscow: Gosgeoltekhizdat, 1968.

Geologiya SSSR. T. I. Leningradskaya, Pskovskayai Novgorodskaya oblasti. Geologicheskoe opisanie. Severo- Zapadnoe GU (USSR Geology. Vol. I: Leningrad, Pskov and Novgorod Regions, Geological Description. North Western Geological Bureau), Moscow: Nedra, 1971. 504 p.

Geologiya SSSR. T. IV Tsentr Evropeiskoi chasti SSSR. Geologicheskoe opisanie(USSR Geology. Vol. IV: Central Part of the USSR European Territory), Moscow: Nedra, 1971. 742 p.

Iosifidi, A., G, Mac Niocaill, C., Khramov, A.N., Dekkers, M.J., and Popov, V.V., Palaeogeographic implications of differential inclination shallowing in Permo-Carboniferous sediments from the Donets basin, Ukraine, *Tectonophysics*, 2010, vol. 490, nos. 3–4, pp. 229–240.

Iosifidi A. G., Mikhailova V. A. Paleomagnetism of Carboniferous Formations of the Russian Platform: Reinvestigation of Old Collections, Izvestiya, Physics of the Solid Earth, 2017, Vol. 53, No. 3, pp. 461–476.

Iosifidi, A.G. and Khramov, A.N., Paleomagnetism of the upper Carboniferous and Lower Permian deposits of the East European platform: a key paleomagnetic pole and kinematics of collision with Urals, *Izv., Phys. Solid Earth*, 2002, vol. 38, no. 5, pp. 389–403.

Khramov, A.N., *Palaeomagnetology*, Heidelberg: Springer, 1987. 330 p.

Khramov, A.N., Earths magnetic field in Late Paleozoic, *Izv. Akad. Nauk SSSR, Fiz. Zemli*, 1967, no. 1, pp. 86–108.

Khramov, A.N. and Komissarova, R.A., Earths magnetic field in the Carboniferous from the data of paleomagnetic studies in the European part of the USSR, in *Magnetizm gornykh porod I paleomagnetizm* (Rock Magnetism and Paleomagnetism*)*, Krasnoyarsk: SO AN SSSR, 1963, pp. 341–351.

Khramov, A.N., Iosifidi, A.G., Popov, V.V., Pytidesiatilatie paleomagnitnykh issledovaniei permskikh otlozeniei Donbassa: dostovernost' rezul'tatov/Materialy Vserosieiskogo seminara po paleomagnetizmu I magnetizmugornykhporod/ V. P. Shcherbakov, ed. Borok. 27–30.10.2011. Yaroslavl': OOO " Servisnyeitsentr". 2011. p. 243-247.

Makhlina M. H., Alekseev A. S., Goreva N. V., Isakova T. N., Drutskoei S. N. Sredniy karbon Moskovskoiy sineklizy(yuznay khast'). t. 1. Stratigrafiya. – M.: Paleontologicheskiy institut RAN, 2001b. 244 s.

Makhlina M. H., Vdovenko M. V., Alekseev A. S., Bysheva T. V., Donakova L. M., Zhulitova V. E., Kononova L. I., Umnova N. I., Shik E. M. Nizniiy karbon Moskovskoiy sineklizy i Voronezskoiy antiklizy. – M.: Nauka, 1993. 221 s.

Makhlina M. H., Alekseev A. S., Goreva N. V., Isakova T. N. Ritmo– I biostratigrafiia karbona Moskovskoiy sineklizy//Izvestiya Vuzov. Seriia: geologiayirazvedka. – 2001a. №5. s. 42–58.

McElhinny, M.W. and Lock, J., IAGA palaeomagnetic databases with Access, *Surv. Geophys.*, 1996, vol. 17, pp. 575–591.

McFadden, P.L. and McElhinny, M.W., Classification of reversal test in palaeomagnetism, *Geophys. J. Int.*, 1990, vol. 103, pp. 725–729.

Opdyke N. D., Khramov A. N., Gurevich E. L., Iosifidi A. G., Makarov I. A. A paleomagnetic study of middleCarboniferous of the Donetz Basin, Ukraine. EOS, Truns. Am. Geophys. Union, Spring Meet., 1993, p.118

Osipova A. M., Vel'skay T. H. Opytlitologo-paleo ekologicheskogo izucheniay vize-namyrskikh otlozeniy Moskovskoiy sineklizy//Litologiay I poleznye iskopaemye.1967.№ 5.s. 118–142.

Pisarevsky, S.A., New edition of the global palaeomagnetic database, *EOS Transact.*, 2005, vol. 66,170 p.

Popov, V.V., 2007, 2008. http://paleomag-ifz.ru//ru/soft.

Popov V.V., Sergienko E.S., Iosifidi A.G., Mikhailova V.A. Paleomagnetic studies of the reference section of the lower Carboniferous of Msta river//11th International Conference "PROBLEMS OF GEOCOSMOS", Book of Abstracts, *St. Petersburg, Petrodvorets, October 3–7, 2016, St. Petersburg*: SPSU, 2016. p. 165.

Stratigrafiya URSR,. - Kyiv, Naukova dumka, 1969. - Vol. 5: Karbon.- 412 p.
The Geologic Time Scale 2012 / Editors Gradstein F.M., Ogg J.G., Schmitz M.D., Ogg G.M., The Boulevard, Langford Lane, Kidlington, Oxford OX5 1 GB, UK Radarweg 29, PO Box 211, 1000 AE Amsterdam, The Netherlands 225 Wyman Street, Waltham, MA 02451. USA. 2012. 1160 p.
Torsvik T. H., Smethurst M. A., Pesonen L. J. GMAP - geographic mapping and paleoreconstruction package. NGU rapport nr 90.019, 1990.
Watson, G.S. and Enkin, R.J., The fold test in palaeomagnetism as a parameter estimation problem, *Geophys. Rev. Lett.*, 1993, vol. 20, pp. 2135–2138.

Chapter 5
Evidence for the Existence of the Gothenburg and Mono Lake Excursions Based on Paleomagnetic Data from Baunt Lake Sediments (Northern Transbaikalia)

M. A. Krainov, E. V. Bezrukova, A. A. Shchetnikov and A. Yu. Peskov

Abstract Paleomagnetic data were obtained from a 13.7 m long-sediment core recovered by drilling at the deepest point 33 m, Lake Baunt in March, 2014. The base of the core was radiocarbon dated to 29 ka BP. The paleomagnetic record from the sediments of Baunt Lake reveals and documents for the first time the Gothenburg and Mono-Lake magnetic excursions for the lakes of the Baikal region. The high sedimentation rate and a reliable age model permit assessment of the age and duration of those excursions. Our evaluation of age yields an estimate of 13.3-13.1 ka BP for the Gothenburg excursion and 26.1-25.9 ka BP for the Mono Lake excursion.

Keywords Transbaikalia · Lacustrine sediments · Pleistocene · Holocene Paleoclimate · Paleomagnetism · Excursions · Lithology

M. A. Krainov (✉) · E. V. Bezrukova · A. A. Shchetnikov
A.P. Vinogradov Institute of Geochemistry SB RAS, Irkutsk, Russia
e-mail: susel_usel@mail.ru

E. V. Bezrukova
Institute of Archaeology and Ethnography SB RAS, Novosibirsk, Russia

A. A. Shchetnikov
Institute of the Earth's Crust SB RAS, Irkutsk, Russia

A. Yu. Peskov
Yu. A. Kosygin Institute of Tectonics and Geophysics FEB RAS, Khabarovsk, Russia

M. A. Krainov · E. V. Bezrukova · A. A. Shchetnikov
Irkutsk Scientific Center SB RAS, Irkutsk, Russia

E. V. Bezrukova
Tyumen Industrial University, Tyumen, Russia

D. Nurgaliev et al. (eds.), *Recent Advances in Rock Magnetism, Environmental Magnetism and Paleomagnetism*, Springer Geophysics,
https://doi.org/10.1007/978-3-319-90437-5_5

Introduction

Lake Baunt is tectonic in origin (Schetnikov 2007) and is located 1060 m a.s.l. in the north of Transbaikalia (Fig. 5.1). Its maximum water depth is 33 m (17 m as the average). The lake has a maximum length of 19 km (from south-west to north-east) and a maximum width of approximately 9 km, and it comprises an area of 111 km^2.

The aim of this study was to obtain the first paleomagnetic record for Lake Baunt from a sedimentary core collected in the lake. Our priority was to obtain the secular variations in the geomagnetic field, i.e., excursions. These will provide a set of reliable age markers for interregional correlation of paleomagnetic and paleogeographic events that could significantly contribute to the modern knowledge of regional and global magnetostratigraphy.

There is still much debate about the existence of Gothenburg and Mono Lake excursions. Some researchers suggest that the Mono Lake event is in fact the Laschamp Excursion (Kent et al. 2002) and that the Gothenburg excursion is a local feature rather than a global phenomenon (Lai and Channel 2007). On the other hand, these excursions are observed in a number of sedimentary records from different regions (Fairbridge 1977; Gus'kova et al. 2012), etc.

This article presents new paleomagnetic data from sedimentary core obtained from Lake Baunt that support the widespread occurrence of the Gothenburg and Mono Lake excursions in the Northern Hemisphere and that provide reliable dates for discriminating between the Mono Lake and Laschamp excursions.

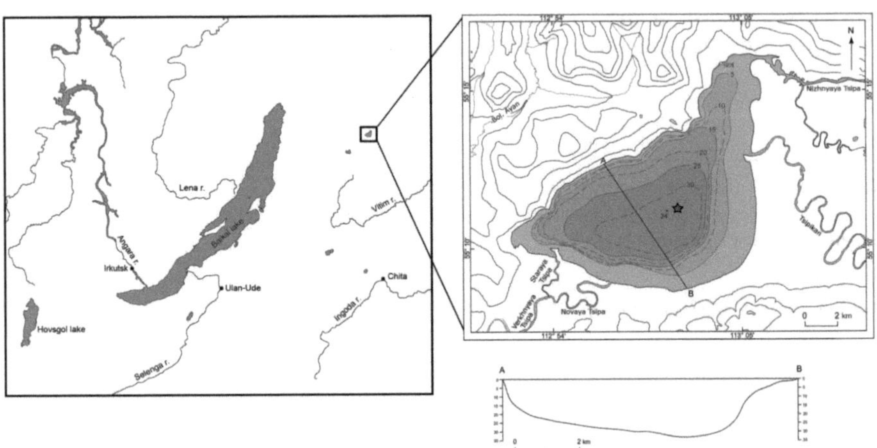

Fig. 5.1 Maps of the Baikal Lake Region with the location of the Baunt Lake (rectangle on left panel) and the Baunt-2014 core position (right panel). The drilling site is marked by an asterisk. The bathymetry scheme is shown in the figure lower right corner

Materials and Methods

Drilling site. In November 2013, prior to drilling, we conducted a bathymetric survey to locate a suitable coring site using a three-dimensional 6-beam digital depth sounder, Humminbird Matrix 748 × 3 D. The map obtained from the processed data is given in Fig. 5.1. The drill site was located at a maximum distance from the shore slopes, where we could expect regular sedimentation rates. The water depth at the drill site turned out to be the maximum for the lake at 33 m.

Drilling technique. The coring operation was performed using a rope-operated Uwitec Piston Corer (Austria) with PVC liners with an inner diameter of 63 mm. The system was installed in March 2014 on the ice at a site located 55° 11′ 15″N and 113° 01′ 45″E. Over a three-day period, we successfully recovered 13.7 m of sediment core in 8 PVC liner tubes. The core recovery amounted to 95%.

Sediment dating. Twelve 14C AMS dates from 200, 400, 600, 692, 800, 950, 1110, 1150, 1172, 1195, 1277 and 1350 cm core depth (Table 5.1) were obtained at the Poznan Radiocarbon Laboratory, Poland (bulk organic matter in sediment). The radiocarbon dates were then converted to calendar dates using the INTCAL 13 calibration curve (Reimer et al. 2013). The calendar dates were calculated using OxCal 4.2 software (Ramsey 2009).

Paleomagnetic and petromagnetic analyses. A total of 259 oriented samples were collected every 5 cm for paleomagnetic and petromagnetic analysis. From the three horizons expected to record the Gothenburg and Mono Lake excursions, we collected an additional 18 samples (6 samples from each layer; every 2 cm) to confirm the discovery of two magnetic excursions.

An Agico spinner magnetometer JR-6 (Czech Republic) was used for remanent magnetization (RM) measurements. Samples were demagnetized in the device using an alternating magnetic field placed in a magnetic vacuum. Then, samples

Table 5.1 Conventional and calibrated radiocarbon ages for the Baunt-2014 core used in this study

Depth in the core (cm)	14C age, years ago	Calibrated age, years ago
200	5775 ± 30	6551 ± 50
400	9000 ± 50	10134 ± 100
600	11620 ± 50	13479 ± 130
692	14090 ± 80	17118 ± 147
800	14930 ± 70	18150 ± 240
950	18220 ± 80	21687 ± 330
1110	18850 ± 120	23453 ± 330
1150	20680 ± 140	24601 ± 270
1172	21670 ± 140	25899 ± 94
1195	21720 ± 140	25989 ± 101
1277	24760 ± 190	28776 ± 203
1350	25350 ± 180	29430 ± 243

were magnetized by an alternating magnetic field placed in the Earth's magnetic field. The saturation remanent magnetization was measured using coercivity spectrometer (Institute of the Earth's Physics, Russia). The magnetic susceptibility was measured on KLY-2 Kappabridge (Agico, Czech Republic).

Natural remanent magnetization (NRM) was measured for all samples in order to determine whether Lake Baunt sediments are suitable for paleomagnetic investigations. Twenty-five samples underwent stepwise demagnetization by alternating magnetic fields of 2, 5, 10, 15, 20, 30, 40, 50, 60, 80, and 100 mT. The rest of the samples were demagnetized by alternating magnetic fields of 10, 20 and 40 mT. With the same field values, the samples were magnetized twice in the Earth's magnetic field (along the z-axis i.e., direction along the core and in the opposite direction). Using the resulting data, anhysteretic remanent magnetization (ARM) was calculated for three values of the alternating magnetic field.

Later, we collected 39 samples to measure the saturation remanent magnetization (SIRM): every 50 cm throughout the core and every 2 cm in places where the Gothenburg and Mono Lake excursions were likely recorded.

Results

Figure 5.2 presents an age-depth model constructed using twelve absolute age dates. All age dates presented in the text are given as calibrated ages. A gradual age increase with core depth in the Baunt-2014 core suggests that the sedimentation was almost continuous during the recorded time span. The average sedimentation rate is 44 cm/kyr, and the bottom core age is estimated at 29.5 ka BP.

The Lake Baunt sedimentary core is composed of biogenic-terrigenous silt and glacial clay (Fig. 5.3). The upper part of the core (0–540 cm) exhibits mainly diatomaceous muds with an enhanced organogenic component; in the middle part of the core (540–1170 cm), diatomaceous muds are interbedded with silty clay; the lower part of the core (1170–1366 cm) is composed of silty clay rich in coarse-grained silt and sand.

The magnetic susceptibility (MS) and anhysteretic remanent magnetization (ARM) data are given in Fig. 5.3. The magnetic susceptibility gradually increases in the upper 550 cm from 10 to 15–20 $\times$ 10^{-5} SI units. Between 550 and 800 cm from the core top the magnetic susceptibility varies markedly, and at this depth range the susceptibility can take either maximum or minimum values. Between 800 and 950 cm, the magnetic susceptibility consistently ranges between 15 and 20 $\times$ 10^{-5} SI units. At the depth range between 950 and 1020 cm, the core exhibits a significant decrease in magnetic susceptibility; further down the core, the magnetic susceptibility reaches values up to $\sim$15–20 $\times$ 10^{-5} SI units. From a depth of 1020 cm to the core bottom, the magnetic susceptibility gradually decreases to 10 $\times$ 10^{-5} SI units.

The values of anhysteretic remanent magnetization in the upper 550 cm are within 10–20 $\times$ 10^{-4} A/m; between 550 and 650 cm, they significantly decrease to

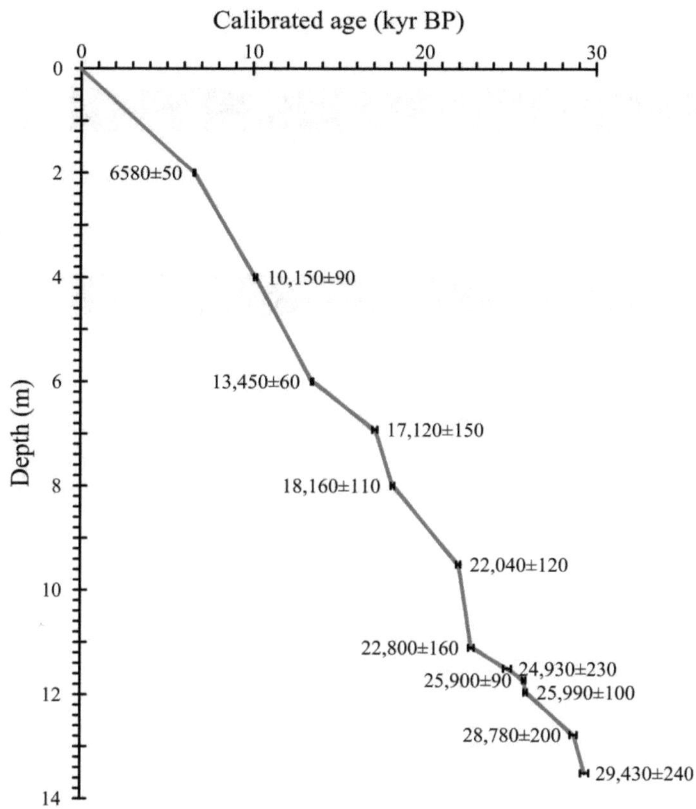

Fig. 5.2 Depth-age model applied to the Baunt Lake sedimentary record

1–2 × 10^{-4} A/m, and remain at that level to the core bottom, except for a short interval of increase (to 10–12 × 10^{-4} A/m) in the depth range between 900 and 1000 cm.

The saturation remanent magnetization in the upper part (to 550 cm from the core top) records an increase from 11 to 23 × 10^{-4} A/m; then, it significantly decreases between 650 and 850 cm, where it reaches approximately 10 × 10^{-4} A/m. In the depth range from 900 to 1000 cm, the saturation remanent magnetization values exhibit a short increase to 16–18 × 10^{-4} A/m, then they markedly decrease to 4–6 × 10^{-4} A/m. The decrease is observed to the core bottom. The triangles in Fig. 5.3 show SIRM values during the excursions.

An example of the detailed step-wise demagnetization by alternating magnetic field is shown in Fig. 5.4. The Zijderwald diagram shows that a weak viscous component is completely removed by a field measuring 10 mT. Only the component of the magnetic field recorded during sedimentation and deposition is left. The plots of natural remanent magnetization show that 15–30% of natural remanent magnetization is destroyed by the alternating magnetic field for samples collected

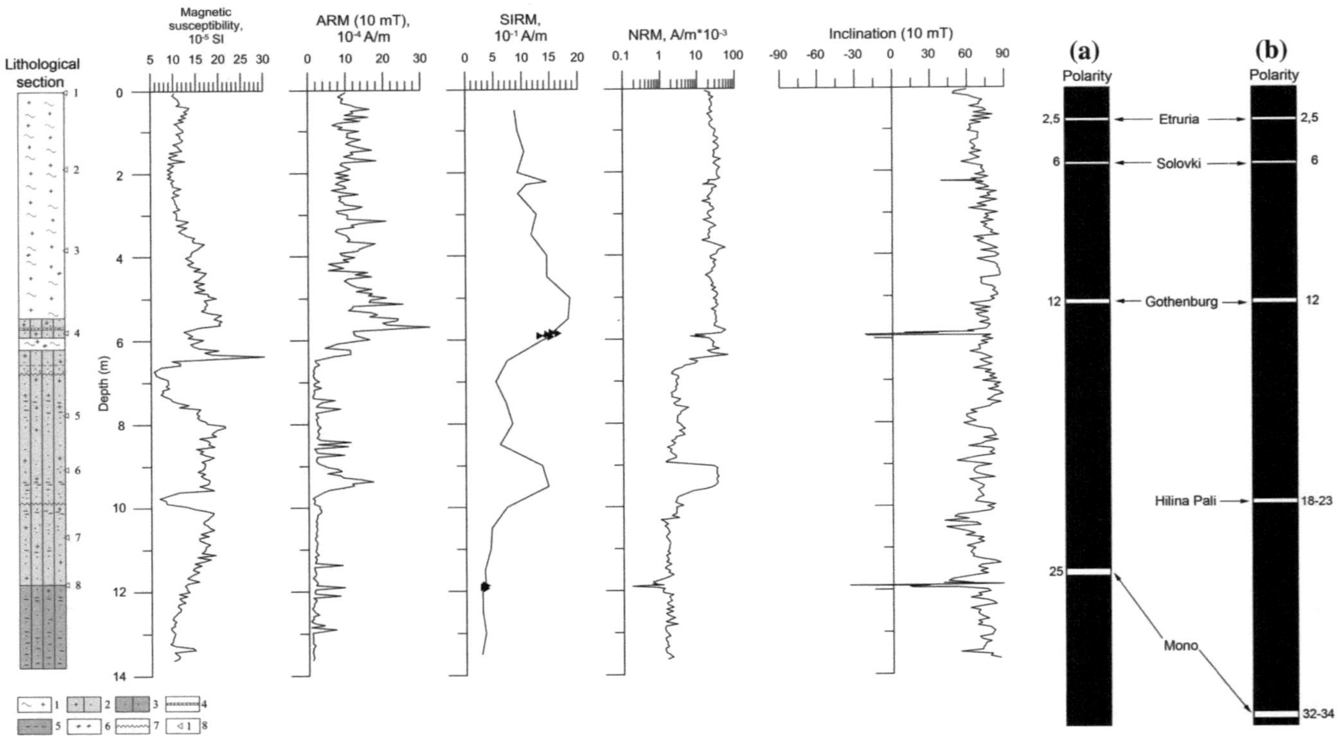

Fig. 5.3 Summary plot for petro- and paleomagnetic values in the Baunt-2014 core plotted together with core lithology scheme against core depth. Former and newly developed global magnetostratigraphic scales (small a and b letters, respectively) (Resolutions of Interagency Stratigraphic Board and its Regular Commissions 2013, 2016) are shown in the rightmost column, where the numbers from 2.5 to 32–34 indicate age of the worldwide recognized excursions in calibrated thousands of years before present. 1. Diatom silt. 2. Silty clay with diatoms. 3. Silty clay. 4. Clayish silt interbeds. 5. Hydrotroilite ($FeS * nH_2O$). 6. Charred plant remains. 7. Core gaps. 8. Drive numbers. SIRM values for the paleomagnetic excursions are marked by triangles

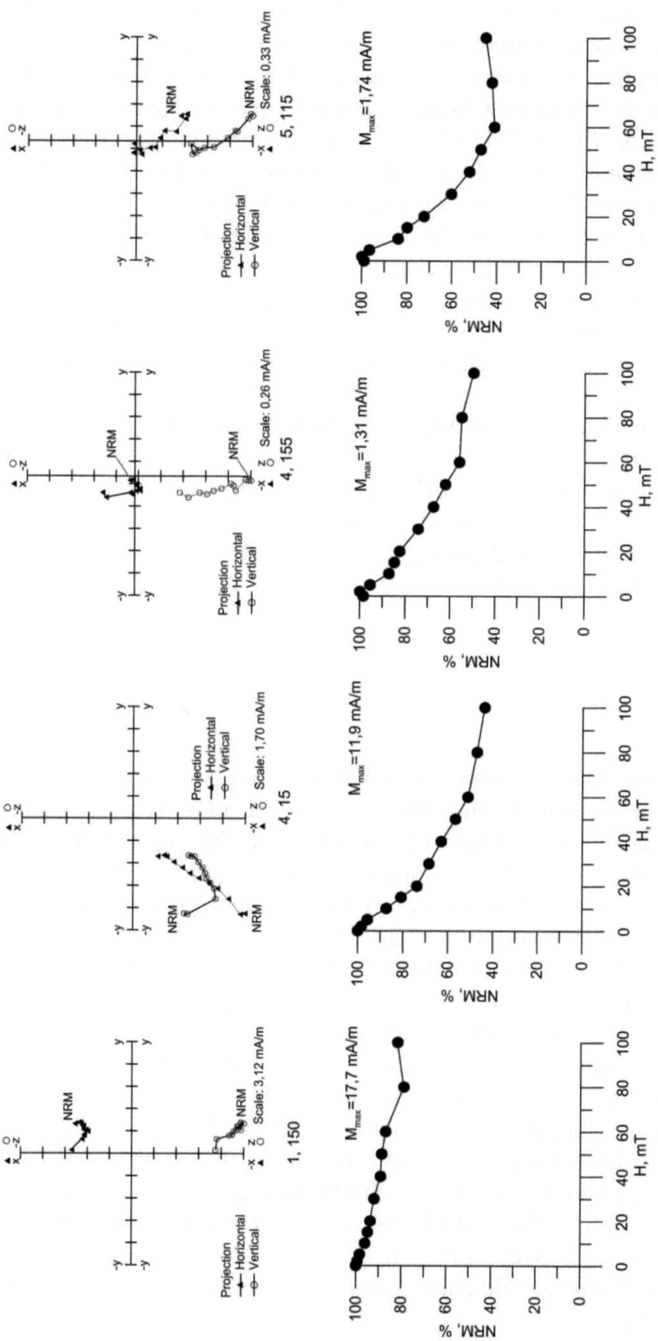

Fig. 5.4 Examples for detailed demagnetization of selected samples. Drive numbers and sample depth in the core are shown

from the upper part of the core (above 600 cm from the core top). For the samples collected lower than 600 cm, the same field destroys approximately 70% of NRM.

Most of the samples throughout the length of the core have inclinations within 60°. These positive inclinations are interpreted as a normal polarity zone corresponding with the Brunhes Chron. Anomalous paleomagnetic inclinations were recorded at 225, 590 and 1190 cm; these are preliminarily interpreted as the Solovki, Gothenburg, and Mono Lake excursions, respectively. The first two excursions were identified in one sample, and the Mono Lake excursion occurred in two samples. In order to confirm these excursions, 6 additional samples were collected from three layers of the archive core every 2 cm. The measurements shown in Fig. 5.3 document the reliable identification of two magnetic events: one at 590 cm (5 samples) and one at 1190 cm (10 samples). The Solovki excursion has not been verified: only a single indication of this event was found at a depth of 225 cm.

The NRM module varies broadly throughout the core: the values range between 20 and 40 $\times$ 10^{-3} A/m in the depth ranges between 0–650 cm and 890–950 cm i.e., in the periods of the stable polarity; between 650 and 890 cm and deeper than 950 cm, these values decrease to 1–3 $\times$ 10^{-3} A/m. During the excursion at a depth of 590 cm, NRM values decrease to 8–15 $\times$ 10^{-3} A/m, while at a depth of 1190 cm, NRM module reduces to 0.2–0.6 $\times$ 10^{-3} A/m (Fig. 5.3).

Discussion

The results show two confirmed excursions that can be reliably distinguished and identified as the Gothenburg and Mono Lake excursions. The first of these paleomagnetic events is characterized by a decrease in the magnetic field inclination between 593 and 583 cm. Five samples from this depth interval have inclinations less than 60°, and two of them record negative polarity. According to the age model (Fig. 5.2) this event dates to 13.3-13.1 ka BP. This appears during generally accepted interval for the Gothenburg excursion (12.35—13.75 ka BP) (Morner 1977; Gus'kova et al. 2012).

The validity of the Gothenburg excursion is verified by a significant (3–5 times) decrease in the natural remanent magnetization value. It has been previously established that excursions occur when the Earth's magnetic field strength drops (Valet et al. 1993; Pospelova and Sharonova 1999). Moreover, in this depth range, there are no visible changes in the petromagnetic characteristics of the sediment (magnetic susceptibility, ARM, and SIRM during the excursions do not show significant variation, and a marked decrease in ARM value is observed lower than 6 m). The decline in NRM value in this depth range is better explained by a decrease in the magnetic field during sedimentation than a change in features of the sediment material.

The second paleomagnetic event related to a reduction in magnetic inclination is found between 1197 and 1172 cm. Nine samples have magnetic inclination lower than 60°; one sample demonstrates a negative polarity. In addition, one sample shows an unusually high inclination value (88.7°). Since the excursion is characterized primarily by unstable polarity induced by a sharp drop in dipole moment, the direction of the vector of the magnetic field during such events is written as the sum of higher order moments, mainly the quadrupole (Valet et al. 2005). Thus, the increased inclination value surrounded by decreased values should be considered to correspond to excursion conditions. In this case, we also observe a decrease in NRM module, even stronger than in the first case (10 times). The petromagnetic characteristics at this depth interval have almost constant values, providing further support for the existence an excursion.

Based on several 14C AMS dates obtained around the excursion, two of which correlate to a polarity change, the age of the second episode can be estimated at 26.1–25.9 ka BP, and this event can be recognized as the Mono Lake excursion (Denham and Cox 1971). In our view, the change of Mono lake sediments dating (e.g., Kent et al. 2002; Zimmerman et al. 2006) doesn't mean that Mono Lake event was absent in principle and there was only Laschamp event. The youngest apparent date reported for the Laschamp event is not younger than 40 Ka BP (Resolutions of Interagency Stratigraphic Board and its Regular Commissions 2016). The excursion recorded in the sediment core retrieved from Lake Baunt has a reliable age of approximately 26 Ka BP and thus cannot be interpreted as the Laschamp excursion.

Conclusion

In 2014, a 13.7 m long lacustrine sediment core was retrieved by drilling at Lake Baunt. The core bottom age of the core is estimated at 29.5 ka BP. An average sedimentation rate amounts to 44 cm/kyr. A combination of high sedimentation rates and the high resolution of the paleomagnetic record obtained serve as the basis for revealing age intervals correlating with the Gothenburg and Mono Lake excursions. A reliable age-depth model further supports the existence of the excursions and provides an estimation of their duration.

The Gothenburg and Mono Lake excursions documented in the sedimentary record from Lake Baunt suggest:

- the Gothenburg excursion can be considered a paleomagnetic event; according to our data, the duration of the excursion can be estimated in the age interval between 13.3 and 13.1 ka BP;
- the Mono Lake excursion is dated at 26.1–25.9 ka BP, which reliably discriminates this event from the Laschamp excursion that occurred 10–20 ka earlier.

The results of this study serve as a basis for further regional studies related to the timing, frequency, duration and character of magnetic excursions. The results also have implications for the influence of magnetic excursions on the evolution of the natural climatic system of Transbaikalia and associated adaptive strategies.

While we were treated and analyzed first paleomagnetic results from the Lake Baunt record presented in this paper, and then wrote the article, the improved magnetostratigraphic scale was adopted (Fig. 5.3b). This new scale opens up additional possibilities for the interpretation of our data, age determination of which is based on the results of the AMS ^{14}C dating only unlike the dating methods used in the new scale.

Acknowledgements The authors are grateful to the team and leadership of the International Baikal-Hokkaido archaeological project (BHAP) for the support to obtain the Uwitec drilling equipment (Austria). This study was performed within the frame of the state assignment No 0350-2016-0026 and was supported by the Russian Science Foundation via project 16-17-10079 (petro- and paleomagnetic analysis), and by Integration Program "Basic researches and breakthrough technologies as a basis of priority development of the Baikal region and its interregional relations" (0341-2016-001).

References

Denham C.R. and Cox C.R. Evidence that the Laschamp polarity event did not occur 13,300–30,400 years ago // Earth Planet. Sci. Lett., 13, p.p. 181–190, 1971.

Fairbridge R.W. Global climate change during the 13,500 BP Gothenburg geomagnetic excursion // Nature, 1977, 292, p.p. 670–674.

Gus'kova E.G., Raspopov O.M., Dergachev V.A., Iosifidi A.G., Sinitsyna G.V. Manifestation of the Gothenburg geomagnetic field excursion in sediments on the Northern Central Russian upland // Geomagnetism and aeronomy, 2012, Vol. 52, N 5, p.p. 675–683.

Kent D.V., Hemming S.R., Turin B.D. Lashamp Excursion at Mono Lake? // Earth and Planetary Science Letters, 2002, vol. 197, p.p. 151–164.

Lai C., Channell J.E.T. Geomagnetic Excursions. Chapter in book Treatise on geophysics. 2007. Elsevier. p. 376.

Mörner N. The Gothenburg Magnetic Excursion // Quaternary Research, 1977, vol. 7, issue 3, p. p. 413–427.

Pospelova G.A., Sharonova Z.V. Geomagnetic Intensity between 53 and 22 ka // Izvestiya, Physics of the Solid Earth. 1999, Vol. 35, No. 10, p.p. 861–870.

Ramsey B. Bayesian analysis of radiocarbon dates. // Radiocarbon, 2009, 51(1), 337–360.

Reimer P.J., Bard E., Bayliss A., Beck J.W., Blackwell P.G., Bronk R.C., Buck C.E., Cheng H., Edwards R.L., Friedrich M., Grootes P.M., Guilderson T.P., Haflidason H., Hajdas I., Hatté C., Heaton T.J., Hoffmann D.L., Hogg A.G., Hughen K.A., Kaiser K.F., Kromer B., Manning S.W., Niu M., Reimer R.W., Richards D.A., Scott E.M., Southon J. R., Staff R.A., Turney C.S.M.; van der Plicht J. IntCal13 and Marine13 radiocarbon age calibration curves 0–50,000 years cal BP // Radiocarbon, vol.55, N 4, 2013, p.p. 1869–1887.

Resolutions of Interagency Stratigraphic Board and its Regular Commissions. 2013. Is. 42–64 pp. (in Russian).

Resolutions of Interagency Stratigraphic Board and its Regular Commissions. 2016. Is. 44–68 pp. (in Russian).

Shchetnikov A.A. Morphotectonics of Lacustrine Basins: The Baikal Rift Zone as an Example // Russian Journal of Pacific Geology, 2007, Vol. 1, No. 2, pp. 120–129. https://doi.org/10.1134/s1819714007020029.

Valet J-P., Meynadier L. Geomagnetic field intensity and reversals during the past four million years // Nature. 1993. V. 366. № 6452. P. 234–238.

Valet J-P., Meynadier L., Guyodo Y. Geomagnetic dipole strength and reversal rate over the past two million years // Nature. 2005. V. 435. P. 802–805. https://doi.org/10.1038/nature03674.

Zimmerman S.H., Hemming S.R., Kent D.V. and Searle S.Y. Revised chronology for late Pleistocene Mono Lake sediments based on paleointensity correlation to the global reference curve // Earth and Planetary Science Letters. 2006. V. 252. P. 94–106.

Chapter 6
Intrusions of the Kulumbe River Valley, NW Siberian Traps Province: Paleomagnetism, Magnetic Fabric and Geochemistry

A. V. Latyshev, N. A. Krivolutskaya, P. S. Ulyahina, Ya. V. Bychkova and B. I. Gongalsky

Abstract We present the new geochemical, paleomagnetic and anisotropy of magnetic susceptibility data from the Kulumbe river area (the Siberian Traps Large Igneous Province). The majority of the studied intrusions have the normal polarity, belong to the Katangsky complex and can be correlated with the upper part of the volcanic section of the Norilsk region. Furthermore, the first intrusive analogues of the Gudchikhinsky Formation are found. The only intrusion showing the reverse magnetic polarity has the geochemical features similar to the ore-bearing Norilsk-type intrusions. Based on the study of the magnetic fabric of the intrusions, we suggest that they were fed from the magma-conducting Imangda-Letninskiy fault.

Keywords Paleomagnetism · Geochemistry · Siberian traps · Anisotropy of magnetic susceptibility · Cu-Ni-PGE deposits

A. V. Latyshev (✉) · P. S. Ulyahina · Ya. V. Bychkova
Geological Faculty, Lomonosov Moscow State University, Moscow, Russia
e-mail: anton.latyshev@gmail.com

A. V. Latyshev · P. S. Ulyahina
Schmidt Institute, Physics of the Earth, Russian Academy of Sciences, Moscow, Russia

N. A. Krivolutskaya
Vernadsky Institute of Geochemistry and Analytical Chemistry, Russian Academy of Sciences, Moscow, Russia

B. I. Gongalsky
Institute Geology of Ore Deposits, Petrography, Mineralogy and Geochemistry, Russian Academy of Sciences, Moscow, Russia

D. Nurgaliev et al. (eds.), *Recent Advances in Rock Magnetism, Environmental Magnetism and Paleomagnetism*, Springer Geophysics,
https://doi.org/10.1007/978-3-319-90437-5_6

Introduction

The problem of the origin of Large Igneous Provinces (LIPs) and associated mineralization is one of the major open questions in geology (Dobretsov 1997; Ernst 2014). The Siberian Traps contain the world-class PGE-Cu-Ni Norilsk deposits, which are the only example of the unique ores within a young ($\sim$251 Ma, Kamo et al. 2003; Burgess and Bowring 2015) large igneous province, while the bulk of similar deposits is related to large Proterozoic basic-ultrabasic plutons. Mineralization is associated with the Permian-Triassic sill-like intrusions of the NW part of the Siberian platform while cognate intrusive bodies are widespread around the trap province. Therefore, the question of the place of the Norilsk ore-bearing intrusions in the LIP evolution has both fundamental scientific and applied importance. In particular, it directly contributes to the successful prospecting of the new deposits not only within the Siberian platform, but also within other LIPs around the world.

The Kulumbe (Kulyumber) river area holds the key position in the Siberian Traps province due to its location inside the intermediate zone between the Norilsk-Igarka paleorift zone and the Tunguska syncline. According to the geophysical data, this zone overlays a branch of the Norilsk-Igarka paleorift. The Kulumbe district comprises intrusive complexes both of the northern and southern parts of the Siberian province. Some intrusions with sulfide mineralization are found in this area. The largest Dzhaltul massif is referred to the Kureysky complex. It is important to estimate the place of ore-bearing intrusions in the magmatic evolution of the region in order to find criteria for the rich mineralization prospecting.

Here we present the results of a paleomagnetic and magnetic fabric study of intrusions from the middle reaches of the Kulumbe river. We have also obtained the rare elements data from these intrusions and ore-bearing Dzhaltul massif and compared them with the data from the Norilsk region.

Methods

The paleomagnetic samples were taken as hand blocks oriented using a magnetic compass. Ten to fifteen hand blocks were collected from each site. The local magnetic declination was calculated using the IGRF model (12th generation, revised in 2014). Paleomagnetic research and measurements of the anisotropy of magnetic susceptibility were carried out in the Paleomagnetic laboratory of Schmidt Institute of Physics of the Earth (IPE RAS, Moscow, Russia). All samples were subjected to the stepwise thermal treatment up to the complete demagnetization in 10–17 steps. The majority of samples were demagnetized up to 580–600 °C, in few cases—up to 630 °C. The size of demagnetization steps was changed from 50 to 100 °C at low temperatures to 15–20 °C at high temperatures depending on the demagnetization pattern. For heating we used MMTD-80 non-magnetic ovens (Magnetic Measurements Ltd, Aughton, U.K.) with internal residual fields of about 5–10 nT. The remanent magnetization of samples was measured using JR-6 spinner magnetometer (AGICO, Brno, Czech Republic) and 2G Enterprises cryogenic magnetometer. Isolation of natural remanent magnetization (NRM) components was performed with Remasoft (Chadima and Hrouda 2006) or Enkin's (1994) paleomagnetic software packages using principal component analysis (Kirschvink 1980). Analysis of paleomagnetic directions was carried out using Fisher statistics (Fisher 1953). Anisotropy of magnetic susceptibility (AMS) was measured by MFK-1FA kappa-bridge (AGICO, Brno, Czech Republic). Processing of the results was performed with Anisoft42 software using Jelinek statistics (Jelinek 1978). Geochemical study of intrusions included determinations of the major components using X-ray fluorescence analysis (Institute of Geology of Ore Deposits, Petrography, Mineralogy and Geochemistry RAS, analyst A. I. Yakushev) and rare elements using the inductively coupled plasma method (ICP-MS, analyst Y. V. Bychkova). In addition, volcanic rocks of the Gudchikhinsky Formation and gabbro of the Talnakh and Dzhaltul intrusions have been analysed.

Geological Background

The study area is located in the middle reaches of the Kulumbe river, 150 km southeast from Norilsk city. The area is situated in the northwestern part of the Siberian platform, at the contact of the Khantaysky-Rybninsky rampart-like uplift and the Tunguska syncline. Ordovician—Lower Carboniferous (O-C_1) carbonate-terrigenous rocks are exposed in the area. Carboniferous deposits of the Tunguska Group overlap the O-C rocks flatly dipping to the east (Fig. 6.1). The main fault structure of this area is the regional Imagda-Letninsky fault of NE strike. The sedimentary rocks are covered by volcanic rocks of the Siberian Traps lava pile (P_3-T_1) representing the southwestern part of the Nirungda Trough.

According to the geological map of Ltd. Norilskgeologiya, the majority of intrusive bodies are referred to as the Katangsky complex, although intrusions of the Kureysky and Ergalakhsky complexes have also been noted. Their diagnostics was carried out in the field conditions only. Meanwhile, their accurate attribution to intrusive complexes is important both for the reconstruction of the magmatic evolution of this area, and for the prospecting of new deposits. New geochemical data on these intrusions are discussed below.

The investigated intrusions are exposed in the Kulumbe river valley and along the creek Khalil (Fig. 6.1). They are mainly located in the sandstones of the Tunguska Group rocks, rarely in the Devonian dolomites referred to as the Nakakhosky Formation. Intrusions usually have sill-like morphology; their thickness varies considerably from the couple of meters to 80–100 m, the visible length is about the first hundreds of meters. Position of intrusions is usually concordant or

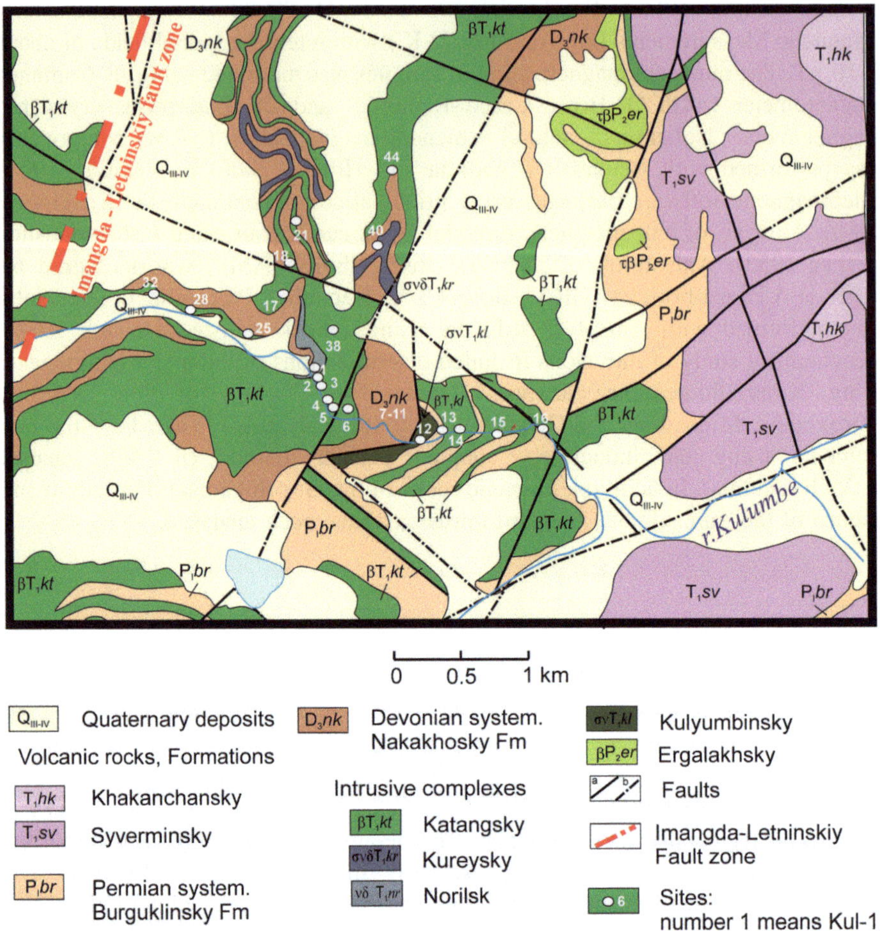

Fig. 6.1 Geological map of the middle reaches of the Kulumbe river valley (after Ltd. Norilskgeologia data with authors' corrections)

sub-concordant with the host rocks which generally form a monocline dipping to the east, with mean dip angles of 10°–15°.

As a rule, intrusive rocks are represented by massive fine-grained gabbro-dolerites without visible differentiation. The exception is the intrusion in site KUL-16 which is characterized by coarse-grained porphyritic texture, where olivine forms large porphyry grains in fine-grained groundmass.

Paleomagnetic Results

We obtained paleomagnetic directions for nine sites from the middle reaches of the Kulumbe river. The quality of the paleomagnetic signal varies from site to site, nevertheless, we were able to calculate the mean directions for all of them. In all sites, we isolated the low-temperature component (LTC) destroyed at 200–300 °C. Since the directions of this component are close to the modern geomagnetic field, we consider that the LTC has a viscous origin. The high-temperature component is also clearly distinguished in all sites. This component is usually destroyed in the temperature range from 350–400 °C to 560−600 °C and decays to the origin (Fig. 6.2a). The directions of this component are close to expected for the Permian-Triassic Siberian Traps (Pavlov et al. 2007, 2015) and we consider it as characteristic. The site-mean directions of the HTC are presented in Table 6.1. Finally, an intermediate middle-temperature component (MTC) occasionally appears in several sites. The temperature range of this component varies from 300 to 500 °C and its directions are scattered. The origin of this component remains unclear.

The HTC in all studied intrusive bodies has the normal polarity except for a single site with the reverse polarity. The concentration parameter K for all directions is slightly higher in the geographic coordinate system than in the stratigraphic system. Moreover, tilt-corrected directions have unusually shallow inclinations compared to those typical for the Siberian Traps. This may imply that the dislocation of the host sedimentary rocks took place before or simultaneously with the

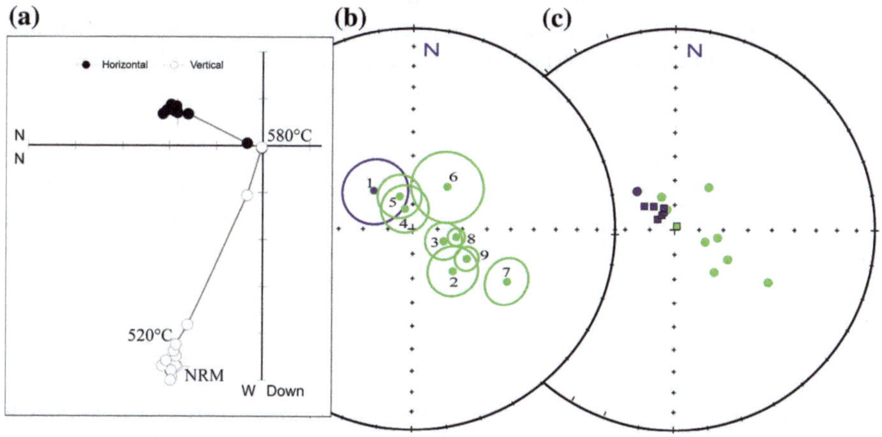

Fig. 6.2 Results of the paleomagnetic investigation. **a**—a typical orthogonal plot for the studied samples. Sample 68, site KUL-14. **b**—site-mean paleomagnetic directions in the geographic coordinate system. Green is normal polarity, blue is reverse polarity. Paleomagnetic sites: 1—KUL-1; 2—KUL-2; 3—KUL-6; 4—KUL-14; 5—KUL-16; 6—KUL-17; 7—KUL-21; 8—KUL-38; 9—KUL-44. **c**—Comparison of our results with the data from Pavlov et al. (2007) which are shown by squares

Table 6.1 The site-mean paleomagnetic directions

Site	N	Dg (°)	Ig (°)	Ds (°)	Is (°)	K	α95
KUL-1	8	314.9	−68.5	301.8	−60.6	18.5	13.2
KUL-2	10	136.4	66.3	124	58	23.9	10.1
KUL-6	14	111.3	76.6	97.9	52.2	29.1	7.5
KUL-14	8	338.5	81.3	39.8	79.5	34.5	9.6
KUL-16	13	337.8	75.8	18.7	76.1	23.2	8.8
KUL-17	13	39.2	67.9	53.6	60.6	9.2	14.5
KUL-21	14	119.2	45.6	115.2	36.7	20.2	9.1
KUL-38	13	100.5	72.1	96.9	62.2	140.5	3.5
KUL-44	14	118.9	65	111.4	55.9	69	4.8
Mean for normal polarity (geographic)	8	99.7	75.6			16.9	13.9
Mean for normal polarity (stratigraphic)	8			95.5	63.6	16.0	14.3

N—number of samples; *Dg*/*Ig*—declination/inclination in situ; *Ds*/*Is*—declination/inclination after tilt correction; K, α95—Fisher statistic parameters

traps emplacement. Our interpretation is consistent with the conclusions of Pavlov et al. (2007) who studied the intrusions in the lower reaches of the Kulumbe river and assumed that the deformations took place at the intermediate stage of the traps emplacement. Therefore, in the discussion below, we consider the site-mean directions in the geographic coordinate system. The distribution of site-mean directions "in situ" is shown in Fig. 6.2b.

The majority of normally magnetized intrusions have steep inclinations, and paleomagnetic directions of the main group are close to those from the upper part of the Norilsk volcanic sequence (the Morongovsky Formation and above; Heunemann et al. 2004). Moreover, the intrusion Norilsk-2 of the Norilsk type also yielded similar direction (Latyshev 2013). According to the geological map of Ltd Norilskgeologiya, the studied intrusions belong to the Katangsky complex widely distributed in the Tunguska syncline.

Site KUL-21 yielded the normal polarity too, but the mean direction is distinct and has significantly shallower inclination (45°) compared to other sites. Such inclinations are unusual for the Siberian Traps and, in the Norilsk region, were reported only for the Syverminsky and Gudchikhinsky Formations from the lower part of the volcanic section (Heunemann et al. 2004; Pavlov et al. 2015). As suggested in these works, this part of the volcanic section was erupted during the reversal of the geomagnetic field within several thousand years and recorded the transitional field state.

Finally, site KUL-1 is the only that yielded the reverse polarity. The site-mean direction is virtually antipodal to that of the main group of normally magnetized intrusions. According to the geological map of Ltd. Norilskgeologiya this intrusion has been referred to as the Katangsky complex as well as the majority of the adjacent sills before, but its paleomagnetic and geochemical (see below) features are distinct.

We compared our results with the data obtained by Pavlov et al. (2007) from the intrusions of the lower reaches of the Kulumbe river. The whole set of site-mean paleomagnetic directions is shown in Fig. 6.2c. As seen from the Figure, the majority of directions reported by Pavlov et al. (2007) has the reverse polarity, in contrast to our data. The reversal test (McFadden and McElhinny 1990) performed for the whole set of site-mean directions is positive: $\gamma/\gamma_{cr} = 8.0°/13.9°$. It argues, firstly, in favor of the primary origin of NRM in studied samples and, secondly, of the averaging of the geomagnetic field secular variations recorded during the intrusions emplacement.

The intrusions sampled by Pavlov et al. (2007) cut Silurian-Devonian sediments, located in deeper horizons of the Siberian platform cover than those studied in this work. Since the normal polarity is predominant in our intrusions, while sills studied by Pavlov et al. (2007) are reversely magnetized (except for one site), we can assume that different magmatic events took place at different levels of the sedimentary cover of the Siberian platform.

AMS Results

We measured the anisotropy of magnetic susceptibility of 15 studied intrusions (Table 6.2). All sites demonstrate the low degree of anisotropy $Pj < 1.035$. The shape parameter T in the majority of sites (11 out of 15) is below zero, corresponding to the prolate shape of AMS ellipsoid. Such values of Pj and T are common for the basic rocks with the primary magmatic magnetic fabric, resulted from the magmatic flow (e.g. Tarling and Hrouda 1993).

Eight sites have so-called normal magnetic fabric, when the minimal axis K3 of AMS ellipsoid is orthogonal to the contact and steep in sills. In this case two other axes K1 and K2 lie in the sill plane and are shallow (Fig. 6.3a). The most common interpretation of the normal magnetic fabric with the low degree of anisotropy is that the magnetic lineation (K1) corresponds to the magma flow direction (Knight and Walker 1988; Tarling and Hrouda 1993; Hrouda et al. 2015). Within all sites with the normal fabric the maximal axes are clustered around the mean shallow axis, and their confidence ellipses do not overlap those of the other axes. Therefore, for the sites with normal magnetic fabric, we interpret the orientation of the maximal axis K1 as the magma flow lineation.

In two other sites, we found the inverse magnetic fabric, when the maximal axis is normal to the contact and steep in sills (Fig. 6.3b). The possible causes of the inversion might be the predominance of single-domain magnetite or titanomagnetite grains (Potter and Stephenson 1988), thermal contraction during the basaltic columns formation (Hrouda et al. 2015) or magnetostatic interaction (Borradaile and Jackson 2010). Within the Siberian Traps province the inverse magnetic fabric was found in sills of the Daldyn-Alakit region (Eastern periphery of the Tunguska syncline) by Konstantinov et al. (2014), who explained the inversion by the contact

Table 6.2 Results of the AMS measurements

Site	Complex	n	Type AMS	Pj	T	K1			K2			K3		
						D	I	Ci	D	I	Ci	D	I	Ci
Kul1	Nr	15	R	1.023	−0.06	49.5	75.2	38.8/22.5	235.9	14.7	48.7/35.7	145.5	1.6	48.8/27.9
Kul2	Kt	8	R	1.023	−0.198	280.0	81.1	17.0/10.8	101.1	8.9	42.2/16.9	11.1	0.2	42.3/10.0
Kul6	Kt	10	N	1.033	0.242	293.6	37.7	43.4/22.3	26.8	4.1	44.4/17.2	122.0	52.0	23.2/21.2
Kul12	Kl	15	S	1.011	−0.100	138.9	52.4	51.7/36.5	355.0	31.9	72.1/50.9	253.5	17.7	72.4/27.2
Kul14	Kt	15	N	1.027	−0.118	309.5	32.0	24.8/10.8	52.7	20.0	31.3/16.4	169.2	50.9	31.4/19.0
Kul16	Kt	16	N	1.022	−0.081	104.4	11.4	38.8/14.0	195.8	6.7	48.5/37.5	315.7	76.7	48.1/11.6
Kul17	Kt	13	S	1.030	−0.123	344.4	7.5	50.5/44.5	81.2	42.1	62.3/48.5	246.3	47.0	62.9/40.2
Kul21	Kl?	11	N	1.017	0.204	273.3	7.5	45.6/28.0	10.3	42.7	62.6/39.5	175.4	46.3	61.5/28.0
Kul24	?	10	N	1.022	−0.115	103.1	5.5	52.9/16.8	194.1	10.5	53.9/26.7	346.0	78.1	32.3/17.4
Kul25	?	15	S	1.023	−0.130	154.9	76.2	83.5/40.7	327.8	13.7	83.6/42.4	58.2	1.6	46.3/41.1
Kul28	?	15	S	1.019	−0.054	357.4	4.7	64.3/40.8	266.1	15.5	65.0/51.6	103.9	73.7	58.5/37.7
Kul32	?	14	N	1.026	0.012	122.7	27.4	21.3/13.0	221.5	16.4	41.3/13.0	338.8	57.3	41.1/20.2
Kul38	Kt	12	N	1.028	0.388	80.0	20.4	34.7/11.9	183.3	31.8	37.1/22.6	322.8	50.8	27.8/10.4
Kul40	?	14	S	1.027	−0.012	39.4	61.2	58.7/36.6	178.6	22.7	76/7/55.7	275.9	16.9	76.6/38.6
Kul44	Kt	9	N	1.028	−0.279	146.7	9.4	13.0/11.2	238.2	8.6	30.5/12.6	9.8	77.2	30.6/10.5

Complex abbreviations: *Kt*—Katangsky, *Nr*—Norilsk, *Kl*—Kulyumbinsky. *n*—number of samples. Types of AMS ellipsoid: *N*—normal, *R*—inverse, *S*—scattered. *Pj*—degree of anisotropy; *T*—shape parameter; *K1*—maximal axis; *K2*—intermediate axis; *K3*—minimal axis. *D*—declination, °; *I*—inclination, °; *Ci*—confidence intervals influence of the younger intrusions. Thus, additional rock-magnetic and petrographic studies are necessary to solve this problem. Finally, in five sites the dispersed magnetic fabric was observed.

Based on the orientation of the K1 axis in sites with normal magnetic fabric, we determined the predominant lineation of the magma transport. In six out of eight sites, the maximal axis has NW orientation (Fig. 6.3c). This lineation is approximately orthogonal to the regional Imangda-Letninsky fault, which is situated westward from the study area and has NE strike. We suggest that this fault could act as the magma feeding zone during the traps emplacement. This idea is consistent with the results of Callot et al. (2004) who studied the upper part of the volcanic section of the Norilsk region (Morongovsky Formation and above). Based on the AMS measurements, Callot et al. (2004) concluded that the lava eruptions were fed from the rift zone situated in the NW border of the Siberian platform and similar to the regional Norilsk-Kharaelakh fault which is parallel to the Imangda-Letninsky fault and controls the location of the Norilsk ore-bearing intrusions.

Geochemical Data

The study of the intrusions chemical composition shows their clear subdivision into three types (Fig. 6.4). Most of the intrusive bodies have similar geochemical characteristics (Table 6.3). All analyses correspond to gabbro (46.99—51.17 wt% SiO_2) with titanium contents of 1.20–1.56% TiO_2 and sodium predominance over potassium. Porphyritic olivine gabbro from the KUL-16 outcrop are distinguished by higher magnesium content (8.5 wt% MgO), but their other petro- and geochemical characteristics are similar to the other intrusions.

Three types of intrusions can be distinguished from the distribution of rare elements in rocks. The patterns of the main group (Fig. 6.4a) differ by a moderate accumulation of rare elements, clearly negative Ta-Nb and positive Pb and Sr anomalies. These patterns are close to those of the volcanic rocks of the main trap magmatic stage within the Siberian platform, mainly to the Morongovsky-Samoedsky Formations. According to titanium concentrations in the rocks, they could be compared with the Mokulaevsky and Kharaelakhsky Formations. Most of the intrusions belong to the Katangsky complex. Two intrusions have rare elements patterns distinct from the main group (samples KUL-1 and KUL-12—Fig. 6.4b). The first of them has less sharp Ta-Nb anomaly than Katangsky complex intrusions and resembles the rocks of the Norilsk complex [Fig. 6.5a, samples OUG-2/1151.7 and OUG-2/1191 of the Talnakh intrusion; see description and data for intrusions of the Norilsk area in Krivolutskaya (2016)]. Moreover, sulfide mineralization was found in the intrusion KUL-1 (up to 3% sulfides).

However, the most interesting is the sample KUL-12 which is similar to the lavas of the Gudchikhinsky Formation (Fig. 6.5b), which represents the unique example of typical mantle magma in the northwest of the Siberian Traps province (Sobolev et al. 2009). Figure 6.5b shows the spectra for samples of the Gudchikhinsky

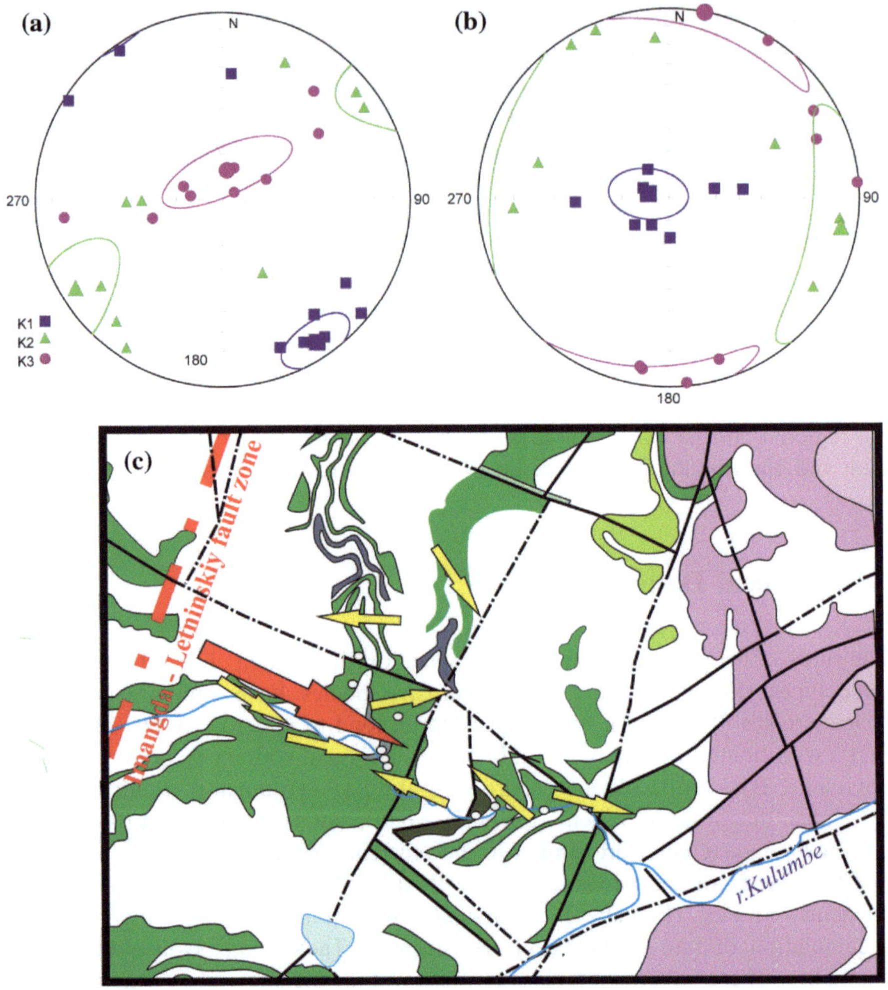

Fig. 6.3 Results of the AMS investigation. **a–b**. Examples of the AMS ellipsoid. Geographic coordinate system, equal-area projection. K1—maximal axis; K2—intermediate axis; K3—minimal axis. **A**—site 44, normal magnetic fabric. **b**—site 2, inverse magnetic fabric. **c**—magma flow directions during the intrusions emplacement. Yellow arrows are the site-mean magnetic lineation in the sites with the normal magnetic fabric. Red arrow is the suggested general magma flow direction. Explanatory notes see in caption to Fig. 6.1

Formation taken from the core of borehole OB-36 drilled within the Vologochan Trough in the west of the Norilsk region. These samples have small Ta-Nb and Pb anomalies. Meanwhile, it was demonstrated earlier that the Gudchikhinsky Formation rocks change their composition from west to east, showing the more primitive composition lacking Ta-Nb and Pb anomalies (Sobolev et al. 2009). The spectrum of the KUL-12 sample is similar to these primitive picrites.

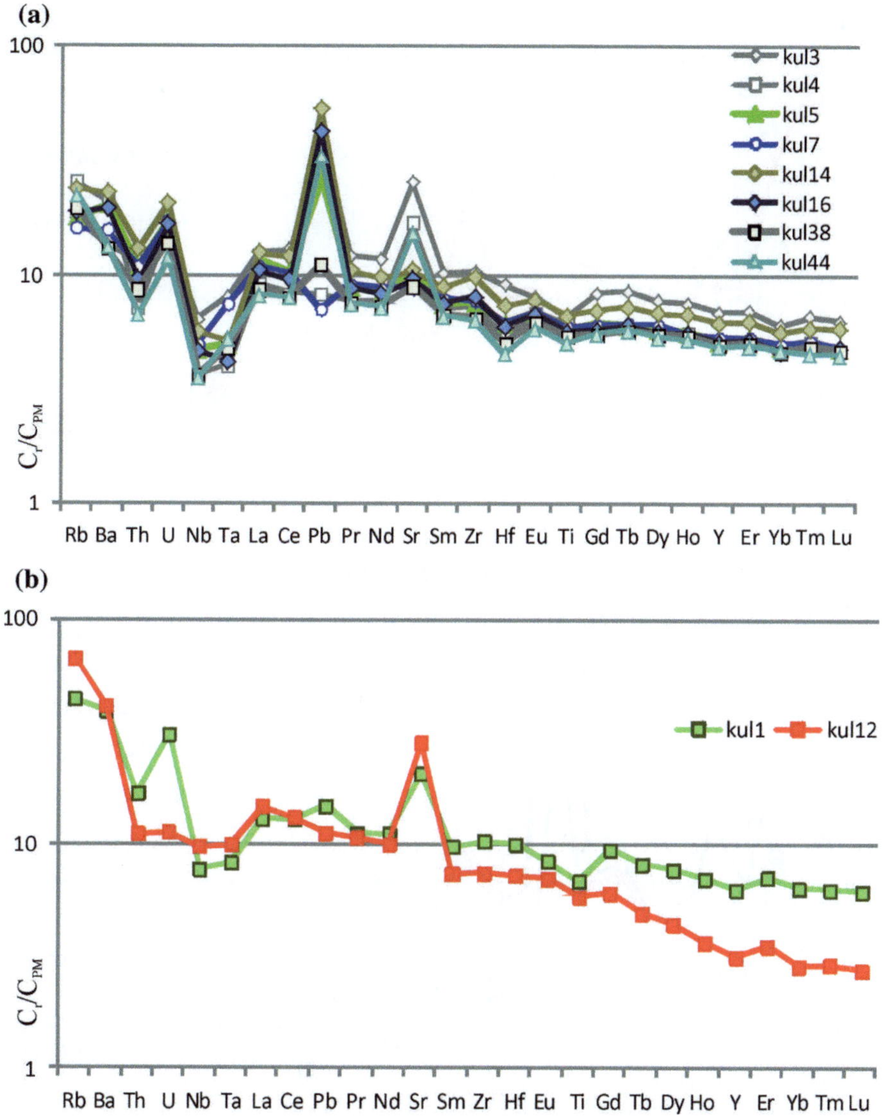

Fig. 6.4 Spider-diagrams for intrusions of the Kulumbe river valley. **a**—Katangsky complex; **b**—other intrusive bodies. Here and in Fig. 6.5 C_r/C_{PM}—content in rock/content in Primitive Mantle (after Hofmann 1988). Data are listed in Table 6.3

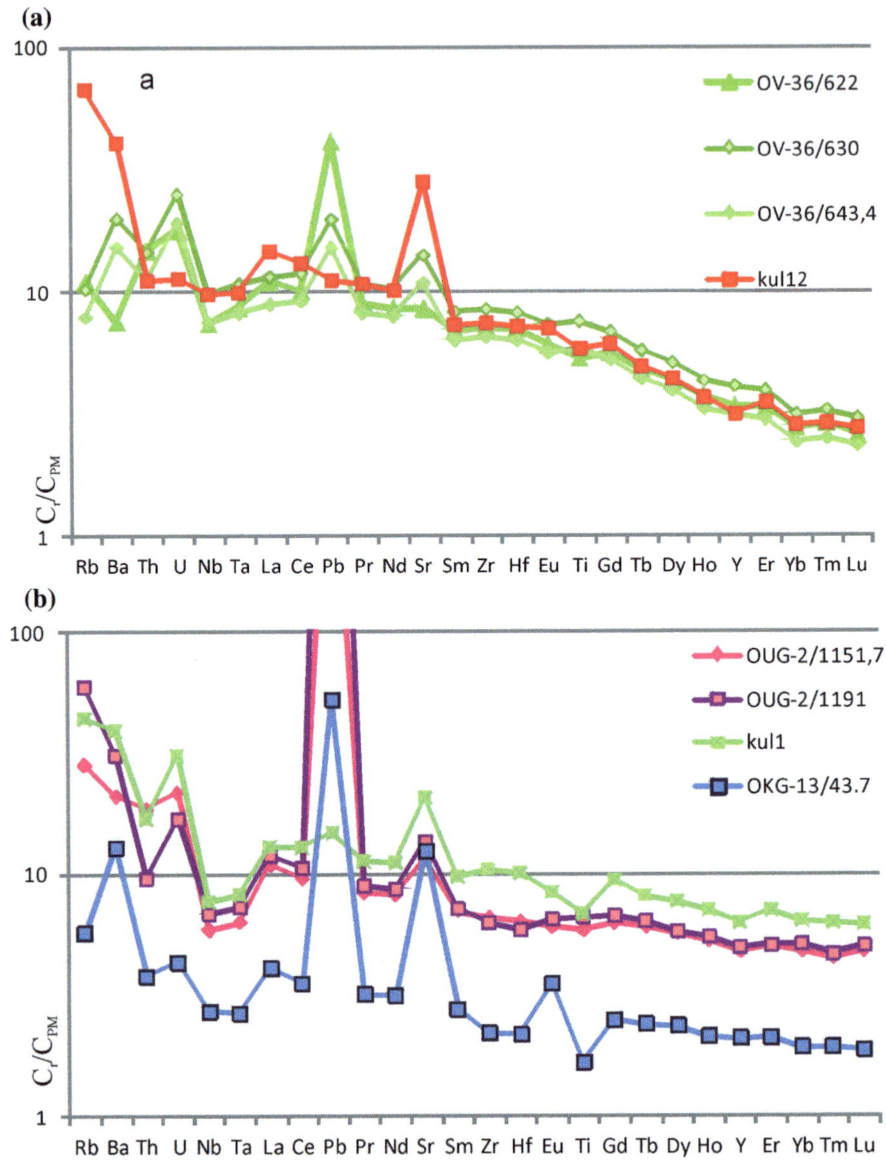

Fig. 6.5 Comparison of spider-diagrams for intrusions and lavas of the Kulumbe river valley and the Norilsk region. Samples, No: OV-36—basalts of the Gudchikhinsky Formation; OUG-2—the Talnakh intrusion (Norilsk complex), OKG-13—Dzhaltul intrusion (Kureysky intrusive complex)

Table 6.3 Chemical composition of the magmatic rocks

N	1	2	3	4	5	6	7	8	9	10	11	12	13	14	15	16
Sample	1151.7	1191	43.7	622	630	643.4	kul3	kul4	kul5	kil7	kul14	kul16	kul38	kul44	kul1	kul12
SiO_2	49.96	49.56	49.66	46.31	48.75	48.25	46.92	46.99	48.18	48.42	48.67	47.59	48.01	48.14	52.17	49.01
TiO_2	1.26	1.42	0.37	1.14	1.62	1.24	1.50	1.43	1.33	1.39	1.56	1.30	1.23	1.20	1.48	1.26
Al_2O_3	14.16	13.35	17.34	8.41	10.77	9.41	14.15	14.21	15.01	14.94	14.69	14.91	15.62	14.98	13.75	14.15
FeO	12.12	12.34	7.59	13.19	13.09	13.17	9.16	13.20	12.88	12.63	13.78	14.11	12.72	9.39	12.87	8.47
MnO	0.21	0.39	0.12	0.18	0.19	0.20	0.12	0.22	0.18	0.17	0.20	0.20	0.18	0.18	0.18	0.11
MgO	7.29	6.73	8.81	19.21	14.47	19.12	6.74	7.34	7.30	7.43	6.23	8.25	7.36	7.27	5.59	6.46
CaO	12.19	11.62	14.02	9.50	9.09	6.98	14.91	9.48	10.04	9.95	10.49	9.99	10.33	11.12	7.18	9.74
Na_2O	2.22	2.84	1.94	1.00	1.17	0.81	3.02	3.50	2.96	2.75	2.59	2.32	2.56	3.53	3.35	2.69
K_2O	0.50	1.06	0.20	0.14	0.11	0.25	0.54	0.69	0.48	0.41	0.56	0.48	0.43	0.57	1.36	1.84
P_2O_5	0.13	0.17	0.03	0.14	0.17	0.14	0.14	0.12	0.14	0.15	0.17	0.14	0.12	0.10	0.15	0.14
LOI							2.49	2.46	1.20	1.55	0.75	0.30	1.18	3.27	1.13	4.81
S	0.01	0.00					0.10	0.13	0.09	0.05	0.08	0.09	0.07	0.04	0.08	0.03
Total	100.07	99.51	100.09	99.23	99.44	99.58	99.79	99.77	99.79	99.84	99.77	99.68	99.81	99.79	99.29	98.72
Rb	17.9	37.9	3.6	6.9	6.5	5.0	11.1	16.1	11.4	10.2	15.1	11.8	12.4	14.0	27.8	42.7
Sr	242	288	262	179	297	228	535	354	220	186	220	202	188	318	434	594
Y	21.7	22.6	9.5	15.4	18.7	14.4	31.2	23.4	23.3	24.5	28.0	23.1	22.7	22.1	28.7	14.4
Ba	147	217	90	52	139	106	94	148	140	109	159	137	91	93	273	285
La	7.47	8.12	2.80	7.55	7.93	6.08	8.73	6.26	8.00	7.46	8.55	7.18	5.87	5.62	8.93	10.13
Ce	16.9	18.8	6.3	17.7	21.1	16.2	23.0	14.8	18.1	18.0	21.3	16.9	14.3	14.1	22.8	23.3
Pr	2.32	2.45	0.87	2.46	2.96	2.27	3.35	2.13	2.45	2.53	2.87	2.42	2.09	2.07	3.09	2.95
Nd	11.2	11.7	4.3	11.6	13.9	10.7	15.8	10.4	11.2	11.9	13.1	11.2	10.0	9.8	15.0	13.6
Sm	3.11	3.19	1.23	3.05	3.67	2.82	4.49	3.18	3.35	3.51	3.94	3.33	3.00	2.94	4.32	3.26
Eu	1.02	1.09	0.59	1.01	1.23	0.94	1.34	1.15	1.16	1.15	1.30	1.14	1.05	0.99	1.41	1.18
Gd	3.77	4.00	1.49	3.35	4.08	3.13	4.97	3.48	3.67	3.62	4.17	3.43	3.27	3.33	5.63	3.63

(continued)

Table 6.3 (continued)

N	1	2	3	4	5	6	7	8	9	10	11	12	13	14	15	16
Tb	0.65	0.69	0.26	0.52	0.62	0.47	0.92	0.64	0.65	0.66	0.78	0.65	0.62	0.62	0.88	0.53
Dy	4.19	4.27	1.73	3.16	3.78	2.90	5.69	4.32	4.30	4.46	5.03	4.14	4.10	3.95	5.69	3.26
Ho	0.87	0.90	0.35	0.61	0.71	0.54	1.23	0.93	0.93	0.91	1.10	0.91	0.89	0.86	1.15	0.60
Er	2.44	2.45	1.01	1.63	1.89	1.45	3.34	2.44	2.56	2.56	3.00	2.50	2.43	2.35	3.39	1.69
Tm	0.33	0.35	0.14	0.21	0.24	0.19	0.49	0.38	0.37	0.38	0.43	0.35	0.36	0.33	0.47	0.21
Yb	2.36	2.52	0.95	1.38	1.55	1.19	2.96	2.42	2.41	2.47	2.76	2.29	2.28	2.31	3.14	1.42
Lu	0.36	0.37	0.14	0.19	0.23	0.17	0.46	0.35	0.36	0.36	0.43	0.36	0.35	0.33	0.46	0.20
Pb	80	2410	3.71	2.91	1.40	1.07	3.02	0.59	1.86	0.50	3.82	3.02	0.79	2.35	1.05	0.79
Th	1.58	0.82	0.32	1.24	1.23	0.95	0.99	0.61	0.93	0.93	1.10	0.81	0.73	0.58	1.43	0.94
U	0.46	0.35	0.09	0.37	0.52	0.40	0.30	0.34	0.36	0.36	0.43	0.35	0.29	0.25	0.65	0.24
Sc	48.1	48.0	42.1	22.7	31.1	23.8	35.1	41.2	35.9	37.4	37.1	31.5	36.2	34.1	36.6	23.7
Ti	7610	8608	2175	6936	9830	7544	8560	7971	7456	7705	8600	7384	7020	6547	8899	7578
V	332	355	224	226	335	257	213	295	262	276	301	250	270	253		
Cr							180	214	151	148	126	119	171	153	149	285
Mn	1794	3323	1101	1522	1817	1394	975	1674	1439	1321	1499	1542	1450	1398	36	21
Co	59	117	53	91	121	93	30	46	47	46	48	56	50	30	69	113
Ni	136	148	304	1058	1296	995	83	132	125	111	99	262	147	109	336	82
Cu	399	4047	361	112	67	51	69	140	115	73	248	659	141	85	62	30
Zn	168	559	52	133	131	100	99	53	70	20	194	47	74	81	20	17
Zr	74	70	25	80	95	73	116	82	85	87	110	90	73	71	116	83
Nb	4.22	4.90	1.93	5.26	6.94	5.33	4.57	2.59	3.34	3.58	4.00	3.30	2.63	2.53	5.51	6.93
Hf	1.97	1.81	0.67	2.18	2.53	1.94	2.79	1.67	1.89	1.91	2.25	1.85	1.56	1.40	3.11	2.23
Ta	0.26	0.30	0.11	0.35	0.44	0.34	0.33	0.17	0.20	0.31	0.21	0.17	0.20	0.22	0.34	0.41

Oxides are given in wt%, elements—in ppm. Empty cell—element was not analyzed. Samples with N = 1–6 are from boreholes (depth in meters is shown in line “sample”): 1–2—OUG-2 (Talnakh intrusion); 3—OKG-13 (Dzhaltul intrusion). 4–6—OV-36 (Gudchikhinsky Formation, basalts)

Discussion

The studied intrusions predominantly belong to the Katangsky intrusive complex which differs from the Norilsk complex by elevated titanium contents and a smaller Ta-Nb anomaly; these intrusions have paleomagnetic directions and geochemical features similar to lavas of the Mokulaevsky-Kharaelakhsky Formations and can be coeval with them. The rocks of the Kureysky complex which was mapped earlier in this area, were not detected (for comparison their geochemical characteristics are given in Table 6.3 and are shown in Fig. 6.5b; see sample OKG-13/43.7 of the Dzhaltul massif of the Kureysky intrusive complex).

For the first time, an intrusive analogue of the rocks of the Gudchikhinsky Formation was discovered. Until now, only the rocks of the Fokinsky massif in the Norilsk region have been compared with the Gudchikhinsky Formation based on high-Mg rock composition. But these rocks were not analyzed using modern geochemical methods, so their comagmatic nature with the Gudchikhinsky rocks was not proved. Besides, the paleomagnetic directions of intrusion KUL-21 point out that its emplacement can be coeval with the Syverminsky-Gudchikhinsky Formations eruption. As the intrusive analogues of the Gudchikhinsky Formation are primarily found in the Kulumbe river valley, we propose to refer to these intrusions as the Kulyumbinsky complex.

KUL-1 intrusion is the only site with the reverse polarity in the studied area and it also differs from other intrusions in the rare element spectra. The geochemical features of sample KUL-1 are similar to those of the Talnakh intrusion.

Our results of the AMS measurements are consistent with the hypothesis of the magma-feeding activity of the regional Imangda-Letninskiy fault of NE strike.

Acknowledgements Authors are very grateful to geologists of Ltd. Norilskgeologia S. Erykalov, V. Rad'ko, V. Sitnikov, A. Lapkovsky, I. Sidorenko and students of Lomonosov Moscow State University M. Nesterenko, D. Korshunov and E. Ostrovsky, Tomsk State University, for their help during the field trip.

This work was funded by RFBR (grants № 16-35-60114, 17-05-01121, 15-05-09250) and the Ministry of Education and Science RF (grant № 14. Z50.31.0017).

References

Borradaile G.J., Jackson M. Structural geology, petrofabric and magnetic fabric (AMS, AARM, AIRM). Journal of Structural Geology, 32 (2010) 1519–1551.

Burgess S.D., Bowring S. A. High-precision geochronology confirms voluminous magmatism before, during, and after Earth's most severe extinction. Science Advances. 2015. Vol. 1, no. 7, e1500470. https://doi.org/10.1126/sciadv.1500470.

Callot J.-P., Gurevitch E., Westphal M., Pozzi J.-P. Flow patterns in the Siberian traps deduced from magnetic fabric studies // Geophys. J. Int. 2004. V. 156. P. 426–430.

Chadima, M., Hrouda, F. 2006. Remasoft 3.0 a user-friendly paleomagnetic data browser and analyzer. Travaux Géophysiques, XXVII, 20–21.

Dobretsov N.L. Permian-Triassic magmatism and sedimentation in Eurasia as reflection of superplume. Doklady Earth Sci. 1997, 354, pp. 220–223.

Enkin R.J. A computer program package for analysis and presentation of paleomagnetic data. Pacific Geoscience Centre, Geological Survey of Canada. 1994. P. 16.

Ernst R.E. Large Igneous Provinces. Cambridge University Press, 2014, 653 p.

Fisher, R. 1953. Dispersion on a Sphere. Proceedings of the Royal Society of London, Series A, Mathematical and Physical Sciences, 217(1130), 295–305.

Heunemann C., Krasa D., Soffel H., Gurevitch E., Bachtadse V. Directions and intensities of the Earth's magnetic field during a reversal: results from the Permo-Triassic Siberian trap basalts, Russia. Earth and Planetary Science Letters. 2004. V. 218. P. 197–213.

Hofmann A. W. Chemical Differentiation of the Earth: Relationship between Mantle, Continental Crust and Oceanic Crust. Earth Planet Sci Lett. 1988. V. 90. pp. 297–314.

Hrouda F., Burianek D., Krejci O., Chadima M. Magnetic fabric and petrology of Miocene sub-volcanic sills and dikes emplaced into the SW Flysch Belt of the West Carpathians (S Moravia, Czech Republic) and their volcanological and tectonic implications. Journal of Volcanology and Geothermal research. 290 (2015) 23–38.

Jelínek, V., 1978. Statistical processing of anisotropy of magnetic susceptibility measures on groups of specimens. Studia geophisica et geodetica 22, 50–62.

Kamo S.L., Czamanske G.K., Amelin Yu. et al. Rapid eruption of Siberian flood-volcanic rocks and evidence for coincidence with the Permian-Triassic boundary and mass extinction at 251 Ma. Earth Planet Sci Lett. 2003. 214, pp. 75–91.

Kirschvink, J.L., 1980. The least-squares line and plane and the analysis of palaeomagnetic data. Geophysical Journal International 62 (3):699–718.

Knight, M.D. & Walker, G.P.L., 1988. Magma flow directions in dikes of the Koolau complex, Oahu, determined from magnetic fabric studies, J. geophys. Res., 93, 4301–4319.

Konstantinov K.M., Mishenin S.G., Tomshin M.D., Kornilova V.P., Kovalchuk O.E. Petromagnetic heterogeneities of the Permo-Triassic traps of the Daldyn-Alakit diamond province (Western Yakutia). Lithosphere. 2014. №2. P. 77–98 (in Russian).

Krivolutskaya N.A. Siberian Traps and Pt-Cu-Ni Deposits in the Noril'sk Area. Springer. 2016. 361 p.

Latyshev A.V. Paleomagnetism of the Siberian Traps: the estimation of the duration and the intensity of magmatic activity in the Norilsk region and Angara-Taseeva depression. PhD thesis. Moscow, 2013 (in Russian).

McFadden P. L., McElhinny M. W. Classification of the reversal test in palaeomagnetism // Geophys. J. Int. 1990. V. 103. P. 725–729.

Pavlov V.E., V. Courtillot, M.L. Bazhenov, R.V. Veselovsky. Paleomagnetism of the Siberian traps: New data and a new overall 250 Ma pole for Siberia. Tectonophysics. V. 443. 2007. pp. 72–92.

Pavlov V., Fluteau F., Veselovskiy R., Fetisova A., Latyshev A., Elkins-Tanton L.T., Sobolev A. V. and N.A. Krivolutskaya (2015). Volcanic pulses in the Siberian Traps as inferred from Permo-Triassic geomagnetic secular variations. Chapter 5 in "Volcanism And Global Environmental Change", A. Schmidt, K.E. Fristad and L. Elkins-Tanton ed., Cambridge University Press, pp. 63–78.

Potter, D.K., Stephenson, A., 1988. Single-domain particles in rocks and magnetic fabric analysis. Geophysical Research Letters, 15, 1097–1100.

Sobolev, A.V., Krivolutskaya N.A., and Kuzmin, D.V. Petrology of the parental melts and mantle sources of Siberian trap magmatism // Petrology, 2009. 17 (3), 253–286.

Tarling, D.H., Hrouda, F., 1993. The Magnetic Anisotropy of Rocks. Chapman, Hall, London, 217 pp.

Chapter 7
Paleomagnetism of Basalts from the Mid-Atlantic Ridge and the Bouvet Triple Junction

V. I. Maksimochkin and A. N. Tselebrovskiy

Abstract The magnetic properties of 18 basalt samples dredged from the south of the Mid-Atlantic Ridge (MAR), the region with Lat = (54.0°–55.0°) S, Long = (0.0°–2.0°) W and from the Bouvet Triple Junction (BTJ) with Lat = (53.5°–54.5°) S, Long = (3.0°–5.0°) E was investigated. The basalts age was determined from spreading rate, distribution of anomalous geomagnetic field (Bulychev et al. in Seafloor chronology in the region of the Bouvet lithospheric plate triple junction. Pleiades Publishing, 2000) at the sampling region and magnetochronological scale (Gradstein et al. in The Geologic Time Scale, 2012). Their ages fluctuate from 0.1 to 5.2 Ma, i.e. covered the Brunhes (9 samples), Matuyama (5 samples), Gauss (3 samples) and Gilbert (1 sample) Chrones. The growth of Curie temperatures from 90 up to 480 °C with the increase of basalts ages, and the decrease in natural remanent magnetization (NRM) due to low-temperature oxidation of titanomagnetite (TM) were observed. The average values of basalts NRM of the Brunhes, Matuyama, and Gauss Chrones was calculated. For MAR basalts NRM_{mid} = 18.7, 11.0 and 2.8 A/m, respectively, and for the basalts of the BTJ—NRM_{mid} = 18.5, 9.2, 3.2 A/m, respectively. There has been a tendency for V_r parameter—NRM normalized to the residual saturation magnetization (M_{rs}), to decrease, as the basalts age approaches the Brunhes-Matuyama boundary. The paleointensity of the geomagnetic field (H_{pl}), determined from no more than 3.5 Ma basalts NRM by the Thellier-Coe procedure, ranged from 16 to 42.5 A/m. For the most of the samples, a satisfactory correlation between V_r and the H_{pl} was found. The virtual axial dipole moment (VADM) values we calculated were fairly close to the averaged values of the VADM from the World Paleomagnetic Bank (WPB) of the Borok Observatory (http://www.brk.adm.yar.ru/palmag/index.html). The obtained results testify to the fact that NRM and V_r variations of basalts of the south of the MAR and the BTJ from the Brunhes and Matuyama Chrones can be caused not only by the formation conditions, but also by variations of the geomagnetic field. The definition of H_{pl} can be used to clarify the age of basalts and the geodynamic features of the spreading in the areas of mid-oceanic ridges.

V. I. Maksimochkin (✉) · A. N. Tselebrovskiy
Faculty of Physics, Lomonosov Moscow State University, Moscow, Russia
e-mail: maxvi@physics.msu.ru

D. Nurgaliev et al. (eds.), *Recent Advances in Rock Magnetism, Environmental Magnetism and Paleomagnetism*, Springer Geophysics,
https://doi.org/10.1007/978-3-319-90437-5_7

Keywords Basalt · paleointensity · rock magnetism · spreading oceanic ridges

Introduction

According to paleomagnetic data, the polarity of the Earth's magnetic field has repeatedly changed its sign (Gradstein et al. 2012). For construction of the Earth's crust magnetoactive layer (MAL) models of oceanic rift zones satisfactory match between the observed and model curves of anomalous magnetic field (AMF) is needed. Magnetization and, sometimes, power of the individual blocks, relating to the epochs of direct or reverse polarity, are chosen for making this match. And only the magnetization decrease with ages due to low-temperature oxidation of TM is counted. For example, in the MAL model of the south of the MAR and the BTJ (Bulychev et al. 2000), the calculated magnetization turned out to be maximal for the central block of directly magnetized rocks, i.e. for the spreading axis, and decreases with distance from it. At the same time, in the previous work (Trukhin et al. 2002) it was shown that a significant variation in the NRM of basalts from the both, direct and reverse, polarity of the geomagnetic field, and the calculated magnetizations in the model in some cases differed significantly from the measured values. There is still no satisfactory explanation for this fact. The NRM value spread can be explained by the difference in the conditions for the formation of magnetic minerals in basalts, for example, the rate of their cooling. Obviously, the thermoremanent part of magnetization in NRM of the MAL should also depend on the magnitude of the magnetic field that exists during the lava cooling. Currently, there is a large number of results that indicate that the H_{pl} also changes with time even in periods of the same polarity and significantly decreases during periods of polarity reversal (Channell et al. 2009; Shcherbakov et al. 2014; Mazaud et al. 2012). By variations in H_{pl} has been explained a significant spread of the NRM basalts of the Reykjanes Ridge, selected from the region of direct polarity of the geomagnetic field and its sharp drop for basalts over 0.7–1.1 Ma, i.e. near the inversion boundary (Maksimochkin et al. 2016). In our opinion, when constructing MAL models and interpreting NRM measurement data, it is necessary to take into account not only geomagnetic field inversions, but also variations in its magnitude.

In this paper, in order to obtain information on the intensity of the geomagnetic field during the Brunhes, Matuyama and Gauss Chrones and to establish the relations between the NRM of oceanic basalts and the geomagnetic field, which exists during their formation as well as to clarify the information on the structure of MAL of the Bouvet triple junction, the magnetic properties of basalts were studied. The basalts samples were lifted from the ocean floor by the "Academic Strakhov" ship of the P. P. Shirshov Institute of Oceanology of the Russian Academy of Sciences (IO RAS). Samples from 5 sampling sites from the south of the MAR from the area with coordinates Lat = (54.0°–55.0°) S, Long = (0.0°–2.0°) W and samples from 13 sampling sites in the area of the BTJ with coordinates Lat = (53.5°–54.5°) S,

Long = (3.0°–5.0°) E were kindly provided to us by the professor A. A. Shreider from the IO RAS.

Results of the Experiments and Discussion

The Age Determination of Basalt Samples

The sampling points and the distribution of the AMF in the region are shown in Fig. 7.1. Samples of basalts №№ S18-48, 51, 52, 63 were selected from the region of the central anomaly (Trukhin et al. 2002) near the axis of the ridge, their age refers to the Brunhes period (cron C1n) (Gradstein et al. 2012). To clarify the age of basalts, the following data were used: the spreading rate, the position of the maximum and the zero magnetic isoline of the AMF according to (Bulychev et al. 2000). It was assumed that the age of the ocean floor rocks lifted from the area of the zero isoline is equal to changes of the geomagnetic field polarity of the Brunhes-Matuyama (0.781 Ma), Matuyama-Gauss (2.581 Ma), Gauss-Gilbert (3.596 Ma) Chrones (Gradstein et al. 2012). The spreading axis was defined as in the other work (Trukhin et al. 2002) for the maximum value of AMF. Due to the low spreading rate (0.7–1.5 cm/year), the resolution of basalt age determination within the axial anomaly is no better than 0.1 Ma.

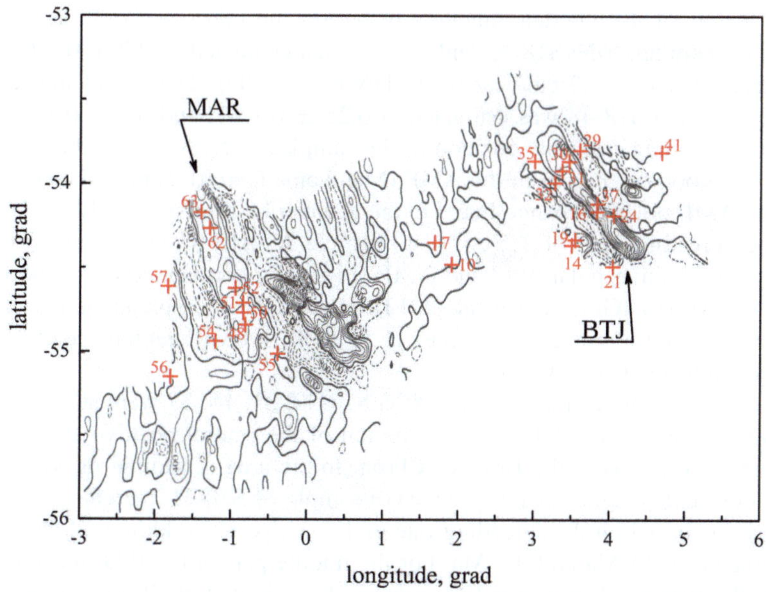

Fig. 7.1 AMF isolines (Bulychev et al. 2000) and sampling sites of the south of the MAR and the BTJ

Table 7.1 Site coordinates and age of samples of the south of the MAR and the BTJ

Sample	Longitude, grad	Latitude, grad	Distance from ridge (km)	Spreading rate (cm/year)	Age (Ma)
S18-14			26[a]	0.8	3.2 ± 0.2
S18-16			1[a]	0.7	0.15 ± 0.1
S18-19			22[a]	0.8	2.75 ± 0.2
S18-21	4.038	−54.493	20	1.2	1.7 ± 0.2
S18-24	4.049	−54.189	7	1.0	0.8 ± 0.2
S18-29	3.603	−53.801	13	1.3	1.0 ± 0.2
S18-30	3.463	−53.867	2	1.3	0.15 ± 0.1
S18-31	3.372	−53.922	1	1.3	0.1 ± 0.1
S18-32	3.276	−53.994	18	1.3	1.4 ± 0.2
S18-35	3.008	−53.865	21	1.3	1.6 ± 0.2
S18-37	3.833	−54.118	2	0.7	0.3 ± 0.1
S18-41	4.688	−53.813	68	0.9	5.2 ± 0.4
S18-48	−0.809	−54.841	4	1.2	0.3 ± 0.1
S18-51	−0.840	−54.710	3	1.2	0.25 ± 0.1
S18-52	−0.942	−54.623	2	1.2	0.2 ± 0.1
S18-54	−1.201	−54.938	26	1.5	1.7 ± 0.2
S18-57	−1.831	−54.612	52	1.5	3.5 ± 0.2
S18-63	−1.398	−54.171	2	1.2	0.1 ± 0.1

[a]direction calculated from AMF (Bulychev et al. 2000), Fig. 7.1

The results of the age determination of basalts are presented in Table 7.1. The age of the samples №№ S18-63 and S18-52 was estimated at 0.2 ± 0.1 Ma with the spreading rate of 1.2 cm/year (Bulychev et al. 2000). The age of the samples №№ S18-51 and S18-48 was estimated at 0.25 ± 0.1 Ma and 0.3 ± 0.1 Ma with the same spreading rate. Site position of the samples № S18-54 is in the middle of the negative polarity region of the AMF (Matuyama Chron), approximately 14 km from the AMF polarity change line, i.e. the age can be estimated at 1.7 ± 0.1 Ma, with the spreading rate of 1.5 cm/year. Sample № S18-57 is selected from the region with positive polarity of the AMF (Fig. 7.1), corresponding to the Gauss Chron (C2An.1n) (Gradstein et al. 2012). If we take the spreading rate on the western flank as 1.5 cm/year (Bulychev et al. 2000), we will get the age of 3.5 Ma, which fits into the Gauss Chron.

Figure 7.1 illustrates that samples №№ S18-30, 31, 16, 37 were selected in the northwestern part of the BTJ from the region of the central positive anomaly, i.e. their age corresponds to the Brunhes Chron. In the area of sample № S18-30, the anomalous field is higher than in the area of sample № S18-31, therefore, the age of these basalts based on the spreading rate of 1.2 cm/year (Trukhin et al. 2002) was determined as 0.15 Ma and 0.1 Ma. For the middle part of the BTJ, the spreading rate for the Brunhes Chron was 0.7 cm/year (Bulychev et al. 2000), so the age of sample № S18-16 was determined as 0.15 ± 0.1 Ma, and the sample S18-37 as 0.3 ± 0.1 Ma (Table 7.1).

The sample № S18-24 is taken from the region of positive polarity of the AMF and it is fairly close to the zero isoline. We estimate the spreading rate in this area as 1.1 cm/year, because the sampling site is in the region between the spreading rate of 0.7 cm/year and 1.3 cm/year (Bulychev et al. 2000). Distance from the spreading axis of 7 km gives us an age of 0.8 Ma. Samples №№ S18-29, S18-32 (Fig. 7.1) are selected from the regions with negative polarity of the AMF, and, therefore, the time of their formation is related to the Matuyama Chron. The site's distance from the spreading axis is approximately 13 km and 18 km, which at the spreading rate of 1.3 cm/year (Bulychev et al. 2000) gives the age of 1.0 ± 0.2 Ma and 1.4 ± 0.2 Ma. At the site of samples №№ S18-35 and S18-21, the AMF is also negative, and the samples age is 1.6 Ma and 1.7 Ma, respectively. Samples №№ S18-14 and S18-19 refer to the Gauss Chron, AMF is positive. Since the average spreading rate for the 0-C2An.1n interval is 0.8 cm/year (Bulychev et al. 2000), their age was estimated at 3.3 ± 0.2 Ma and 2.8 ± 0.2 Ma, respectively. The sample № S18-41 was taken from the region, which corresponds to the C3n (the Gilbert Chron), so its age can be estimated at 5.2 Ma, although with the distance from the axis ridge at 68 km and the average spreading rate of 0.9 cm/year (Bulychev et al. 2000) the age is 7.5 Ma. Probably, this discrepancy is caused by the error in determining the spreading rate (Bulychev et al. 2000), or by the error in the designation of the Chrones of the AMF polarity in the sampling region.

Basalts NRM of the South of the MAR and the BTJ

On the samples of the collection were measured NRM, magnetic susceptibility (k), hysteresis characteristics: coercive force (H_c), remanent coercive force (H_{cr}), M_{rs} after 1 T magnetic field and saturation magnetization (M_s) in the field of 0.4 T. All these measurements were carried out on samples of a cubic shape with the rib size of 1 cm. The remanent magnetization was measured on AGICO JR-6a spinner magnetometer and the magnetic susceptibility was measured on AGICO kappabridge MFK-1A. The hysteresis characteristics and the dependence of the magnetization in the field of 0.24 T were measured on a VM-1 vibration magnetometer.

The NRM values of the basalts are presented in Table 7.2. There is quite a large scatter in the NRM values. Analysis of the NRM dependence on age shows that there is a general decrease of NRM with age. The mean values of samples NRM with age related to the Brunhes, Matuyama and Gauss Chrones calculated by the MAR basalts were NRM_{mid} = 18.7, 11.0 and 2.8 A/m, respectively. Similar calculations for the basalts of the BTJ—NRM_{mid} = 18.5, 9.2 and 3.2 A/m were fairly close to the MAR values (Fig. 7.2).

The revealed decrease in NRM with their age is likely to be due to the oxidation of TM, the carrier of the NRM. It is known (Stephenson 1972) that Curie temperature (T_c) of TM increases with the process of low-temperature oxidation. To estimate the phase state of ferrimagnetic grains of basalts the Curie temperature was determined. Determination was made by the dependence of magnetization in the

Table 7.2 Magnetic characteristics of the basalts of the south of the MAR and the BTJ

Site	Sample №	NRM (A/m)	$k \times 10^3$ SI	M_{rs} (A/m)	M_{rs}/M_s	H_{cr} (mT)	H_{cr}/H_c	NRM/M_{rs}	T_c, °C by $k(T)$	T_c, °C by $M_s(T)$	Age, (Ma)
S18-21	29[a]	9.91	5.5	396				0.025		330	1.7
	21	5.74	4.01								
S18-24	26	12.74		290	0.250	49	1.42	0.044			0.8
S18-29	71	28.52	11.34								1.0
	72	29.14	11.55	550	0.241	44.5	1.89	0.053	95	180	
S18-30	18-3	4.17	12.04	183	0.136	22.2	2.23	0.046	90	200	0.15
	18-1	5.09	10.89								
	26	8.13	16.97	95.2	0.139	12.9	1.95	0.043	70	90	
	40	13.49	31.58	259	0.097	16.3	2.90	0.051	160	150	
S18-31	46	21.21	8.28	229	0.361	75.1	1.05	0.046			0.1
	54	5.66	4.57	92.1	0.373	15.6	1.00	0.031		70	
s18-32	11	2.85	63.1	467	0.07	29.5	2.81	0.003	410, 595		1.4
S18-37	21-2	27.92	4.71	375	0.285	30.9	1.33	0.075	190		0.3
	21-1	31.76	5.82								
	9-2	34.54	3.84	664	0.379	47.8	1.21	0.052	330, 475	240	
	9-1	43.76	5.54						220. 465		
	9-4	34.53	3.86								
S18-41	1-4	3.31	1.44	148	0.331	76.2	1.37	0.014	481		5.2
	1-1	3.05	1.52							290	
S18-48	4	8.21	9.27	328.5	0.212	26.4	1.40	0.013	215	200	0.3
	38	9.21	7.52	330	0.220	28.1	1.48				
	33[b]	6.54	1.39	409				0.016		185	
S18-51	26[b]	19.28	3.35	459				0.042	145		0.25
S18-52	7	15.65	4.55	348	0.268	25.2	1.22	0.045	155	140	0.2
	3	14.36	3.5								
S18-54	35	11.13	4.23	448.2	0.275	27.3	1.29	0.025	225		1.7
	29	10.14	4.12						185		
	48	11.41	4.5								
	19[b]	15.67	4.6	422	0.45		1.33	0.026		265	
	24[b]	9.32	4.17	345	0.43		1.39	0.026		245	

[a]Maksimochkin and Tselebrovskiy 2015, [b]Maksimochkin et al. 2016

field of 0.24 T on the temperature $M_s(T)$ in the air, which was recorded on a VM-1 vibrational magnetometer upgraded with the ZET-Lab module. Zet-Lab module allows us to use the program of automatic management and data collection (https://www.zetlab.com). The magnetic susceptibility as a function of temperature $k(T)$ in argon was measured at the MFK-1A. The results are shown in the Fig. 7.3 and in the Table 7.2. It was found that the Curie temperature of the most of the samples ranged from 70 to 290 °C, and the T_c data for $k(T)$ and $M_s(T)$ were close to each other. On the sample № S18-37/9, quite a large discrepancy of these values was found

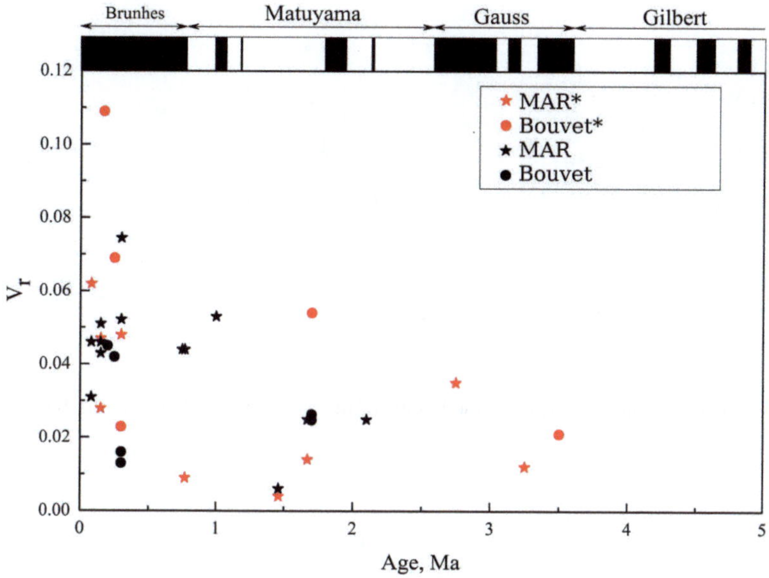

Fig. 7.2 The ratio between the basalts age and V_r = NRM/M_{rs} for the south of the MAR and the BTJ samples, *Trukhin et al. 2002

(Table 7.2). Probably, more reliable is the Curie temperature determined from the $M_s(T)$ dependence—T_c = 230 °C, since on the other sample of this number S18-37/21 the value of T_c = 190 °C was obtained from the susceptibility dependence.

Considering data in Fig. 7.4 (Trukhin et al. 2002) an increase in the Curie temperature with increasing their age (Fig. 7.4) is observed on the samples from the south of the MAR and the BTJ, probably due to low-temperature oxidation of TM.

Since laboratory studies were carried out on samples with the volume of 1 cm^3 the difference in the NRM value of the samples can also be not only due to the difference in the level of oxidation of TM, but also to the variation in the concentration of ferrimagnetic grains. The ratio of NRM to the M_{rs} should not depend on the concentration of ferrimagnetic grains. From Fig. 7.3 and Table 7.2 show that the scatter of the value V_r over the samples also turned out to be quite large. It is also evident that within the Brunhes Chrone the value of V_r decreases with the age of the basalts. Also, a low value of normalized NRM had samples with the age of 1.4 million years.

Determination of the Geomagnetic Field Paleointensity

The variability of V_r and NRM basalts of different ages may be due to the difference in physico-chemical conditions for the formation of ferrimagnetic minerals, as well as variations in the geomagnetic field in which the primary thermally-based

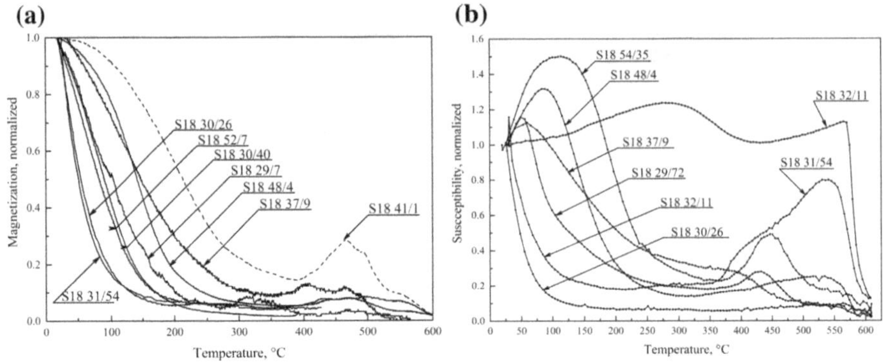

Fig. 7.3 Thermomagnetic analysis of the basalts of the south of the MAR and the BTJ: **a** $M_s(T)$, **b** $k(T)$

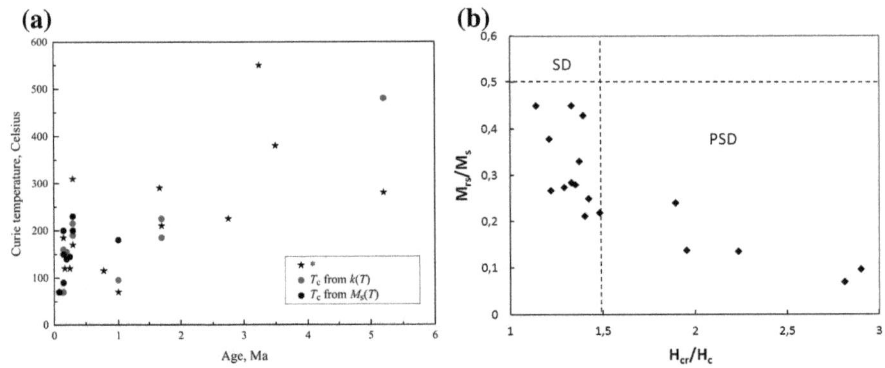

Fig. 7.4 **a** Change in the Curie temperature of the MAR basalts and the BTJ with age, *Trukhin et al. 2002, **b** Hysteresis characteristics of the basalts of the south of the MAR and the BTJ

magnetization of basalts was formed. To verify this assumption, the geomagnetic field paleointensity was determined by the Thellier–Coe procedure (Coe 1978). The validity of the application of this method is confirmed by the hysteresis characteristics of the samples. The M_{rs}/M_s and H_{cr}/H_c ratios (Table 7.2 and Fig. 7.5) indicate that the ferrimagnetic grains of most basalt samples according to the Day criteria (Day et al. 1977) are in the SD-PSD state, i.e. sufficiently small. Most of the samples we studied had $T_c \leq 230$ °C, which suggests that the oxidation level of TMs is low, and NRM has preserved information on ancient magnetic field.

The examples of the Thellier-Coe procedures are shown in the Arai-Nagata (Nagata et al. 1963) and Zijderveld (1967) diagrams in Fig. 7.5. The NRM of a part of the samples is represented by two components of viscous and thermo-stable nature (№№ S18-51/26, 37/9, 29/71, 52/7). On the samples 19 and 24 from the site

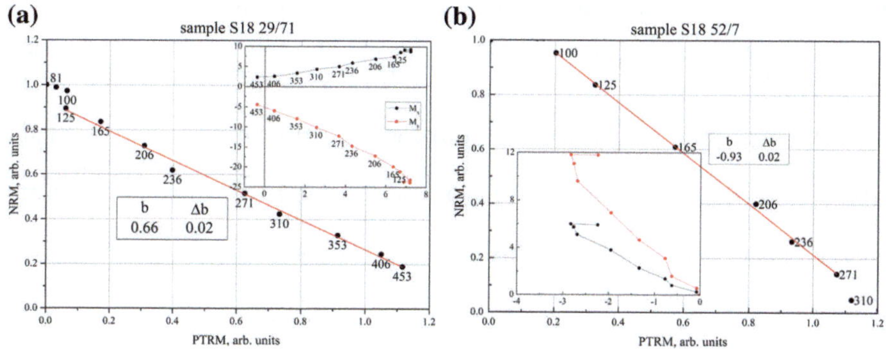

Fig. 7.5 Examples of determining the geomagnetic field paleointensity on samples of basalts of the south of the MAR and the BTJ

S18-54 there were identified three components of NRM. The component of viscous nature was destroyed after heating up to 80–100 °C. The second component is likely to be thermoremanent magnetization (TRM), and the third component, which was destroyed at higher temperatures, was determined as chemical remanent magnetization (CRM). To separate the thermal and chemical components of NRM, the technique proposed in related works (Maksimochkin and Tselebrovskiy 2015; Gribov et al. 2017) was used.

The results of the H_{pl} determination from the basalts NRM of the south of the MAR and the BTJ and VADM values calculated from these data are presented in Table 7.3. H_{pl} varies over quite large ranges: from 18.0 to 42.5 A/m. The quality factor (q) of H_{pl} determination, calculated by the method (Coe 1978), for most of the samples proved to be sufficiently high, q > 9. It was also found that on the samples with high Curie temperatures (T_c > 300 °C) and, accordingly, with the chemical component of NRM, the quality factor was low: for № S18-30/44 (T_c = 348 °C), № S18-54/19 (T_c = 265 °C) and № S18-54/24 (T_c = 245 °C) q < 3, for № S18-21/29 (T_c = 330 °C)—q < 7.

If the TM has a low degree of oxidation, and therefore NRM is basically thermally stable, and the paleointensity is correctly determined and corresponds to the magnetic field during basalt formation, the value of V_r should increase with increasing H_{pl}. As can be seen from Fig. 7.6, such a correlation is observed if we exclude from the analysis the data obtained on samples of basalts S18-30/44 and S18-21/29 with high Curie points. The reasons for excluding these data from the analysis are as follows. Data on sample S18-30/44 should be deleted, as it can be seen from Table 7.3 that the results of the determination of H_{pl} on samples №40 and №44 taken from one site S18-30 were fairly close, and the value of V_r differs more than 2 times. It can be assumed that the NRM of S18-30/44 is significantly distorted by the chemical changes of TM, since the Curie temperature (see Table 7.3) is considerably higher for this sample than for S18-30/40 (T_c = 150 °C) and CRM was not recognized by us.

Table 7.3 The results of the determination of geomagnetic field paleointensity

Sample	NRM (A/m)	T_c, (°C)	b	q	H_{lab} (A/M)	H_{pl} (A/M)	VADM × 10^{22} (A M^2)	Age (Ma)
S18-21/29[a]	9.91	330	0.70 ± 0.05	6.64	40	28.0 ± 2.0	5.2 ± 0.4	1.67
S18-24/26[a]	12.74		0.54 ± 0.02	19.5	40	21.6 ± 0.8	4.1 ± 0.2	0.77
S18-30/40	13.49	150	0.72 ± 0.02	30	40	28.8 ± 0.8	5.4 ± 0.1	0.15
S18-30/44[a]	8.33	348	0.85 ± 0.06	2.43	40	34.1 ± 2.4	6.4 ± 0.5	0.15
S18-37/9	34.54	230	0.58 ± 0.01	33	40	23.2 ± 0.4	4.4 ± 0.1	0.3
S18-29/72	29.14	180	0.71 ± 0.02	24	40	28.4 ± 0.8	5.4 ± 0.1	1
S18-51/26[b]	19.28	145	0.51 ± 0.03	9.76	40	20.4 ± 1.2	3.8 ± 0.2	0.25
S18-52/7	15.65	135	0.93 ± 0.02	37	40	37.3 ± 0.8	7.0 ± 0.2	0.2
S18-54/24[b]	8.98	245	0.40 ± 0.04	2.07	40	16.0 ± 1.6	3.0 ± 0.3	1.70
S18-54/19[b]	10.98	265 (Coe 1978)	0.61 ± 0.04	3.81	40	24.0 ± 1.6	4.5 ± 0.3	1.7
S18-48/33[b]	6.54	185	1.06 ± 0.01	25.9	40	42.5 ± 0.4	7.9 ± 0.1	0.33
S18-31/37[c]			0.80		40	32	6.00	0.08
S18-51/3[c]			0.83		40	33	6.20	0.25
S18-48/8[c]			0.50 ± 0.11		80	40	7.50	0.33
S18-54/34[c]			0.43 ± 0.14		80	34.4	6.40	1.7
S18-57/16[c]			0.65 ± 0.13		40	21.3	4.00	3.5

[a]Maksimochkin and Tselebrovskiy 2015, [b]Maksimochkin et al. 2016, [c]Maksimochkin et al. 2010

As for the sample S18-48/33, it was shown in (Maksimochkin and Tselebrovskiy 2015) that the part of the NRM with unlocking temperatures greater than T_c determined by $I_s(T)$ is of chemical nature, and H_{pl} was determined for NRM of thermal nature.

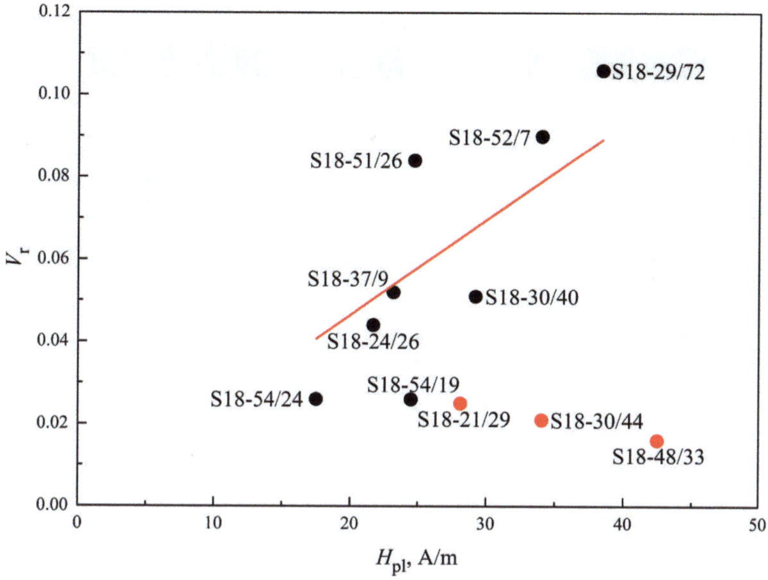

Fig. 7.6 The ratio between the V_r and the geomagnetic field paleointensity. Red points are excluded from approximation

M_{rs}, which is created in the laboratory on a modified phase, can also be larger than in the original state, and causes a low value of V_r. Probably, this reason can explain the understated V_r value of the sample S18-48/33, as well as S18-21/29.

The VADM values calculated for the majority of the samples, as can be seen from Fig. 7.7, were fairly close to the averaged values of the H_{pl} from the WPD (http://www.brk.adm.yar.ru/palmag/index.html).

This is also true for data obtained from the samples S18-48/33 (VADM = 7.9×10^{22} A m^2, age = 0.33 Ma) and S18-21/29 (VADM = 5.2×10^{22} A m^2, age = 1.7 Ma).

From the results presented in Fig. 7.6, we can see that VADM determined by the samples № S18-37/9 (age 0.3 Ma) and № S18-51/26 (0.25 Ma) were approximately half the size of averaged VADM from the WPD and from the results obtained by us on other samples of approximately the same age. This is probably due to the fact that the NRM of these samples contains CRM, which we did not recognize in Thellier-Coe procedures. This is evidenced by the presence in the sample № S18-37/9 of two phases with Curie points T_{c1} = 220–240 °C and T_{c2} = 330–460 °C (Fig. 7.3; Table 7.2). In this case, as was shown in (Maksimochkin and Tselebrovskiy 2015; Gribov et al. 2017), the Thellier-Coe procedure on ocean basalts gives an underestimation of H_{pl}. The results of the approximation $V_r = f(H_{pl})$ in Fig. 7.6, also confirm this assumption: the point on the figure is deflected towards smaller values of the magnetic field in comparison with the approximation curve (Pearson's correlation coefficient is high—0, 82).

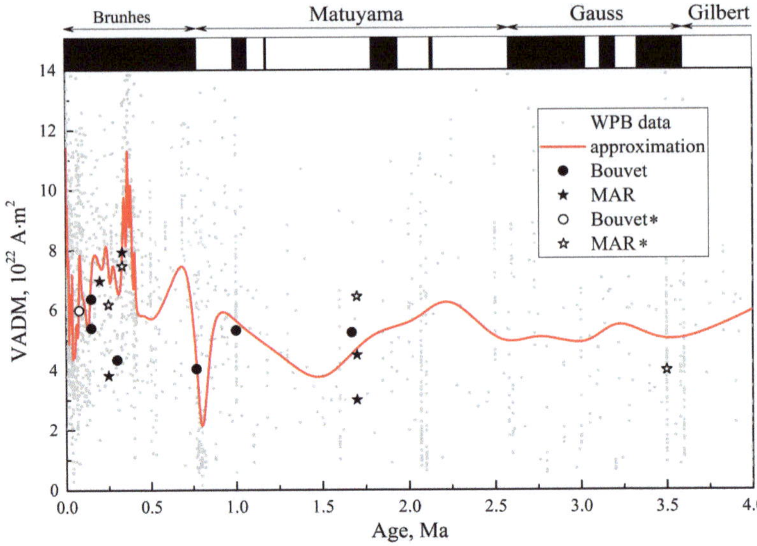

Fig. 7.7 Changes in the VADM of the Earth with time: gray points (http://www.brk.adm.yar.ru/palmag/index.html), solid curve—average data, circles and stars—the results of the VADM determination on the MAR and BTJ basalts, *Maksimochkin et al. 2010

The satisfactory correspond between VADM determined by us from the samples of the south of the MAR and the BTJ, and the WPD, indicates that for most of the samples correct results of H_{pl} determinations by the Thellier-Coe procedure and age of basalts were obtained on the basis of the analysis of linear magnetic anomalies and spreading theory. Figure 7.7 also shows that both the averaged WPD and our data show that 0.35 million years ago the Earth had geomagnetic field of sufficiently high intensity, and during the Brunes-Matuyama boundary its intensity was significantly decreasing.

From the analysis of the results of V_r and H_{pl} determinations obtained from the samples of the MAR and the BTJ, it can be concluded that the V_r parameter, which does not depend on the variation of the ferrimagnetic concentration, reflects the H_{pl} during the basalts formation and the presence of oxidized TM.

Therefore, the decrease in the V_r with the samples age approaches to the boundary of polarity reversal can be explained not only by the basalts oxidation increase with increasing their age, but also by the decrease in the H_{pl}.

Conclusions

1. The results of paleomagnetic studies of the south of the MAR and the BTJ showed that if titanomagnetite of ocean basalts has a low degree of single-phase

oxidation, than the parameter V_r and the geomagnetic field, paleointensity determined by this NRM has a positive correlation.

2. The low NRM and V_r values of basalts from the south of the MAR and the BTJ may be due to the decrease in geomagnetic field with their age approaches to the Brunhes-Matuyama boundary.
3. The definition of geomagnetic field paleointensity can be used to interpret geomagnetic anomalies, to clarify the age of basalts, and geodynamic features of the spreading process in the mid-oceanic ridges.

Acknowledgements This work was supported by the Russian Foundation for Basic Research (the grant № 16-05-00144).

References

Seafloor chronology in the region of the Bouvet lithospheric plate triple junction. **Bulychev A. A., Gilod D. A., Kulikov E. Yu., Shreider Al. A., Shreider A. A.** 2, s.l.: Pleiades Publishing, 2000, Vol. 40.

Gradstein F. M., Ogg J. G., Schmitz M. D., Ogg, G. M. *The Geologic Time Scale.* 2012.

Borok Geophysical Observatory IPE RAS. [Online] http://www.brk.adm.yar.ru/palmag/index.html.

Geomagnetism of the Bouvet triple junction. **Trukhin V. I., Zhilyaeva V. A., Shreider A. A.** 8, s. l.: Izvestiya. Physics of the Solid Earth, 2002, Vol. 38, pp. 628–648.

Stacking paleointensity and oxygen isotope data for the last 1.5 Myr (PISO-1500). **Channell J., Xuan C., Hodell. D. A.** 283, s.l.: Earth and Planetary Science Letters, 2009, pp. 14–23.

Comparison of the Brunhes epoch geomagnetic secular variation recorded in the volcanic and sedimentary rocks. **Shcherbakov V. P., Sycheva N. K., Khokhlov A. V.** 2, s.l.: Pleiades Publishing, 2014, Vol. 50, pp. 222–228.

Relative paleointensity and environmental magnetism since 1.2 Ma at IODP site U1305 (Eirik Drift, NW Atlantic). **Mazaud A., Channell J.E.T., Stoner J. S.** s.l.: Earth and Planetary Science Letters, 2012, Vols. 357–358, pp. 137–144.

Evolution of the geomagnetic field of the Brunhes-Matuyama epochs over the basalts of the Mid-Atlantic ridge (russian). **Maksimochkin V. I., Tselebrovsky A. N., Shreider A. A.** 3, s. l.: Scientific notes of the Physics Department of Moscow State University, 2016, p. 163910.

The Influence of the Chemical Magnetization of Oceanic Basalts on Determining the Geomagnetic Field Paleointensity by the Thellier Method. **Maksimochkin V. I., Tselebrovskiy A. N.** 6, s.l.: Moscow University Physics Bulletin, 2015, Vol. 70, pp. 566–576.

Spontaneous Magnetization Curves and Curie Points of Cation Deficient Titanomagnetites. **Stephenson A.** s.l.: Geophys. J. R. astr. Soc., 1972, Vol. 29, pp. 91–107.

Zet-Lab. [Online] https://www.zetlab.com.

Geomagnetic Paleointensities From Radiocarbon-Dated Lava Flows on Hawaii and the Question of the Pacific Nondipole Low. **Coe R. S.** s.l.: J. Geophys. Res., 1978, Vol. 83, pp. 1740–1756.

Hysteresis properties of titanomagnetites: Grain-size and compositional dependence. **Day R., Fuller M., Schmidt V. A.** s.l.: Physics of the Earth and Planetary Interiors, 1977, Vol. 13, pp. 260–266.

Secular variation of the geomagnetic total force during the last 5000 years. **Nagata T., Arai Y., Momose K.** 18, s.l.: J. Geophys. Res., 1963, Vol. 68, pp. 5277–5281.

A.C. demagnetization of rocks: analysis of results. In: D.W. Colllnson, K.M. Creer and S.K. Runcorn (Editors), Methods in Paleomagnetism. **Zijderveld J. D. A.** 5803—5806, Amsterdam: Geophys. Res., 1967, Vol. 82, pp. 254–286.

Experimental Modeling of the Chemical Remanent Magnetization and Thellier Procedure on Titanomagnetite-Bearing Basalts. **Gribov S. K., Dolotov A. V., Shcherbakov V. P.** 2, s.l.: Pleiades Publishing, 2017, Vol. 53, pp. 274–292.

Paleointensity of the Geomagnetic Field in the Last Half-Million Years in Regions of the Red Sea and South of the Mid-Atlantic Ridge. **Maksimochkin V. I., Mbele J. R., Trukhin V. I., Shreider A. A.** 6, 2010, MOSCOW UNIVERSITY PHYSICS BULLETIN, Vol. 65, pp. 531–538.

Chapter 8
Archaeomagnetic Studies of the Material of the Archaeological Monument Dmitrievskaya Sloboda II of the Second Millennium B.C

O. V. Pilipenko, I. E. Nachasova, S. K. Gribov and O. V. Zelentsova

Abstract An archaeomagnetic study of ceramic material from the archaeological site Dmitrievskaya Sloboda II (Murom district, Vladimir region) was conducted. The site is archaeologically dated to the middle of II millennium B.C. Nine geomagnetic field intensity determinations were obtained. The geomagnetic field intensity varies between 40 and 75 μT with an average value of about 55 μT. Comparing the field intensity data sets obtained in this study and in that on the material from the Sakhtysh-I monument (IV-III millennium B.C., Nachasova et al. 2018) shows an increase in the geomagnetic field intensity variation and the mean value of the geomagnetic field intensity in the II millennium B.C. against the level of the field in the previous two millennia. This might be a manifestation of the 8-thousand-year cycle in the field intensity variation.

Keywords Archaeomagnetic intensity · Geomagnetic field variations Thellier method · Arai-Nagata diagrams

Introduction

Acquisition of new data about the Earth's magnetic field makes it possible to advance the studies of geomagnetic variations and clarify their regularities, and thereby to determine the key elements in the process of generation of the geo-

O. V. Pilipenko (✉) · I. E. Nachasova
Schmidt Institute of Physics of the Earth RAS, Moscow, Russia
e-mail: pilipenko@ifz.ru

S. K. Gribov
Geophysical Observatory "Borok", Yaroslavl Region, Russia
e-mail: gribov@borok.yar.ru

O. V. Zelentsova
Institute of Archeology of RAS, Moscow, Russia
e-mail: olgazelentsova2010@yandex.ru

D. Nurgaliev et al. (eds.), *Recent Advances in Rock Magnetism, Environmental Magnetism and Paleomagnetism*, Springer Geophysics,
https://doi.org/10.1007/978-3-319-90437-5_8

magnetic field. The study of the fine structure of the geomagnetic field, of its individual elements and their interrelations forms a natural part of comprehensive research of the processes taking place in the Earth's core and shells.

Archaeomagnetic research makes it possible to obtain data about the intensity of the geomagnetic field in recent millennia by studying thermoremanent magnetisation of burnt material from archaeological monuments. Most important, data on absolute intensity of the ancient geomagnetic field are recovered, which cannot be obtained by investigating other types of remanence.

Examination of geomagnetic field intensity variations carried out on the materials from archaeological monuments of Eurasia (in the south-east of Europe (Kovacheva 1980), in the Caucasus, in Central Asia and Siberia (Burlatskaya 1965; Nachasova 1998), in Japan (Sakai and Hirooka 1986) showed that a gradual change in the field intensity occurs with a characteristic time of about 8000 years.

An analysis of the world data of geomagnetic field intensity for the last eight millennia (Nachasova 1998) led to the conclusion that the change in the field intensity can be largely represented as a superposition of a number of oscillations with periods from ranging 300 to 8000 years. The characteristic feature of these oscillations is their drift; moreover, oscillations with different characteristic times show a drift in both western and eastern directions. At the same time, these studies led to a significant increase of the information about the geomagnetic field intensity for Eurasia.

The longest time series of archaeomagnetic intensity determinations for the last 13 ky (Nachasova et al. 2015), constructed for Siberia, reveals new peculiarities of the "basic" oscillation of the geomagnetic field intensity, suggesting the existence of its eastern drift.

The greatest amount of data on the geomagnetic field intensity was obtained for the latitudinal belt 40°–45°N. The distribution on the time scale of the data is very irregular, most of the determinations belonging to the last two millennia. Therefore the task of extending the archaeomagnetic database for the time B.C. is very relevant. This work is a part of the research aiming to obtain information about the geomagnetic field intensity in the European part of Russia, located to the north of the above mentioned latitudinal belt, in the time interval from the Neolithic to the boundary of eras. It is especially interesting to obtain data for the time interval between 4000 and 2000 years B.C., for which there are much less data for Eastern Europe (Tema and Kondopoulou 2011) compared to other time intervals. This work studies the material from the Dmitrievskaya Sloboda II archaeological monument dated at the second millennium B.C.

The Object of Study and Age Estimations

The Dmitrievskaya Sloboda II settlement was discovered in the Murom district of the Vladimir region on the northern periphery of the Murom town, Russia, (φ = 55° 34′, λ = 42° 03′), Fig. 8.1. It occupies a low-pitched side of the first terrace of the left bank of the Oka river, and it is single-layered. The ceramic fragments are well

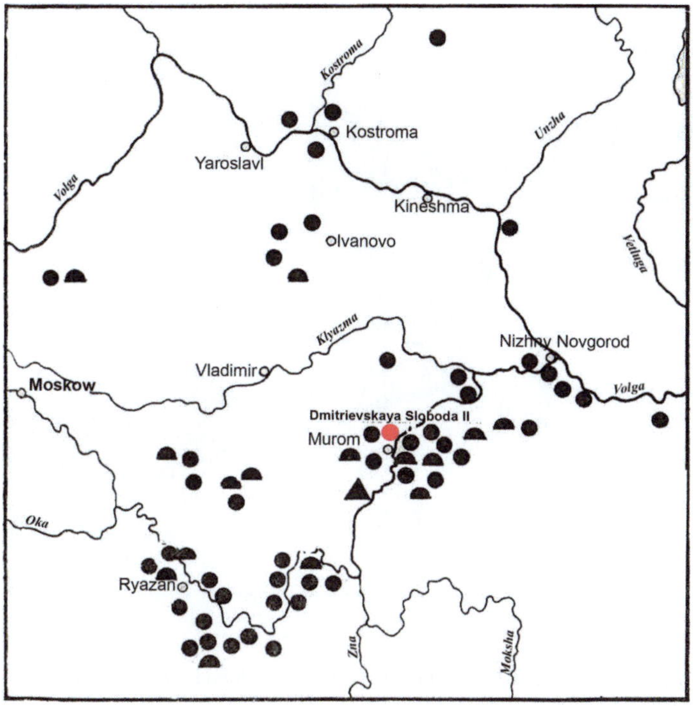

Fig. 8.1 Archaeological sites of Pozdnyakov culture (after Bader 1987). Bold circles are settlements, semicircles are burial mounds, triangles are archaeological sites, red circle is the Dmitrievskaya Sloboda II monument

burned pottery scraps, which are of brick or yellowish color outside, and dark brown or dark gray inside (Fig. 8.2). In general, the ceramic complex and implements look quite uniform in chronological and cultural terms. The monument belongs to the Pozdnyakov culture which dates to middle of II—beginning of I millennium B.C. (Sulerzhitsky 1993). For the dating of the monument, a series of radiocarbon dates were obtained for soil, coal and ceramics (Voronin 2013). The determinations made for coal have a calibrated age of ~1750–1500 years B.C. The dates obtained for ceramics are more variable (Saprykina et al. 2010). The most probable interval of the calibrated age of dates for ceramics is ~2050–1750 B.C. These dates correspond to the time of the transition from the Middle Bronze to the Late Bronze period and are related to the early stage of the formation of the Pozdnyakov culture.

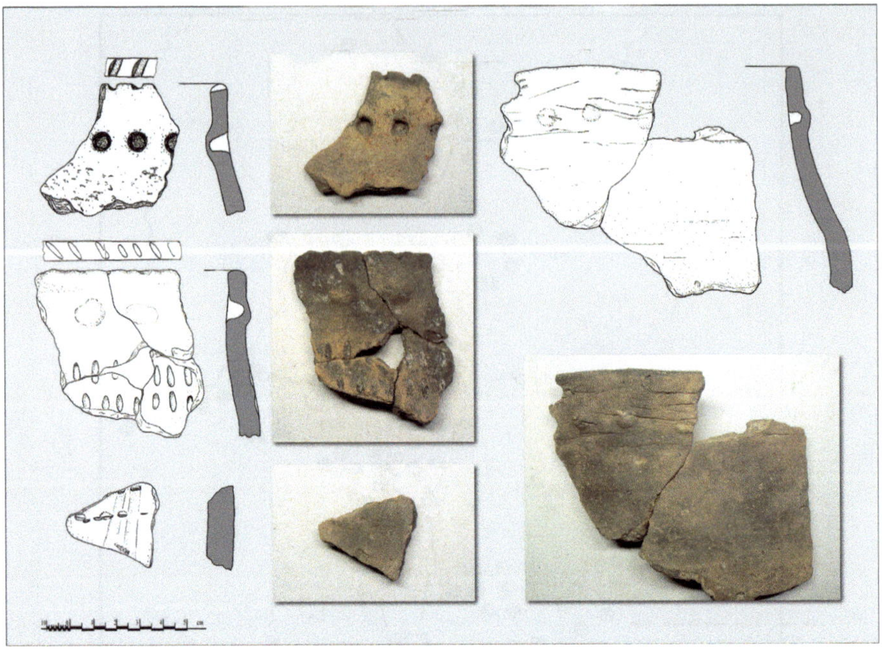

Fig. 8.2 Fragments of ceramics from the Dmitrievskaya Sloboda II settlement

Methods of Investigation and Equipment

Fourteen ceramic fragments were subjected to an archaeomagnetic study. The composition of the ferromagnetic fraction was studied on a pilot collection, consisting of seven samples (G-7/2-1694, B-6/3-2419, A-6/3-2639, A-5/3-2646, B-6-2667, $G_2$8-3922, VV10/1-4259) by the method of powder X-ray diffraction. Before X-ray diffractometry, we dispersed each sample in water by ultrasound and extracted a ferromagnetic fraction along the test tube wall with a neodymium magnet. We performed this procedure several times in order to improve separation. Since the initial data were inconclusive, we separated a sufficient amount of the magnetic fraction to reliably identify magnetic minerals by X-ray diffraction. The latter procedure was performed at room temperature with a STADI-MP multifunctional diffractometer (STOE, Germany) with an arched germanium monochromator crystal (reflection 111) which maintains a strictly monochromated CoK_α emission.

Thermomagnetic analysis (TMA) was carried out measuring the dependence of the saturation magnetization on temperature $Js(T)$ in the field of $\sim$0.4 T with an analyzer of ferromagnetic fraction (ORION, Russia).

The natural remanent magnetization (*NRM*) and thermoremanent magnetization (*TRM*) were measured on a spinner magnetometer JR-6 (AGICO, Czech Republic) in three rotation positions.

The intensity of the ancient geomagnetic field has been determined using the modified double heating Thellier method (Thellier and Thellier 1959; Coe 1978). For this purpose we used a non-magnetic oven MMTD80 (Magnetic Measurements, UK) and a DC magnetic field source. To exclude the effect of magnetic anisotropy, the samples in the furnace were placed with the maximum axes parallel to the direction of the magnetic field in the furnace. To eliminate the cooling rate effect, the samples in the furnace cooled at a natural rate, without turning on a cooling fan. For all samples, *pTRM* check-points were measured at temperatures of 300, 400 and 500 °C (Coe 1967; Paterson et al. 2014). Repeated field-free, *pTRM* tail-checks were also measured at temperatures of 150, 250, 350, 450 and 550 °C to test composition changes of magnetization carriers at blocking temperatures above the *TRM* creation temperature (Bolshakov and Shcherbakova 1979; Riisager and Riisager 2001; Shcherbakov and Shcherbakova 2002). For each sample, the Arai-Nagata diagram was used to determine the angular coefficient *K* of the slope of the straight line and to calculate the ancient geomagnetic field intensity according to the formula $B_{an} = K B_{lab}$, where $B_{lab} = 50$ μT (Fig. 8.3). We used for paleointensity estimations only that points, when difference between tail-check and *NRM* were less than or equal 10% and when difference between check-point and *pTRM* were less than or equal 10%.

The whole experiment was carried out at the Schmidt Institute of Physics of the Earth RAS and Geophysical observatory "Borok".

Rock- and Archaeomagnetic Results

On the base of the powder X-ray diffraction method, the component analysis of spinel and hexagonal reflexes indicates that two phases are present in all cases. The elementary cell parameters calculated for both phases are given in Table 8.1. The period *a* of a spinel phase crystal lattice is close to those of magnetite ($a = 8.396$ Å) or maghemite ($a = 8.33$ Å). The parameters of a hexagonal phase correspond well to periods *a* and *c* of the hematite crystal lattice ($a = 5.034$ Å, $c = 13.75$ Å). We conclude therefore that in the investigated magnetic fraction the X-ray phase analysis shows the presence of magnetite, in most cases single-phase oxidized to maghemite, as well as of hematite.

Thermomagnetic analysis shows that the *Js(T)* curves of the first and second heating have a bend in the temperature interval of ~570–600 °C (Fig. 8.4) suggesting that the main carriers of magnetization in the studied samples are magnetite and/or maghemite. The mineralogical composition does not change much after the heating.

Attempt to determine the geomagnetic field intensity has been carried out for fourteen ceramic fragments. After applying the "*pTRM* check-point" and "*pTRM* tail-check" criteria five ceramic fragments, whose experiments results do not satisfy these criteria, were excluded. Thus, the archaeointensity data obtained from the rest nine ceramic fragments were taken into consideration (Table 8.2).

The obtained geomagnetic field intensity data vary from 39 to 77 μT, with the average value of 54.6 ± 5.1 μT.

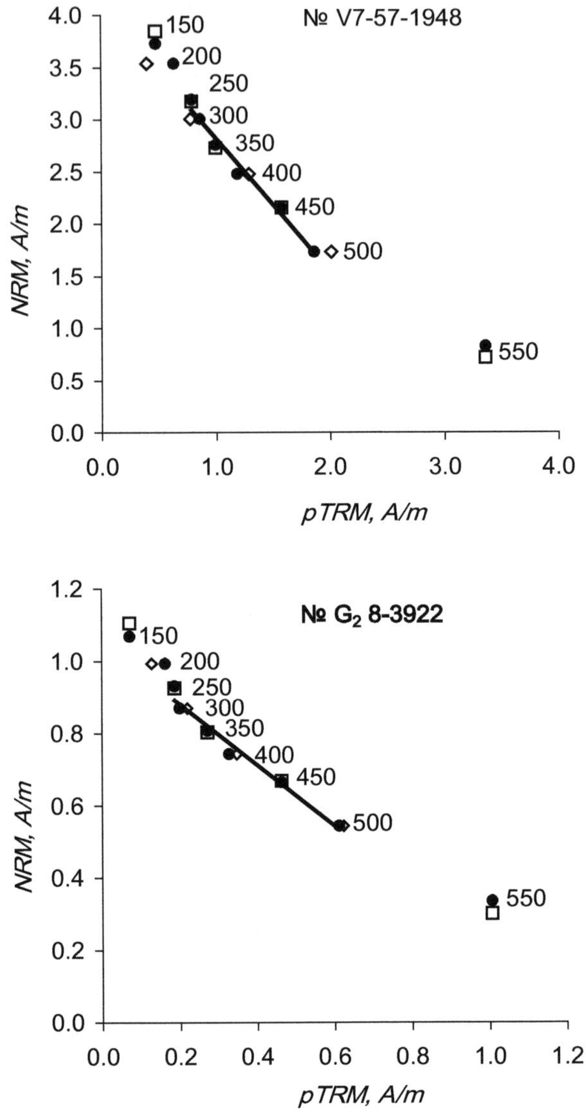

Fig. 8.3 The Arai-Nagata diagrams. Circles denote the results of studies using the modified Thellier method, hollow diamonds—*pTRM* check-points, hollow squares—*pTRM* tail-checks. The numbers near the symbols indicate the heating temperatures

Discussion and Conclusions

Data of the geomagnetic field intensity in the past obtained for different regions of Eurasia (Spain, the Caucasus, Central Asia and Siberia) for the last 7000–8000 years (Nachasova 1998; Nachasova and Akimova 2015; Nachasova et al. 2015) allowed us to obtain a picture of the magnetic field intensity variations in time and in space. The minimum of the “basic” variation (8 thousand years) falls at

Table 8.1 Crystallographic parameters of the magnetic fraction separated from the samples of the pilot collection

No. of the sample	Spinel phase	Hexagonal phase
G-7/2-1694	a = 8,41(3) Å V = 593.8(37) Å^3	a = 5026(7) Å c = 13.73(3) Å V = 300.5(10) Å^3
B-6/3-2419	a = 8.357(5) Å V = 583.7(6) Å^3	a = 5.029(5) Å c = 13.71(3) Å V = 300.3(9) Å^3
A-6/3-2639	a = 8.33(3) Å V = 578.1(34) Å^3	a = 5.03(3) Å c = 13.70(10) Å V = 300.2(39) Å^3
A-5/3-2646	a = 8.349(8) Å V = 581.9(10) Å^3	a = 5.031(4) Å c = 13.728(15) Å V = 300.9(5) Å^3
B-6-2667	a = 8.340(8) Å V = 580.0(9) Å^3	a = 5.0344(20) Å c = 13.747(7) Å V = 301.7(3) Å^3
$G_2$8-3922	a = 8.341(4) Å V = 580.2(5) Å^3	a = 5.033(4) Å c = 13.752(8) Å V = 301.7(4) Å^3
VV10/1-4259	a = 8.323(11) Å V = 576.5(14) Å^3	a = 5.0360(13) Å c = 13.7550(20) Å V = 302.11(13) Å^3

Note a, c are the periods of the crystal lattice; V—volume of the crystal lattice, numbers in the brackets show the experimental error

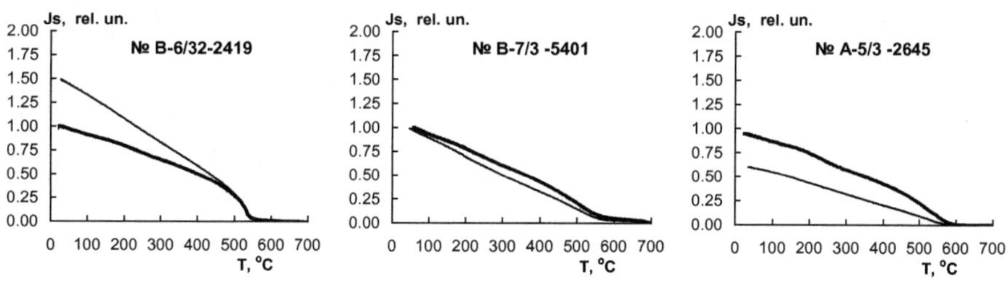

Fig. 8.4 Thermomagnetic analysis results. The bold line corresponds to the first heating, the thin line to the second heating

Table 8.2 Paleointensity of the geomagnetic field the ceramics from the Dmitrievskaya Sloboda II settlement

No. of the sample	$B_{an} \pm \sigma$ (μT)	Calculation temperature range (°C)
V7-57-1948	76.7 ± 0.9	250–450
B-6/2-2641	58.2 ± 0.8	200–450
A-5/3-2646	48.7 ± 3.6	250–450
B-6-2667	39.4 ± 1.3	250–450
B-6/6-2743	47.2 ± 0.7	400–550
G_2 8-3922	41.9 ± 1.3	250–500
G_2 8-3923	51.8 ± 5.1	250–450
VV 10/1-4259	54.8 ± 0.3	250–450
B-7/3-5401	67.9 ± 1.8	250–450

the time interval 5500–3500 B.C. The average level of the field intensity in this time interval is approximately two times less than the average level of the 8-thousand-year variation maximum. The variations of smaller periods are superimposed on the main variation. Figure 8.5 shows as an example a pattern of the change of the geomagnetic field intensity obtained from Iberian Peninsula archaeological monuments.

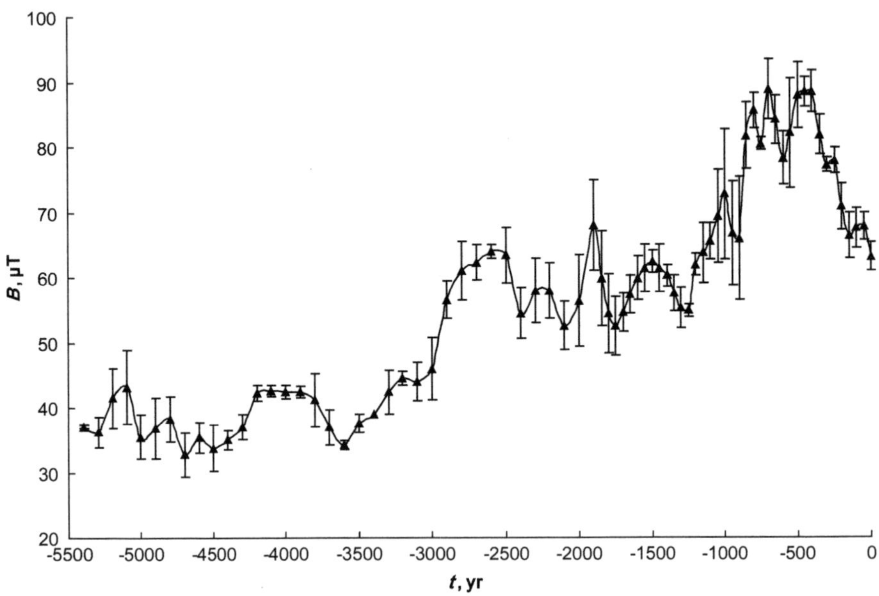

Fig. 8.5 The values of the intensity of the geomagnetic field B (μT), average for the time periods. The averaging interval is 200 year for the period from the sixth to third millennia B.C., 100 year for the period from the second millennium B.C. to the boundary between the eras

Variations of different periods show different drift direction, resulting in discrepancies in the detailed patterns of geomagnetic field intensity variations in regions from different longitude sectors. In this regard, a direct comparison of the obtained data on the geomagnetic field intensity makes sense if the territories for which the data are obtained are regions whose difference in longitude is <30°. The data on the geomagnetic field intensity obtained in the present study can thus be compared with the data obtained for the southeastern Europe.

Previously obtained and new data of the geomagnetic field intensity in south-east Europe (in the longitude sector 26 ± 3°E) are compiled (Tema and Kondopoulou 2011). Despite the relatively small amount of data for the II millennium B.C., one can conclude that the average level of field intensity in the II millennium B.C. is noticeably higher than the level of field intensity in the III millennium B.C. A rapid change in the field intensity occurs in the middle of the II millennium B.C.

Data of the geomagnetic field intensity in the II millennium B.C. were also obtained for Egypt and western Asia (Aitken et al. 1984). One can note the rapid growth of the field intensity 1750–1500 year. B.C. and the change in the field intensity during the millennium by a factor of 2, although there are time intervals for which there are almost no data (XVII and XV centuries).

The picture of geomagnetic field intensity variations in the II millennium B.C. was also obtained on the material from Georgia archaeological monuments, having the longitude 44 ± 3°E (Nachasova and Burakov 1987). The main trend of geomagnetic field intensity variations in the II millennium B.C. is an increase in the field intensity by a factor of about 2. Against the background of this increase, variations of several centuries duration occur. In general, the patterns of the geomagnetic field intensity variations in the II millennium B.C. according to the data obtained in the cited works is very similar to that observed in Dmitrievskaya Sloboda II. In the time interval at which the Dmitrievskaya Sloboda II monument is dated, values of the field intensity from 45 to 96 μT were obtained on the material from Georgia (Nachasova and Burakov 1987). Average value (56.7 μT ± 8.3 μT) obtained on the material from Georgia, dated at XVII–XV centuries B.C., agrees well from the average value obtained at Dmitrievskaya Sloboda II (54.6 ± 5.1 μT). According to the data obtained from the material of Georgia, in the XX–XIX centuries B.C. the geomagnetic field intensity was relatively low (about 25 μT). Later in the XVIII–XVII centuries B.C. there is an increase in the geomagnetic field intensity. Such a change in the field intensity makes it possible to use the data obtained in this study to assign the time interval of the manufacture of the Dmitrievskaya Sloboda II ceramics as the XVIII–XV centuries B.C.

In a previous work, ceramic material of the Neolithic monument of Sakhtysh-I was studied. It is located in the same region as the Dmitrievskaya Sloboda II monument. Archaeointensity data were obtained for the time interval between V—III thousand years B.C. (Pilipenko et al. 2016; Nachasova et al. 2018). The geomagnetic field intensity varies mostly within a range of 30–60 μT. The average level of field intensity is about 50 μT in the second half of the V–beginning IV millennium B.C decreasing to about 35 μT in the IV—first half of the III millennium B.C. The limits of the change in the geomagnetic field intensity and the average value of the geomagnetic field intensity obtained for Dmitrievskaya Sloboda II are noticeably higher than for the Sakhtysh-I monument. This may indicate a tendency that the geomagnetic field intensity in the II millennium B.C. increased in comparison to the field level in the previous three millennia, being a manifestation of the 8-thousand-year variation of the field intensity.

In all, according to the data obtained for the different regions of Eurasia, in the II millennium B.C. the average level of field intensity appears much higher than the field level in the previous three millennia. The average value of the geomagnetic field intensity, obtained for Dmitrievskaya Sloboda II material in II millennium B. C., is also noticeably above the average level of geomagnetic field intensity in the IV–III millennium B.C. New data about the limits of variation and the average level of geomagnetic field intensity in Eastern Europe in the II millennium BC allow to improve the models of the geomagnetic field intensity variations in this time interval.

As a result of the performed research new data for the geomagnetic field intensity in the middle of the II millennium B.C are obtained. The geomagnetic field intensity varies within the limits of $\sim$40–75 μT with an average value of about 55 μT.

Acknowledgements This work was supported by the State Task of IPE RAS, the Russian Foundation Research (project no. 16-05-00378) and by the Government of the Russian Federation (project no. 14.Z50.31.0017).

References

Aitken M.J., Allsop A.L., Bussel G.D., Winter M.B. Geomagnetic intensity in Egipt and western Asia during the second millennium BC // Nature. 1984. V.310. Pp. 305–312.

Bader O.N. Pozdnyakovskaya culture // Archeology of the USSR. The bronze epoch of the forest belt of the USSR. M. 1987. Pp. 131–135.

Bolshakov A.S., Shcherbakova V.V. Thermomagnetic criterion for determining the domain structure of ferromagnetics // Izv. Academy of Sciences of the USSR, ser. Physics of the Earth, 1979. № 2. Pp. 38 – 47.

Burlatskaya S.P. Archeomagnetism. Investigation of the Earth's magnetic field in the past eras. 1965. Moscow. Nauka. 127 pp.

Coe R.S. Paleointensities of the Earth's magnetic field determined from tertiary and quaternary rocks // J. Geophys. Res. 1967. Vol. 72, Pp. 3247–3262.

Coe R.S. Geomagnetic paleointensity from radiocarbon-dated flows on Hawaii and the question of the Pacific nondipole low // J. Geophys. Res. 1978. Vol. 83. Pp. 1740–1756.

Kovacheva M. Summarized results of the archaeomagnetic investigations of the geomagnetic field variation for the last 8000 yr in south-eastern Europe // Geophys.J.R.Astr.Soc. 1980. V.61. Pp. 57–64.

Nachasova I.E. Characteristics of the geomagnetic intensity variations from archaeomagnetic data, Extended Abstract of Doctoral (Phys.-Math.) Dissertation, Schmidt Institute of Physics of the Earth of the Russian Academy of Sciences. 1998. 65 pp.

Nachasova I.E., Akimova S.V. The geomagnetic field intensity variations in the Iberian peninsula during the last millennium // Izv., Phys. Solid Earth. 2015. Vol.51. №. 5. Pp. 709–715.

Nachasova I.E., Burakov K.S. Variation of the geomagnetic field intensity in the second millennium BC in Georgia // Geomagn. and aeronom. 1987. № 5. Pp. 869–872.

Nachasova I.E.,Burakov K.S., Pilipenko O.V. Variations in the intensity of the geomagnetic field in Siberia during the last 13000 years // Izv., Phys. Solid Earth. 2015. Vol.51. № 1. Pp. 44–50.

Nachasova I.E., Pilipenko O.V., Markov G.P., Gribov S.K., Tsetlin Yu.B. (2018) Geomagnetic Field Intensity during the Neolith in the Central East European Plain. Geomagnetism and Aeronomy 58 (3):438–447

Paterson G.A., Tauxe L., Biggin A.J., Shaar R., Jonestrask L.C. Standard Paleointensity Definitions. 2014. Vol. 1.0. 43 p. https://earthref.org/PmagPy/SPD/DL/SPD_v1.0.pdf.

Pilipenko O.V., Nachasova I.E., Markov G.P., Gribov C.K., Tsetlin U.B. New archaeomagnetic research of pottery from Neolithic settlement Sahtysh –I (Ivanovo region, Russia) c. 5000–2000 BC //11th International Conference and School Problems of Geocosmos. October 3-7, 2016. St.-Ptersburg, Russia. Book of Abstracts. P. 164.

Riisager P., Riisager J. Detecting multidomain magnetic grains in Thellier palaeointensity experiments // Physics of the Earth and Planetary Interiors. 2001. Vol. 125. Pp. 111–117.

Saprykina I., Zelentsova O., Voronin K. What is it that we date when dating pottery? View of archaeologist // 38th International Symposium on Archaemetry. Tampa, Florida, USA, 2010. Program and Abstracts.

Sakai H., Hirooka K. Archaeointensity determinations from Western Japen //J.Geomagn. and Geoelectr. 1986. V.38. №12. Pp. 1323–1331.

Shcherbakov V.P., Shcherbakova V.V. Dependence of the errors in the determination of paleo direction from the domain structure of ferrimagnetic grains of rocks // Izv., Phys. Solid Earth. 2002. № 5. Pp. 404–411.

Sulerzhitsky L.D. Radiocarbon dates of archaeological monuments of the Middle Oka basin. Ancient monuments of the Oka basin. Ryazan. 1993. Pp. 42–55.

Tema E., Kondopoulou D. Secular variation of the Earths magnetic field in the Balkan region during the last 8 millennia based on archaeomagnetic data // Geophys. J. Int. 2011. Vol. 186. Pp. 603–614.

Thellier E., Thellier O. Sur l'intensité du champ magnéttique terrestre dans le passé historique et géologique // Ann. Geophys. 1959. Vol. 15. Pp. 285–378.

Voronin K.V. Bronze Age Complex Settlements- Pesochnoe -1 and Dmitrievskaya Sloboda II // Tverskoy archaeological collection. Issue 9. Tver: The Triad. 2013. Pp. 329–344.

Part II
Rock and Environmental Magnetism

Chapter 9
Influence of Magnetostatic Interaction on Magnetic Characteristics of Decay Products of Nanodisperse Titanomagnetites

S. V. Anisimov, L. L. Afremov and I. G. Iliushin

Abstract A theoretical analysis of the effect of the magnetostatic interaction on the temperature dependence of the hysteresis characteristics and various types of remanent magnetization of the decay products of the titanomagnetite nanoparticle system is carried out. It is established that the magnetic interactions between nanoparticles with an effective field of less than the coercive field leads to an insignificant decrease in the temperature dependence of the hysteresis characteristics. An increase in the degree of decay of titanomagnetite (the size of the magnetite core of the nanoparticle) lowers the dependence of the hysteresis characteristics on temperature. In addition, it is shown that an increase in the intensity of the magnetic interactions leads to a drop in the thermo-remanent and chemical remanent magnetizations, regardless of the degree of decay of titanomagnetite. The thermal and chemical remanent magnetizations to the ideal ratios (R_t and R_c) increases with increasing magnetostatic interaction, while the increase in the degree of decay leads to a decrease in the values of R_t lying in the range $0.8 \leq R_t \leq 1.4$ and an increase in R_c, which is limited by the relation $0 < R_c \leq 0.9$.

Keywords Titanomagnetite · Core-shell nanoparticle · Model
Magnetic properties · Decay · Hysteresis characteristics

S. V. Anisimov (✉) · L. L. Afremov · I. G. Iliushin
Far Eastern Federal University, Vladivostok, Russian Federation
e-mail: ahriman25@gmail.com

L. L. Afremov
e-mail: afremov.ll@dvfu.ru

I. G. Iliushin
e-mail: futted@gmail.com

D. Nurgaliev et al. (eds.), *Recent Advances in Rock Magnetism, Environmental Magnetism and Paleomagnetism*, Springer Geophysics,
https://doi.org/10.1007/978-3-319-90437-5_9

Introduction

Success of solving multiple paleomagnetic tasks is dependent on the precision of the assessment of the primary component of natural remanent magnetization I_n. The basis for a method of defining of primary component of I_n is the difference of values and stabilities of various types of magnetization, coercive spectra, hysteresis characteristics (Sholpo 1977; Belokon et al. 1973; Hramov 1982; Petrova 1977; Özdemir and Dunlop 2014) and the relationships of different types of remanent magnetization to an ideal [see (Ivanov et al. 1981; Ivanov and Sholpo 1982; Nguen 1985a; Nguen 1985b; Gribov et al. 2017)]. All of the abovementioned characteristics are greatly dependent on a chemical composition and structure of a material [see (Kudryavceva 1988)], magnetic properties and size of the magnetic grains and density of distribution in a non-magnetic rock matrix. Note, that process of magnetization of rocks where magnetic properties are defined by the homogeneous nanoparticles, is well described with the model of interacting single-domain nanoparticles [please, see (Belokon 1985; Scherbakov and Scherbakova 1977; Afremov and Panov 2004; Goroshko et al. 2010)]. In these references studies of an effect of magnetostatic interaction on the properties of various types of remanent magnetization and ratios between them can be found, as well as studies of its effect on the coercive spectra and hysteresis characteristics. It is showed that increase in concentration of magnetic nanoparticles (magnetostatic interaction enhancement) results in a fall of hysteresis characteristics and increase of the variance of a distribution of coercive spectra. Thermo-remanent magnetization to ideal magnetization ratio $R_t = I_{rt}/I_{ri}$ depends on the intensity of magnetostatic interaction. In a case of weak interaction ratio it is $R_t < 1$. And in a case of strong interaction it is $1 < R_t \leq 2.7$ for all materials, which agrees with an experimental results (Borisova and Sholpo 1985). In the references (Nguen 1985a; Haig 1962; Avilova et al. 1985) results of a study of an effect of chemical reaction on the properties of the various types of magnetization are demonstrated. Thus, as a result of the formation of magnetite from pyrite, a chemical magnetization arises, the ratio of which to the ideal $1 \leq R_c \leq 1.4$ (Nguen and Pechersky 1984, 1985). Measurement of R_c for hematite formed as a result of the hydrogoetite transformation (Avilova et al. 1985), showed that $R_c \ll 1$, while for the chemical magnetization that arose on the decay products of titanomagnetite $0.6 \leq R_c \leq 0.9$ (Nguen and Pechersky 1987).

Calculations of the chemical magnetization lead to results that are in agreement with the experimental data noted above. In addition, the simulation of chemical magnetization conducted within the model of single-domain particles showed that an increase in the intensity of the magnetostatic interaction leads to an increase in the ratio R_c (Afremov et al. 2010). The main drawback of theoretical constructions is the assumption of the homogeneity of magnetic nanoparticles, it is not observed in the presence of chemical transformation during the "life" of the rock and also by temperature heating.

In this paper, an attempt was made, within the framework of the theory of two-phase interacting particles developed by us (Afremov et al. 2018), to model the effect of the magnetostatic interaction on various types of remanent magnetization and the coercive field of decay products of nanodispersed titanomagnetite (TM).

Magnetization of an Interacting Core-Shell Nanoparticles

To model the effect of magnetic interaction on the residual magnetization of nanodispersed TM, we shall use the theory of two-phase interacting particles described in detail in the references (Afremov et al. 2015, 2018). According to this theory, the magnetization of the system N_0 of two-phase magnetic nanoparticles uniformly distributed in a non-magnetic matrix of volume V_0 is determined by the following expression:

$$I(t) = \int \Phi(h,I,B)dh \int \mathbf{c}(a)\Big((1-\varepsilon)I_s^{(1)} + \varepsilon I_s^{(2)}\Big)[(n_1(t,a,H+h) - n_3(t,a,H+h)) + \Big((1-\varepsilon)I_s^{(1)} - \varepsilon I_s^{(2)}\Big)(n_2(t,a,H+h) - n_4(t,a,H+h))\Big] F(a)da. \tag{9.1}$$

Here $W(h)$ is the density of the distribution function over interaction fields h, the form of which depends on the volume concentration of two-phase nanoparticles c (Scherbakov 1979).

$$\Phi(h,I,B) = \begin{cases} \frac{1}{\pi}\frac{B}{B^2 + \left[h - \widehat{h}(M)\right]^2}, c < 0.1, \\ \frac{1}{\sqrt{\pi B^2}} exp\left(-\frac{\left(h - \widehat{h}(M)\right)^2}{B^2}\right), c > 0.1 \end{cases} \tag{9.2}$$

where, depending on the volume concentration of magnetic nanoparticles, the effective interaction field B is defined as the Cauchy distribution parameter or as the variance of the normal distribution:

$$B = 5\int \mathbf{c}(a) \int \Big[\Big((1-\varepsilon(a))I_s^{(1)} + \varepsilon(a)I_s^{(2)}\Big)(n_1(t,h) + n_3(t,h)) + \left|(1-\varepsilon(a))I_s^{(1)} - \varepsilon(a)I_s^{(2)}\right|(n_2(t,h) + n_4(t,h))\Big]\Phi(h,I,B)dhF(a)da, \mathrm{c} < 0.1,$$

$$B^2 = \int c(a) \int \Big[\Big((1-\varepsilon(a))I_s^{(1)} + \varepsilon(a)I_s^{(2)}\Big)^2 (n_1(t,h) + n_3(t,h)) + \Big[\Big((1-\varepsilon(a))I_s^{(1)} - \varepsilon(a)I_s^{(2)}\Big)^2 (n_2(t,h) + n_4(t,h)) + \Big]\Phi(h,I,B)dhF(a)da, \mathrm{c} < 0.1 \tag{9.3}$$

In the Eqs. 9.1–9.3 $\mathbf{c}(a) = N_0V(a)/V_0$ is the volume concentration of TM nanoparticles of volume $V(a)$, $I_s^{(1,2)}$ are spontaneous magnetizations of the shell (1-st phase) and the core (2-nd phase), $\varepsilon(a)$ is the relative volume of the nanoparticle's core, $\boldsymbol{n}(t) = \{n_1(t), n_2(t), n_3(t), n_4(t)\}$ stands for the normalized population vector of states of two-phase nanoparticles $(n_1(t) + n_2(t) + n_3(t) + n_4(t) = 1)$, three components of which are determined by means of a matrix exponent (Afremov et al. 2010, 2018)

$$\begin{pmatrix} n_1(t) \\ n_2(t) \\ n_3(t) \end{pmatrix} = \exp(\mathcal{W}t) \cdot \begin{pmatrix} n_{01} \\ n_{02} \\ n_{03} \end{pmatrix} + \int_0^t \exp(\mathcal{W}(t-\tau))d\tau \cdot \begin{pmatrix} W_{41} \\ W_{42} \\ W_{43} \end{pmatrix}, \tag{9.4}$$

where the matrix elements of the matrix $\mathcal{W}$ are expressed in terms of the matrix elements of the transition matrix $W_{ik} = f_0\, exp(-E_{ik}/k_B T)$:

$$\mathcal{W}_{ik} = \begin{cases} -\sum\limits_{j \neq i, j=1}^{4} W_{ij} - W_{4i}, & i = k \\ W_{ki} - W_{4i}, & i \neq k \end{cases}, \tag{9.5}$$

E_{ik} is the height of a potential barrier between an i-th and k-th states, $f_0 = 10^{10} s^{-1}$ is a frequency factor, $F(a)da$ is the size distribution function and H is an external magnetic field.

Modeling of Thermal Dependence of Hysteresis Characteristics

We use the model of the decay process, described in detail in (Afremov et al. 2015; Anisimov and Afremov 2018), according to which the decay of the primary homogeneous TM $Fe_{3-x_0}Ti_{x_0}O_4$ occupying the entire volume of nanoparticles produces two-phase nanoparticles in which titanomagnetite cores ($Fe_{3-x_{core}}Ti_{x_{core}}O_4$) depleted of titanium ($x_{core} < x_0$) are covered by the TM ($Fe_{3-x_{shell}}Ti_{x_{shell}}O_4$) with the concentration $x_{shell} > x_0$. The concentration of titanium in the shell is related to the model parameters as follows: $x_{shell} = (x_0 - \varepsilon x_{core})/(1 - \varepsilon)$.

Modeling of various types of remanent magnetization was carried out on a TM with $x_0 = 0.3$. It was believed that as a result of the decay of the TM in an ellipsoidal nanoparticle, whose minor semi-axis size is 50 nm, a magnetite core ($x_{core} = 0$) with a minor semi-axis r, coated with a TM shell $Fe_{3-x_{shell}}Ti_{x_{shell}}O_4$. The decay process was considered as a discrete process of increasing the size of the core of magnetite and a corresponding decrease in the thickness of the shell, the titanium concentration in which increases with increasing relative volume of the core as follows: $x_{shell} = x_0/(1 - \varepsilon)$.

Further we will characterize the degree of decay of TM in a nanoparticle using the size of the minor semi-axis of the core of magnetite r.

For the calculations, we used the results of measurements of the dependence of the spontaneous magnetization $I_s(a, x)$ on the sizes of the nanoparticles a

(Nikiforov et al. 2017) and the titanium concentration x (O'Reilly et al. 1984) which were approximated by the following expressions:

$$I_s(a,x) = 480\left(1-\frac{a_0}{a}\right)(1-1.22x)G, \quad a_0 = 4.07\times 10^{-7}\text{cm}. \tag{9.6}$$

The same way was used to calculate the dependence of the magnetic crystallographic anisotropy constant on the titanium concentration (O'Reilly et al. 1984) and the size of the nanoparticles a (Gubin et al. 2005):

$$K_A(a,x) = \left(K_V + \frac{K_S}{a}\right)\left(1+4.54x-15.70x^2+9.85x^3\right). \tag{9.7}$$

Here $K_V = -1.06\times 10^5$ erg/cm^3 (Krupichka 1976), $K_S = 0.029$ erg/cm^2 (Perez et al. 2008).

The temperature dependence of the spontaneous magnetization $I_s(a,x,T)$ and the crystallographic anisotropy constant $K_n(a,x,T)$ were estimated using the well-known relations

$$\frac{I_s(a,x,T)}{I_s(a,x)} = \left(1-\frac{T}{T_c(a,x)}\right)^{1/2}, \quad \frac{K_n(a,x,T)}{K_n(a,x)} = \left(1-\frac{T}{T_c(a,x)}\right)^{n(2n+1)/2}.$$

The dependence of the Curie temperature of TM on nanoparticle sizes (Wang et al. 2011) and titanium concentration (O'Reilly et al. 1984) was determined on the basis of experimental data:

$$T_c(a,x) = 860\left(1-\frac{a_1}{a}\right)^{1/v}(1-0.87x)K, \quad a_1 = 0.51\times 10^{-7}cm, \; v = 0.82. \tag{9.8}$$

The Modeling Results

Figures 9.1 and 9.2 shows the temperature dependences of the coercive field H_c and the remanent saturation magnetization I_{rs} of interacting and noninteracting systems of magnetic TM nanoparticles, which differ in the degree of decay (sizes of the magnetite core r).

It follows from the figures that at room temperature the values of H_c and I_{rs} of TM nanoparticles of size $r = 20$ nm are higher than for nanoparticles of size $r = 40$ nm. This is due to the dependence of the potential barriers on the size of the magnetite core (see Fig. 9.3). The increase in the size of the core leads to an increase in the heights of the potential barriers E_{ik} separating the first (fourth) state of the two-phase nanoparticles from the second (third) E_{12} (E_{34}), and also to the reduction of the barriers between the first (second) state and the fourth (third)

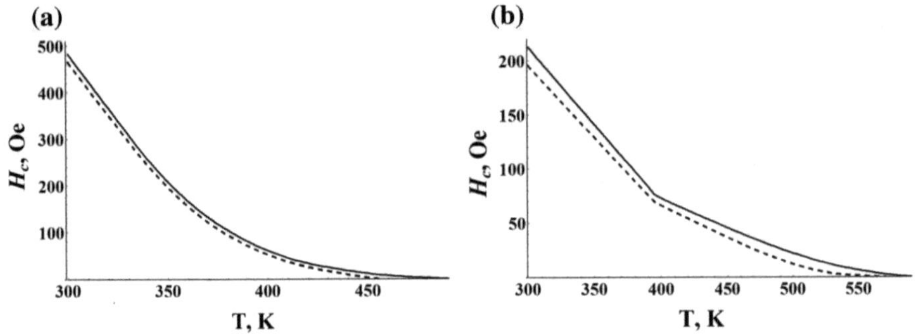

Fig. 9.1 Temperature dependences of the coercive field H_c of noninteracting (solid line) and interacting (dashed line) TM nanoparticles with $r = 20$ nm (**a**) and $r = 40$ nm (**b**)

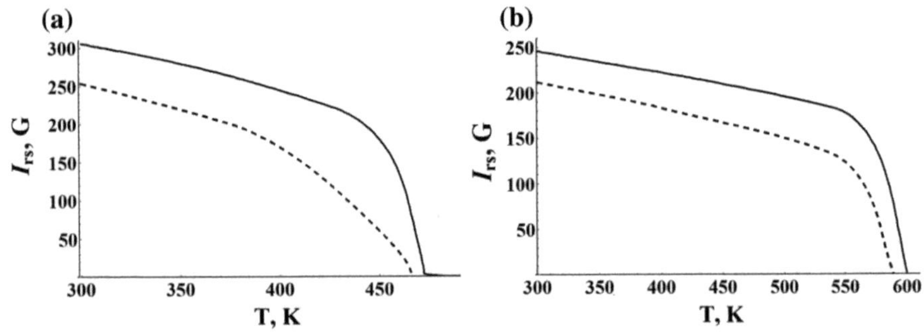

Fig. 9.2 Temperature dependences of the remanent saturation magnetization I_{rs} of noninteracting (solid line) and interacting (dashed line) TM nanoparticles with $r = 20$ nm (**a**) and $r = 40$ nm (**b**)

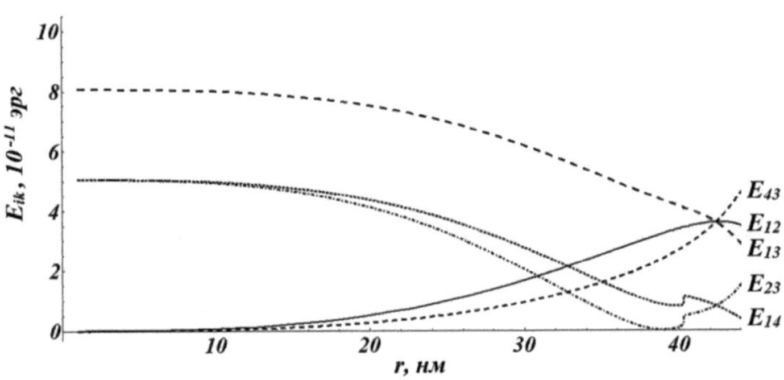

Fig. 9.3 Dependence of potential barriers E_{ik} on the degree of decay of titanomagnetite $Fe_{2,7}Ti_{0,3}O_4$ (size of minor semi-axis of magnetite core r)

E_{14} (E_{23}). At $r = 20$ nm, the barriers are higher than at $r = 40$ nm, which in the presence of thermal fluctuations of the magnetic moments of the phases leads to the aforementioned relationship between the coercive fields. The difference in the values of the remanent magnetizations is due to the relation described above between potential barriers. For nanoparticles with $r = 20$ nm, the smallness of the barriers E_{12} and E_{34} leads to the alignment of the first and second $n_1 = n_2$ (or the third and fourth $n_3 = n_4$) states. The remanent magnetization of such particles can be estimated as follows: $I_{rs}(20) \sim (n_1 - n_3)\left((1-\varepsilon)I_s^{(1)} + \varepsilon I_s^{(2)}\right) + (n_2 - n_4)\left((1-\varepsilon)I_s^{(1)} - \varepsilon I_s^{(2)}\right) = 2(n_1 - n_3)(1 - \varepsilon(20))I_s^{(1)}$. In the case of $r = 40$ nm $n_2 = n_3$ and $n_1 = n_4$, which leads to $I_{rs}(40) \sim 2(n_1 - n_3)\varepsilon(40)I_s^{(2)}$. Since $\varepsilon(20) \approx 0.06, \varepsilon(40) \approx 0.5,\ I_s^{(1)} = 306G, I_s^{(2)} = 480G$ then $I_{rs}(20)/I_{rs}(40) \approx 1.2$. The resulting ratio coincides with that shown in Fig. 9.2.

Note that even a sufficiently large volume concentration of TM nanoparticles c = 0.2, the magnetostatic interaction with the effective field B = 176 Oe weakly affects the hysteresis characteristics. This is due to the excess of the coercive field over the effective field $H_c > B$. That is why the magnetostatic interaction is more affected by low-coercivity nanoparticles with $r = 40$ nm. The weak dependence of the hysteresis characteristics on the magnetic interaction of nanoparticles at $B < H_c$ was noted in the reference (Afremov et al. 2018) in which it was shown that the interaction significantly affects H_c and I_{rs} only in fields $B > H_c$.

Thermo-Remanent and Chemical Magnetizations

The dependences of the thermo-remanent I_{rt} and chemical I_{rc} magnetizations on the effective interaction field B, calculated with the aid of relations 9.1–9.7, are presented in Table 9.1. It was assumed that the magnetizations are formed in the systems of TM nanoparticles with different degrees of decay in an external magnetic field $H = 1$ Oe.

As one would expect, an increase in the magnetostatic interaction between TM nanoparticles leads to a drop in the thermo-remanent and chemical remanent magnetizations of nanoparticle systems with different sizes of magnetite core (the

Table 9.1 The dependences of the thermo-remanent I_{rt} and the chemical remanent magnetization I_{rc} on the effective interaction field B and the degree of decay of r titanomagnetite $Fe_{2,7}Ti_{0,3}O_4$

B, Oe	$r = 10$ nm		$r = 20$ nm		$r = 30$ nm		$r = 40$ nm	
	I_{rt}	I_{rc}	I_{rt}	I_{rc}	I_{rt}	I_{rc}	I_{rt}	I_{rc}
5	55.8	0	54.7	3.35	54.0	11.31	44.2	26.80
10	27.5	0	24.3	2.39	22.6	8.07	20.3	19.13
15	24.3	0	22.8	1.77	20.6	5.99	19.0	14.82
20	19.3	0	18.6	1.12	17.1	4.85	15.4	12.32
25	16.2	0	15.3	1.12	14.9	3.76	13.0	10.41

degree of decay), which is related to the randomization of the magnetic moments of nanoparticles by random fields of magnetostatic interaction (Belokon 1985; Scherbakov and Scherbakova 1977; Afremov and Kharitonsky 1988). Moreover, the value of the thermo-remanent magnetization formed at room temperature $I_{rt}(300)$ is practically independent of the degree of decay up to $r = 30$ nm. The decay effect on I_{rt} is realized only at $r \approx r_{max} = 44$ nm. Thus, at $r = 40$ nm, the thermo-remanent magnetization decreases by almost 20%. The observed decrease in I_{rt}, as well as the increase in the chemical remanent magnetization I_{rc}, is due to the above-described peculiarities of the change in potential barriers with increasing magnetite core size r (see Fig. 9.3).

Figures 9.4 and 9.5 show the results of modeling the effect of the magnetostatic interaction and the degree of TM decay on the ratios of the thermo-remanent magnetization ($R_t = I_{rt}/I_{ri}$) and chemical remanent magnetization ($R_c = I_{rc}/I_{ri}$) to the ideal remanent magnetization I_{ri}. The dependence of R_t and R_c on the effective interaction field repeats the course of the curves calculated within the model of single-domain particles (Afremov et al. 1989, 2010). In the region of small interaction fields, the ratio R_t depends on the relation between B and the fluctuation field at the blocking temperature $H_f = 25kT/vI_s$. As the calculation with the increase in the degree of decay of the TM has shown, the field of fluctuations increases from $H_f \approx 6$ Oe to $H_f \approx 7$ Oe. At $B \leq H_f$, the thermally stable and ideal residual magnetizations are practically independent of the interaction, which leads to the independence of R_t from the effective field B in the initial section of the curves depicted in Fig. 9.4.

Since I_{rt} is determined mainly by high-coercivity particles, and I_{ri} is determined by both high and low coercive particles, the randomizing effect of the magnetic interaction leads to a more significant decrease in the ideal remanent magnetization

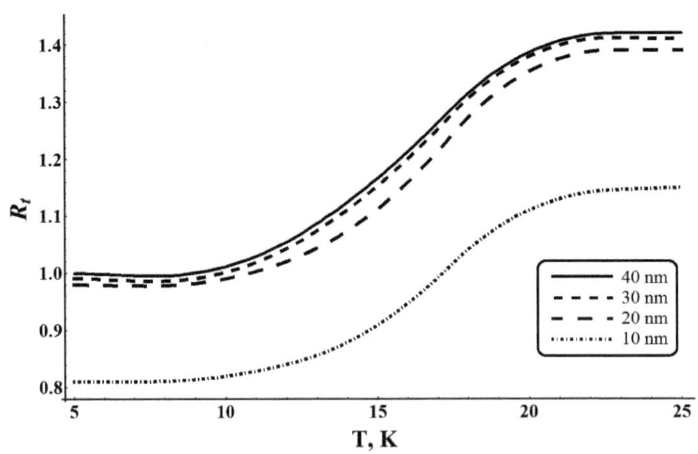

Fig. 9.4 Dependences of the ratio of the thermo-remanent magnetization to the ideal remanent magnetization R_t on the effective interaction field B and the decay rate r of titanomagnetite $Fe_{2,7}Ti_{0,3}O_4$

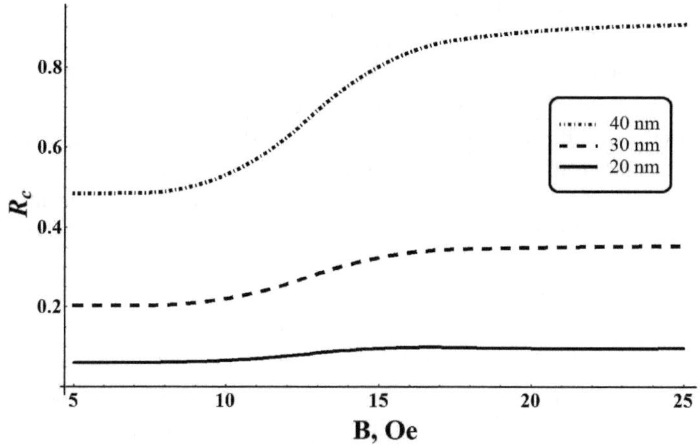

Fig. 9.5 Dependences of the ratio of the chemical remanent magnetization to the ideal remanent magnetization R_c on the effective interaction field B and the decay rate r of titanomagnetite $Fe_{2,7}Ti_{0,3}O_4$

than the thermo-remanent one. This leads to an increase in the ratio R_t. An increase in the interaction field leads to an equalization of the chaotic effect on both types of magnetization, which corresponds to the independence of the ratio R_t from the effective interaction field. We note that the ratio of the thermo-remanent magnetization to the ideal magnetization lies in the range $0.8 \leq R_t \leq 1.4$. The calculated values for R_t are consistent with the results of experiments carried out on titanium-containing rocks (Borisova and Sholpo 1985) and on fine-dispersed titanomaghemites (Nguen 1985b).

The effect of the magnetostatic interaction on the ratio of chemical remanent to ideal remanent magnetization is similar to the effect of interaction on R_t (see Fig. 9.5). However, depending on the degree of decay, the ratios of R_c are limited ($0 < R_c \leq 0.9$) and do not exceed R_t. The noted limitations for R_c are consistent with the experimental values $0.6 \leq R_c \leq 0.9$ (Nguen and Pechersky 1987).

Conclusion

Theoretical analysis of the influence of magnetostatic interaction on the coercive field and various types of remanent magnetization of the decay products of the system of TM nanoparticles carried out within the framework of the theory of two-phase interacting nanoparticles showed that:

- an increase in the degree of decay of TM (the size of the magnetite core or the concentration of titanium in the shell of a two-phase nanoparticle) leads to a drop in the coercive field and remanent saturation magnetization at any temperature up to the Curie temperature;

- magnetic interaction between nanoparticles with an effective field of less than the coercive field slightly reduces the hysteresis characteristics;
- the growth of the effective interaction field leads to a decrease in the thermo-remanent and chemical remanent magnetization;
- the degree of TM decay has little effect on the value of the thermo-remanent magnetization I_{rt} in the size range of the magnetite core $r \leq 30$ nm and only about 20% near $r_{max} = 44$ nm at $r = 40$ nm;
- the chemical remanent magnetization increases with the degree of decomposition to values not exceeding the thermo-remanent one;
- the value of the ratio of the thermo-remanent magnetization to the ideal R_t increases with the increase of the magnetostatic interaction and lie in the range $0.8 \leq R_t \leq 1.4$, which agrees with the experimental data (Borisova and Sholpo 1985);
- the ratio of the chemical remanent magnetization to the ideal increases with increasing magnetic interaction and the degree of decay and has bounded values $0 < R_c \leq 0.9$, which does not contradict the results of the experiments (Nguen and Pechersky 1987).

References

Afremov L., Anisimov S., Iliushin I., Size effect on the hysteresis characteristics of a system of interacting core/shell nanoparticles. *JMMM*, 2018, 447, pp 88–95.

Afremov L.L., Belokon V.I. Kirienko Yu.V., Nefedev K.V., Magnetic properties of nanodispersed magnets. *Vladivostok: Publisher FEFU*, 2010, 120 p.

Afremov L.L., Iliushin I.G., Anisimov S.V., Modeling the implications of chemical transformations for the magnetic properties of a system of titanomagnetite nanoparticles. *Izv., Phys. Solid Earth,* 2015, 51, pp. 613–621.

Afremov L.L., Kharitonsky P.V., On the magnetostatic interaction in an ensemble of growing single-domain grains. *Izvestiya AN USSR, Fizika Zemli,* 1988, 2, pp. 101–105.

Afremov L.L., Kharitonsky P.V., Some diagnostic features of the thermo-sufficient and chemical magnetizations of an ensemble of single-domain particles. *In the book "Magnetism of rocks"*, Moscow, 1989, pp. 37–45.

Afremov L.L., Panov A.V., Remanent magnetization of ultradispersed magnets. *Vladivostok: Publisher FEFU*, 2004, 192 p.

Anisimov S.V., Afremov L.L., Thermoremanent and chemical magnetization of exsolution products of nanosized titanomagnetites. *Izvestiya, Physics of the Solid Earth*, 2018, 54, pp. 1–6.

Avilova T.E., Bagin V.I. Gendler T.S., Chemical remanent magnetization and structural-sensitive characteristics of hematite. *Izvestiya AN USSR, Fizika Zemli*, 1985, 4, pp. 67–77.

Belokon V.I., Kochegura V.V., Sholpo L.E., Methods of paleomagnetic studies of rocks. *Leningrad Nedra,* 1973, 248 p.

Belokon V.I., On the ratio of some types of remanent magnetization of an ensemble of single-domain interacting particles. *Izvestiya AN USSR, Fizika Zemli,* 1985, 2, pp. 55–64.

Borisova G.P., Sholpo L.E., On the possibility of statistical estimates of paleo magnetic field strength. *Izvestiya AN USSR, Fizika Zemli*, 1985, N7, p. 71.

Goroshko O.A., Ivanov V.A., Soppa I.V., Observation of interaction assembly of single domain particles. *JMMM*, 2010, 322, pp. 3385–3390.

Gribov C.K., Dolotov A.V. Scherbakov V.P., experimental modeling of chemical magnetization and Telie method for the titanomagnetite-containing basalts. *Fizika Zemli,* 2017, 2, pp. 109–128.

Gubin S.P., Koksharov Yu.A., Khomutov G.B., Yurkov G.Yu., Magnetic nanoparticles: preparation, structure and properties. *Russian Chem. Rev.*, 2005, 74, pp. 489–520.

Haig G., The occurrence of remanent magnetization in chemical changes. *In the book "Paleomagnetism" under the redaction of Petrova G.N.*, Moscow, 1962, pp. 67–86.

Hramov A.N., Paleomagnetics. *Leningrad Nedra,* 1982, 312 p.

Ivanov V.A., Haburzaniya I.A., Sholpo L.E., The use of Preisoch diagram for the diagnosis of single- and multi-domain grains in rock samples. *Izvestiya AN USSR, Fizika Zemli,* 1981, 1, pp. 55–65.

Ivanov V.A., Sholpo L.E., Quantitative criteria for single- and multi-domain states of ferromagnetic minerals of rocks. *Izvestiya AN USSR, Fizika Zemli,* 1982, 8, pp. 84–90.

Jun Wang, Wei Wu, Fan Zhao, and Guo-meng Zhao, Curie temperature reduction in SiO_2-coated ultrafine Fe_3O_4 nanoparticles: Quantitative agreement with a finite-size scaling law. *Applied physics letters*, 2011, 98, 083107.

Krupichka S., Physics of ferrites and related magnetic oxides. Moscow, 1976, 2, 504 p.

Kudryavceva G.P., Ferrimagnetism of natural oxides. *Moscow Nedra*, 1988, 232 p.

Nguen T.K.T., Chemical remanent magnetization of magnetite, formed during the decay of titanomaghemite. *VINITY RAS* № 4886–85, 1985a, 17 p.

Nguen T.K.T., On the stability and the ratio between the values of crystallographic, ideal and thermo-remanent magnetization of magnetite formed during the decay of pyrite. *Izvestiya AN USSR, Fizika Zemli,* 1985b, 7, pp. 100–104.

Nguen T.K.T., Pechersky D.M., An experimental study of the crystallization magnetization formed in the oxidation of pyrite. *Izvestiya AN USSR, Fizika Zemli*, 1984, N5, pp. 48–62.

Nguen T.K.T., Pechersky D.M., Experimentally obtained features of the crystallization remanent magnetization of magnetite-bearing rocks. *Izvestiya AN USSR, Fizika Zemli*, 1985, N8, pp. 48–62.

Nguen T.K.T., Pechersky D.M., Signs of the chemical magnetization of magnetite formed during the decay of titanomagnetite. *Izvestiya AN USSR, Fizika Zemli,* 1987, N5, pp. 69–76.

Nikiforov V.N., Ignatenko A.N., Irkhin V. Yu., Size and surface effects on the magnetism of magnetite and maghemite nanoparticles. *J. Exp. Theor. Phys.*, 2017, 124, pp. 304–310.

O'Reilly W., Rock and mineral magnetism. *New York*. 1984. 230 p.

Özdemir Ö., and Dunlop D. J., Hysteresis and coercivity of hematite. *J. Geophys. Res. Solid Earth,* 2014, 119, 2582–2594.

Perez N., Guardia P., Roca A.G., Morales M.P., Serna C.J., Iglesias O., Bartolome F., Garcia L. M., Batlle X., Labarta A., Surface anisotropy broadening of the energy barrier distribution in magnetic nanoparticles. *Nanotechnology,* 2008, 19, 475704.

Petrova G.N., Laboratory methods in paleomagnetic studies. *Geomagnetic studies,* 1977, 19, pp. 40–49.

Scherbakov V.P., On the distribution function of molecular fields in systems with randomly distributed interaction centers. *FMM,* 1979, 48(6), p. 1134.

Scherbakov V.P., Scherbakova V.V., To the calculation of the thermo-sufficient and ideal magnetization of an ensemble of interacting single-domain grains. *Izvestiya AN USSR, Fizika Zemli,* 1977, 6, pp. 69–83.

Sholpo L.E., The use of the rock magnetism to solve the geological tasks. *Leningrad Nedra,* 1977, 182 p.

Chapter 10
An Estimate of the Remanent Magnetization in the Case of a Ferromagnet Transformation Accompanied by a Change in the Curie Temperature

V. I. Belokon, E. V. Chibiriak and O. I. Dyachenko

Abstract The remanent magnetization of a system of interacting one-domain particles arising during the transformation of a ferromagnet accompanied by a change in the Curie point is estimated. The case is considered when the initial Curie temperature is lower than the temperature at which the structure of the ferromagnet changes. The investigation was carried out on an Ising model in the framework of random interaction fields method.

Keywords Remanent magnetization · Titanomagnetite · Ising model
Diffusion · Curie point · Ferromagnetism

Introduction

A possible significant change in the primary remanent magnetization of igneous rocks is associated with the features of its formation and the duration of action of a relatively weak geomagnetic field. Diffusion can be one of the factors of the change in the remanent magnetization. It determines, for example, the mechanism of the single-phase oxidation (diffusion of vacancies) or the mechanism of the decay (diffusion of impurity atoms, which contributes to the increase the local concentration of exchange interacting atoms (ions) and, respectively, the Curie point)

V. I. Belokon · E. V. Chibiriak
Department of Theoretical and Nuclear Physics, the School of Natural Sciences, Far Eastern Federal University, Vladivostok, Russia

O. I. Dyachenko (✉)
Department of General and Experimental Physics, the School of Natural Sciences, Far Eastern Federal University, Vladivostok, Russia
e-mail: dyachenko.oi@dvfu.ru

D. Nurgaliev et al. (eds.), *Recent Advances in Rock Magnetism, Environmental Magnetism and Paleomagnetism*, Springer Geophysics,
https://doi.org/10.1007/978-3-319-90437-5_10

(Belokon et al. 2013, 1993). Another factor is the reordering of cations in two-sublattice magnets, such as titanomagnetite (Bowles et al. 2013).

The influence of diffusion is considered in the framework of the random interaction fields method. It is the development of the molecular field method. The method of an effective (molecular) field in the theory of magnetism is simple, but unfortunately yields an overestimate to the Curie point T_c. For example, Smart (1996) uses dimensionless units, where the Curie temperature T_c is expressed in units of kT and for the sake of convenience the Boltzmann constant k, the magnetic spin moment m_0, and the exchange integral J are equal to one. The exact solution of Onsager shows that the dimensionless Curie temperature is 2.27 for a planar square lattice with coordination number $z = 4$ in the framework of Ising model. In the Bragg-Williams approximation, the simplest version of the mean-field theory, $T_c = z$. The reason for this discrepancy is that the theory of Bragg-Williams does not take into account correlations between the directions of the magnetic moments of neighboring atoms. An approximate account of such correlations was proposed by Oguchi, Bethe, Peierls (Smart 1996). An example of the successful use of the effective field theory in many-sublattice magnets can be found in (Fabian et al. 2015), in which the magnetic properties of hematite-ilmenite systems are investigated.

In this article, we estimate the remanent magnetization arising in a system of single-domain interacting particles by a change in Curie temperature in the interval from $T_1 < T$ to $T_1 > T$, where T_1 is the temperature of the structural transformation of the ferromagnet.

Influence of the Diffusion of Atoms on the Curie Point

In the case of a random distribution of exchange-interacting ions in the crystal lattice, the distribution function of the random fields of the exchange interaction has the form (Belokon et al. 2001):

$$W(H) = \frac{1}{\sqrt{\pi}B}\exp\left(\frac{-[H - H_0(\alpha - \beta)]^2}{B^2}\right). \tag{10.1}$$

It is assumed that the atom creating a molecular field H (ferromagnetic atoms), has a random coordinate for amorphous magnetic material or a random lattice site. In addition, a magnetic moment has also random interaction. Moments of the distribution function:

$$H_0 = p\sum_k \varphi_k, \tag{10.2}$$

$$B^2 = 2p\left[1 - (\alpha - \beta)^2 p\right]\sum_k \varphi_k^2, \tag{10.3}$$

where $\varphi_k = \varphi_k(\mathbf{m}_k, \mathbf{r}_k)$ describes the law of interaction between particles, α is the relative probability of the spin being up, β is the relative probability of the spin being down, p is the concentration of the magnetic particles.

The mean magnetization is defined as

$$\langle M \rangle = \int \mathrm{th}\left[\frac{m_0 H}{kT}\right] W(H)\,\mathrm{d}H, \tag{10.4}$$

where $\langle M \rangle = \langle \alpha - \beta \rangle$. Here … denotes thermodynamic and configurational averaging, m_0 is the magnetic moment per one ion, k is the Boltzmann constant, and T is the temperature.

Then the angle brackets will be omitted.

The substantial simplification of the first equation can be achieved by the replacing of distribution function on the rectangle function (Belokon et al. 2001, 2015, 2016):

$$\widetilde{W}(x) = \begin{cases} 0, & -B > x, B < x, \\ \frac{1}{2B}, & -B \le x \le B. \end{cases} \tag{10.5}$$

Near the Curie point such substitution practically does not affect the dependence of M on T.

After integration of expression of the magnetization can be determined as:

$$M = \frac{1}{2B} \int\limits_{-B}^{B} \mathrm{th}\left[\frac{m_0(H + MH_0)}{kT}\right] \mathrm{d}H, \tag{10.6}$$

For small M can be represented in the form:

$$M^2 = \frac{3\left(\frac{H_0}{B}\,\mathrm{th}\left[\frac{m_0 B}{kT}\right] - 1\right)}{\frac{H_0^3}{Bk^2T^2}\left[\mathrm{th}\left[\frac{m_0 B}{kT}\right] - th^3\left[\frac{m_0 B}{kT}\right]\right]}. \tag{10.7}$$

It follows from Eq. 10.7 that the solution with nonzero (ferromagnetism) becomes possible only under the condition

$$\frac{H_0}{B}\,\mathrm{th}\left[\frac{m_0 B}{kT}\right] > 1. \tag{10.8}$$

Since the limiting value of the hyperbolic tangent is one, the solution can be obtained under the condition that $\frac{H_0}{B} > 1$.

For direct exchange $\varphi_k = m_0 J$ (J is the exchange integral), therefore it's necessary to carry out the summation over nearest neighbours. Hence,

$$\gamma = \frac{H_0}{B} = \frac{p_c z}{\sqrt{2p_c z}} = 1, \quad p_c = \frac{2}{z}, \tag{10.9}$$

where z is the number of nearest neighbours, p_c is critical concentration at which ferromagnetism can be established. The critical concentrations p_c calculated using the formula (10.9), have good agreement with known results given in works on the percolation theory (Kirpatrick 1973; Strauffer 1985).

It means that the ordering is possible in the case when a flowing cluster appears. For $p < p_c$ there are clusters of finite dimensions, which may have a different magnetic moment, but the total magnetic moment of the system is zero.

We consider a binary alloy with coordination number z consisting of exchange-interacting ferromagnetic atoms and non-ferromagnetic impurity atoms at a concentration p of ferromagnetic atoms with $p < 1$. If we admit diffusion, then three states are possible for an arbitrary site of the crystal lattice in the Ising model:

1. The site is occupied by a "ferromagnetic atom" with its spin oriented "up";
2. The site is occupied by a "ferromagnetic atom" with its spin oriented "down";
3. The site is occupied by a "non-ferromagnetic atom".

Then the grand partition function has the form:

$$Z = 1 + \lambda \exp\left\{\frac{m_0 H}{kT}\right\} + \lambda \exp\left\{-\frac{m_0 H}{kT}\right\}, \tag{10.10}$$

where $\lambda = \exp[\mu/T]$, μ is the chemical potential.

The average number of ions per site:

$$\overline{p} = \frac{2\lambda \mathrm{ch}\left\{\frac{m_0 H}{kT}\right\}}{1 + 2\lambda \mathrm{ch}\left\{\frac{m_0 H}{kT}\right\}} = \frac{\mathrm{ch}\left\{\frac{m_0 H}{kT}\right\}}{\beta + \mathrm{ch}\left\{\frac{m_0 H}{kT}\right\}}. \tag{10.11}$$

Here $\overline{\ldots}$ denotes thermodynamic averaging, $\beta = 1/2\lambda$.

The average value of the magnetic moment per site:

$$\overline{pM} = \frac{2\lambda m_0 \mathrm{sh}\left\{\frac{m_0 H}{kT}\right\}}{1 + 2\lambda \mathrm{ch}\left\{\frac{m_0 H}{kT}\right\}} = \overline{p}\frac{2\lambda m_0 \mathrm{sh}\left\{\frac{m_0 H}{kT}\right\}}{2\lambda \mathrm{ch}\left\{\frac{m_0 H}{kT}\right\}} = \overline{p} m_0 \mathrm{th}\left\{\frac{m_0 H}{kT}\right\}. \tag{10.12}$$

To simplify the equations, in the following the normalizations $m_0 = 1$, $k = 1, J = 1$ are used.

The configuration averaging over random interaction fields leads to the following equation:

$$\langle pM \rangle = \frac{1}{2B} \int_{-B}^{B} \frac{\mathrm{sh}\left\{\frac{H + pMH_0}{T}\right\}}{\beta + \mathrm{ch}\left\{\frac{H + pMH_0}{T}\right\}} \, \mathrm{d}H, \tag{10.13}$$

Here H_0 does not contain the factor p, which is included in the definition of thermodynamically and configurationally average magnetic moment per one site.

As a result of the expansion in a series up to a value $\langle pM^3 \rangle$, we obtain after integration:

$$\langle pM^2 \rangle = \frac{6\left(\frac{H\mathrm{sh}\left[\frac{B}{T}\right]}{B\left(\beta+\mathrm{ch}\left[\frac{B}{T}\right]\right)} - 1\right)}{\frac{H^3\left(2-\beta^2+\beta\mathrm{ch}\left[\frac{B}{T}\right]\right)\mathrm{sh}\left[\frac{B}{T}\right]}{BT^2\left(\beta+\mathrm{ch}\left[\frac{B}{T}\right]\right)^3}} \tag{10.14}$$

After configurational averaging of value $\overline{p}$, we have the equation:

$$\langle p \rangle = \frac{1}{2B}\int_{-B}^{B} \frac{\mathrm{ch}\left\{\frac{H+pMH_0}{T}\right\}}{\beta+\mathrm{ch}\left\{\frac{H+pMH_0}{T}\right\}}\,\mathrm{d}H. \tag{10.15}$$

Equation (10.14) allows us to determine the critical temperature at which $\langle pM^2 \rangle = 0$:

$$\frac{H\mathrm{sh}\left[\frac{B}{T_c}\right]}{B\left(\beta+\mathrm{ch}\left[\frac{B}{T_c}\right]\right)} = 1. \tag{10.16}$$

For concentration $p \geq 1/3$ at $\beta < 2$ the corresponding temperature is the Curie point T_c. At conditions that $p \leq 1/3$ and $T = T_c$ we have that $\langle pM^2 \rangle = 0$. It means the disappearance of a flowing (infinite) cluster, which can be restored with increasing temperature and a more uniform distribution of exchange-interacting atoms.

By the way in Eq. (10.15),

$$\langle p \rangle = \frac{1}{2B}\int_{-B}^{B} \frac{\mathrm{ch}\left\{\frac{H+pMH_0}{T}\right\}}{\beta+\mathrm{ch}\left\{\frac{H+pMH_0}{T}\right\}}\,\mathrm{d}H = p, \tag{10.17}$$

because the configuration averaging is performed over all the sites of the crystal lattice, and the number of magnetic atoms in the system does not change.

Equation (10.17) in the zero-order approximation of $\langle pM \rangle$ (near the critical temperature T_c) makes it possible to relate the parameter β to the density p:

$$p = 1 - \frac{2T\mathrm{arctn}\left[\frac{\sqrt{1-\beta^2}\mathrm{th}\left[\frac{B}{2T}\right]}{1+\beta}\right]}{B\sqrt{1-\beta^2}}. \tag{10.18}$$

The joint solution Eqs. (10.16) and (10.18) gives us the dependence $T_c = T_c(p)$.

The graphs of the dependence of critical temperature T_c on the concentration of the "ferromagnetic" atoms p are shown in Fig. 10.1 for the lattice with $z = 8$. The dotted curve corresponds to the case of the absence of the diffusion, the solid curve- the diffusion takes place. In the absence of the diffusion the critical concentration $p_c = 2/z = 0.25$.

From Fig. 10.1 follows that the diffusion process can cause an increase in the Curie point, which is especially noticeable at low concentrations of ferromagnetic atoms p. We note that this result does not take into account the possibility of the existence of a "crystal" field, which facilitates the separation of ferromagnetic and impurity atoms. In particular, such a separation can occur at temperatures above the Curie point due to the crystal field. To some extent, this effect is realized, for example, in titanomagnetite (heterophase or secondary oxidation of titanomagnetite in the terminology of Butler). The primary titanomagnetite of the intermediate composition is replaced by iron-rich titanomagnetite with a higher Curie temperature and a higher saturation magnetization. In addition, the particle size decreases, because the original particle is subdivided into essentially smaller particles separated by the paramagnetic ilmenite. The Curie point can also change as a result of the low-temperature oxidation ($T < T_c$).

Figure 10.2 shows the dependence of the Curie point on the degree of the oxidation, given in (Readman et al. 1972). If we turn to basalts, titanomagnetites have a titanium concentration $p = 0.6$, then it can be seen the Curie point can vary from $T_c \approx 150$ °C to $T_c \approx 350$ °C with an increase in the degree of oxidation. If the oxidation process occurs due to the secondary warming at a temperature of 200 °C, then the situation of the transition of the Curie point throw the temperature of the transformation of the ferromagnet will be realized (it is natural that such a process can be realized in the laboratory conditions). A simple estimate of this magnetization is made in the next section.

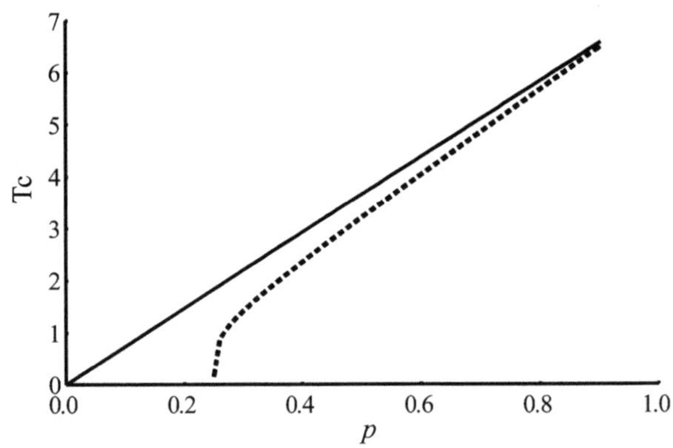

Fig. 10.1 The graphs of the dependence of the Curie point T_c on the concentration of the "ferromagnetic" atoms p for the lattice with $z = 8$. The dotted curve corresponds to the case of the absence of the diffusion, the solid curve- the diffusion takes place

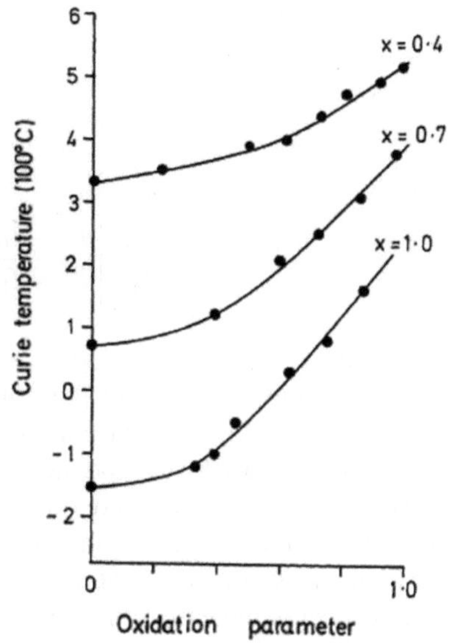

Fig. 10.2 Variation of Curie temperature as a function of oxidation for titanomagnetites $x = 0.4, x = 0.7$ and 1.0. (Readman et al. 1972)

Diffusion Thermal Magnetization?

Perhaps this name is not universally accepted and, as a variant of this kind of magnetization can be called thermochemical remanent magnetization (TCRM), as used in (Fabian 2009).

It is obvious that there is a fundamental possibility of the appearance of an "isothermal" magnetization in the field due to the change in the value of the Curie point from $T_{c,1} < T$ to $T_{c,2} > T$. In this case, the mechanism for the formation of the magnetization I_{rdt} ("diffusion thermal magnetization") is the same as the mechanism for the formation of the thermal magnetization I_{rt} (TRM). The difference is connected with the fact that the Curie point changes, and not the temperature. The simplest comparative estimate I_{rt} and the ideal magnetization I_{ri} (ARM) is given by us, for example, in work (Belokon et al. 2001), and it is based on a comparison of regions on the Preisach-Neel diagram. In this case, $I_{ri} \sim h/b(T_0)$, where h is the external field ($h \ll b$), $b(T_0)$ is the maximum value of the interaction field of clusters at temperature T_0, $b \sim I_s$, I_s is the spontaneous magnetic moment of a unit volume. In the framework of the theory of the effective field $I_s(T) \sim (T_c - T)^{1/2}$. The evaluation of the thermal magnetization is based on the fact that the impact of thermal fluctuations is the same as the impact of alternative field, and the result should be average in the range $T_0 \leq T \leq T_b$, where T_b is the blocking temperature, it is close to the temperature T_c of the highest-coercive particles.

$$I_{rt} \sim \left\langle \frac{h}{b} \right\rangle = h \frac{1}{T_b - T_0} \int_{T_0}^{T_b} \frac{1}{b(T)} \mathrm{d}T$$
$$= \frac{h}{b(T_0)} \frac{1}{T_b - T_0} \int_{T_0}^{T_b} \frac{I_s(T_0)}{I_s(T)} \mathrm{d}T \tag{10.19}$$
$$= \frac{2h}{b(T_0)} \left[\frac{T_c - T_0}{T_b - T_0} - \frac{(T_c - T_0)^{1/2} (T_c - T_b)^{1/2}}{T_b - T_0} \right]$$

In many cases, T_b is close to T_c, so $I_{rt} \approx 2I_{ri}$. For many rocks, the experimentally determined ratio I_{rt}/I_{ri} is in the range $2 \div 3$, so the above estimate is plausible. A large volume of relevant experimental data is given in (Borisova et al. 1985). Figure 10.3 shows the dependence I_{rt}/I_{ri} as a function of T_b. The volume of grain V is assumed to be fixed and unchanged. The main role in the blocking of the magnetic moment of the grain is played by the sharp growth of I_s as the difference between the Curie temperature and the temperature at which the magnetization process occurs increases. In turn, the blocking temperature is determined from the ratio $I_s V H_c \approx 50$ kT, where H_c is critical cluster remagnetization field. The blocking of grains for a given V is determined by the formula $I_s H_c \approx I_s^2$.

Naturally, as applied to the real ferromagnet, it is necessary to translate the conditional temperature values into Kelvin degrees in Fig. 10.3. The dependence of the result on the volume of the ferromagnetic grains lies in the blocking temperature. If the magnetic moment of the grain is smaller, than the difference between the temperature T_c and T_b is greater and the ratio I_{rt}/I_{ri} is closer to unity. In addition, these estimates make sense at a sufficiently high concentration of the ferromagnet, when the interaction field b (T) appreciably exceeds the applied external field.

We consider the case when the magnetization arises in the field h at the process of change T_c at a constant temperature T from $T_{c,1} < T$ to $T_{c,2} > T$. In this case, the ratio h/b should be averaged as follows:

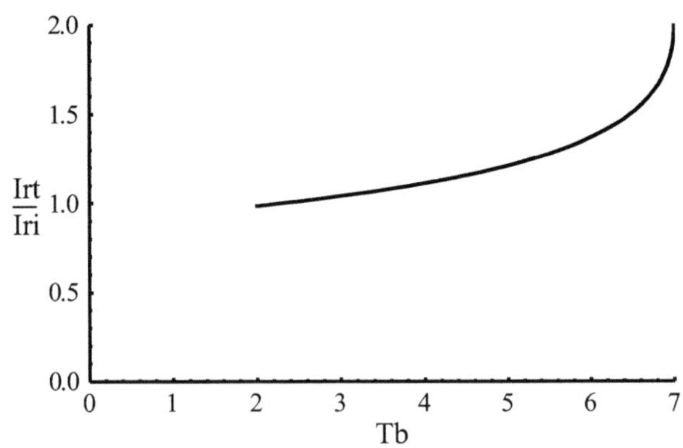

Fig. 10.3 The dependence I_{rt}/I_{ri} as a function of T_b

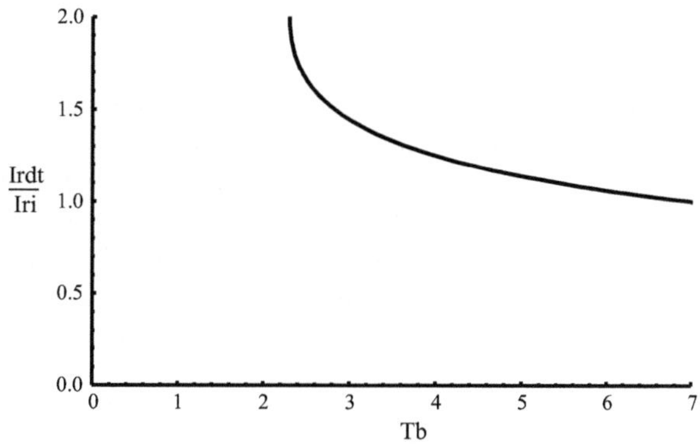

Fig. 10.4 The dependence I_{rdt}/I_{ri} as a function of $T_b nn$

$$\left\langle \frac{h}{b} \right\rangle = \frac{1}{T_{c,2} - T_b} \int_{T_b}^{T_{c,2}} \frac{1}{b(T_c)} \mathrm{d}T_c, \tag{10.20}$$

where $T_b \geq T$, $b(T_c) \sim (T_c - T)^{1/2}$, $b(T_{c,2}) \sim (T_{c,2} - T)^{1/2}$.

So,

$$\begin{aligned} \left\langle \frac{h}{b} \right\rangle &= \frac{h}{b(T_{c,2})} \frac{1}{T_{c,2} - T_b} \int_{T_b}^{T_{c,2}} \frac{(T_{c,2} - T_b)^{1/2}}{(T_c - T)^{1/2}} \mathrm{d}T_c \\ &= \frac{2h}{b(T_{c,2})} \left[\frac{T_{c,2} - T}{T_{c,2} - T_b} - \frac{(T_{c,2} - T)^{1/2} (T_b - T)^{1/2}}{T_{c,2} - T_b} \right]. \end{aligned} \tag{10.21}$$

At $T_b \approx T, I_{rdt}/I_{ri} \approx 2$. Thus, I_{rdt} in terms of the limit of the value at $T_b \geq T$ is comparable with the thermal magnetization.

The formulas (10.19) and (10.21) are valid under the condition that the maximum interaction field is larger than the external field practically on the entire temperature range, that is the concentration of particles is quite large. In other words, practically throughout the entire temperature range, there is a linear dependence of I_{ri} on the constant external field to a maximum value exceeding the field h.

Figure 10.4 shows the dependence I_{rdt}/I_{ri} as a function of T_b.

Note that the dependence I_{rdt}/I_{ri} as a function of T_b mirrors the ratio I_{rt}/I_{ri} as a function of T_b, which is to be expected due to the mechanism of the formation I_{rdt}.

Conclusion

1. The diffusion process at $p \geq 1/3$ can lead to an increase in the Curie point due to an increase in the local concentration of ferromagnetic atoms.
2. At $p \leq 1/3$, the tendency to unite the exchange-interacting atoms leads to the destruction of the flowing cluster. In this case, the configuration and thermodynamically averaged magnetic moment of the system M turns to zero at $T \leq T_c$.
3. The motion of the Curie point in the diffusion process under certain conditions can lead to the appearance of remanent magnetization similar to TRM.
4. For rocks, the main carrier of magnetization is titanomagnetite, the direct application of the results obtained can raise doubts due to the fact that titanomagnetite is a ferrimagnet. However, the data given in Bowels' article indicate the possible movement of T_c, and applying of Eqs. 10.19, 10.20 and 10.21 does not imply knowledge of the specific mechanism responsible for this movement.

Acknowledgements This work was financially supported by the State Program of the Ministry of Education and Science of the Russian Federation № 3.7383.2017/8.9. Authors are thankful to Karl Fabian for the very helpful review.

References

Belokon V.I., Dyachenko O.I., Kapitan V.Yu. Izvestiya, Physics of the Solid Earth. 2015. **51**(5). P. 622.

Belokon V.I., Kapitan V.Yu., Dyachenko O.I Journal of Magnetism and Magnetic Materials. 2016. **401**. P. 651.

Belokon V.I., Nefedev K.V., Dyachenko O.I. Phase transitions in the binary alloys with annealed magnetic impurities. Applied Mechanics and Materials. 2013. **328**. P. 789–793.

Belokon V.I., Nefedev K.V. Journal of experimental and theoretical physics. 2001. **120** (1(7)). P. 156.

Belokon V., Semkin S. The method of a random field in the theory of ferromagnetism of binary alloys. JETP. 1993. **104** #5 (11). P. 3784–3791.

Borisova G.P, Sholpo L. Ye. On the possibility of statistical estimates of the paleointensity of the geomagnetic field. Izvestiya AN SSSR., Fizika Zemli. 1985. #7. P. 71.

Bowles, J.A., M.J. Jackson, T.S. Berquó, P.A. Solheid, and J.S. Gee. Inferred time- and temperature-dependent cation ordering in natural titanomagnetites. *Nature Commun. 2013.*

Fabian K., Shcherbakov V.P., McEnroe S.A., Robinson P., Burton B.P. Magnetic mean-field modelling of solid solutions: theoretical foundations and application to the hematite–ilmenite system. GJI. 2015. **202** #2. P. 1029–1040.

Fabian, K., Thermochemical remanence acquisition in single-domain particle ensembles: A case for possible overestimation of the geomagnetic paleointensity, Geochemistry Geophysics Geosystems 10, 1–10, https://doi.org/10.1029/2009gc002420, 2009.

Kirpatrick S. Percolation and Conduction. *Rev. Mod. Phys.* 1973. **45**. P. 547–586.

Strauffer D. Introduction to percolation theory. Taylor&Francis. 1985. 74 p.

Smart J. Effective field theories of magnetism. Saunders. 1996. P. 188.

Readman P.V, O'Reilly V. Magnetic properties of oxidized (cation-deficient) titanomagnetites. J. Geomag.Geoelekctr. 1972. 24. P. 69–90.

Chapter 11
A 13,000-Yr Record of Environmental Change from Tschuchye Lake in Northeast Yakutia

S. S. Burnatny, A. N. Naumov and Yu. A. Korzun

Abstract Tschuchye Lake is located in the central part of the Oimyakon Upland in eastern Yakutia (63.351465°N, 140.990055°E). A multi-proxy investigation of a sediment core from the lake provides evidence of abrupt changes in paleolimnological conditions during the Holocene. The chronology of the 347 cm lacustrine sediment sequences is delimited by four AMS ^{14}C ages from plant macrofossils. Three distinct lithological units are identified marking changes in sedimentation. Four pollen zones represent shifts in biocoenosis during the Holocene. Clear geochemical zonation occurs in the core. Coarse detrital material represented by silty sand and sandy silt are characterized by (1) low content of TiO_2, Al_2O_3, MgO, K_2O, Rb, Fe_2O_3, and LOI; (2) high content of SiO_2, Zr and Sr; and (3) low values of magnetic susceptibility, saturation magnetization, and saturation remanence. Laminated silt shows the highest content of Fe_2O_3, P_2O_5, MnO, MS, Jrs and Js, which implies autigenic or diagenetic processes. Magnetic particles are predominantly single domain, superparamagnetic and rarely multi domain and possibly are of bacterial origin.

Keywords Holocene · Lacustrine sediments · Rock magnetism
Inorganic geochemistry · Palynology

Introduction

Lacustrine sediments represent one of the best archives for paleoenvironmental information because of their continuous deposition and low-level of disturbance. Multidisciplinary research on lakes in central and eastern Yakutia has rarely been conducted and has focused so far on palynology and vegetation reconstruction (Andreev et al. 1997; Anderson et al. 2002). The use of magnetic and geochemical

S. S. Burnatny (✉) · A. N. Naumov · Yu. A. Korzun
North East Interdisciplinary Scientific Research Institute n.a.
N.A. Shilo of FEB RAS, 16, Portovaya St, Magadan, Russia 685000
e-mail: tesfarkon@gmail.com

D. Nurgaliev et al. (eds.), *Recent Advances in Rock Magnetism, Environmental Magnetism and Paleomagnetism*, Springer Geophysics,
https://doi.org/10.1007/978-3-319-90437-5_11

properties of lake sediments for interpreting environmental changes in this region has received less attention, although they are widely used as environmental proxies (Boyle 2002; Evans and Heller 2003; King and Peck 2002). In summer 2005, two lakes, Tschuchye and Ulu, were cored in eastern Yakutia. In addition standard palynological analyses, the sediments of Tschuchye Lake were studied to examine the variability and the origin of sediment magnetic and geochemical properties, with the aim to evaluate their sensitivity and usefulness for environmental reconstructions, to assess the relationship of biotic and abiotic indicators and the natural environment.

Study Area

Lake Tschuchye is located in the Oimyakon Upland of eastern Yakutia between the Kobyuma and Suntar Rivers (Fig. 11.1). The lake has a length and width of 950 and 360 m, respectively. The maximum water depth is 6.1 m. The lake is connected with the larger Ulu Lake by a shallow channel. The lake has a shallow discharge into the Tiryakh-Yuryakh River. The distance between the both lakes is about 800 m.

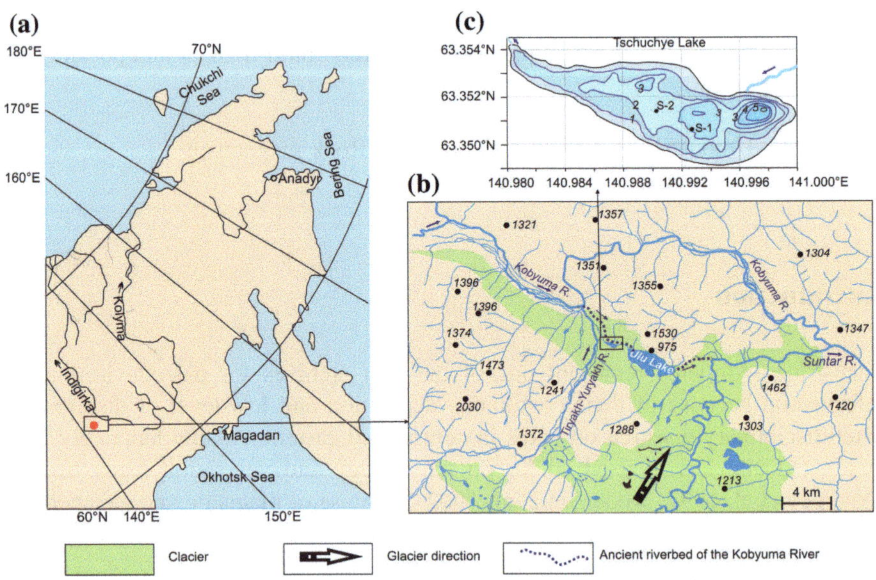

Fig. 11.1 Maps showing details of the study area, the location is indicated by the a red dot in a square in (**a**); drainage patterns, elevation, and locations of Tschuchye Lake (**b**); bathymetric map of Tschuchye Lake with depths shown in 1 m intervals (**c**). Coring locations are indicated on the bathymetric maps by a dark circle

Tschuchye and Ulu Lakes are of glacial origin. The glacier moved from the south along the valley of the Suntar River which originates on the northern slope of the Suntar-Hayat Range (Fig. 11.1). The damming effect of the glacier caused the ancient Kobyuma River to change direction. It is assumed that on the site of Tschuchye and Ulu lakes there is a buried river valley. Tschuchye and Ulu lakes are located in the moraine field. The deposits of the latter are distributed to the east and south of the lakes.

The climate of Oimyakon is subarctic with extremely severe and long winters, and only little precipitation (snow). The annual average January temperatures are—46.4 °C, and the annual average July temperatures are +14.9 °C. The absolute minimum winter and maximum summer temperatures measures are −65.4 and 34.6 °C, respectively.

The average annual humidity is 71% and the annual precipitation is 212 mm (http://www.pogodaiklimat.ru/climate/24688.htm).

The Oimyakon region is classified within the North Taiga subzone (Karaveav 1958). The light coniferous forest is dominated by *Larix cajanderi* Mayr. Other trees within the larch forests include birch (*Betula lanata* V. Vass., *B. platyphylla* Sucacz.), *Chosenia arbutifolia* (Pall.) A. Skvorts and poplar (*Populus suaveolens* Fisch.). The understory in the larch forests usually is a mix of *Betula middendorffii* Trautv. & C.A. Mey., *B. exilis* Sucacz., *Pinus pumila* (Pall.) Regel, various species of *Salix, Duschekia fruticosa* (Rupr.) Pouzar, *Rhododendron adamsii* Rehd., *Rh. parvifolium Adams*, Ledum.

Method

Two cores (Fig. 11.1) were taken in the central part of the Tschuchye Lake using a modified Livingstone piston sampler (Wright et al. 1984). Only the sediments of one core (S-1) were studied. Samples for rock magnetic analyses were cut every 1 cm from this core, yielding a total of 327 samples. The samples were dried at room temperature during a few days.

Mass-specific low-field magnetic susceptibility (MS) was measured on a MFK1-FA Kappabridge (AGICO Ltd., Brno, Czech Republic) having a sensitivity of 3×10^{-8} SI. Hysteresis parameters, including saturation magnetization (Js), induced magnetization (Ji), saturation remanence (Jrs), coercive force (Bc), and remanence coercivity (Bcr) were measured by an automatic coercive spectrometer (Burov et al. 1986). The sensitivities of remanent (Jr) and induced (Ji) magnetic moment are 1×10^{-8} Am^2 and 1×10^{-6} Am^2, respectively. The maximum specimen volume is 1.92 cm^3. The paramagnetic susceptibility (MSp) was recorded from the high-field slopes of the hysteresis curves.

The temperature-dependent susceptibility (MS-T) of the sediments was measured continuously from room temperature up to 700 °C and back to room temperature using a MFK1-FA Kappabridge equipped with a CS-3 high temperature furnace (AGICO Ltd., Brno, Czech Republic) having a sensitivity of 1×10^{-7} SI and a maximum specimen volume of 0.25 cm^3. The heating and cooling rates were 10–12 °C min^{-1}. The samples were heated continuously from room temperature to 700 °C and then cooled back to room temperature in air, and for a few samples in argon. The Curie temperature was determined using the Hopkinson peaks and the temperature dependence of inverse susceptibility values (Lattard et al. 2006; Petrovský and Kapička 2006; Fabian et al. 2013).

Major and trace elements were analyzed in 95 samples using a S4 Pioneer X-ray fluorescence spectrometer (Bruker, Germany) and a VRA-30 XRF spectrometer (Germany). The elemental compositions were determined using the fundamental parameters method (Borkhodoev 2002). The relative content of organic material was estimated using loss on ignition (LOI) after heating the samples to 500 and 1000 °C (Heiri et al. 2001).

Palynological samples were prepared using standard procedures for organic-poor sediments (PALE 1994). A minimum of 300 known arboreal and non-arboreal pollen grains was identified in each sample. Pollen calculations and diagrams were done using Tilia and Tilia-graph (https://museum.state.il.us./pib/grimm/tilia). Pollen zonation was done qualitatively based on percentage changes in major taxa.

The ^{14}C dates were obtained from plant macrofossil samples and carried out at the Center for Accelerator Mass Spectrometry, Lawrence Livermore National Laboratory (USA). Radiocarbon and tephra dates were converted into calibrated ages with IntCal 13 (Reimer et al. 2013). An age-depth model was then constructed using Bacon (Blaauw and Christen 2011).

Micrographs of magnetic extract were obtained by D.V. Fomin in Far East Scientific Center (Vladivostok) using microscope SIGMA HD (Carl Zeiss Microscopy GMBH).

Lithology and Chronology

The sediments from core S-1 include three lithological units.

Unit 1 (315–347 cm) is composed of massive homogeneous sandy silt with layers (0.5-cm-thick) of sand between 320 and 325 cm.

Unit 2 (250–315 cm) is characterized by laminated silt. Layering is due to the alternation of light gray and gray silts. The unit includes layers of sandy silt (a, b, c, d) at 294–298, 274–275, 260–261, and 251–252 cm.

Unit 3 (0–250 cm) consists mainly of a homogeneous and massive gray silt. The unit includes layers (a, b, c, d, e) of silty sand at 213–225, 182–187, 112–124, 43–48, and 15–22 cm, as well as a tephra (1.5–2 mm thick) at 244 cm. The tephra is correlated with the 7600 ^{14}C-yearBP eruption of the Kurile Lake caldera on Kamchatka. It is a valuable stratigraphic marker for southern Kamchatka, the Sea of Okhotsk, and a large part of the northern Asian mainland, where it has been identified in terrestrial and lake sediments (Ponomareva et al. 2004).

The chronology of lake sediments is based on AMS ^{14}C dates and the Kuril tephra (Table 11.1) (Korzun 2017).

Table 11.1 Radiocarbon dating

Depth (cm)	Lab no.	14C age	Calibrated age	Material dated
101	CAMS 174,820	4315 ± 35	4900 ± 45	Plant macrofossil
143	CAMS 174,821	4955 ± 45	5690 ± 50	Plant macrofossil
187	CAMS 174,818	6140 ± 35	7060 ± 75	Plant macrofossil
244	Tephra	7650 ± 50	8465 ± 50	
329	CAMS 174,822	11,770 ± 30	13,665 ± 110	Plant macrofossil

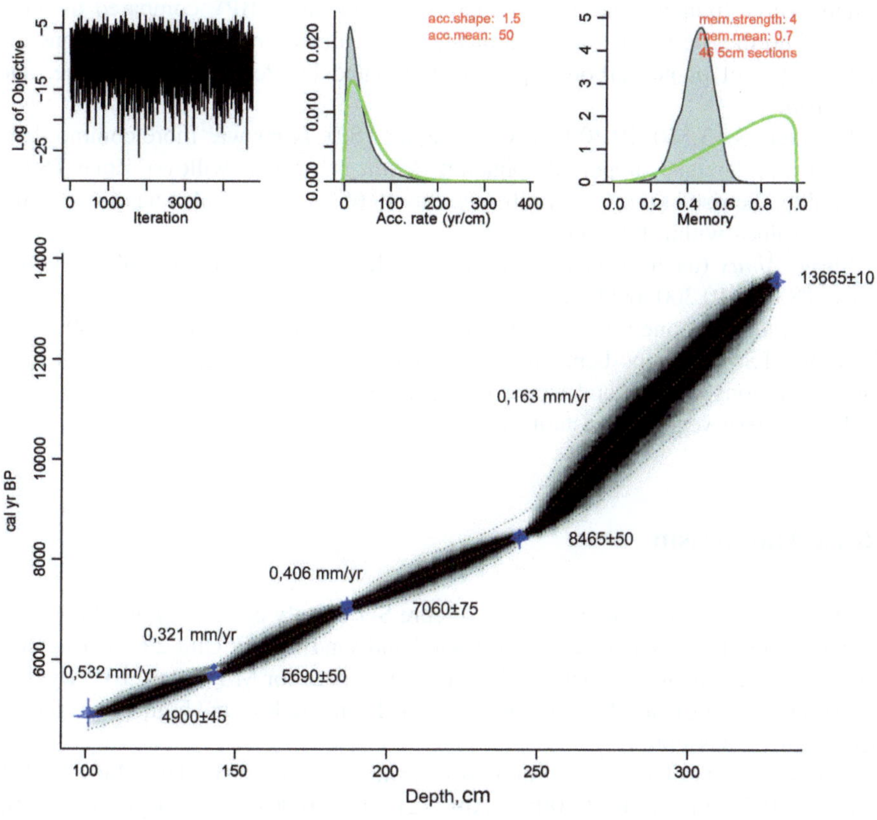

Fig. 11.2 Age-depth model for Tschuchye Lake sediments produced by the Bacon software package (Blaauw and Christen 2011). Dotted lines show the 95% confidence

Figure 11.2 shows age-depth model for Tschuchye Lake. Mean sedimentation rates were calculated for intervals between adjacent dates. The sedimentation rates vary from 0.163 to 0.532 mm/yr. The lowest sedimentation rate corresponds to unit 2, which is characterized by laminated silt.

Palynology

Four pollen zones trace the major changes in vegetation near Tschuchye Lake. The pollen concentration in the basal 12 cm of the core was too low for counting. One of the most striking characteristics of this record are the consistent and relatively high *Larix* pollen percentages. In contrast to many other pollen records from West Beringia, the Tschuchye data indicate the presence of tree environments since ~13,800 cal yr. BP.

Open *Larix* (larch) forests mixed with shrub *Betula* (birch) established exceptionally early (pollen zone TS1 ~13,800–13,300 cal yr BP), compared to other areas of northeastern Siberia and indicate warm, dry climates. Zone TS1 has the highest content of non-arboreal pollen, dominated by Poaceae, Cyperaceae, and *Artemisia*.

Between ~13,300–10,300 cal yr BP (zone TS2), *Larix* was more common and the forest understory supported abundant *Betula* and *Salix* (willow). Zone TS2 is marked by the rise in tree and shrub sums (up to 65%). *Salix* pollen has some of the highest values within the core.

Shrub *Alnus* (alder) became important in the region around 10,300 cal yr BP (zone TS3 ~ 10,300–6900 cal yr BP).

Pinus pumila (stone pine) was the last major taxa and occurred at ~6900 cal yr BP (zone TS4), perhaps being most abundant ~5600–4900 cal yr BP, and indicates conditions warmer and wetter than present.

The modern vegetation established ~700 yr ago.

Rock Magnetism

Shifts in the main magnetic properties of core S-1 closely correspond to lithological changes (Fig. 11.3). The layers of silty sand and sandy silt in Unit 2 (a, b, c, d) and Unit 3(a, b, c, d, e) are characterized by average values for MS (8.9×10^{-8} kg/m^3), Js (0.0045 Am^2/kg), and Jrs (0.0016 Am^2/kg). There are low and high values for the paramagnetic susceptibility.

The sediments of Unit 1, composed of sandy silt, show similar values for MS (9.67×10^{-8} kg/m^3), Js (0.0055 Am^2/kg), Jrs (0.0025 Am^2/kg), and MSp (5×10^{-8} kg/m^3) but have higher values for Bc (30.34 mT) and Bcr (54.16 mT). These magnetic zones are shown by colored bars in Fig. 11.3.

Sediments of Unit 2 are clearly distinguished in the core, because of their strong variations in magnetic parameters. The sediments are characterized by maximum average values for MS (23.8×10^{-8} kg/m^3), Js (0.019 Am^2/kg), and Jrs (0.0085 Am^2/kg). In contrast to Unit 1, Bc and Bcr correspond to enhanced values of MS, Js and Jrs.

The magnetic parameters of the sediments in the upper part of Unit 1, excluding the sandy layers, vary slightly throughout the unit: MS = 13.6×10^{-8} kg/m^3,

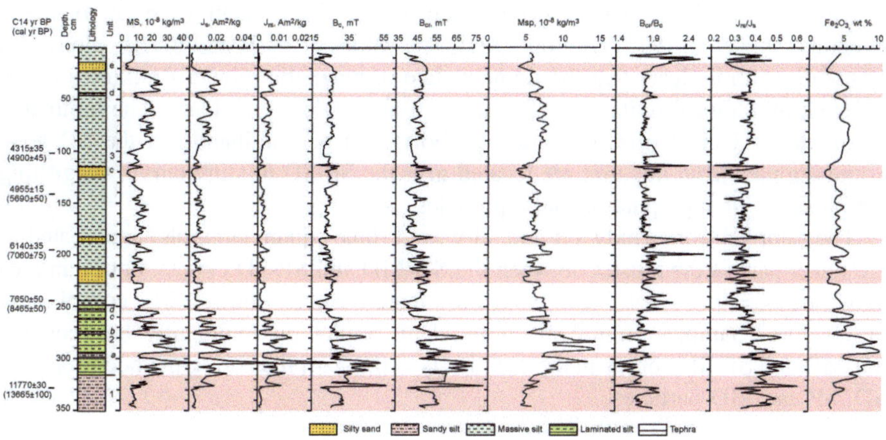

Fig. 11.3 Magnetic susceptibility (MS), saturation magnetization (Js), saturation remanence (Jrs), coercive force (Bc), remanence coercivity (Bcr), paramagnetic susceptibility (MSp), ratios Bcr/Bc, Jrs/Js, and Fe_2O_3 contents from core S-1 plotted versus depth. The lithological units described in the text are given to the right of the diagram

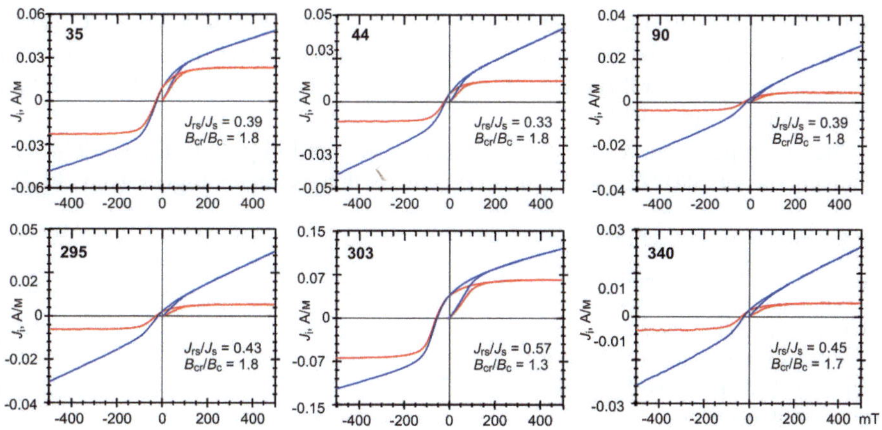

Fig. 11.4 Hysteresis loops of representative samples. The blue and red curves represent the data before and after the correction for the paramagnetic component, respectively. The sample numbers correspond to depths and are marked in bold font. Jrs/Js and Bcr/Bc ratios were calculated after correction for the paramagnetic component

Js = 0.0082 Am^2/kg, Jrs = 0.0030 Am^2/kg. Higher values of MS, Js and Jrs characterize the sediments from 20 to 100 cm.

Figure 11.4 shows hysteresis loops of representative samples. In all samples, the raw and dia/para-corrected loops are apparently different because of the paramagnetic or diamagnetic contributions. Higher diamagnetic/paramagnetic contributions are observed in silty sand (sample 340, 295) and sandy silt (sample 44).

Day-plot has been widely used to determine the domain state of magnetic minerals (Day et al. 1977; Dunlop 2002). The values of Jrs/Js and Bcr/Bc plotted on a Day diagram indicate that the domain state of magnetic minerals mixtures in the different lithological units from Tschuchye Lake varies (Fig. 11.5). The laminated sediments of Unit 2 (green rhombi symbols) plot predominantly in the SD area. Some samples from this unit are located near the SP-SD mixture curve, suggesting the presence of superparamagnetic particles.

The samples from sandy silt and silty sand (blue square symbols) are located in the lower part of the cluster between the SD-MD and SP-SD curves, but in an area above Jrs/Js = 0.2. These locations approximately correspond to a MD contribution of 20–40%. Samples from Unit 1 and Unit 3 plot predominantly between the SD-MD and SP-SD curves in the area Jrs/Js = 0.3–0.4, indicating the occurrence of SD, SP and MD particles.

High-Temperature Dependence of Magnetic Susceptibility

Selected samples of lacustrine sediments, tephra, and pebbles were subjected to susceptibility vs. temperature measurements in air. The temperature was increased continuously from room temperature to 700 °C and cycled back to 50 °C; some samples were measured in argon (Fig. 11.6).The behavior of the samples can be divided into four general types of thermomagnetic curves. All curves from the first heating run are irreversible.

The first type of curve shows a decrease in susceptibility up to the Curie points of magnetite at 580 °C, with a cooling curve that is much higher (Fig. 11.6k)

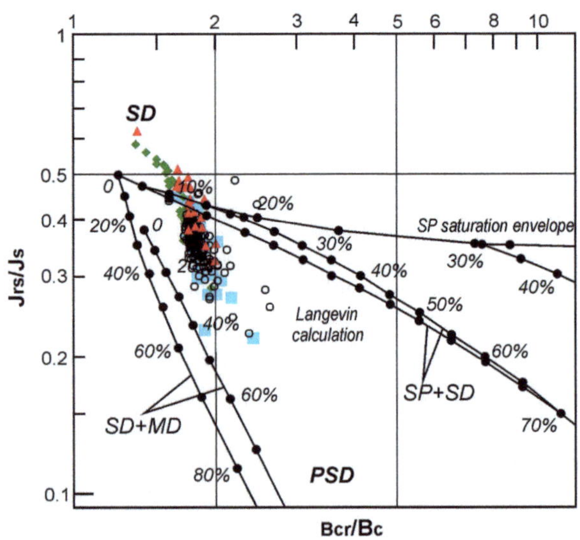

Fig. 11.5 The Day diagram for sediments of core S-1. SD, SP, PSD, and MD are single domain, superparamagnetic, pseudo-single domain, and multi domain particles, respectively. The SD–MD and SD–SP mixing lines with black dots refer to calculations from Dunlop (2002). The green rhombi, blue squares, red triangles, and open circles represent samples from laminated silt of Unit 2, and sandy silt and silty sand from Unit 1 and Unit 3, respectively

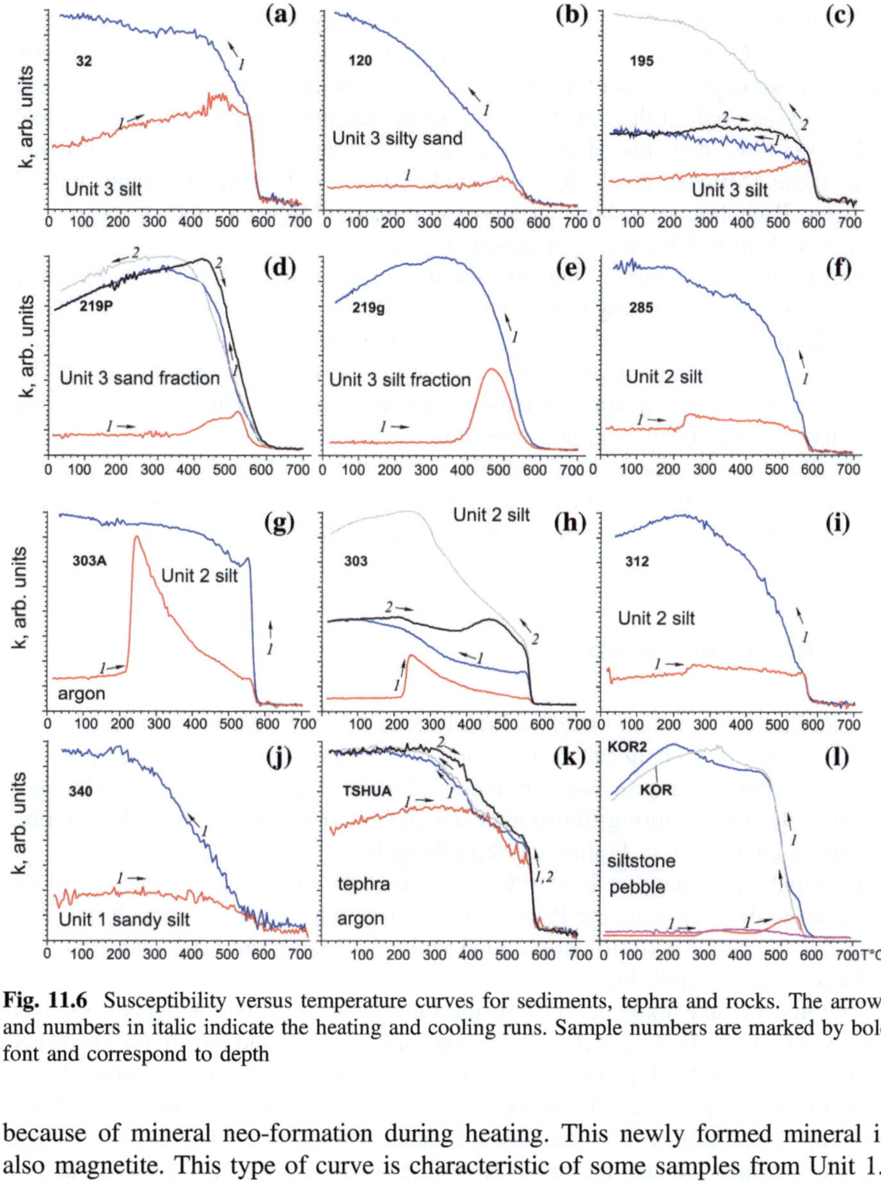

Fig. 11.6 Susceptibility versus temperature curves for sediments, tephra and rocks. The arrows and numbers in italic indicate the heating and cooling runs. Sample numbers are marked by bold font and correspond to depth

because of mineral neo-formation during heating. This newly formed mineral is also magnetite. This type of curve is characteristic of some samples from Unit 1.

The second type shows an increasing magnetic susceptibility at T ~220 °C, followed by a decrease during further heating to ~580 °C (Fig. 11.6f, h, i). Heating in argon also shows an increase in MS at T ~220 °C (Fig. 11.6g). This behavior supposedly is associated with the presence of lepidocrocite, which transforms into maghemite–hematite (Gehring et al. 1990; Gehring and Hofmeister 1994; Gendler et al. 2005; Hanesch et al. 2006). Upon cooling, increases in susceptibility are seen at the Curie temperature of magnetite. Such curves are typical for the laminated sediments of Unit 2.

The third type displays an increase in susceptibility at 380–420 °C and a sharp drop at 580 °C (Fig. 11.6a, b, c, d, e). Cooling curves all show a considerable increase in susceptibility starting at 580 °C. This type of curve reflects the influence of organic material on the transformation of Fe-bearing minerals, such as goethite, ferrihydrite and hematite (Hanesch et al. 2006) as well as biotite, vivianite, hypersthene (Minyuk et al. 2011). Sample 219 was divided into sand and silt fractions. Both fractions show an increase in susceptibility at 380 °C but with a sharper peak in MS for the silt fraction (Fig. 11.6d, e).

The forth type of curve characterizes the tephra (Fig. 11.6). The sample was heated in argon. The heating curve shows a small but distinct decrease near 547 °C and a sharp drop at ~590 °C, marking the presence of low-titanium magnetite and magnetite. The cooling curves display a strong increase in susceptibility at 425 °C, indicating the formation (transformation) of a new mineral. This mineral is stable. All curves of the second run are reversible.

Thermomagnetic curves from pebbles of siltstone and sandstone are variable. The second and third types of curves typify these samples (Fig. 11.6m). All samples finish with susceptibility hundreds of times higher than the initial values.

Inorganic Geochemistry

The geochemical characteristics of lacustrine sediments depend on many factors, including the: (1) chemical composition of the provenance; (2) physical and chemical weathering processes in the catchment area; (3) tectonic and eolian activity; (4) sorting during transport and sedimentation; and (5) post-depositional diagenetic changes (e.g. Fralick and Kronberg 1997).

Principal component analysis (PCA) was used to reduce the dimensions of the multivariate data set using the PAST software program (Hammer et al. 2001). This analysis was performed on a correlation matrix of major and trace elements, LOI and magnetic susceptibility.

The first PC axis explains 52% of the total variance. It is positively correlated with Al_2O_3, TiO_2, Fe_2O_3, MgO, Cr, Rb, Ba, MS, and LOI and negatively correlated with SiO_2, Zr, and Sr (Fig. 11.7). The second PC axis explains an additional 18% of the variability and is characterized by positive loadings of P_2O_5, MnO, MS, CaO, MS, Fe_2O_3, and LOI. It is negatively correlated with Al_2O_3, TiO_2, Cr, Ni, Rb, and Ba. These results indicate the presence of three main data groups. SiO_2, Sr, and Zr are clearly related to silty sand and sandy silt, where as Al_2O_3, TiO_2, Cr, Ni, and Ba correspond to silt. P_2O_5, MnO, MS, CaO, MS, Fe_2O_3, and LOI are associated with sediments of high authigenic processes and post-depositional diagenetic changes.

The elemental record from core S-1 can be divided into two groups or geochemical zones based on the variability of the inorganic compounds and LOI (Figs. 11.8 and 11.9). These zones coincide with the rock magnetic zones. Different environmental conditions are reflected in the variations in the sediment deposition. Sandy silt and silty sand (colored bars in Figs. 11.8 and 11.9) represent dynamic

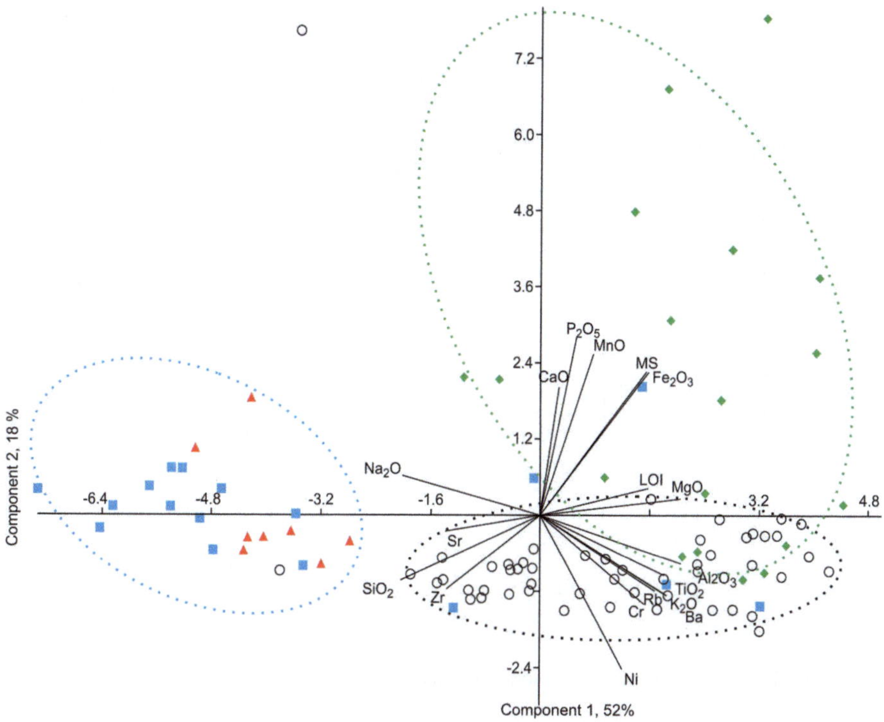

Fig. 11.7 Principal component analysis of the Tschuchye Lake sediments. Green rhombi, blue squares, red triangles, and open circles represent samples from laminated silt of Unit 2, sandy silt and silty sand from Unit 1 and Unit 3, respectively

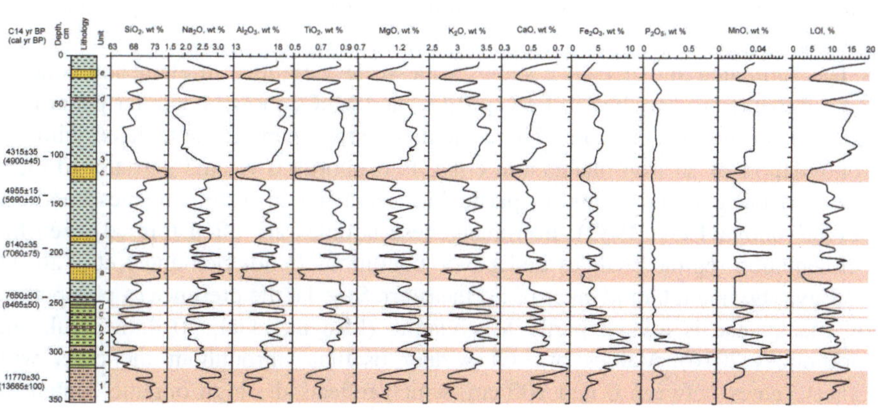

Fig. 11.8 Graphs of major elements plotted versus depth

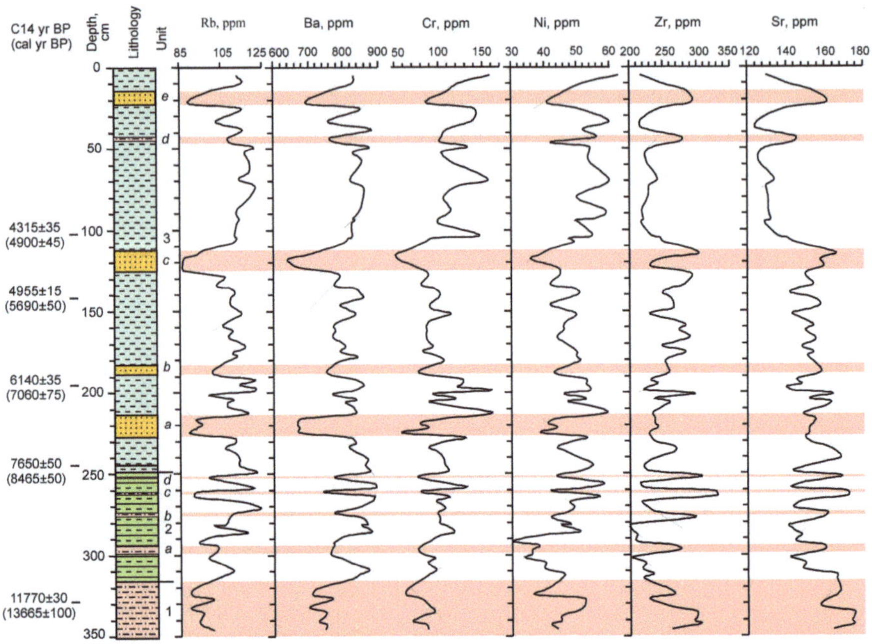

Fig. 11.9 Graphs of trace elements plotted versus depth

facies and reflect a high input to the lake basin of coarse detrital material. It was noted that coarse detrital sediments, such as sand, are enriched with SiO_2, Sr, and Zr and have low LOI. These sediments have a low TiO_2/Zr index, indicating dynamic sorting of the material (Yudovich and Ketris 2011). Petrographic analyses of thin sections of the sand indicate the presence of quartz, feldspar and siltstone-sandstone grains. In turn, siltstone-sandstone grains consist of quartz, feldspar and clay minerals.

The laminated silt of Unit 2 shows wide-scale variations in elemental content. The sediments are enriched in Fe_2O_3, P_2O_5, and MnO, showing similar patterns of variation throughout the core. These elements are apparently associated with the accumulation of an iron-manganese crust. Phosphorus is related partially to manganese in oceanic sediments, in particular, occurring in ferromanganese nodules (Sevast'yanova 1981, 1982). Iron-manganese nodules are often formed when the sedimentation regime changes under oxic conditions (Yudovich and Ketris 2011). In El'gygytgyn and Grand lakes of northeastern Siberia, the presence of phosphorus and manganese is related partly to vivianite (Minyuk et al. 2014; Minyuk and Borkhodoev 2016). Values of LOI widely oscillate throughout the core, with highest values between 0 and 100 cm, which reflects the high organic content in this part of the record.

Discussion

The very low values of the magnetic parameters that characterize the sediments of Tschuchye Lake reflect the low magnetic properties of the catchment material. The main input into the lake comes from Permian sedimentary rocks, located along the northern shore, and moraine deposits from the southern and eastern shores of the lake. The magnetic susceptibility of the pebbles (18 samples) collected near the lake has low values—3.3–10.8 (5.6) $\times 10^{-8}$ kg/m^3, which are comparable with the MS of sandy silt and silty sand (average of 8.9×10^{-8} kg/m^3). Regarding the Tschuchye Lake, low magnetic properties of the sediment mark an intense influx of coarse detrital material enriched with quartz and feldspars; thus mainly dia- and paramagnetic minerals which explain the low susceptibility, Js and Jrs values. The presence of sandy silt and silty sand layers is apparently associated with flood events.

The magnetic parameters of sediments correlate with the content of total iron. A positive correlation exists between MS, Js, Jrs and MSp with iron (Fig. 11.10). This is the first lake from the northeastern Siberia, where this pattern is revealed. In other lakes from this region (e.g., Lake El'gygytgyn, Chukotka (Minyuk et al. 2014), Lake Grand, northern Priokhotye (Minyuk and Borkhodoev 2016) no correlation of the iron content with the MS, Js, Jrs is observed. The main source of iron in the sediments of those lakes is in the paramagnetic minerals. The content of total iron is lower in coarse, weakly magnetic detrital material and higher in silt in the sediments of Tschuchye Lake, suggesting that iron influx to the basin is primarily due to clay particles and organic matter (Davison 1993). Silt enriched with Fe_2O_3, Al_2O_3, TiO_2, MgO, and increased LOI, which agrees with the geochemical data on the study of various granulometric fractions of the diluvium near Lake Grand,

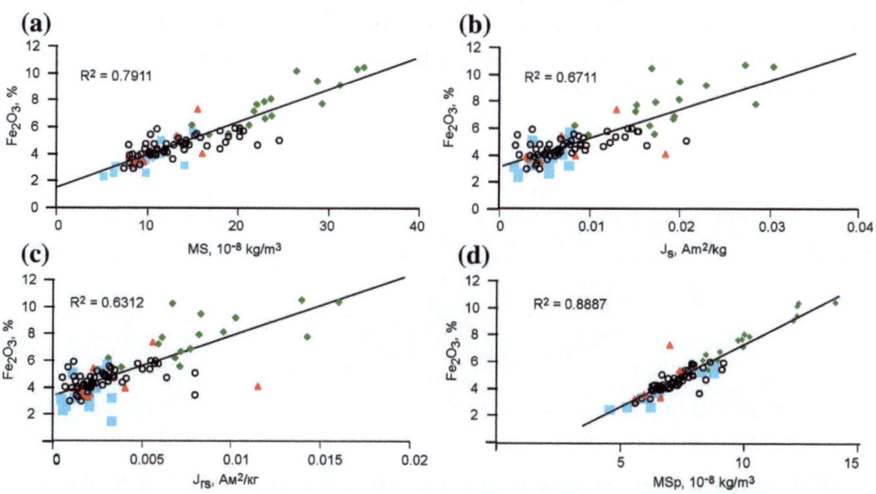

Fig. 11.10 Fe_2O_3–MS, Fe_2O_3–Jrs, Fe_2O_3-Js, Fe_2O_3-MSp diagrams for sediments

where silt and clay fractions also are enriched in these elements (Minyuk and Borkhodoev 2016).

The iron in the sediment is structure-bound in clay minerals and/or forms amorphous iron hydroxide that available for reduction to ferrous iron (Davison 1993). High positive correlations occur between Fe_2O_3 and MgO (r = 0.79), TiO_2 and MgO (r = 0.76) suggesting that part of iron is associated with the presence of chlorite ($Mg_{3.5}Fe_{1.5}Al_2Si_3O_{14}$) (Boyle 2002), or biotite ($(K_{0.9}Na_{0.01})(Mg_{1.5}$ Fe1.1) $(Al_{0.05}Ti_{0.2})$ $[Si_{2.9}Al_{1.1}O_{10}](OH)_2$ (e.g., Just and Kontny 2012).

The strong positive correlation between Fe_2O_3 and MS (r = 0.89) indicates that part of the iron is incorporated also into detrital or/and chemical (biogenic) ferromagnetic minerals.

Because of the high mineralogical transformations during heating, it is not possible to determine the composition of detrital magnetic minerals by the thermomagnetic method. On the other hand, the extremely low magnetic properties of sediments do not allow us to separate the magnetic extract for mineralogical study. Magnetite was determined in some samples from Unit 1, where there were no new magnetic minerals formed during heating.

Hysteresis data indicate the dominance in sediments of single-domain and superparamagnetic particles. Micrographs of magnetic extract from sediment of Unit 2 (depth 303 cm) show the needle-like morphology particles and rounded shape particles, that are interpreted as lepidocrocite and magnetite, respectively (Fig. 11.11) (Webster et al. 2012; Song et al. 2015). The size of the rounded particles is up to several tens of nanometers as reported for synthesized abiotic and biotic magnetite (e.g. Li et al. 2013).

Principal component analysis shows that MS, Fe_2O_3, MnO, and P_2O_5 form a separate cluster on the PCA diagram (Fig. 11.7). The positive correlations between Fe_2O_3 and P_2O_5 (r = 0.67), Fe_2O_3 and MnO (r = 0.63), P_2O_5 and MnO (r = 0.64) are supposed to be caused by the formation of autigenic Fe–Mn oxyhydroxides or/and vivianite ($Fe_3(PO_4)_2$ $8H_2O$). Visible vivianite as a blue spots was found in the sediments of Unit 2 and Unit 3. Sediments from Unit 2, enriched with Fe_2O_3, MnO, P_2O_5, include also lepidocrocite.

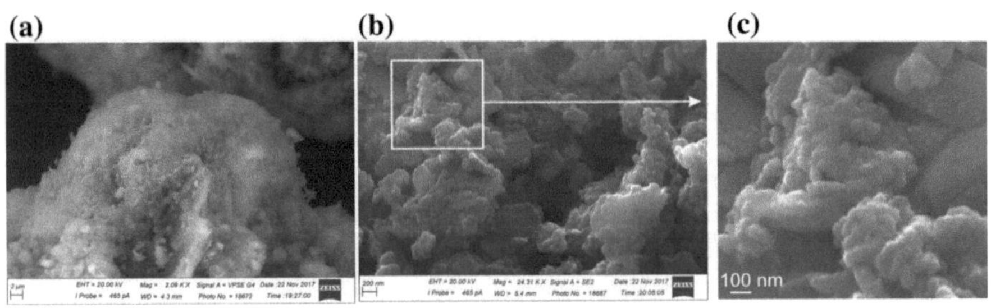

Fig. 11.11 Micrographs of magnetic extract from the sediments of Unit 2: **a** lepidocrocite?, **b**, **c** magnetite?

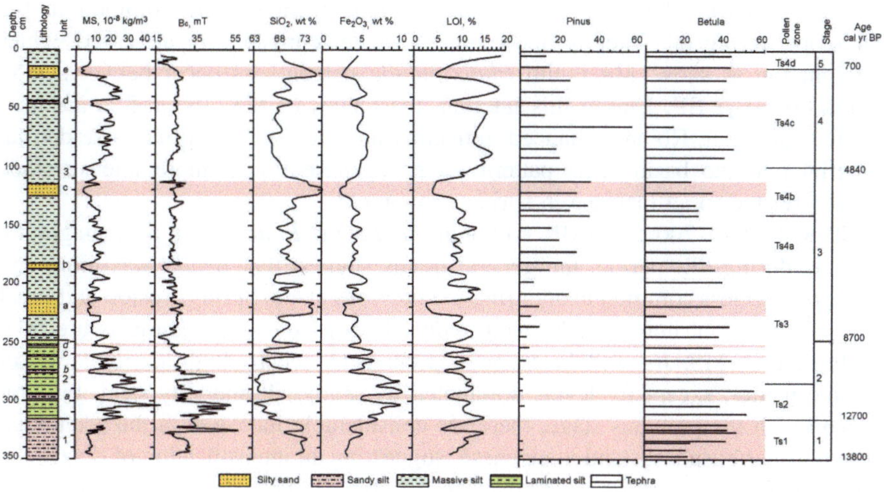

Fig. 11.12 Selected rock magnetic, geochemical, and palynological proxies plotted versus depth

The magnetic and geochemical characteristics, lithological units, and palynological zones indicate that the palaeolimnological and palaeoenvironmental history of Tschuchye Lake can be divided into five stages (Fig. 11.12).

Stage 1, 13,800–12,700 cal. yr BP. This stage marks initial formation of the lake following climate warming, glacial degradation and establishment of open *Larix* forest-*Betula* shrub tundra in the valley. This interval is also characterized by dynamic sedimentation within the Tschuchye basin, which is associated with an input of weak magnetic (low MS, Js, Jrs) detrital material that is enriched in quartz, feldspars, and clay minerals. High values of Bc and Bcr reflect the accumulation of Fe-oxyhydroxides.

Stage 2, 12,700–8780 cal yr BP. During this stage, the core lithology changes to laminated silt, which suggests seasonal variations in sedimentation. The accumulation of coarse material decreases from the previous stage, reflecting a decline in erosion within the catchment area. During this interval, the climate was likely as warm as or warmer than in present times, but relatively dry. *Larix* forest was common with abundant *Betula* and *Salix. Alnus* became a major landscape component ~11,200 cal yr BP. The high contents of iron, phosphorus and manganese indicate the occurrence of intensive digenetic and autigenic processes and the possible accumulation of a Fe–Mn crusts and of lepidocrocite. These results suggest that oxic conditions characterized the lake. The lowest contents of Zr and SiO_2 in laminated silt suggest the decreasing of detrital input to the lake. High values of MS, Jrs, Js, Bcr, Bc, and Jrs/Js indicate the presence of predominantly single domain particles, possible, of magnetite of chemical or bacterial origin. In most cases, magnetite-producing magnetotactic bacteria were found in and above the oxic-anoxic transitional zone (Bazylinski, 1996).

Stage 3, 8700–4840 cal yr BP. *Pinus pumila* became more common in the landscape during this interval, indicating of wetter climate with increased winter precipitation as snow. Maximum *Pinus pumila* percentages occur between 5655 and 4840 cal yr BP. During this interval, the values of MS, Jrs, and Js are low, whereas SiO_2, Sr, Rb are enhanced, reflecting an increased input of coarse detrital material into the basin. This pattern seems consistent with more intense slope erosion related to an increase in spring snow-melt.

Stage 4, 4840–700 cal yr BP. Although *Betula* and *Pinus* pollen percentages are variable during this period, the general trends suggest that *Betula* was more and *Pinus pumila* less common than during the previous stage. Spring erosion and input of coarse material into the lake were reduced. Consequently, sediments are relatively high in MS, Jrs, and Js. Stage 4 is characterized by high accumulation of organic matter, except for levels dominated by sandy silt and silty sand. In comparison with the previous stage, there are considerable increases in the accumulation of iron, phosphorus and manganese, suggesting an intensification of diagenetic and authigenic processes but less than during stage 2. The accumulation of Cr and Ni is also increased. Possibly these elements were absorbed by organic matter indicating an increase in biological production in the basin (e.g., Sharma et al. 2004).

Stage 5, 700–0 yr BP. This stage is characterized by the establishment of the modern vegetation. It is also characterized by the highest accumulation of organic matter but with a low input of coarse detrital material (low content of SiO_2). The values of MS, Jrs, Js are decreasing, presumably due to dilution by organic matter, while the contents of Cr and Ni are increased.

Conclusion

A multi-proxy investigation of a sediment core from Tschuchye Lake provides evidence of abrupt changes in limnological conditions during the Holocene.

Three distinct lithological units are identified, marking changes in sedimentation.

Coarse detrital material is characterized by (1) low content of TiO_2, Al_2O_3, MgO, K_2O, Rb, and Fe_2O_3, and low LOI values; (2) high content of SiO_2, Zr and Sr; and (3) low values of magnetic susceptibility, saturation magnetization, and saturation remanence.

Laminated silt shows the highest content of Fe_2O_3, P_2O_5, MnO, MS, Jrs and Js, which implies the occurrence of autigenic or diagenetic processes. The magnetic mineral grain size fraction consists predominantly of single domain, super-paramagnetic and rarely of multi domain and possibly are of bacterial origin.

Five stages of paleolimnological and paleoenvironmental history of Tschuchye Lake are defined based on the above changes.

Acknowledgements These studies were funded by the Far East Branch Russian Academy of Sciences (15-I-2-067).

References

Anderson, P.M., Lozhkin, A.V. & Brubaker, L.B. Implications of a 24,000-yr palynological record for a Younger Dryas cooling and for boreal forest development in Northeastern Siberia. Quaternary Research. 2002. Vol. 57. P. 325–333.

Andreev, A.A., Klimanov, V.A. & Sulerzhitsky, L.D. Younger Dryas pollen records from central and southern Yakutia. Quaternary International. 1997. Vol. 41/42. P. 111–117.

Bazylinski, D.A. Controlled biomineralization of minerals by magnetotactic bacteria. Chemical Geology. 1996. Vol. 132. P. 191–198.

Blaauw, M. and Christen, J.A. Bacon Manual. 2011. Vol. 2.2. http://www.chron.qub.ac.uk/blaauw/bacon.html.

Borkhodoev, V. Ya. Accuracy of the fundamental parameter method for x-ray fluorescence analysis of rocks. X-Ray Spectrom. 2002. Vol. 31. P. 209–218.

Boyle, J. F. Inorganic geochemical methods in paleolimnology, in: Tracking Environmental Change Using Lake Sediments Volume 2: Physical and Geochemical Methods, edited by: Last, W. M. and Smol, J. P., Springer, Berlin, 2002. P. 83–141.

Burov, B.V., Nurgaliev, D.K., and Yasonov, P.G., Paleomagnitnyi analiz (Paleomagnetic Analysis), Kazan: KGU, 1986 (in Russian).

Davison, W. Iron and manganese in lakes. Earth Sci. Rev. 1993. Vol. 34. P. 119–163.

Day, R., Fuller, M., and Schmidt, V.A. Hysteresis Properties of Titanomagnetites: Grain_Size and Compositional Dependence. Phys. Earth Planet. Inter. 1977. Vol. 13. P. 260–267.

Dunlop, D.J. Theory and application of the Day plot (*M*rs/*M*s versus *H*cr/*H*c) 1. Theoretical curves and tests using titanomagnetite data. J. Geophys. Res. 2002. Vol. 107. P.56–60.

Evans, M.E. and Heller, F. Environmental Magnetism. Principles and Applications of Enviromagnetics, edited by Dmowska R., Holton J.R., and Rossby H.T. Academic Press. 2003. 299 p.

Fabian, K., Shcherbakov, V.P., and McEnroe, S.A. Measuring the Curie temperature. Geochem. Geophys. Geosyst. 2013. Vol. 14. P. 947–961.

Fralick, P.W., Kronberg, B.I. Geochemical discrimination of elastic sedimentary rock sources. Sedimentary Geology. 1997. Vol. 113. P. 111–124.

Gehring, A.U. and Hofmeister, A.M. The Transformation of Lepidocrocite during Heating: a Magnetic and Spectroscopic Study. Clays Clay Miner. 1994. Vol. 42. P. 409–415.

Gehring, A.U., Karthein, R., and Reller, A. Activated State in the Lepidocrocite Structure during Thermal Treatment. Naturwissenschaften. 1990. Vol. 77. P. 177–179.

Gendler, T.S., Shcherbakov, V.P., Dekkers, M.J., Gapeev, A.K., Gribov, S.K., and McClelland, E. The Lepidocrocite-Maghemite-Haematite Reaction Chain: I. Acquisition of Chemical Remanent Magnetization by Maghemite, Its Magnetic Properties and Thermal Stability. Geophys. J. Int. 2005. Vol. 160. P. 815–832.

Hammer, Ø., Harper, D. A. T., and Ryan, P. D. PAST: Paleontological statistics software package for education and data analysis Palaeontol. Electron. 2001. Vol. 4. 9 pp.

Hanesch, M., Stanjek, H., and Petersen, N. Thermomagnetic Measurements of Soil Iron Minerals: the Role of Organic Carbon. Geophys. J. Int. 2006. Vol. 165. P. 53–61.

Heiri, O., Lotter, A. F., and Lemcke, G. Loss on ignition as a method for estimating organic and carbonate content in sediments: reproducibility and comparability of results. J. Paleolimnol. 2001. Vol. 25. P. 101–110.

Just, J. and Kontny, A. Thermally induced alterations of minerals during measurements of the temperature dependence of magnetic susceptibility: a case study from the hydrothermally altered Soultz-sous-Forêts granite, France. Int. J. Earth Sci. 2012. Vol. 101. P. 819–839.

Karavaev, M.N. Synopsisof the Flora of Yakutia. USSRAcademy of Sciences, Moscow-Leningrad, 1958 (In Russian).

King, J and Peck, J. Use of palaeomagnetism in studies of lake sediments. In Tracking Environmental Change Using Lake Sediments Volume 1: Basin Analysis, Coring, and Chronological Techniques, edited by: Last, W. M. and Smol, J. P., Springer, Berlin, 2002. P. 371–389.

Korzun, Ju. A Palynological analysis of glacial lake sediments from the upper basin of the Indigirka River. Bulletin of the North-East Scientific Center, RussianAcademy of Sciences Far East Branch. 2017. Vol. 1. P. 24–31 (In Russian).

Lattard, D., Engelmann, R., Kontny A. and Sauerzapf, U. 2006. Curie temperatures of synthetic titanomagnetites in the Fe-Ti-O system. Reassessment of some methodological and crystal chemical effects. J. Geophys. Res. 2006. Vol. 111, B12S28.

Li, J., Benzerara, K., Bernard, S., Beyssac, O. The link between biomineralization and fossilization of bacteria: Insights from field and experimental studies. Chemical Geology. 2013. Vol. 359. P. 49–69.

Minyuk, P. S. and Borkhodoev, V. Ya. Geochemistry of Sediments from Lake Grand, Northeast Russia. Geochemistry International. 2016. Vol. 54 (9). P. 807–816.

Minyuk, P. S., Borkhodoev, V. Y., and Wennrich, V. Inorganic geochemistry data from Lake El'gygytgyn sediments: marine isotope stages 6-11. Climate of the Past. 2014. Vol. 10. P. 467–485.

Minyuk, P.S., Subbotnikova, T.V., and Plyashkevich, A.A. Measurements of Thermal Magnetic Susceptibility of Hematite and Goethite. Izvestiya, Physics of the Solid Earth. 2011. Vol. 47 (9). P. 762–774.

PALE, 1994. Research Protocols for PALE: Paleoclimates of Arctic Lakes and Estuaries. 53 pp. PAGES Workshop Report Series, Bern, Switzerland.

Petrovský, E. and Kapička, A. On determination of the Curie point from thermomagnetic curves. J. Geophys. Res. 2006. Vol. 111. B12S27, https://doi.org/10.1029/2006jb004507.

Ponomareva, V.V., Kyle, P.R., Melekestsev, I.V., Rinkleff, P.G., Dirksen, O.V., Sulerzhitsky, L. D., Zaretskaia, N.E., and Rourke, R., The 7600 (14C) Year BP Kurile Lake Caldera_Forming Eruption, Kamchatka, Russia: Stratigraphy and Field Relationships. J. Volcanol. Geotherm. Res. 2004. Vol. 136. P. 199–222.

Reimer, P.J., Bard, E., Bayliss, A., Beck, J.W., Blackwell, P.G., Bronch Ramsey, C., Buck, C.E., Cheng, H., Edwards, R.I., Friedrich, M., Grootes, P.M., Guilderson, T.P., Halidasson, H., Hajdas, I., Hatte, C., Heaton, T.J., Hoffman, D.I., Hogg, A.G., Hughen, K.A., Kaiser, K.F., Kromer, B., Manning, S.W., Niu, M., Reimer, R.W., Richards, D.A., Scott, E.M., Southon, J. R., Staff, R.A., Turney, C.A. & van der Plicht, J. IntCal13 and marine13 radiocarbon age calibration curves 0-50,000 years cal BP. Radiocarbon. 2013. Vol. 55. P. 1869–1887.

Sevast'tyanova, E. S. Behavior of phosphorus in oceanic ferromanganese micro and macronodules. Lithol. Polezn. Iskop. 1981. Vol. 6. P. 96–101 (in Russian).

Sevast'tyanova, E. S. Correlation of phosphorus and tetravalent manganese contents in pelagic oxidized clays. Okeanologiya. 1982. Vol. 32(6). P. 970–974 (in Russian).

Sharma, S., Joachimski, M., Sharma, M., Tobschall, H. J., Singh, I. B., Sharma, C., Chauhan, M. S., and Morgenroth, G. Late glacial and Holocene environmental changes in Ganga plain, Northern India. Quaternary Sci. Rev. 2004. Vol. 23. P. 145–159.

Song, J., Jia, S.-Y., Yu,B., Wu,S.-H., Han,X. Formation of iron (hydr)oxides during the abiotic oxidation of Fe(II) in the presence of arsenate. Journal of Hazardous Materials. 2015. Vol. 294. P. 70–79.

Webster, N.A.S., Loan, M.J., Madsen, I.C., Knott, R.B., Brodie, G. M., Kimpton, J. A. An in situ synchrotron X-ray diffraction investigation of lepidocrocite and ferrihydrite-seeded $Al(OH)_3$ crystallisation from supersaturated sodium aluminate liquor. Journal of Crystal Growth. 2012. Vol. 340. P. 112–117.

Wright, H.E., Jr., Mann, D.H., and Glaser, P.H. Piston Corers for Peat and Lake Sediments. Ecology. 1984. Vol. 65. P. 657–659.

Yudovich, Ya.E., Ketris, M.P. Geochemical indicators of Lithogenesis. (Ed. Tkachev Yu. A.). Syktytvkar, 2011. 742 p (in Russian).

Chapter 12
Comparison of the Porosity Determination on the Whole Core and Petrophysical Samples

A. V. Fattakhov, V. E. Kosarev, D. L. Melnikova, V. D. Skirda and A. V. Starovoytov

Abstract The paper presents the result of core porosity measurements obtained by traditional (nuclear magnetic resonance method and liquid saturation method using the standard petrophysical samples) and alternative method—nuclear magnetic resonance using a whole core. The goal of the research is to prove the correctness of the porosity values obtained by the nuclear magnetic resonance method using whole core. The comparative analysis of the porosity values was carried out. The analysis shows a high degree of correlation between data obtained by different methods. This allows to state that it is possible to use nuclear magnetic resonance method for the porosity evaluation of a whole core. A great advantage of method is its non-destructive ability and high speed of measurement. At the end of the paper, conclusions were drawn about the applicability and expediency of using the nuclear magnetic resonance method using a whole core in core analysis programs and obtaining filtration-volumetric characteristics of rocks.

Keywords Nuclear magnetic resonance · Dolomites · Whole core
Porosity

Introduction

The method of nuclear magnetic resonance (NMR) has proven itself in geophysical and petrophysical studies (Dzhafarov et al. 2002; Dunn et al. 2002). The porosity obtained by this method is considered reliable in well logging operations and core analysis (Coates et al. 1999). The most widespread NMR effect has in the porosity determination of standard petrophysical samples (Dunn et al. 2002). Comparisons

A. V. Fattakhov (✉) · V. E. Kosarev · A. V. Starovoytov
Institute of Geology and Petroleum Technologies,
Kazan Federal University, Kazan, Russia
e-mail: AVFattahov@kpfu.ru

D. L. Melnikova · V. D. Skirda
Institute of Physics, Kazan Federal University, Kazan, Russia

D. Nurgaliev et al. (eds.), *Recent Advances in Rock Magnetism, Environmental Magnetism and Paleomagnetism*, Springer Geophysics,
https://doi.org/10.1007/978-3-319-90437-5_12

of the porosity values obtained by the NMR method and the traditional liquid saturation method (hydrostatic weighing) using the petrophysical samples have repeatedly demonstrated a high degree of correlation of the results obtained (Dzhafarov et al. 2002; Dunn et al. 2002; Timur 1972). However, the filtration-volumetric characteristics of the petrophysical samples do not always qualitatively and reliably characterize the reservoir's filtration-volumetric characteristics (Fattakhov et al. 2016). The solution to such problems is to work with a whole core (full-size core, full-diameter core) (Musin 2012). At present, the evaluation of petrophysical and lithological characteristics of the whole core is presented in core analysis programs in a very limited scope. At the same time, the definition of the reservoir porosity and permeability of a material such whole core is a non-trivial task, for the solution of which a special approach both in hardware and in the methodic of work is required. The main purpose of the work was to compare the porosity values of a whole core and standard petrophysical samples cut from this core.

Equipment

One of the main reasons for the weak using of the whole core to assess the porosity of the reservoir is related to the hardware limitations. There are very few instruments available to conduct this type of research in the all world. If it is talked about the NMR method, at present there are only two special installations that allow to determine the values of the porosity of a whole core correctly by this method. The first such device is «GeoSpec» from Oxford Instruments (https://www.oxford-instruments.com/OxfordInstruments/media/industrial-analysis/magnetic-resonance-pdfs). The second solution is the installation «NMR-Core», created by the employees of the Kazan Federal University in conjunction with the specialists of «TNG-Group» Ltd (Nurgaliev et al. 2012).

The Oxford Instruments installation can work only in laboratories, it allows the use of a core with a diameter of up to 6 inches (15.24 cm), but the length of the sample is limited to 10 cm, and the resulting petrophysical characteristics are generalized for the entire volume of the sample (https://www.oxford-instruments.com/OxfordInstruments/media/industrial-analysis/magnetic-resonance-pdfs). If a detailed analysis of the porosity of the whole core is required (measurement step less than 10 cm), the Geospec installation cannot be optimal, since it is necessary to make additional breakdown of the core material. This approach does not always justify itself in relation to the large time expenditure and the forced destruction of full-diameter samples.

An essential limitation of the NMR method is the mandatory filling of the porous space with fluid. Otherwise, the sensors of the device do not allow to determine the hydrogen content in the rock and, accordingly, to obtain the correct value of the core porosity. Full saturation of the whole core is a nontrivial problem that can be solved only under certain conditions (Fattakhov et al. 2017). The main problem for carrying out the saturation procedure is preliminary extraction. As practice shows, extraction of a whole core does not give the desired result; in addition, after this procedure, there is a possibility of changing the wettability of the core, which in consequence can lead to incorrect filtration-volumetric characteristics (Mikhailov and Polischuk 2014).

To solve such problem, it is desirable to carry out measurements directly near the well immediately after extracting the whole core from the core recovery tube. At this moment, the core has saturation, close to its natural. However, the installation «GeoSpec» is a laboratory equipment not adapted to work in the field.

In 2012, the Kazan Federal University introduced a new facility designed to determine the filtration-volumetric characteristics of a whole core by nuclear magnetic resonance directly on the well after the core is extracted from the core recovery tube. Developed equipment has the following characteristics:

- the maximum diameter of the core sample can be up to 116 mm;
- the length of the sample (or the length of the sample set) is up to 1000 mm;
- the accuracy of the positioning of the core relative to the measuring system is 1 mm;
- the spatial resolution along the core axis is 10 mm;
- the measurement time of 1 m of the core varies depending on the mode of investigation and is at least 20 min (Nurgaliev et al. 2012).

The described features and characteristics of the «NMR-Core» device allow to get rid of the above-mentioned problems when performing NMR measurements. First, the mobility of the installation allows you to work with the core directly on the well before losing of its fluid saturation. Secondly, the device, thanks to a unique system of magnets, ensures the homogeneity of the field in the volume under investigation, which makes it possible to obtain a porosity profile along the core axis without generalizing this characteristic for the entire volume of the sample (Nurgaliev et al. 2012).

Methods and Object of Research

The most commonly used method for determining the porosity from the core material today is liquid saturation method (hydrostatic weighing) (USSR State Committee for Standards and Product 1985). This method has proved to be the most accurate for many years of use by petrophysical laboratories.

The liquid saturation method is the following sequence of actions:

(1) drying and extraction until the saturable fluid is completely removed;
(2) weighing of the dry sample (m_1);
(3) saturation of the sample by a reservoir fluid until full saturation;
(4) weighing of saturated sample in air (m_3) and in saturating liquid (m_2).

Further, void volume (K_p) is determined according to the formula:

$$K_p = \frac{m_3 - m_1}{m_3 - m_2} * 100 \tag{12.1}$$

Another popular method for determining porosity in core analysis programs is nuclear magnetic resonance using the petrophysical samples. The principle of the method is based on recording the relaxation curve, which describes the reduction in the magnetization of the protons in the fluid. The number of hydrogen nuclei involved in the resonance characterizes the initial amplitude of such curve. The number of protons in its turn is proportional to the volumetric hydrogen content (Dzhafarov et al. 2002). After calibration of the equipment, the resulting hydrogen content is converted into a porosity factor (K_p). The main advantage of the NMR method is the independence of the resulting porosity from the mineral skeleton (Coates et al. 1999). Value of the total porosity can be obtained regardless of the geological section type as a result of the measurements.

However, both methods described above are carried out on standard petrophysical samples. Such samples by the time of beginning the investigation have already lost their "natural" saturation. They require drying and extraction, followed by a full saturation, preferably a reservoir fluid model. During the preparatory and experimental work, the samples can partially change their petrophysical parameters, which negatively affects on the determined filtration-volumetric characteristics (Kovalev et al. 2017). An additional disadvantage of working with petrophysical samples is the large process duration of their preparation (API RP 40 1998). Working with a whole core directly on the well is devoid of these shortcomings.

When carrying out whole core measurements with the «NMR-Core» device, there is no universal measurement mode. It must be selected depending on the type of rock, the type of fracture and the composition of the fluid filling the rocks. The main indicators that needed to be selected are:

(1) The distance between radio frequency pulses is the time interval between the "bursts" of pulses.
(2) Number of measurements of the noise path and the useful signal at each measurement point.
(3) The duration of the pulse sequence is the time during which the precession of the nuclei is recorded.
(4) The rest time is the time interval between each measurement.

In order for constructing the function between the relaxation time and the porosity of the studied core, a calibration sample with a known hydrogen content and porosity is placed at the beginning of each measured interval. Through this approach, the construction of the function takes place in real time.

The object of the study was the core from the well «Universitetskaya-1». The well is educational and is located on the territory of the Institute of Geology and Petroleum Technologies of Kazan Federal University. The drilling of this well was completed in early 2015. The total depth of the well is 300 m. During the drilling from a hundred meters deep and down to the bottom, there was a constant core sampling for educational and scientific purposes. The diameter of the core was 92 mm to a depth of 190 m and 67 mm from a depth of 190 m to the bottom, the real thickness of the core is about 187 m. Lithologically, the section is composed of anhydrite, gypsum and dolomite. Stratigraphically the rocks are confined to the Cisuralian section of the Permian system and the upper section of the Carboniferous system.

Results and Discussion

The choice of the object of investigation is due to the possibility of carrying out on it the studies of porosity by all the methods described above: NMR method and liquid saturation method on standard petrophysical samples, NMR method on a whole core. During the drilling, on the core material from the well «Universitetskaya-1» the porosity was determined by the NMR method on a whole core in 1 cm steps along the core axis. Based on the measurement results, the curve of porosity values changing was obtained for the entire sampling interval (200 m). By the core analysis program, the standard petrophysical samples were cut and the porosity was determined by the liquid saturation method. This operation was carried out in the core library of Kazan Federal University (KFU). Points for sampling were selected by «NMR-Core» readings. Then the porosity of saturated petrophysical samples was determined by NMR method. The measurements were carried out on the equipment of Center of Shared Facilities for Physical and Chemical Research of Substances and Materials in KFU spectrometer "AVANCE 400 III TM", optimized for a solid, self-diffusion and microtomography.

The porosity of 30 petrophysical samples was determined by NMR method. Figure 12.1 shows a cross plot of porosity values of petrophysical samples by NMR method and porosity values of the whole core by NMR method. When comparing the values obtained can be noticed a high degree of correlation of porosity values.

With the same samples, we measured the porosity by the liquid saturation method. Figure 12.2 shows the correlation between the NMR porosity readings on a whole core and the liquid saturation method on petrophysical samples.

When performing the correlation analysis, a high degree of correlation was found. Approximation veracity is 0.85–0.88 (Figs. 12.1 and 12.2). This indicates a high degree of convergence of the porosity values, determined by different methods

The main reason why there is no complete coincidence of the determined porosity values is the differences in the samples with which the studies were carried out. The "NMR-Core" installation does not work with standard petrophysical samples, but with a whole core. Several factors affect the work with this kind of material. The first factor is the lack of consolidation of the whole core. The gaps between the fragments extracted from the inner core tube influence the determined filtration-volumetric characteristics. The second factor is the penetration of the drilling mud into the surface part of the whole core. It distorts the real petrophysical characteristics of the rock samples. The third and, in our opinion, the main factor is

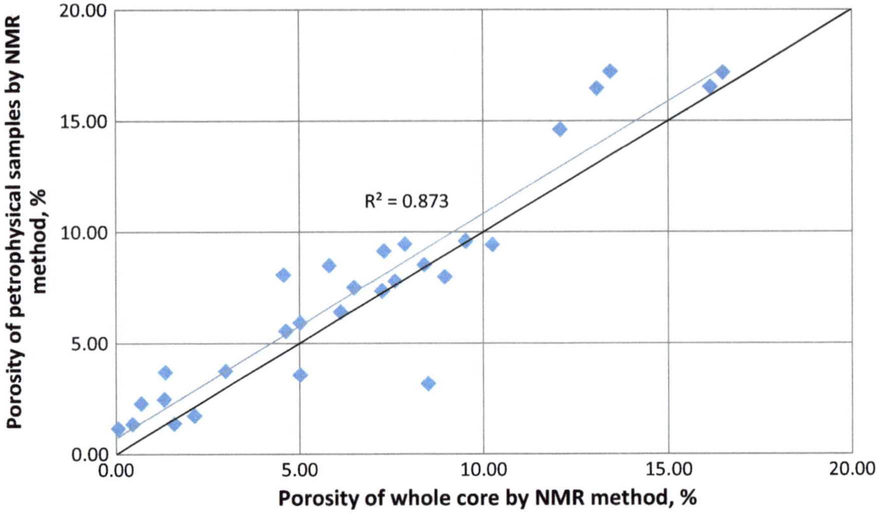

Fig. 12.1 Cross plotter of whole core porosity by NMR method and porosity indications of petrophysical samples by NMR method

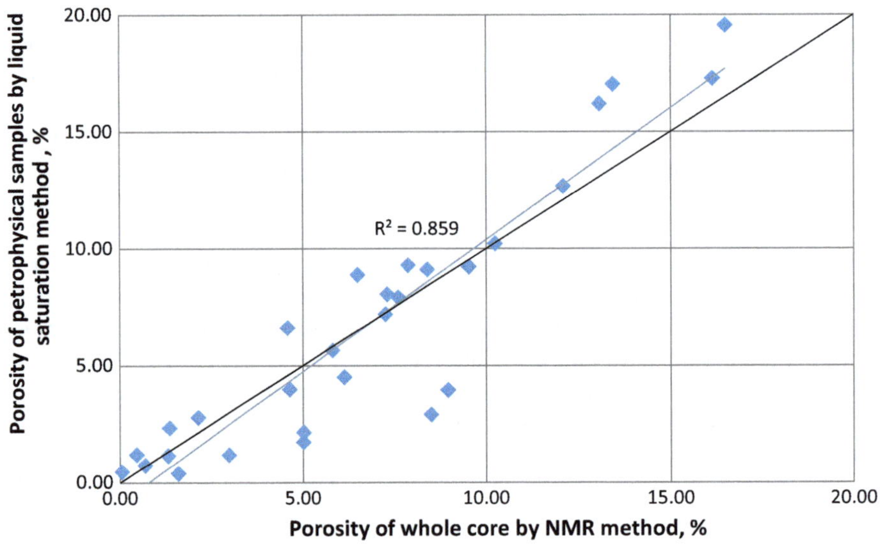

Fig. 12.2 Cross plotter of whole core porosity by NMR method and porosity readings of petrophysical samples by liquid saturation method

the size of the samples. Standard petrophysical samples have the diameter of 3 cm and are cut from the central part of the whole core. The measurements average the readings of porosity by three centimeters, but only the axial part of the full-length core is evaluated. Working with a whole core involves analyzing the entire extracted rock volume. This approach allows to state that the result obtained on whole core correctly characterizes the reservoir properties than the properties obtained on petrophysical samples.

For greater clarity, based on the results of the measurements, a log plot was created (Fig. 12.3). Here is 3 curves on log plot: a curve obtained from the results of

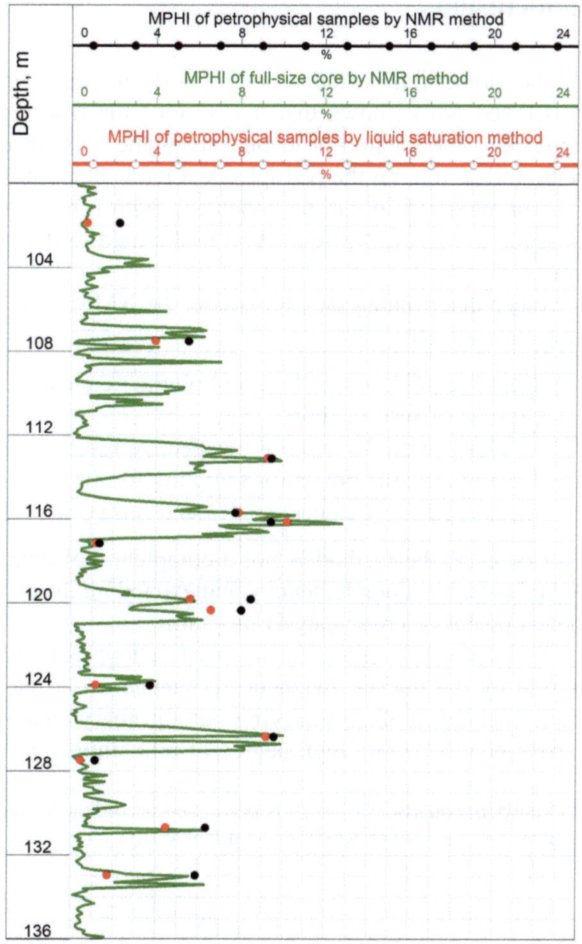

Fig. 12.3 Fragment of log plot with all methods of porosity determination

the porosity measurements on whole core by the NMR method (green curve), as well as point clouds obtained from porosity measurements on petrophysical samples by NMR (black points) and liquid saturation method (red points). This log plot also demonstrates good reproducibility of the results obtained on a whole core and on petrophysical samples.

Conclusion

At the moment, the whole core is poorly involved in core analysis programs. The performed work shows that a new tool for working with full-diameter samples appeared—the method of nuclear magnetic resonance, which shows a good result both in determining the petrophysical parameters of the rock and in the part of the measuring speed. It can be concluded with a high degree of confidence that it is possible to use the NMR method in the field to determine the porosity of a whole core.

When determining the porosity by the NMR method directly at the well, several tasks are solved:

(1) Rapid acquisition of filtration-volumetric characteristics of rocks already in the process of well drilling.
(2) Analysis and allocation of perspective reservoir for further development work.
(3) Determining the points for sampling for the purpose of further detailed analysis of petrophysical characteristics and reservoir properties.

However, the method is not a complete replacement of standard petrophysical techniques, it is complementary and expanding the capabilities of laboratories in the development of core analysis programs.

The studies described in the article allow us to state that the porosity determination by the nuclear magnetic resonance using the whole core is a reliable and rapid method. It allows carrying out a primary analysis of the filtration-volumetric characteristics with a high degree of reliability even at the drilling stage.

Acknowledgements The work is performed according to the Russian Government Program of Competitive Growth of Kazan Federal University.

References

API RP 40, Recommended Practices for Core Analysis, second edition. 1998. Washington, DC: API

Coates, G.R., Xiao, L., Prammer, M.G. NMR Logging Principles and Applications. Houston: Hullibarton Energy Services, 1999. 335 p.

Dunn, K.J., D.J. Bergman, and G.A. LaTorraca. Nuclear Magnetic Resonance Petrophysical and Logging Applications. New York: Pergamon, 2002. 312 p.

Dzhafarov, I.S., Syngaevskii, P.E., Khafizov, S.F. Application of the method of nuclear magnetic resonance to determine the structure and distribution of bed fluids. Moscow, 2002. 439 p. (in Russian).

Fattakhov, A.V., Kosarev, V.E., Shakirzyanov, R.A. Porosity evaluation of the "archival" full-size core using nuclear magnetic resonance. Conference Paper. Horizontal Wells 2017: Challenges and Opportunities; Kazan; Russian Federation; 15 - 19 May 2017. (in Russian). https://doi.org/10.3997/2214-4609.201700472

Fattakhov, A.V., Kosarev, V.E., Skirda, V.D., Doroginitskii, M.M. Study of drying process in full-size core by nuclear magnetic resonance method. Neftyanoe khozyaystvo - Oil Industry, 2016, Issue 3, 61–63 pp. (in Russian).

Kovalev, K.M., Grishin, P.A., Gabsiya, B.K., Kurochkin, A.D., Kolesnikov, M.V., Levchenko, A. S. Carbonate core: features, complexities, research prospects. 2017 SPE Russian Petroleum Technology Conference, 2017, SPE-187872-RU. (in Russian).

Mikhailov N.N., Polischuk V.I. Study of the statistical regularities of manifestation of scale effects in implex carbonate reservoirs. Open scientific journal Georesursi. Geoenergetica, geopolitica - Georesources, geoenergetics, geopolitics, 2014, №1, Issue 9. http://oilgasjournal.ru/vol_9/mikhailov.pdf (in Russian).

Musin, K.M., Abdullin, T.R. Integrated full-size core analysis. Neftyanoe khozyaystvo - Oil Industry, 2012, Issue 9, 52–54 pp. (in Russian).

Nurgaliev, D.K., Kosarev, V.E., Murzakaev, V.M, Tagirov, M.S., Skirda, V.D., Tyurin, V.F., Gizatullin, B.I. The nuclear magnetic resonance equipment for the research in laboratory and field conditions of full-sized core samples. Georesursy – Georesurces, 2012, № 4, Issue 46, 16–18 pp. (in Russian).

Timur, A. Nuclear magnetic resonance study of carbonate rocks. SPWLA 13th Ann. Logging Symp., Tulsa, OK, 1972, May 7 - 10:15 pp.

USSR State Committee for Standards and Product QUALITY MANAGEMENT. 26450.1-85 The Rocs. The method of opened porous coefficient determination by liquid saturation]. Moscow, 1985. (in Russian).

GeoSpec manual. Core analysis. Oxford Instruments. https://www.oxford-instruments.com/OxfordInstruments/media/industrial-analysis/magnetic-resonance-pdfs [electronic resource]

Chapter 13
Magnetic Properties of Soils from the Volga-Kama Forest-Steppe

L. A. Fattakhova, L. R. Kosareva and A. A. Shinkarev

Abstract We study the magnetic properties of virgin forest-steppe soils which formed on original vertical uniform unconsolidated parent material. In this work, profile samples of virgin dark-gray forest light loamy soil developed on a Permian (Kazanian stage) siltstone and virgin leached medium-thick fat light-loam chernozem developed on the Quaternary deluvial loam were used. Both soils are characterized by accumulative type of magnetic susceptibility and its components. The accumulative type refers to a profile where the maximum accumulation of substances takes place at the surface with a gradual fall in their content with depth. The calculation of frequency-dependent magnetic susceptibility (F-factor) shows that in studied soils F-factor reflect changes in their ferromagnetic fraction. In the humus part of the soil profile magnetic materials are present predominantly as fine mineral particles, and their content in the <2.5 μm fraction decreases toward the soil-forming rock. Estimates of relative contributions of dia -/paramagnetic, superparamagnetic, and ferromagnetic components obtained from the coercive spectra shows that the increase in the magnetic susceptibility in organogenic horizons of forest-steppe soils is due to the contribution of the ferromagnetic component.

Keywords Virgin dark-gray forest soil · Virgin chernozem · Magnetic susceptibility · Soil profile distribution

Introduction

Starting with the Devonian period, formation, destruction, demolition and redeposition of soil material or their burial are an integral part of the sedimentary process. Therefore, study of magnetism of sedimentary rocks motivated also the

---

L. A. Fattakhova (✉) · L. R. Kosareva · A. A. Shinkarev
Department of Geophysics and Geoinformation Technologies,
Kazan Federal University, Kazan, Russia
e-mail: L.a.fattakhova@yandex.ru

D. Nurgaliev et al. (eds.), *Recent Advances in Rock Magnetism, Environmental Magnetism and Paleomagnetism*, Springer Geophysics,
https://doi.org/10.1007/978-3-319-90437-5_13

studies of the magnetic properties of soils which formed as a result of complex processes of energy and mass transfer in the boundary layer between rocks and the atmosphere. According to modern concepts the causes of ferrimagnetic minerals accumulation in the soil humus horizons can be very diverse and not confined by an in situ pedogenic formation of ferrimagnetic minerals (Blundell et al. 2009; Torrent et al. 2010). However, despite the widespread use of magnetic measurements in the study of soils, loess–palaeosols, as well as fluvial, limnic and marine sediments and other natural objects containing a material from soil profiles at some stage of development, the mechanisms of magnetic mineral formation in soils remain a subject of debate. So the development of novel and improvement of existing methods of rock magnetism as applied to soil science is a very active area of study. In particular, a routine use of coercive spectrometry (Jasonov et al. 1998; Nurgaliev and Jasonov 2009) complemented by results obtained with other physical methods would allow to enhance an informative value and expand the interpretative potential of magnetic methods.

Data on soil magnetic properties show that magnetic characteristics in the zonal series of the Russian plain soils vary naturally with depth, showing a close relationship to the type and intensity of the soil-forming process (Babanin et al. 1995). Magnetic properties depend primarily on the iron content, its phase composition and dispersion (Vodyanitskiy 2003). In most modern automorphic soils in the northern hemisphere, formed in the intensively altered unconsolidated Quaternary sediments ferromagnetic minerals can be detrital and authigenic. The majority of authigenic magnetic minerals in soil are formed in the organogenic horizons where chemical and biological weathering is most intense (Babanin et al. 1995). According to modern concepts, ferromagnetic minerals can accumulate in the soil horizons due to various causes, which are not confined to the in situ pedogenic processes (Blundell et al. 2009; Torrent et al. 2010).

Information about the mechanisms and the kinetic parameters of the magnetic susceptibility enhancement in automorphic and well-drained temperate soils most consistently and thoroughly summarized in the chemical kinetic model proposed by Boyle et al. (2010). The magnetic susceptibility enhancement in this model is uniquely associated with the weathering duration main controlling variables being the primary ferrous silicate concentration in the soil-forming rocks and the mean annual precipitation, mean annual temperature playing a subordinate role. The kinetics of the process consists of two stages. The first stage (the characteristic time of $\sim 10^4$ years) involves active production of secondary magnetic minerals from Fe^{2+} released by weathering of less resistant primary minerals (e.g., chlorite). As their reserves in the soil forming rock are depleted the enhancement rate of the magnetic susceptibility is reduced. The second stage (the characteristic time of 10^5 years) involves passive enhancement of secondary ferrimagnetic minerals concentration due to loss of less reactive minerals (e.g.,—feldspars) from the soil by weathering. The contribution of dissimilatory Fe-reducing microorganisms to magnetic susceptibility enhancement is ignored in this model. The chemical kinetic model has some uncertainty, - however, following its logic it can be assumed that the formation of the magnetic profile of Holocene soil is currently at the stage of active formation of secondary magnetic minerals. To check and clarify the main provisions of the above model it is necessary to extend the circle of the studied types of soil and carry out complex studies of the magnetic profiles formation and accumulation of ferromagnetic minerals.

Nevertheless, despite the widespread use of magnetic measurements in the study of soils the mechanisms of magnetic mineral formation in soils remain a subject of discussions. Anyway, we can confidently state that the magnetic profile of a soil results from complex processes of energy and mass transfer, leading to vertical differentiation of the original parent rock's composition and properties. This article presents the results of a pilot study of the magnetic properties of virgin dark-gray forest soil and virgin leached chernozems which were initially formed on vertically homogeneous loose soil-forming rocks. Soils were chosen at random, considering only the fact that they are virgin, zonal and automorphic. The main goal of the research was to study the variation of the magnetic properties with depth in these soils.

Materials and Methods

Experiments were conducted on a virgin dark-gray forest light loamy soil developed on a Permian (Kazanian stage) siltstone (Fig. 13.1) and on a virgin leached medium thick fat light-loam chernozem developed on a Quaternary deluvial loam (Fig. 13.1). The samples were taken in Alekseyvsky district located at the shore of the Kuibyshev reservoir.

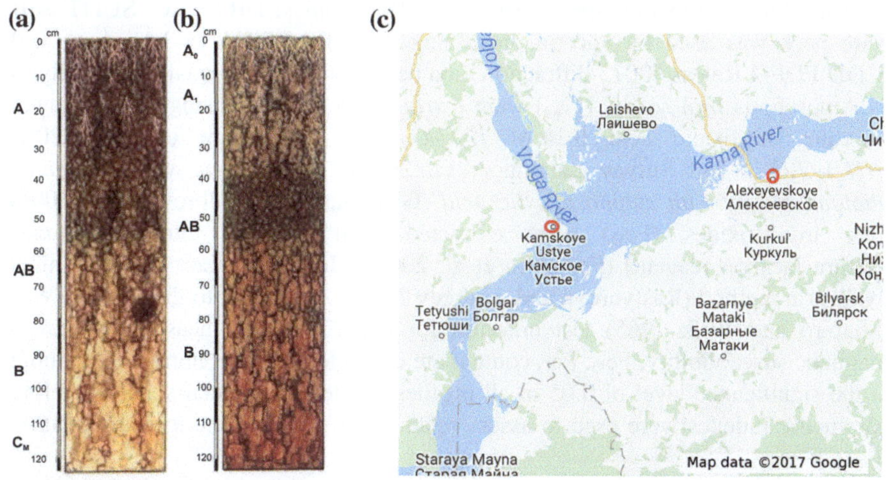

Fig. 13.1 Profile of leached chernozem (**a**), dark-gray forest soil (**b**), and geographic map showing locations of studied soils (**c**)

Initial vertical parent material uniformity of these soils was established using the concentration ratios of Ti, Zr, and Y (Shinkarev et al. 2010). Preparation of soil samples was carried out according to the recommendations of the International Organization for Standardization (ISO 1146).

To remove carbonates, samples were treated with 1 mol/l CH_3COOH solution for 24 h and washed off with distilled water. Then, to remove organic matter (OM), the samples were treated with 30% H_2O_2 at room temperature for 15–20 days. The precipitate was separated by centrifugation, then dried and homogenized. Carbonates and OM were removed prior to separation of particles with the Stokes diameter of <2.5 μm and of heavy mineral fractions with density >2.9 g × $cm^{-3.}$ The clayey fraction was separated by elutriation after sedimentation in a liquid column. The heavy mineral fraction was separated by repeated centrifugation in bromoform.

Magnetic susceptibility values were determined at frequency 0.465 kHz for χ_{lf} and 4.65 kHz for χ_{hf} using a Bartington MS2–B magnetic susceptibility meter. To do this, the original samples were crushed in an agate mortar and mass normalized.

Thermomagnetic analysis (TMA) was performed using a Curie express balance (Burov et al. 1986). Intensity of induced magnetization was measured as a function of temperature up to 800 °C (heating rate 100 °C/min) in the constant magnetic field of 0.2 T. The TMA curves to 800 °C were recorded twice. Coercive spectra of normal magnetization in the magnetic fields of up to 0.5 T were obtained by means of a coercive spectrometer («J_meter») (Jasonov et al. 1998; Burov et al. 1986), which allows separate recording of remanent and induced magnetization at room temperature. Original samples were magnetized from their natural state.

Bruker Corporation D2 Phaser Powder Diffractometer (Bragg-Brentano θ-θ geometry, 30 kV and 10 mA on X-Ray tube, Ni-filtered CuKα, goniometer radius —141.4 mm, fixed slits, 1-dimensional LYNXEYE detector, sample rotation) was used to obtain diffraction spectra from powder samples. DIFFRAC. SUITE software pack was used for spectra processing. DIFFRAC.EVA (v-3.1) along with ICDD PDF-2 Release 2013 diffraction data base were used for qualitative analysis.

Amorphous *iron oxides* (Fe_O) were *extracted* in the dark using Tamm reagent (mixture of oxalic acid 0.14 M $H_2C_2O_4$ and ammonium oxalate 0.20 M $(NH_4)_2C_2O_4$) (Soil Survey Laboratory … 2004). *Iron content* was *determined photometrically with sulphosalicylic acid* (Glebova 2006; Fedorova et al. 2006). Free iron oxides (Fe_D) were extracted with dithionite-citrate-bicarbonate (Mehra-Jackson reagent) (Fedorova et al. 2006). Humus content was determined by Tyurin method (Rastvorova and Andreev 2006). According to the Shapiro-Wilk (Shapiro and Wilk 1965) test criterion for experimentally measured values of magnetic susceptibility, Fe_O, Fe_D content are characterized by a normal distribution at the significance level of 0.05 in all profiles of the investigated soils. Therefore, parametric indexes were used to assess relationship between the measured data.

Results and Discussion

Soil-forming rock of the dark gray forest soil is a Permian siltstone of the Kazanian stage. The diffraction spectra of powder mounts (Fig. 13.2) show that the parent rock contains not only clay minerals, but also quartz, calcite, potassium and sodium feldspars, as well as a small amount of dolomite. Soil-forming rock of the virgin leached medium-thick fat light-loam chernozem is the Quaternary deluvial loam (Fattakhova et al. 2016). The soil-forming rock contains clay minerals, quartz, calcite, dolomite, potassium and sodium feldspars and a small admixture of amphiboles.

Graphical analysis shows that the background virgin dark-gray soil and virgin chernozem are characterized by accumulative type of profile distribution patterns of magnetic susceptibility (Fig. 13.3).

Selective organic matter removal has no effect on the magnetic susceptibility profile differentiation. Somewhat elevated absolute values of χ_{lf} at the top and bottom of the profile may be attributed to removal of organic matter and carbonates, because magnetic susceptibility measurement was conducted on mass normalized samples. In the humus part of the soil profile magnetic materials are present predominantly as fine mineral particles, and their content in the <2.5 μm fraction is reduced toward the soil-forming rock. Accumulative distribution of magnetic susceptibility in the virgins soil formed on the vertically uniform parent material implies the pedogenic origin of magnetic materials. Pedogenic magnetic particles are usually fine grained, and the difference between magnetic susceptibility values

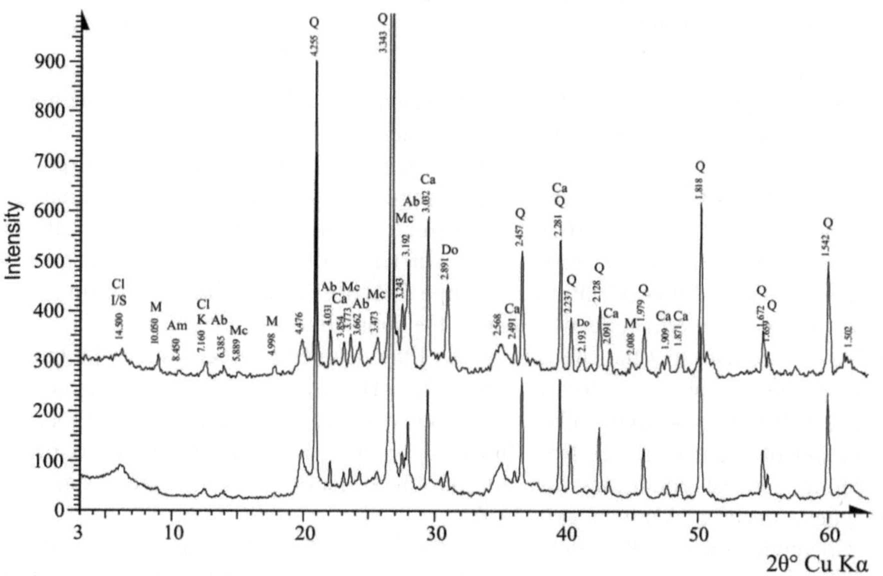

Fig. 13.2 X-ray diffraction patterns of powder preparations of soil-forming rocks of the virgin leached chernozem (**a**) and virgin dark-gray forest soil (**b**)

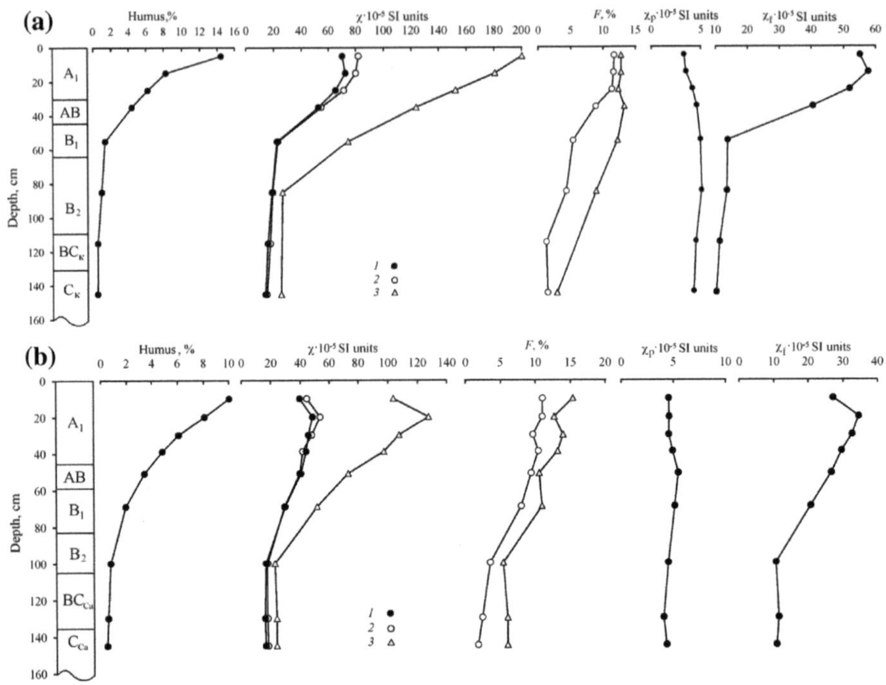

Fig. 13.3 Humus content and magnetic properties as a function of depth in the profile of virgin dark-gray forest soil (**a**) and virgin leached chernozem (**b**): 1—original soil samples; 2—soil samples after the organic matter removal; 3—<2.5 μm fraction isolated after the removal of carbonates and organic matter

(χ) at different frequencies reflecting the presence of superparamagnetic fine grains is considered an indicator of bio-inert interactions (Evans and Heller 2003). Calculation of the frequency-dependent magnetic susceptibility ($F = (\chi_{lf}-\chi_{hf})/\chi_{lf}100\%$) shows that the F-factor values range from ~2 to ~12% in original soil samples and from ~3 to ~16% in <2.5 μm fraction—in the soil-forming rock and the upper parts of profile, respectively. In terms of F-factor, as in the case of magnetic susceptibility, virgin forest-steppe soils are characterized by accumulative type of profile patterns reflecting mainly the single-domain state of ferromagnetics. It is not necessary that the characteristics of the magnetic properties and the humus content functionally depend on each other. However, it should be taken into account that such behavior of the profile curves can partly be mediated by a contribution of organic matter from biogenic organo-accumulative surface horizons creating optimal conditions for heterotrophic microorganisms synthesizing the magnetic material (Vodyanitskiy 2003).

Coercivity spectra allow determining the contribution of dia-/paramagnetic (χ_p) and ferromagnetic component (χ_f) to the magnetic susceptibility (Kosareva et al. 2015). In our case, shape of χ_p and χ_f depth dependences clearly proves that magnetic susceptibility enhancement in organogenic horizons of virgin dark-gray forest soil and virgin chernozem is due to the contribution of ferromagnetic components (Fig. 13.3). A strong correlation (at the significance level $\alpha = 0.001$) between the total magnetic susceptibility and its ferromagnetic component (Fig. 13.4) lends further support to this conclusion.

Magnetic susceptibility in soil profiles is often correlated with the iron content. Strong correlation (at the significance level $\alpha = 0.001$) can be seen between magnetic susceptibility and oxalate-extractable Fe (Fig. 13.5) and between the magnetic susceptibility and Schwertmann's criterion values (relationship between magnetic susceptibility and index of the degree of crystallinity of free Fe in layer samples (Fig. 13.6). There is no correlation between magnetic susceptibility and crystalline Fe oxides content determined by the difference between contents of dithionite-citrate-bicarbonate-extractable Fe and oxalate-extractable Fe (Fig. 13.5). This, however, does not exclude a possibility of magnetic iron oxides actively forming in the surface horizons. Technically, correlation analysis results can be regarded as merely an evidence of dominating contribution of "amorphous" and poorly crystalline Fe oxides to the magnetic properties of humus profile.

Magnetic and mineralogical characteristics of bulk samples and <2.5 μm fraction were studied by differential TMA. Figure 13.7 shows accumulative (in the upper part of the figures) and differential (in the lower part of the figures) TMA curves. Thermomagnetic analysis reveals that magnetite is the main carrier of magnetization of soils. All samples are characterized by the presence of free, bound and crystallization water and complete oxidation of magnetic minerals during thermomagnetic studies. In addition, each sample contains a small amount of hematite. Burnout of organic material is reflected in an increase of magnetization in the temperature range 480–580 °C due to the formation of magnetite. Peaks or

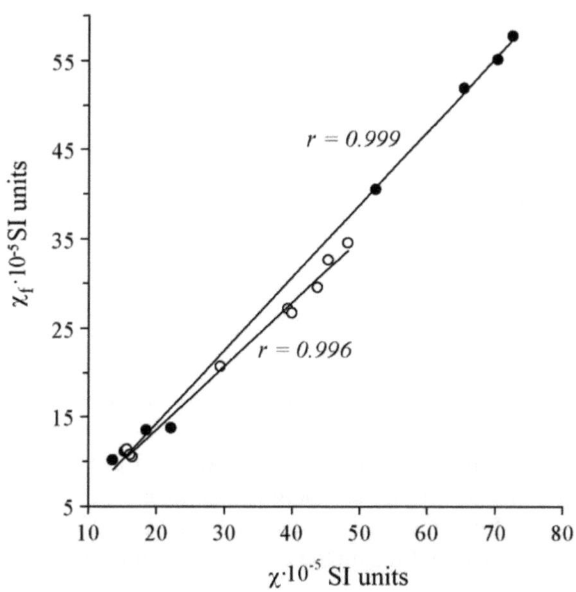

Fig. 13.4 Relationships between the total magnetic susceptibility and its ferromagnetic component in profile samples of dark-gray forest soil (●) and leached chernozem (○)

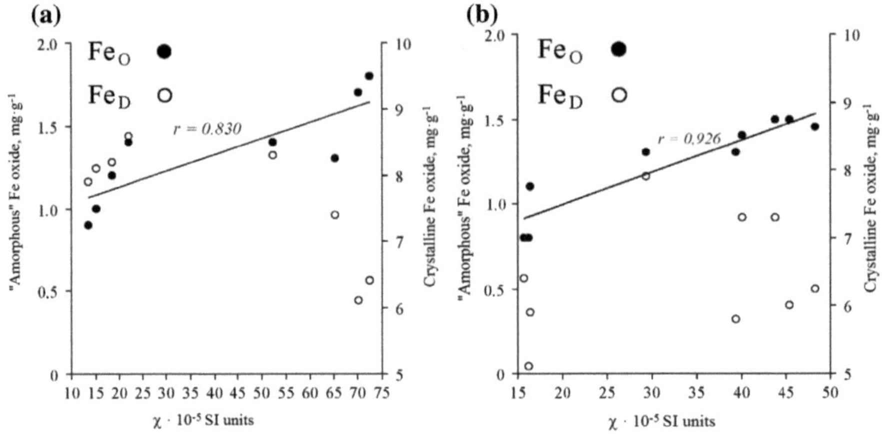

Fig. 13.5 Relationship between magnetic susceptibility and contents of "amorphous" (Fe_O) and crystalline (Fe_D) Fe oxides in layer samples of virgin dark-gray forest soil (**a**) and virgin leached chernozem (**b**)

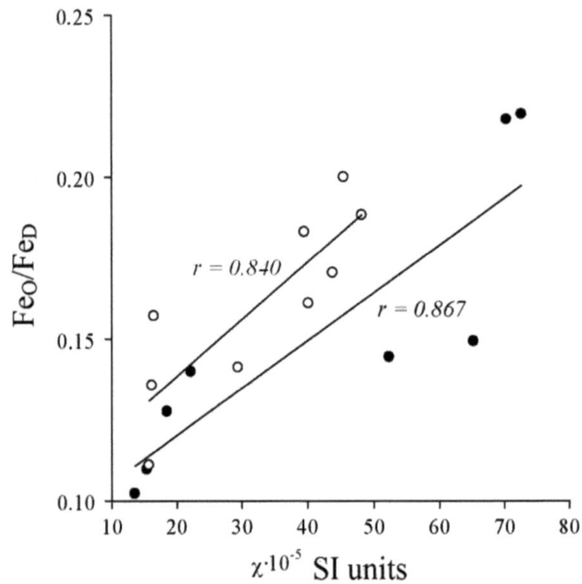

Fig. 13.6 Relationship between magnetic susceptibility and index of the degree of crystallinity of free Fe in layer samples of virgin dark gray forest soil (●) and virgin leached chernozem (○)

inflections on differential TMA curves near 300 °C might be due to a presence of iron sulfides.

The differential TMA curves for <2.5 μm fraction after the organic matter and carbonates removal are more informative (Fig. 13.8). There is a peak at 180–210 °C (apart from the one associated with free and bound water) which is typical for maghemite and magnetite and usually appears due to relieving the stress related to

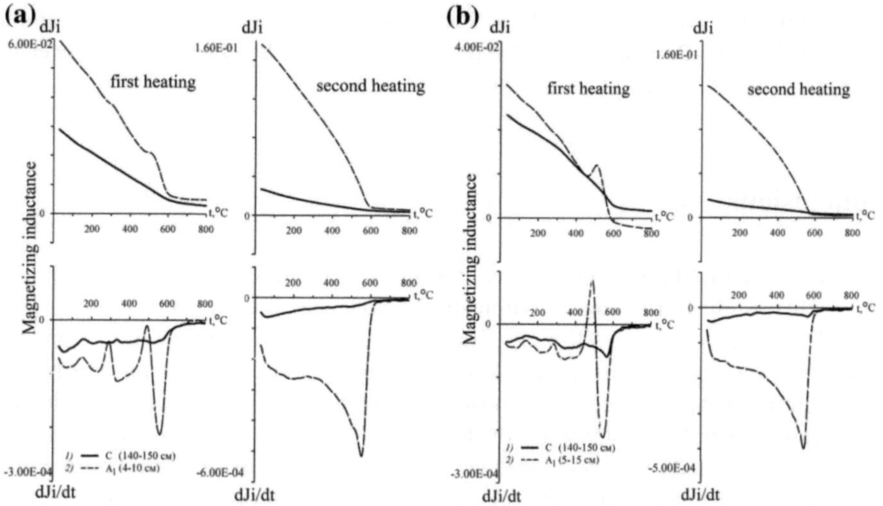

Fig. 13.7 Thermomagnetic analysis results for the parent rock sample and the samples taken from the upper layers of the virgin dark-gray forest soil (**a**) and of the virgin leached chernozem (**b**). Note the different magnetization scales in first- and second-heating graphs for both soils

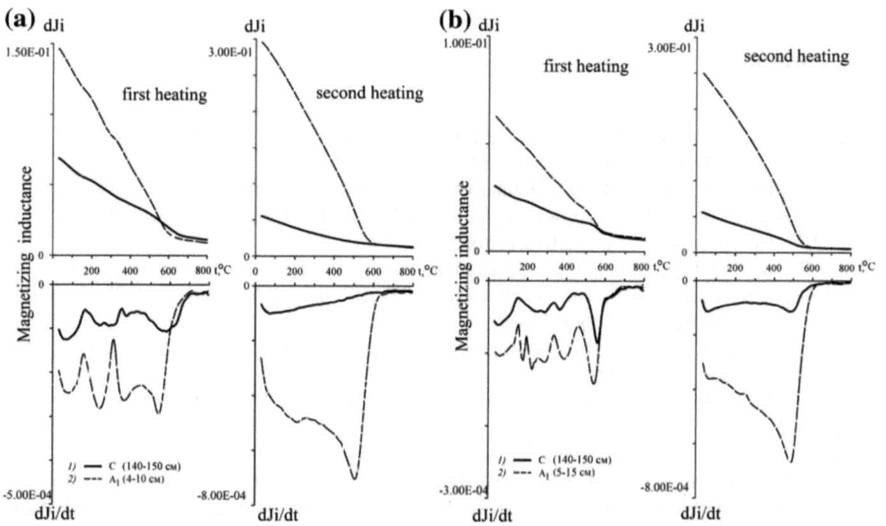

Fig. 13.8 Results of thermomagnetic analysis of <2.5 μm fraction isolated after the removal of carbonates and organic matter from parent rock (1) and upper layers (2) of the virgin dark-gray forest soil (**a**) and of the virgin leached chernozem (**b**)

maghemitization (Burov et al. 1986). The samples also contain iron sulfide, and a slight bend at 250–280 °C is considered as an indicator of lepidocrocite (Burov et al. 1986).

Conclusions

A study was carried out of the magnetic profiles of virgin dark-gray forest soil and virgin chernozem formed on lithologically homogeneous, loose soil-forming rocks typical for the temperate climatic zone. Against the background of a gradual decrease of the organic matter with depth the magnetic susceptibility also decreases consistently. In the presence of the most favorable prerequisites for the formation of the organo-mineral complex the development of soils under meadow and under forest vegetation led to a pronounced accumulative type of the magnetic profile due to the accumulation of magnetite and maghemite concentrated mainly in the clay fraction.

Acknowledgements The work is performed according to the Russian Government Program of Competitive Growth of Kazan Federal University.

References

Babanin V.F., Trukhin V.I., Karpachevskiy L.O., Ivanov A.V., Morozov V.V. Soil Magnetism. Moscow, Yaroslavl, YaGTU Publisher, 1995. 222 p. (In Russian).

Blundell A., Dearing J.A., Boyle J.F., Hannam J.A. Controlling factors for the spatial variability of soil magnetic susceptibility across England and Wales. Earth-Science Rev., 2009. vol. 95. pp. 158–188. https://doi.org/10.1016/j.earscirev.2009.05.001.

Boyle J.F., Dearing J.A., Blundell A. and Hannam J.A. 2010. Testing competing hypotheses for soil magnetic susceptibility using a new chemical kinetic model. Geology. Vol. 38. pp. 1059–1062. https://doi.org/10.1130/g31514.1.

Evans M.E., Heller F. Environmental Magnetism: Principles and Applications of Enviromagnetics San Diego: Acad. Press, 2003. 311 p.

Fattakhova L.A., Shinkarev A.A., Kosareva L.R., Nurgaliev D.K., Shinkarev *A.A. (Jr.), Bagautdinova Y.S. Magnetic properties of different-aged chernozemic soil profiles.* ARPN Journal of Engineering and Applied Sciences, *2016. 11 (19), pp. 11383–11394.*

Fedorova N.N., Kasatkina G.A., Rastvorova O.T. Indexes and methods for estimating element composition in the mineral part of soils (total element determination). Teoriya i praktika khimicheskogo analiza pochv [Theory and Practice of Chemical Analysis of Soils]. Moscow, GEOS, 2006, pp. 141–193. (in Russian).

Glebova G.I. Indexes and methods for estimating group (fractional) composition of chemical elements in soils. Teoriya i praktika khimicheskogo analiza pochv [Theory and Practice of Chemical Analysis of Soils]. Moscow, GEOS, 2006, pp. 248–309. (in Russian).

ISO 11464. 1994. Soil Quality – Pretreatment of samples for physico-chemical analysis. 11 p.

Burov B.V., Nurgaliev D.K., Jasonov P.G. Paleomagnetic analysis. Kazan, KGU, 1986. 166 p. (In Russian).

Jasonov P.G., Nurgaliev D.K., Burov B.V., Heller F. A modernized coercivity spectrometer. Geologica Carpathica, 1998. vol. 49. pp. 224–226.

Kosareva L.R., Nourgaliev D.K., Kuzina D.M., Spassov S., Fattakhov A.V. Ferromagnetic, dia-/paramagnetic and superparamagnetic components of Aral Sea sediments: significance for paleoenvironmental reconstruction. *ARPN Journal of Earth Science* 2015. *Vol. 4, pp. 1–6.*

Nurgaliev D.K., Jasonov P.G. Coercivity Spectrometer. Utility Model Patent No. 81805, Fed. Inst. Ind. Prop. Bul. No. 9, 2009. (In Russian).

Rastvorova O.T., Andreev D.P. 2006. Preparation of soil samples for the analysis and methods of expressing the results of analysis. Theory and Practice of Chemical Analysis of Soils. Moscow. GEOS. pp. 103–111. [in Russian].

Shapiro S.S., Wilk M.B. An analysis of variance test for normality.—Biometrica, 1965, 52, № 3 – pp. 591–611.

Shinkarev A.A., Kornilova A.G., Lygina T.Z., Giniyatullin K.G., Gilmutdinov R.R. Assessing Parent Material Uniformity by Elemental Analysis. Uchenye Zapiski Kazanskogo Universiteta. Seriya Estestvennye Nauki, 2010. vol. 152. no. 4, pp. 78–91. (in Russian).

Soil science. Textbook for students / Kovda V.A. ed., Part 2. Types of soils, their geography and use / etc. - Moscow.: Vysshaya Shkola, 1988. - 368 p.: ill. (In Russian).

Torrent J., Liu Q.S., Barrón V. Magnetic susceptibility changes in relation to pedogenesis in a Xeralf chronosequence in northwestern Spain. European J. Soil Sci., 2010. vol. 61. pp. 161–173. https://doi.org/10.1111/j.1365-2389.2009.01216.x.

Soil Survey Laboratory Methods Manual. Soil Survey Investigations Report No 42, Version 4.0. Lincoln, NE, USDA-NRCS, 2004. 700 p.

Vodyanitskiy Yu. N. Chemistry and Mineralogy of the Soil Iron. Moscow, V.V. Dokuchaev Soil Science Institute of Russian Academy of Agricultural Sciences, 2003. 238 p. (In Russian).

Chapter 14
Magnetic Properties of Artificial CRM Created on Titanomagnetite-Bearing Oceanic Basalts

S. K. Gribov, V. P. Shcherbakov and N. A. Aphinogenova

Abstract Laboratory modeling of the CRM formation on basaltic rocks from the Reykjanes submarine ridge and the Red Sea rift zone containing titanomagnetites (TM) with $T_c \approx 200$ °C was carried out by annealing at different temperatures T_{an} for different time intervals up to 200-hour exposure in the external field $B_{CRM} = 50$ μT. Monitoring the CRM value against annealing time revealed a strong correlation between the CRM intensity and a strong field magnetization M_s. The existence of this correlation implies that the process of CRM acquisition actually takes place only in a relatively short initial time interval of annealing, so that the further CRM changes in time basically follow variations of M_s due to the process of oxidation of TM taking place during the annealing. It is shown that at T_{an} = 350 and 400 °C TM grains are subjected to high degree of single-phase oxidation while annealing at high T_{an} = 450 and 500 °C leads to the oxyexsolution. The Arai-Nagata diagrams for samples bearing the isolated induced CRM can be approximated by two linear segments with a break point at T = (560–570) °C. For the low temperature interval (T_{an}, 560 °C), the calculated field B_{calc} underestimates the true field B_{CRM} by 33–56% at T_{an} = 350 and 400 °C but by 29–42% at T_{an} = 500 °C. Surprisingly, all specimens annealed at 450 °C yield B_{calc} which is very close to the true field B_{CRM}. The observed error in the determination of B_{CRM} tends to decrease with increasing T_{an} probably due to the relative narrowness of the blocking temperature interval, centered near T_c.

Keywords Titanomagnetites · Oxidation · Chemical remanent magnetization
Thellier procedure · Paleointensity

S. K. Gribov (✉) · V. P. Shcherbakov · N. A. Aphinogenova
Geophysical Observatory "Borok" IPE RAS, Borok, Russia
e-mail: gribov@borok.yar.ru

V. P. Shcherbakov
Kazan Federal University, Kazan, Russia

V. P. Shcherbakov
IPE RAS, Moscow, Russia

D. Nurgaliev et al. (eds.), *Recent Advances in Rock Magnetism, Environmental Magnetism and Paleomagnetism*, Springer Geophysics,
https://doi.org/10.1007/978-3-319-90437-5_14

Introduction

The question of reliability of paleomagnetic data is critical for paleomagnetism. For volcanic rocks the necessary condition of the reliability is the requirement that the normal remanent magnetization (NRM) is the thermoremanent magnetization (TRM). However, this fundamental assumption may be not correct if metamorphic processes took place during the rock unit cooling or later at temperatures lower than the Curie temperature T_c of magnetic minerals of rocks because of these processes can induce such secondary magnetizations as chemical (CRM) and thermochemical (TCRM) (Smirnov and Tarduno 2005; Draeger et al. 2006). It has been recognized since long ago that there is often a real danger of confusing CRM or TCRM with TRM and thus obtaining erroneous paleointensity of geomagnetic field B_{anc} (Smirnov and Tarduno 2005; Draeger et al. 2006; Gribov and Dolotov 2016; Gribov et al. 2017; Shcherbakov et al. 2017). Hence, recognition of CRM (or TCRM) in volcanic rocks represents an important task of paleo- and rock magnetism which can be achieved only by a thorough investigation of different magnetic properties of samples carrying CRM and TCRM.

To this moment, a few studies of magnetic properties of an artificial laboratory induced CRM have been performed (Draeger et al. 2006; Gribov and Dolotov 2016; Gribov et al. 2017; Shcherbakov et al. 2017). In particular, the Thellier experiments conducted on rocks carrying the induced CRM showed that a substantial underestimation of B_{lab} occurs when CRM is present instead of TRM. Here, B_{lab} is the external magnetic field used in the Thellier experiments. However, the question of recognition of CRM and selection of rocks carrying this remanence, remains an unsolved problem of rock magnetism. Obviously, in order to approach the solution, first of all the understanding of physical and chemical processes lying behind the CRM and TCRM formation must be achieved.

Since the formation of CRM and TCRM is closely related to the mineralogical and structural transformations of ferrimagnetic grains, it is necessary in parallel to the Thellier experiments to conduct a detailed magnetomineralogical study of the samples used for these experiments. For this purpose, in addition to the Thellier experiments, we carried out a complex of laboratory studies including the creation of CRM on oceanic basalts at different annealing times and temperatures combined with monitoring of thermomagnetic and hysteresis properties performed on different stages of thermal treatment of the samples. To obtain a deeper insight into the physical and chemical processes during the CRM acquisition and Thellier experiments, the study of internal morphology of ferrimagnetic grains, their chemical and phase composition by means of electron microscopy and X-ray diffraction was undertaken.

Samples, Experimental Procedure and Equipment

In the present study, two tholeitic basalt samples were used. The first one was dredged from the bottom of the rift zone of the south of the Red Sea (number P-72/4) during the 30-th voyage (1980) of the research vessel "Akademik Kurchatov". The rock age is <0.5 Ma (Shreider et al. 1982; Trukhin et al. 2006). The second sample with the age <1 Ma ("Rift …" 1990) was collected from the axial part of the rift zone of the Reykjanes oceanic ridge (number 27503-7) during the 4-th voyage (1982) of the research vessel "Akademik Mstislav Keldysh".

Each sample was cut into 20 sister cubes with an edge of 1 cm. The remaining material was used for a complex magnetomineralogy studies comprising the light and scanning electron microscopy, X-ray phase and spectrochemical analyses, thermomagnetic experiments and standard magnetic hysteresis parameters measurements. The Curie points T_c and the thermal stability of magnetic minerals were estimated from thermomagnetic curves of strong-field magnetization M_s during heating–cooling cycles to incrementally higher temperatures T. These measurements were made with a Curie balance in the external magnetic field 450 mT. Hysteresis loops and additional high-field thermomagnetic curves (up to 900 mT) were measured using a variable field translation balance (VFTB) produced by Petersen Instruments company (Germany). To estimate the magnetic hardness and mineralogy of samples, magnetic susceptibility, hysteresis loop parameters, such as coercive force B_c, remanent coercive force B_{cr}, saturation magnetization M_s and residual saturation magnetization M_{rs} were measured.

Monitoring of CRM directly at the temperature of its formation as a function of time t was conducted using a two-component (2D) vibrating thermal magnetometer ORION (Russia) which measures continuously the horizontal components of specimen's remanent magnetic moment. The noise threshold of the magnetometer is 3×10^{-9} Am^2 and the maximum available external field is 0.2 mT (residual field <100 nT).

The scanning electron microscope (SEM) backscatter electron images were taken and chemical composition of separate magnetic grains was determined on the Tescan Vega II LMU scanning electron microscope (TESCAN, Czech Republic) equipped with energy dispersive spectrometer INCA Energy 450 (Inca Oxford Instruments, England). The X-ray diffractometry of the separated magnetic fraction of both the initial and annealed specimens was conducted at room temperature on the STADI–MP powder diffractometer (STOE, Germany).

The isolated CRM was created as follows: first the sample was slowly heated in air while the NRM was continuously monitored until it totally disappeared which occurred at temperatures about 300 °C. Then the sample was heated in air to the chosen reaction temperature T_{an} where the external magnetic field $B_{CRM} = 50$ μT was switched on and the CRM value was monitored as a function of time t. After the giving annealing time, the sample was cooled (in zero field) to room temperature T_r where the imparted CRM was subjected to different experiments. However, natural rocks are cooled in the Earth's magnetic field so in addition to CRM they acquire also a pTRM from T_{an} to T_r so that the total remanence is CRM + pTRM. To understand the role of this pTRM, another set of Thellier experiments was carried out when the external field was applied during both annealing and cooling.

Magnetic Properties and Composition of Magnetic Fraction of Samples

According to the scanning electron microscopy observations of the fresh samples of the collection, the ore fraction is represented only by small grains of TM of skeletal irregular shape (Fig. 14.1) which is typical for oceanic basalts and indicates a high cooling rate of the effluent magma in the aquatic environment. Moreover, in the Red Sea basalts, the upper limit of the size of the TM fraction does not exceed 6 μm, and in the basalt samples of the Reykjanes ridge—4 μm.

The surface images of polished samples obtained in backscattered electrons showed that fresh TM grains are rather homogeneous and they do not exhibit exsolution structures. From the microprobe data, the average content of the ulvospinel component x in the initial TM $Ti_xFe_{3-x}O_4$ in the basalt of the Reykjanes Range is 0.525 while for the Red Sea basalt $x = 0.46$. Also, a small amount of different impurities was detected among which the most abundant are Al ions whose average content is 1.86 and 2.44 at. % in the basalts of the Reykjanes Range and the Red Sea correspondingly. Using these data, the T_c calculated according to Richards et al. (1973), are 158 and 188 °C. These values are close to the measured ones (168 and 200 °C, respectively) determined from the $M_s(T)$ curves of fresh material (Fig. 14.2). The proximity of the measured Curie points to the calculated ones, as well as the homogeneity of the grains, indicate a low degree of single-phase (SPh) oxidation of fresh TM of both basalts. This conclusion is consistent with the X-ray data which give the lattice parameter $a_0 = 8.4639$ and 8.4560 Å, for the basalts of the Reykjanes and Red Sea ridge, respectively.

Fig. 14.1 Red Sea sample. Electron microscopic images of polished sections in backscattered electrons. **a** fresh sample; **b**, **c** after annealing in air at 500 °C for 200 h

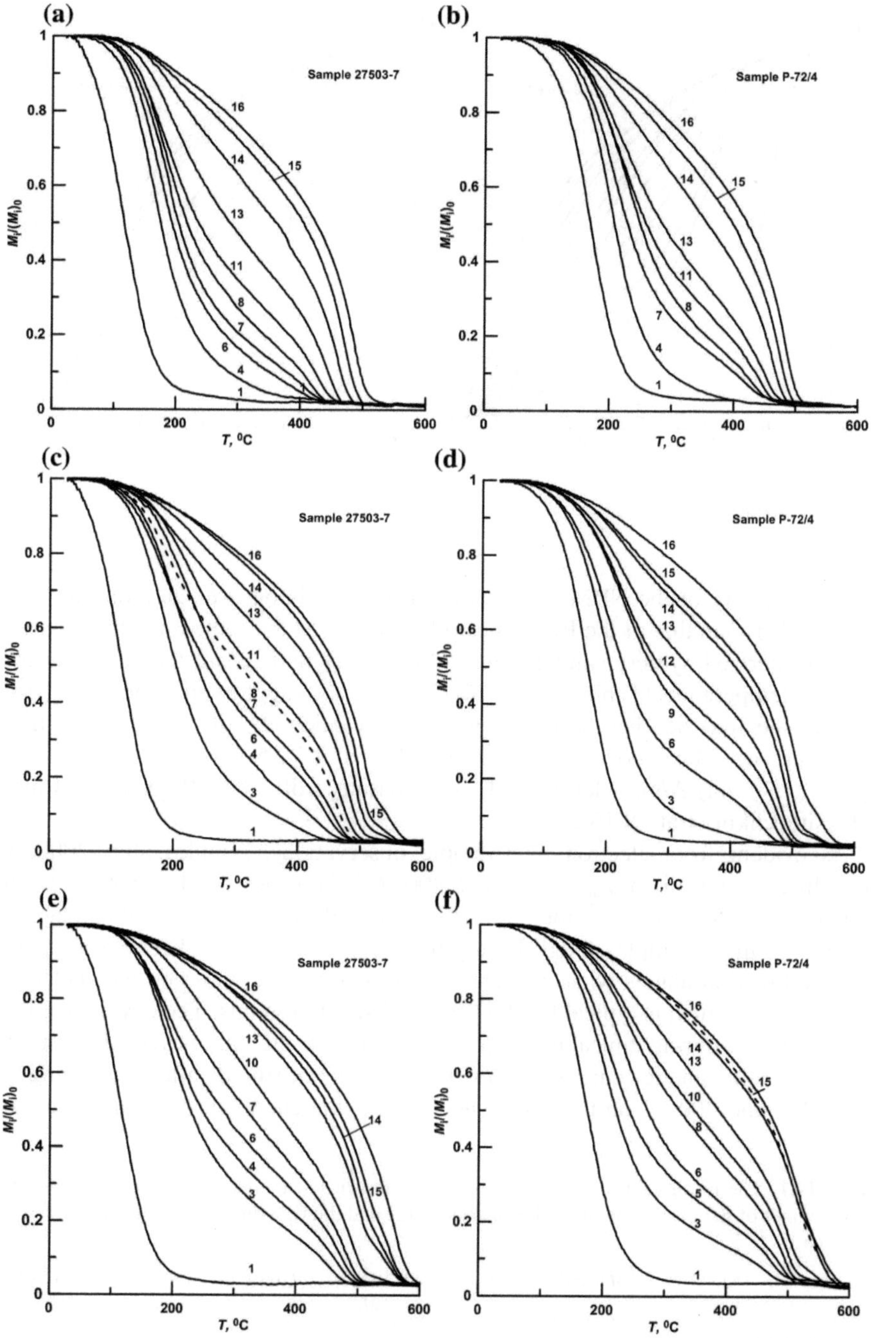

Fig. 14.2 Strong field thermomagnetic curves of samples from Reykjanes Ridge (left) and Red Sea (right) annealed in air at **a** and **b** 350 °C, **c** and **d** 400 °C, **e** and **f** 450 °C, **g** and **h** 500 °C for time intervals: 1—the initial state, 2–5, 3–10, 4–20, 5–25 min, 6–1, 7–2, 8–4, 9–5, 10–6, 11–7, 12–8, 13–15, 14–50, 15–100, 16–200 h

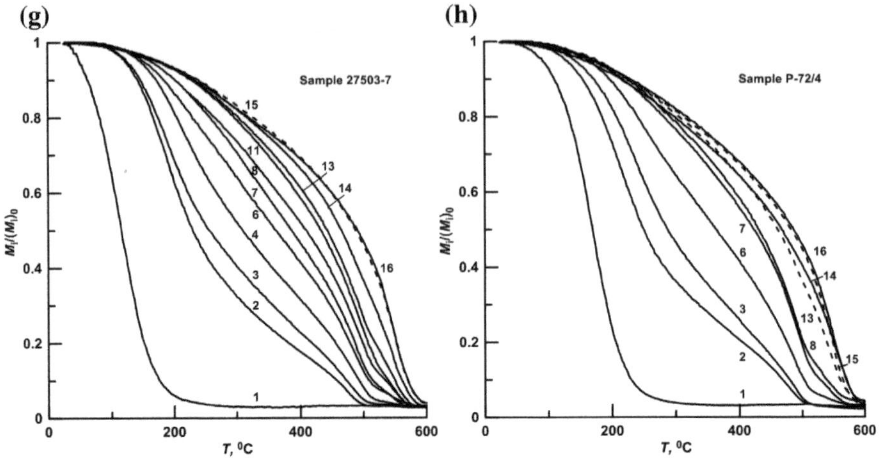

Fig. 14.2 (continued)

The intensity of the NRM for the Red Sea basalt is on average more than two times higher than that of the Reykjanes ridge basalt. At the same time, these basalts are characterized by very similar values of other magnetic characteristics measured at room temperature (Table 14.1). This discrepancy suggests that the high NRM level of the Red Sea basalts is due to the high paleointensity value B_{anc} which, according to Maksimochkin et al. (2010) is ≈77.5 A/m as compared to $B_{anc} = (21 \div 44)$ A/m, determined for basalts of the Reykjanes Range (Maksimochkin et al. 2016).

As evident from electron microscopic observations, samples after 200-hour annealing in the air at T_{an} = 350 and 400 °C, show no visible changes in the structure of TM. At the same time, the X-ray diffractograms of the magnetic fraction of these samples fixed the asymmetry of the spinel reflections from the larger diffraction angles what indicates the decrease in the unit cell parameters of spinel phases which is typical for the SPh oxidized TM grains. The development of SPh oxidation during annealing is in agreement with the data provided by strong field thermomagnetic curves $M_s(T)$ (Fig. 14.2) measured after annealing at T_{an} = 350 and 400 °C on these samples. Indeed, the curves demonstrate a regular

Table 14.1 The mean values of the magnetic characteristics of the basalt samples studied in the northern part of the axial zone of the Reykjanes Ridge and the rift zone of the southern part of the Red Sea

No. of sample	NRM (A/m)	M_s (Am²/kg)	M_{rs} (Am²/kg)	M_{rs}/M_s	B_c (mT)	B_{cr} (mT)	B_{cr}/B_c	T_C (°C)
Reykjanes ridge								
27503-7	15.35	0.725	0.182	0.249	8.13	15.72	1.82	168
Red Sea								
P-72/4	35.43	0.647	0.162	0.250	8.59	15.80	1.84	200

shift of T_c toward higher temperatures with increasing annealing time which is the well-known marker for the presence of SPh oxidation. In the end, after the annealing for 200 h at 350 °C, we observed the phase with $T_c \approx 520$ °C which corresponds to the degree of oxidation $z \approx 1.0$ for the given composition $x \approx 0.5$ (Nishitani and Kono 1983). Note that annealing at 400 °C for 100 and 200 h already produces a tiny magnetic phase with $T_c \approx 570$ °C as indicated by the tails of the corresponding thermomagnetic curves (Fig. 14.2c, d). Most likely the appearance of this phase reflects the onset of initial stage of TM exsolution.

The process of exsolution becomes dominating when the annealing temperature is increased to 450 and 500 °C. As is seen from the Fig. 14.2e–h, exposure to these temperatures for 100 or 200 h leads to appearance of a single magnetic phase with $T_c \approx 580$ °C, which must be attributed to magnetite and this conclusion is confirmed by electron microscopic observations of etched polished sections taken from these samples after this long annealing. Indeed, the images clearly reveal inhomogeneous structures on submicron level, caused by heterophase oxidation (oxyexsolution) expressed sometimes in the form of fine magnetite-ilmenite octahedral grid (Fig. 14.1b). The micro-inhomogeneous structures form sometimes very fine rippled patterns, barely visible under the maximum possible magnification ×50,000 (Fig. 14.1c). Thus, a sufficiently treatment at T_{an} = 450 and 500 °C culminates in a complete disintegration of the initial TM phase with the release of near-magnetite cells. This conclusion is further confirmed by the X-ray diffraction experiments according to which the lattice parameter of the annealed samples is $a_0 \approx (8.38 \div 8.40)$ Å, close to the magnetite lattice parameter. Besides, in addition to the spinel reflections, XRD patterns of these samples display pronounced ilmenite and pseudobrookite reflections. The latter are known to be a product of the high-temperature oxidation of ilmenite indicating a more advanced stage of oxidation.

Variations of hysteresis parameters during isothermal annealing were studied on sister cubes. The annealing at 350 and 400 °C causes a steady increase in value of M_s with time while the value of M_{rs} experiences only slight non-monotonous variations (Fig. 14.3a, b). As far as the annealing promotes SPh oxidation, the increase in M_s seemingly contradicts the common believe that SPh oxidation results in a decreases of the M_s value due to reduction of a number of Fe^{2++} ions per a unit cell. However, besides this effect, the SPh oxidation also causes an increase of the Curie temperature that implies the increase in the interval between T_r and T_c which in turn leads to the increase of M_s measured at T_r. The combination of the two opposite tendencies might increase the $M_s(T_r)$ value in accordance with our observations.

Speaking about the time variations in M_s and M_{rs} values at T_{an} = 450 and 500 °C, we see that they also increases on the initial stages of the annealing but start to decrease after they reach the maximum values at some annealing time (this decrease is visible already on $M_s(t)$ curve recorded for T_{an} = 400 °C). This fall most likely reflects the subsequent oxidation of the magnetite cells.

What concerns the time evolution of the coercive parameters B_c and B_{cr}, at the beginning they decrease at any T_{an} but the decrease changes for a rise with annealing time and the higher T_{an} the sooner. Interestingly, at T_{an} = 500 °C and

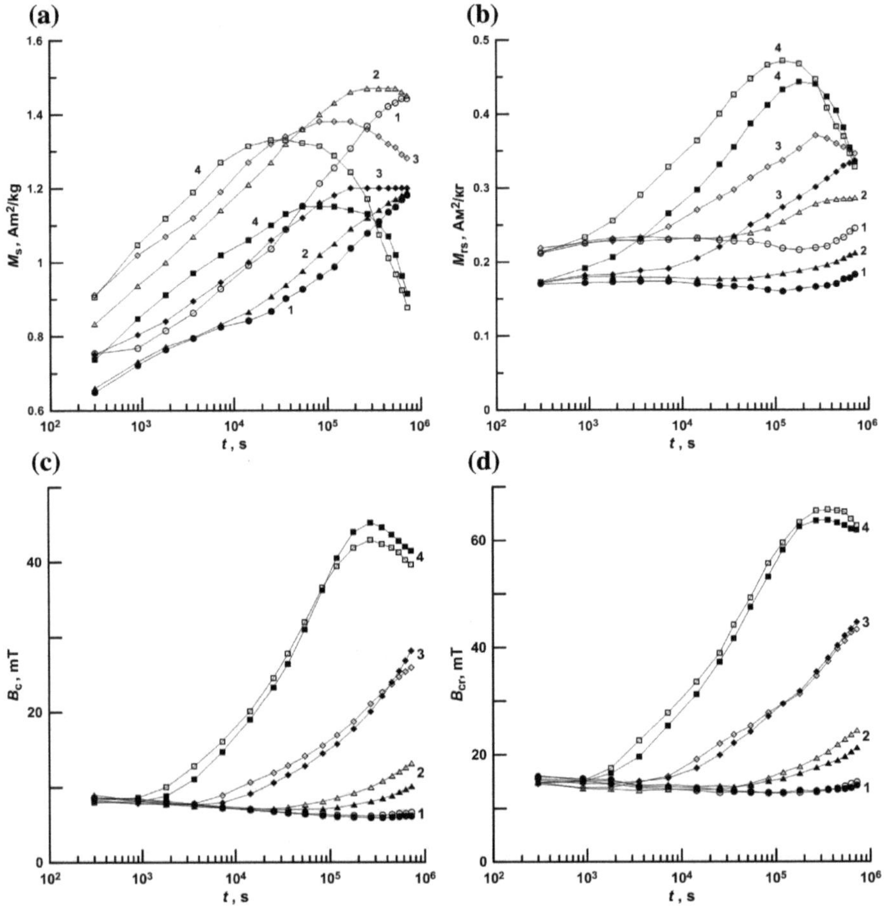

Fig. 14.3 Dependence of the parameters M_s (**a**), M_{rs} (**b**), B_c (**c**) and B_{cr} (**d**) on the duration t and temperature T_{an} of annealing in air of the fresh samples of basalts of the Reykjanes ridge (hollow circles) and the Red Sea (filled circles): 1–350, 2–400, 3–450, 4–500 °C. All measurements are performed at room temperature in 0.65 T field

$t > 2 \times 10^5$ s the values of B_c and B_{cr} return to the trend of decrease, probably due loss of stresses because of the violation of coherence at the interface magnetite cells—ilmenite lamellae with the formation of misfit dislocations.

Taken together, the changes in magnetic parameters cause a noticeable shift of the representing points (in comparison with their original locations) on the Day diagram (Fig. 14.4). First, for samples annealed at 350 °C, we observe a substantial motion towards the multi domain (MD) field what probably indicates diminishing stresses due to the decrease in inhomogeneities in distribution of different ions and vacancies because of enhanced diffusion processes. But annealing at high

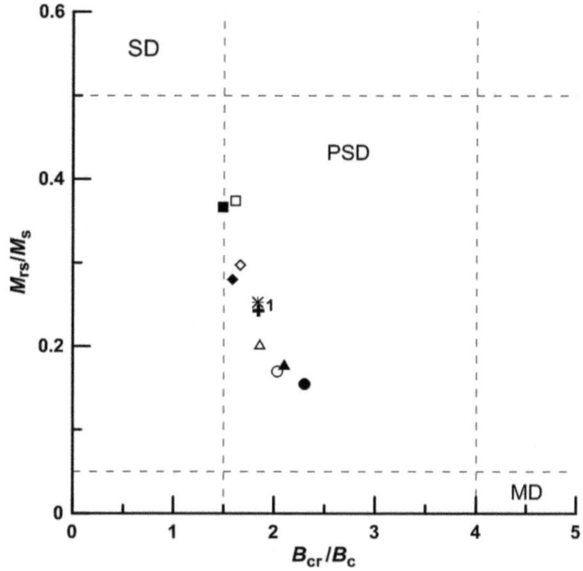

Fig. 14.4 Location of the representative points of samples of basalts of the Reykjanes ridge (hollow figures) and the Red Sea (filled figures) on the Day diagram (Day 1977) after annealing for 200 h at T_{an} = 350 °C (circles), 400 °C (triangles), 450 °C (rhombus), 500 °C (squares). The digit 1 indicates the initial state

temperatures gradually move representative points back to the single domain (SD)—pseudo-single domain (PSD) boundary, evidently as a consequence of continuing breakdown of individual TM grains due to the ongoing SPh oxidation.

From the data shown in Figs. 14.2 and 14.3, it can also be seen that the underwater basalts of the Red Sea are more thermostable in comparison with basalts from the Reykjanes ridge. This difference should be attributed to the larger size of the TM grains contained therein, as well as to the increased concentration of the isomorphous Al impurities which stabilize the crystal lattice of TM with respect to phase transformations (Bowles et al. 2012).

Monitoring of Acquisition of CRM

The intensity of CRM (M_{CRM}) measured directly at temperature of its creation as a function of time is displayed in Figs. 14.5 and 14.6 for the given set of T_{an}. For comparison, the curves $M_s(t)$, measured on sister cubes exposed to exactly the same annealing regime as the cubes used for $M_{CRM}(t)$ are placed on the same graphs. We underline that, in contrast to measurements of M_{CRM} intensities, the M_s values were measured after cooling the sample to room temperature. From these figures, a positive correlation between the corresponding $M_{CRM}(t)$ and $M_s(t)$ curves is clearly visible. This fact has been already noted by us earlier (Gribov et al. 2014, 2017; Gribov and Dolotov 2016) and interpreted by Shcherbakov et al. (2017), who claim that the observed correlation appears because of the process of CRM acquisition takes place only in a relatively short initial time interval of annealing. After this,

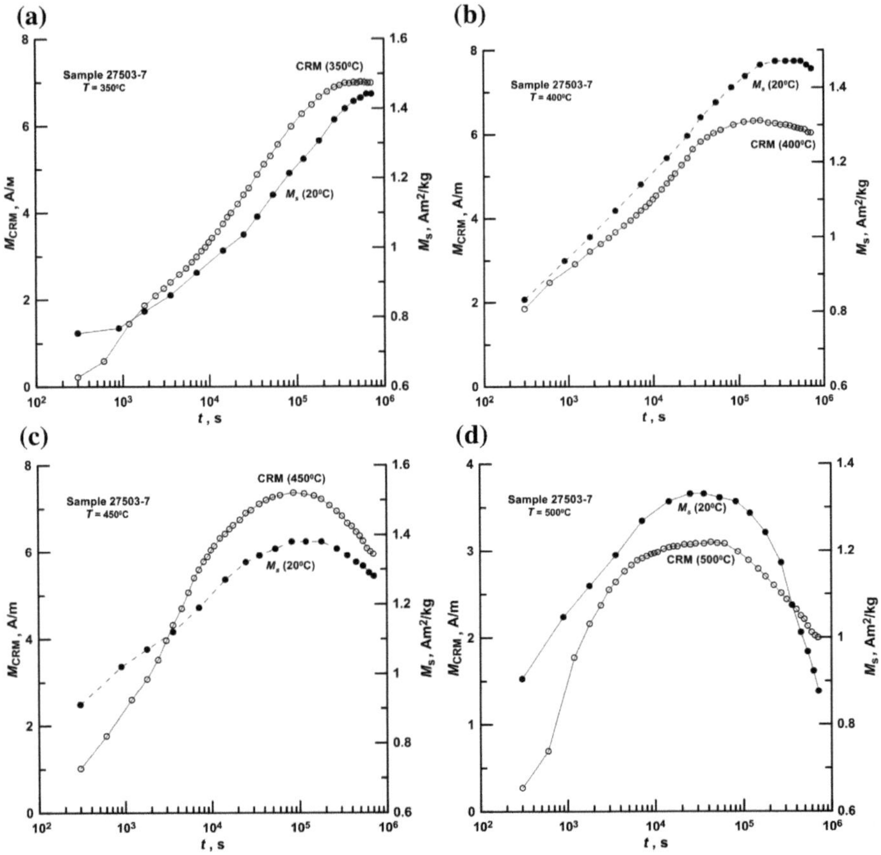

Fig. 14.5 Reykjanes ridge. Changes of M_{CRM} and M_s with time t (logarithmic scale) imparted at different temperatures: **a** 350 °C, **b** 400 °C, **c** 450 °C, **d** 500 °C. Values of M_{CRM} were measured directly at the annealing temperature while the M_s measurements were carried out after cooling each sister cube to room temperature and this heating-annealing-cooling cycle was repeated sequentially for each annealing time interval used for the monitoring of the corresponding CRM

the magnetic moments of grains are blocked and during further annealing changes in CRM basically follows the changes in M_s− with time.

Thellier Experiments

The Thellier experiments (Thellier and Thellier 1959) were carried out according to the double-heating protocol following Coe (1967), completed by pTRM checks (Figs. 14.7, 14.8, 14.9 and 14.10). All magnetizations are normalized to CRM (Figs. 14.7 and 14.8) or CRM + pTRM values (Figs. 14.9 and 14.10). Open circles

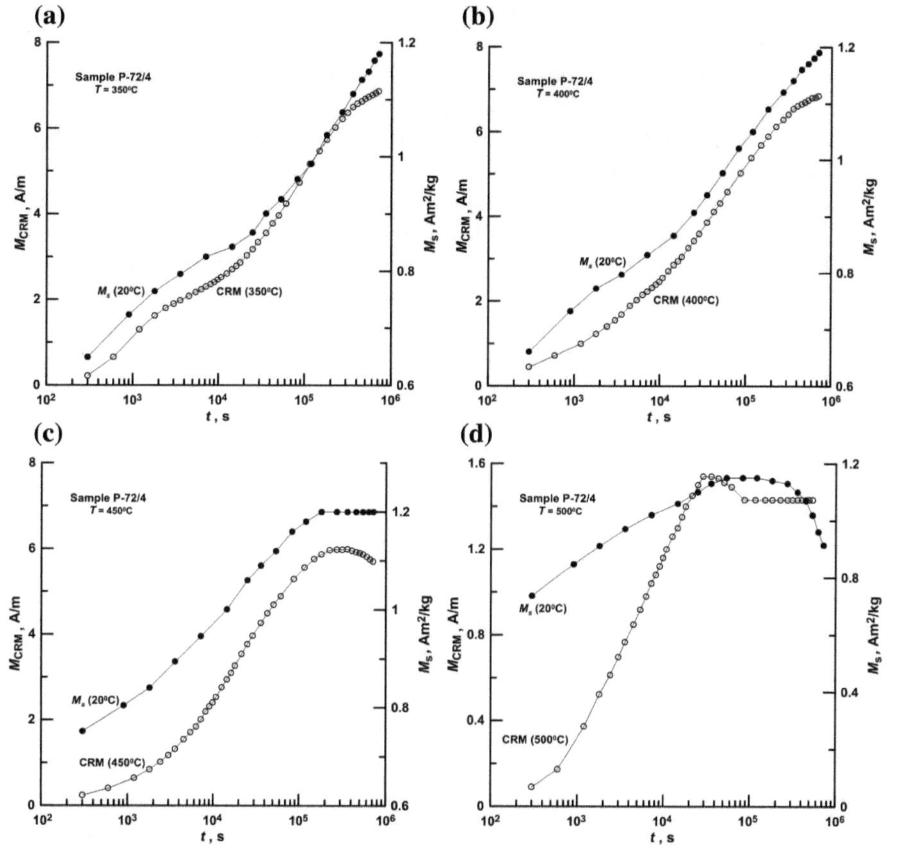

Fig. 14.6 The same as in Fig. 14.5 but for the Red Sea sample

correspond to the representative points for each heating-cooling cycle, the numbers near the circles indicate the heating temperature of the cycle, the solid triangles represent the position of check-points. The straight lines are the fits to the Arai-Nagata diagrams over a particular temperature interval used for the given "paleointensity" determination.

Discussion

The results of our XRD experiments, electron microscopic observations (Fig. 14.1) and a set of thermomagnetic (Figs. 14.2 and 14.3) and magnetic hysteresis measurements (Fig. 14.4) clearly indicates that the final state of samples after 200 h annealing strongly depends on T_{an}, that is, the higher T_{an} is the more SD behavior the sample exhibits.

All measured Arai-Nagata diagrams can be divided into two-three practically straight lines. For an isolated CRM (samples were cooled in zero field), in most cases the diagrams can be approximated by two linear segments characterized by different slopes k with the break point T_{br} ranging from 530 to 570 °C (Figs. 14.7 and 14.8). Accordingly, the diagrams (CRM + pTRM, pTRM) for samples imparted a combination of CRM and pTRM display often three quasi-linear sections with the breaks in the slopes localized at $T = T_{an}$ and $T = T_{br} \approx 560–570$ °C (Figs. 14.9 and 14.10). Because in paleomagnetic practice the thermal pre-history of rocks is usually poorly known, we cannot decide *a priory* which of the segments represents (if at all) the true field and may try to use any of them for paleointensity determinations. Hence, it is worthwhile to find out from controlled experiments, are there grounds to expect some regularities in behavior of slopes for linear segments

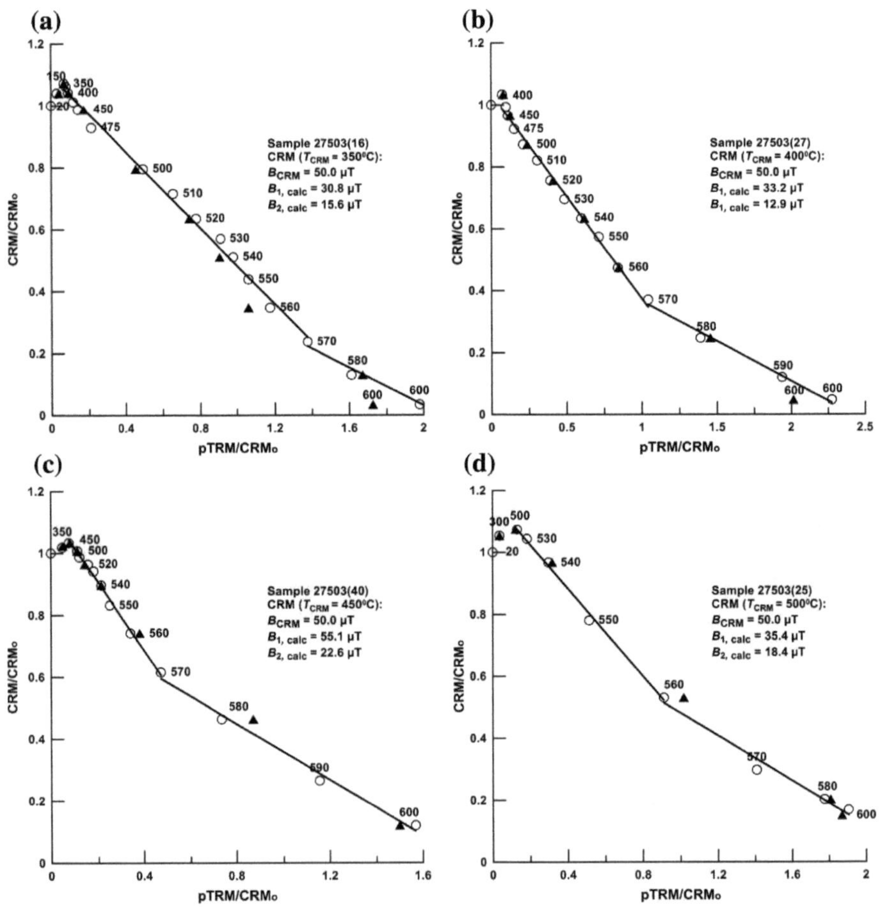

Fig. 14.7 Reykjanes Ridge, isolated CRM. The Arai-Nagata diagrams for T_{an} = 350°C **a**, T_{an} = 400 °C **b**, T_{an} = 450 °C **c**, T_{an} = 500 °C **d**

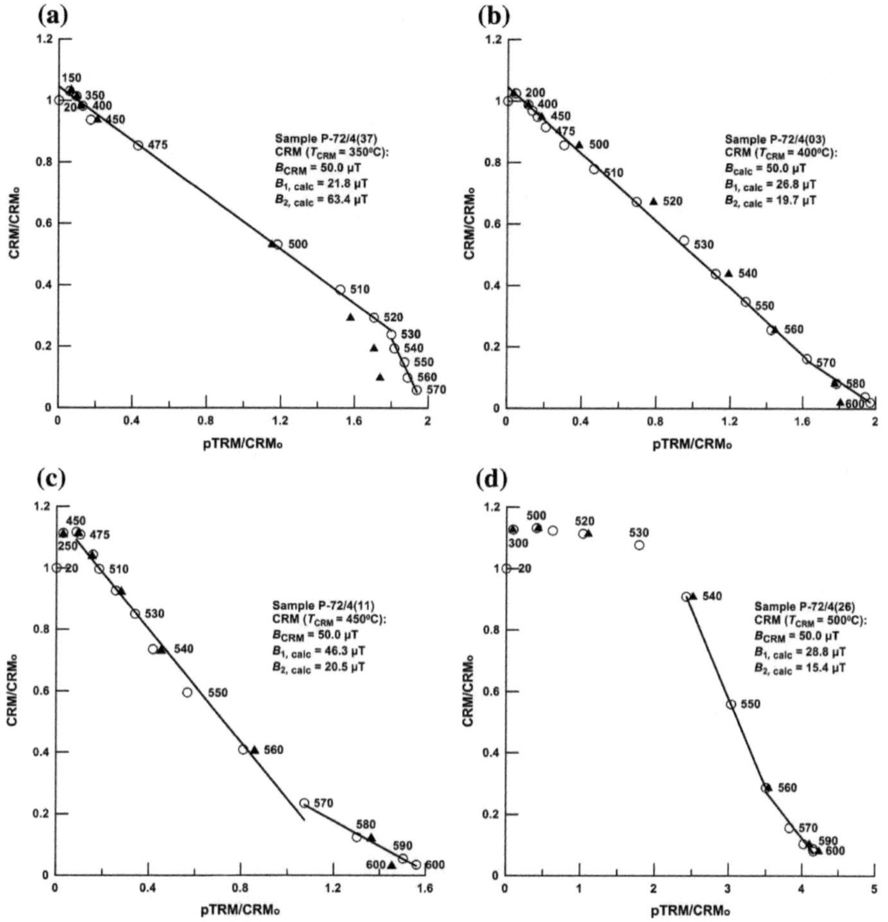

Fig. 14.8 The same as in Fig. 14.7 but for basalt samples of the Red Sea

responding different thermal intervals? For this purpose, we calculated the "paleointensity" B_{calc} according to the relation $B_{calc} = |k| \cdot B_{CRM}$. Results of these calculations for all specimens used in our experiments are summarized in Tables 14.2 and 14.3 altogether with the corresponding parameters characterizing the quality of the determinations.

Isolated CRM. As noted above, the Arai-Nagata diagrams for isolated CRM almost always can be separated into two linear segments with sufficiently different slopes. The low-temperature segment $T_{an} < T < T_{br}$ is of a special interest as it occupies a wide enough temperature interval with successful pTRM-check test judging from rather small values of *DRAT* parameters listed in the Tables 14.2 and 14.3. As is seen from these Tables, the calculated field B_{calc} sufficiently underestimates the true field B_{CRM} except for the case T_{an} = 450 °C. For the

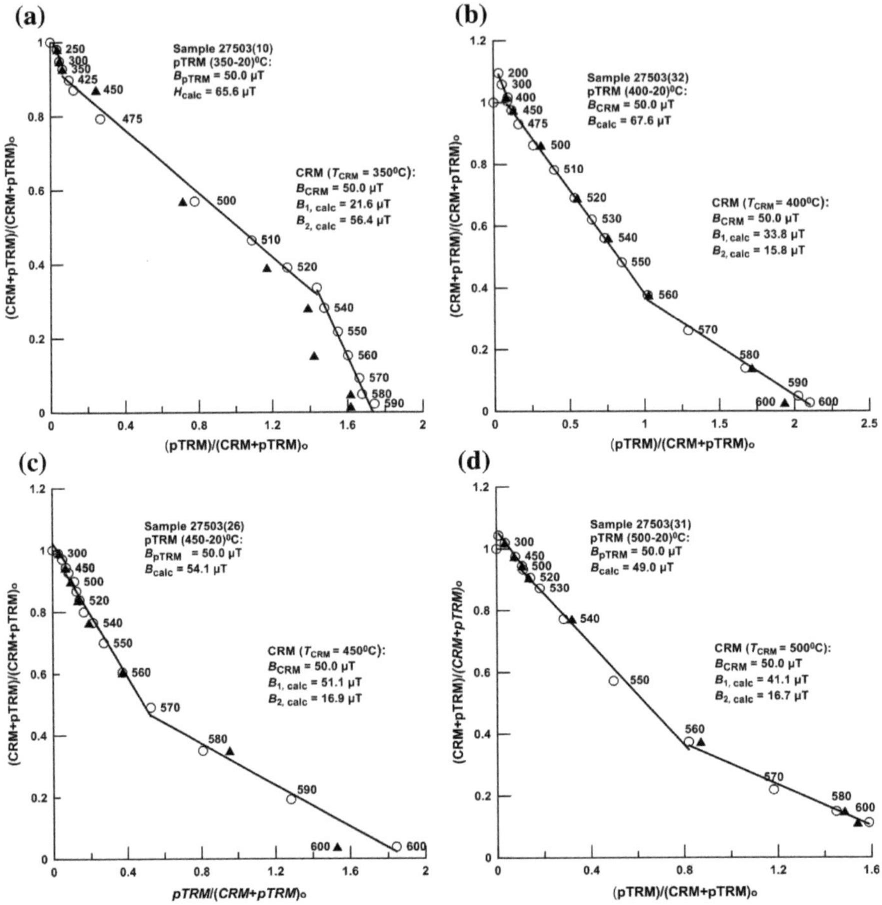

Fig. 14.9 Reykjanes Ridge, CRM + pTRM. The Arai-Nagata diagrams for T_{an} = 350 °C **a**, T_{an} = 400 °C **b**, T_{an} = 450 °C **c**, T_{an} = 500 °C **d**

high-temperature segment (T_{br}, T_c), the data listed in the Tables 14.2 and 3 clearly show that the "paleointensity" determinations performed for this interval always yield too low B_{calc} and exhibit too big *DRAT* to be even formally accepted. Evidently, this rather narrow interval corresponds to the temperatures where the magnetic grains become again thermally unstable and for these grounds we will not discuss this high-temperature component anymore.

Combination of CRM and pTRM. For the samples imparted a combination of CRM and pTRM, the Arai-Nagata diagrams (CRM + pTRM, pTRM) are often approximated by three linear functions, and the breaks in the slopes are localized at $T = T_{an}$ and $T = T_{br} \approx 560$–570 °C (Figs. 14.9 and 14.10). The low-temperature interval (T_r, T_{an}) corresponds to a pTRM acquired during the cooling of the sample and superimposed on the CRM. Because of this, it seems reasonable to expect that

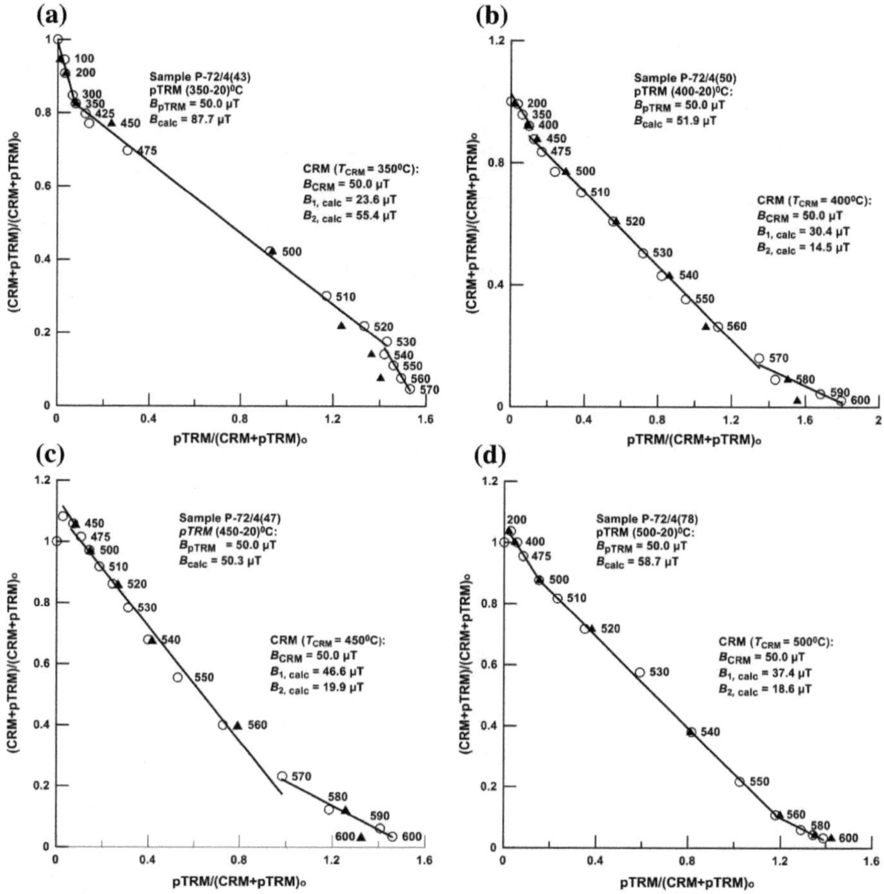

Fig. 14.10 The same as in Fig. 14.9 but for basalt samples of the Red Sea

values of B_{calc} determined for this interval should reproduce the true field B_{CRM}. However, it occurred to be not exactly the case: these values systematically overestimate B_{CRM} though the excess comprises not more than 20%. Although the question remains open, a possible explanation might be the MD state of some grains carrying the pTRM.

Turning to the most important middle-temperature segment ($T_{an} \approx 560$ °C) corresponding to the interval where thermal demagnetization of the majority of CRM takes place, we can just repeat the conclusions of the previous paragraph: the values of B_{calc} noticeably underestimate the true field except for the case when the annealing temperature T_{an} = 450 °C.

Summarizing the data obtained, the following tendency can be noticed. The "paleointensity" B_{calc} calculated for these segments for the specimens cut from the Reykjanes ridge sample annealed at T_{an} = 350, 400 and 500 °C underestimate

Table 14.2 Results of determination of B_{CRM} by the Thellier-Coe method on samples from the Reykjanes Ridge

T_{an}, °C	Specimen	(T_1, T_2), °C	*N*	*g*	*q*	*f*	*k*	*σ*(*k*)	*DRAT* (max)	*DRAT* (mean)	B_{calc}, μT	*B**	Remanence
1	2	3	4	5	6	7	8	9	10	11	12	13	14
350	16	350–570	14	0.89	31.9	0.75	0.62	0.01	4.8	1.7	30.8	38	CRM
		570–600	4	0.51	1.3	0.30	0.31	0.04	17.4	5.3	15.6	69	
	10	350–530	10	0.78	18.2	0.63	0.43	0.01	8.0	1.8	21.6	57	**CRM** + pTRM
		530–600	8	0.83	2.1	0.17	1.13	0.07	41.3	12.1	56.4	13	
		20–350	12	0.84	0.4	0.08	1.31	0.21	33.0	12.1	65.6	31	CRM + **pTRM**
	18	20–350	12	0.86	7.0	0.56	1.13	0.07	2.8	1.3	56.4	13	pTRM
400	27	400–570	12	0.89	17.5	0.63	0.66	0.02	3.5	1.5	33.2	34	CRM
		570–600	4	0.65	4.4	0.51	0.26	0.02	5.1	1.6	12.9	74	
	32	400–560	11	0.88	31.1	0.60	0.67	0.01	4.7	1.4	33.8	33	**CRM** + pTRM
		560–600	5	0.70	6.6	0.51	0.32	0.02	4.6	1.5	15.8	68	
		20–400	14	0.85	0.2	0.08	1.35	0.05	17.5	8.9	67.6	35	CRM + **pTRM**
	21	20–400	14	0.91	11.1	0.70	1.11	0.06	2.3	1.5	55.6	11	pTRM
450	40	450–570	10	0.81	11.5	0.38	1.10	0.03	3.7	1.5	55.1	10	CRM
		570–600	4	0.66	5.7	0.61	0.45	0.03	11.2	1.7	22.6	55	
	23	450–570	10	0.82	5.3	0.46	0.98	0.07	2.6	1.3	49.2	2	CRM
		570–600	4	0.65	5.9	0.70	0.34	0.03	9.1	1.2	16.8	66	
	45	450–570	10	0.84	4.0	0.34	1.17	0.08	8.6	2.6	58.4	17	CRM
		570–600	4	0.64	2.9	0.58	0.34	0.04	9.9	1.7	17.2	66	
	26	450–570	10	0.82	7.9	0.46	1.02	0.05	4.0	1.9	51.1	2	**CRM** + pTRM
		570–600	4	0.66	5.3	0.69	0.34	0.03	10.6	1.6	16.9	66	
		20–450	16	0.87	0.4	0.07	1.08	0.17	25.1	11.8	54.1	8	CRM + **pTRM**
	47	20–450	16	0.91	18.9	0.60	1.08	0.03	2.0	1.1	53.8	8	pTRM

(continued)

Table 14.2 (continued)

T_{an}, °C	Specimen	(T_1, T_2), °C	*N*	*g*	*q*	*f*	*k*	σ(*k*)	*DRAT* (max)	*DRAT* (mean)	B_{calc}, μT	*B**	Remanence
500	25	500–560	7	0.64	11.7	0.47	0.71	0.02	2.1	0.9	35.4	29	CRM
		560–600	5	0.58	3.3	0.44	0.37	0.03	9.8	1.6	18.4	63	
	31	500–560	7	0.70	8.9	0.57	0.82	0.04	4.0	1.1	41.1	18	**CRM** + pTRM
		560–600	5	0.61	3.4	0.41	0.33	0.02	6.3	1.8	16.7	67	
		100–500	17	0.87	0.8	0.09	0.98	0.10	10.8	5.6	49.0	2	CRM + **pTRM**
	22	20–500	18	0.91	43.7	0.53	1.00	0.01	3.7	1.4	49.9	<1	pTRM

The column (T_1, T_2) shows the temperature interval of the Arai–Nagata diagram used for the determination of B_{calc}, *N* is the number of points within this interval used to determine the best linear fit. The parameter *f* represents the fraction of the primary magnetization spanned by the linear fit; the gap factor *g* quantifies the uniformity of the distribution of successive data points in a chosen temperature interval, the quality factor $q = k \times f \times g/\sigma(k)$ quantifies the general quality of a given palaeointensity determination with the coefficient $k = B_{calc}/B_{CRM}$ being the slope of the linear fit to the Arai–Nagata plot and σ(*k*) being its standard error (Coe et al. 1978). The parameters *DRAT* (max) and *DRAT* (mean) (in %) describe, respectively, the estimates of the maximal and average deviations of the pTRM-check-point values from the fitting line in the selected temperature interval (Selkin and Tauxe 2000). In the case of the combined magnetization CRM+ pTRM the values of B_{calc} correspond to the component indicated in boldface. *B** is the error in B_{true} determination (%) of the corresponding remanence (CRM or pTRM)

Table 14.3 The same as in Table 14.2 but for sample from the Red Sea

T_{an}, °C	Specimen	(T_1, T_2), °C	N	g	q	f	K	$\sigma(k)$	$DRAT$ (max)	$DRAT$ (mean)	B_{calc}, μT	B^*	Remanence
1	2	3	4	5	6	7	8	9	10	11	12	13	14
350	37	350–530	10	0.74	36.7	0.73	0.44	0.01	13.1	2.1	21.8	56	CRM
		530–570	5	0.73	0.5	0.07	1.27	0.14	110.0	23.7	63.4	27	
	43	350–530	10	0.73	48.1	0.75	0.47	0.01	14.8	5.8	23.6	53	**CRM** + pTRM
		530–570	5	0.72	0.3	0.07	1.11	0.22	135.8	24.2	55.4	11	
		200–350	7	0.80	1.1	0.09	1.75	0.11	20.5	2.6	87.7	75	CRM + **pTRM**
	77	20–350	12	0.83	5.3	0.56	1.21	0.08	3.6	1.7	60.5	21	pTRM
400	03	400–570	12	0.89	60.3	0.79	0.54	0.01	5.2	1.8	26.8	46	CRM
		570–600	4	0.58	1.1	0.17	0.39	0.03	23.7	7.6	19.7	61	
	50	400–570	12	0.89	24.4	0.80	0.61	0.01	4.0	1.5	30.4	39	**CRM** + pTRM
		570–600	4	0.64	0.7	0.25	0.29	0.07	14.4	5.9	14.5	71	
		20–400	14	0.84	0.5	0.09	1.04	0.14	28.0	11.8	51.9	4	CRM + **pTRM**
	75	20–400	14	0.91	8.5	0.55	1.19	0.07	0.9	0.7	59.5	19	pTRM
450	11	450–570	10	0.84	16.2	0.77	0.93	0.04	2.7	1.1	46.3	7	CRM
		570–600	4	0.58	2.9	0.30	0.41	0.02	11.6	3.8	20.5	59	
	18	450–570	10	0.83	10.9	0.66	0.96	0.05	2.6	1.0	47.9	4	CRM
		570–600	4	0.65	2.7	0.46	0.42	0.05	8.4	2.1	21.0	58	
	22	450–580	11	0.86	32.7	0.82	0.87	0.02	2.5	0.8	43.4	13	**CRM** + pTRM
		580–600	3	0.47	2.3	0.26	0.38	0.02	14.2	3.3	19.1	62	
		20–450	16	0.85	0.2	0.06	1.13	0.32	20.9	7.0	56.6	13	CRM + **pTRM**
	47	450–570	10	0.85	16.0	0.77	0.93	0.04	1.9	0.8	46.6	7	**CRM** + pTRM
		570–600	4	0.59	1.8	0.31	0.40	0.04	13.7	3.6	19.9	60	
		250–450	9	0.81	0.2	0.04	1.01	0.19	34.3	11.6	50.3	<1	CRM + **pTRM**

(continued)

Table 14.3 (continued)

T_{an}, °C	Specimen	(T_1, T_2), °C	N	g	q	f	K	$\sigma(k)$	$DRAT$ (max)	$DRAT$ (mean)	B_{calc}, μT	B^*	Remanence
	21	20–450		0.92	14.1	0.68	1.13	0.05	1.8	1.0	56.7	13	pTRM
500	26	540–560	3	0.49	17.5	0.27	0.58	0.01	7.0	2.3	28.8	42	CRM
		560–600	5	0.58	0.9	0.15	0.31	0.03	12.6	4.8	15.4	69	
	78	500–560	7	0.81	33.6	0.77	0.75	0.01	2.3	0.7	37.4	25	**CRM** + pTRM
		560–600	5	0.59	1.3	0.14	0.37	0.02	13.3	4.5	18.6	63	
		20–500	18	0.70	0.7	0.15	1.17	0.18	8.8	4.0	58.7	17	CRM + **pTRM**
	31	20–500	18	0.91	27.5	0.43	1.07	0.01	2.5	1.6	53.8	7	pTRM

the true field B_{CRM} with the errors (38–57), (33–34) and (18–29) %, correspondingly. Accordingly, for the Red Sea sample the errors are (53–56), (39–46) and (25–42)% for T_{an} = 350, 400 and 500 °C, correspondingly. These observations are in line with results of earlier works (Draeger et al. 2006; Gribov and Dolotov 2016; Gribov et al. 2017; Shcherbakov et al. 2017). The observed tendency suggests that the higher the temperature of CRM acquisition T_{an} is the smaller is the error in B_{true} determination. The same tendency was noted by Draeger et al. (2006) who found that the underestimation error B^* was as large as 200% in experiments performed at T_{an} = 400 °C but it starts to reduce consistently with the T_{an} increase down to 50% at T_{an} = 450 °C and to only 10% at T_{an} = 500 °C due to the relative narrowness of the blocking temperature interval, centered near T_c.

However, our previous experiments with Permo-Triassic Siberian tholeitic basalts (Shcherbakov et al. 2017) and alkaline basalts selected on the island of San Thome (Atlantic Ocean) (Gribov and Dolotov 2016) as well as with Kamchatka basalt samples (Gribov 2017) showed no clear dependence of B^* on T_{an} yielding the underestimation errors confined in 6 to 58% interval with no respect to T_{an}.

Conclusions

1. Thellier experiments with laboratory induced CRM lead to 1.2–3.5-fold underestimation errors with respect to the true induced field value. This conclusion agrees with results of similar experiments performed earlier (Draeger et al. 2006; Gribov 2016, 2017; Gribov et al. 2017; Shcherbakov et al. 2017), and provides experimental support for the hypothesis that low values of the ancient geomagnetic field obtained from rocks carrying predominatly oxidized titanomagnetites may not reflect the actual field level but are artifacts due to unrecognized CRM erroneously taken for the TRM (Smirnov and Tarduno 2005; Draeger et al. 2006).
2. According to our data and the data by Draeger et al. (2006), the underestimation error tends to decrease with increase of the annealing temperature. However, it cannot be considered as a rule as the data by Gribov (2016, 2017), Gribov et al. (2017), Shcherbakov et al. (2017) do not exhibit such the tendency.
3. Surprisingly, all specimens annealed at 450 °C yielded B_{calc} whose values are very close to the true field B_{CRM}. It is quite an intriguing result but we cannot suggest here a reasonable explanation leaving it for future investigations.

Acknowledgements The work was supported by the Russian Foundation for Basic Research (project no. 17-05-00259), by the Russian Federation Ministry of Science and Education (contract no. 14.Z50.31.0017) and by the State Research Task IPE RAS 0144-2014-00117. The authors are grateful to Tsel'movich for the help in producing the electronic micrographs.

References

Bowles J.A., Tatsumi-Petrochilos L., Hammer J.E., Brachfeld S.A. Multicomponent cubic oxide exsolution in synthetic basalts: temperature dependence and implications for magnetic properties // J. Geophys. Res. 2012. V. 117. B03202, https://doi.org/10.1029/2011jb008867.

Coe R.S. The determination of paleointensities of the Earth's magnetic field with special emphasize on mechanisms which could cause nonideal behavior in Thellier method // J. Geomag. Geoelectr. 1967. V. 19, N 3. P. 157–179.

Coe R.S., Gromme C.S., Mankinen E.A. Geomagnetic paleointensities from radiocarbon-dated lava flows on Hawaii and the question of the Pacific nondipole low // J. Geophys. Res. 1978. V. 83, N B4. P. 1740–1756.

Day R. Hysteresis properties of titanomagnetites: Grain-size and compositional dependence // Phys. Earth Planet. Inter. 1977. V. 13, N 4. P. 260–267.

Draeger U., Prévot M., Poidras T., Riisager J. Single-domain chemical, thermochemical and thermal remanences in a basaltic rock // Geophys. J. Int. 2006. V. 166, N 1. P. 12–32.

Gribov S.K. The assessment of the impact of process of single-phase oxidation and subsequent decomposition of titanomagnetites on the results of palaeointensity determination by the Thellier method (results of laboratory simulation) / Materialy mezhdunar. shkoly-seminara po problemam paleomagnetizma I magnetizma gornykh porod [Proc. Int. School-Seminar on Paleomagnetism and Rock Magnetism, St. Peterburg, Petrodvorets, October 3–7, 2016]. Yaroslavl: Filigran, 2016. P. 40–47 (In Russian).

Gribov S.K. The effect of chemical magnetization on the palaeointensity determined by the Thellier method: experimental study on titanomagnetite-containing basalts // Geophys. Research. 2017. V. 18, N 1. P. 37–48 (In Russian).

Gribov S.K., Dolotov A.V. Peculiarities of the behavior of chemical remanent magnetization in the oxidation of natural titanomagnetite under isothermal conditions / Materialy mezhdunar. shkoly-seminara po problemam paleomagnetizma I magnetizma gornykh porod [Proc. Int. School-Seminar on Paleomagnetism and Rock Magnetism, St. Peterburg, Petrodvorets, October 3–7, 2016]. Yaroslavl: Filigran, 2016. P. 35–39 (In Russian).

Gribov S.K., Dolotov A.V., Shcherbakov V.P. Experimental modeling of the chemical remanent magnetization and Thellier procedure on titanomagnetite-bearing basalts // Izv. Phys. Solid Earth. 2017. V. 53. N 2. P. 274–292.

Gribov S.K., Dolotov A.V., Tsel'movich V.A. Peculiarities of magnetomineralogical transformation of natural titanomagnetites in air under osothermal conditions // Uch. Zap. Kazan. Univer., Ser. Estestv. Nauki. 2014. V. 156, book 1. P. 64–78.

Maksimochkin V.I., Mbele J.R., Trukhin V.I., Schreider A.A. Paleointensity of the geomagnetic field in the last half-million years in regions of the Red Sea and south of the Mid-Atlantic Ridge // Moscow Univ. Phys. Bull. 2010. V. 65. N 6. P. 531–538.

Maksimochkin V.I., Tselebrovskiy A.N., Shreider A.A. The Brunhes-Matuyama epoch magnetic field evolution from the Mid-Atlantic Ridge basalts // Memoirs of the Faculty of Physics. 2016. N 3. 163910.

Nishitani T., Kono M. Curie temperature and lattice constant of oxidized titanomagnetite // Geophys. J. R. Astr. Soc. 1983. V. 74. P. 585–600.

Richards J.C.W., O'Donovan J.B., Hauptman Z., O'Reilly W., Creer K.M. A magnetic study of titanomagnetite substituted by magnesium and aluminium // Phys. Earth Planet. Inter. 1973. V.7. N 4. P. 437–444.

Rift zone of the Reykjanes Ridge (edited by A.P. Lisitsyn, L.P. Zonenshain) Moscow: Nauka, 1990. 238 p. (in Russian).

Selkin P.A., Tauxe L. Long-term variations in palaeointensity // Philos. Trans. R. Soc. London, ser. A. 2000. V. 358, N 1768. P. 1065–1088.

Shreider A.A., Trukhin V.I., Sychev V.A., Rimsky-Korsakov. Detailed geomagnetic studies of the rift zone in the southern part of the Red Sea // Okeanologiya. 1982. V. 22. Issue 3. P. 439–445 (In Russian).

Smirnov A.V., Tarduno J.A. Thermochemical remanent magnetization in Precambrian rocks: Are we sure the geomagnetic field was weak? // J. Geophys. Res. 2005. V. 110, N B06103.

Shcherbakov V.O., Sycheva N.K., Gribov S.K. Experimental and numerical simulation of the creation of chemical remanent magnetization and the Thellier procedure // Izv. Phys. Solid Earth. 2017. V. 53. N 5. P. 645–657.

Thellier E., Thellier O. Sur l'intensité du champ magnétique terrestre dans le passé historique et géologique // Ann. Geophys. 1959. V. 15. P. 285–376.

Trukhin V.I., Maksimochkin V.I., Zhilyaeva V.A., Kurochkina E.S., Shreider A.A., Kashintsev. Magnetic properties of basalts and geodynamics of the rift zone in the southern Red Sea // Izv., Phys. Solid Earth. 2006. V. 42. N 11. P. 928–941.

Chapter 15
Blocking Temperature and Hysteresis Characteristics of Nanoparticles of Oxidated Magnetite

Ilia Iliushin and Leonid Afremov

Abstract An effect of a single-phase oxidation process on the hysteresis characteristics and blocking temperature of magnetite has been carried out within the framework of the model of core-shell nanoparticle. It has been shown that an increase of the degree of oxidation of magnetite grains results in a decrease of the spontaneous magnetization and slight change of coercive field and remanent saturation magnetization to spontaneous magnetization ratio. Increase of the portion of maghemite lead to decrease of the blocking temperature. All results are in agreement with an experimental data.

Keywords Oxidation · Nanoparticles · Blocking temperature · Magnetization

Introduction

The main source of a natural remanent magnetization of oceanic basalts are the magnetite oxides and titanomagnetite oxides, which can alter due to the natural low-temperature chemical processes. During these processes secondary magnetization can emerge. This complicates determination of the ancient geomagnetic field using the natural remanent magnetization. For instance, in the basalts of the upper-layer oceanic crust titanomagnetite $Fe_{3-x}Ti_xO_4$ is the most common magnetic material ($x \approx 0.6$, Stephenson 1972), and magnetite oxides in the loess, red sandstone and soils (Prevot et al.1981; Liu et al. 2004). It is known, that at low temperature formation of the magnetite, maghemite or titanomaghemite during the oxidation of the titanomagnetite is possible (Smith and Banerjee 1986; Ryall and Hall 1979; Carvallo et al. 2004). Primary thermo-remanent magnetization (I_{rT}) replaced with the chemical remanent magnetization (I_{rc}). Direction of chemical magnetization of grains formed during the decay or oxidation of cation-deficient

I. Iliushin (✉) · L. Afremov
Department of Theoretical and Nuclear Physics, Far Eastern Federal University, Vladivostok, Russia
e-mail: iliushin.ig@dvfu.ru

D. Nurgaliev et al. (eds.), *Recent Advances in Rock Magnetism, Environmental Magnetism and Paleomagnetism*, Springer Geophysics,
https://doi.org/10.1007/978-3-319-90437-5_15

single-domain or pseudo single-domain titanomagnetite grains may not coincide with the direction of the external field applied during the process (Walderhaug 1992). Some investigators think that the data of paleointensity of the geomagnetic field obtained for the ferromagnetic minerals prone to conversion to secondary minerals should be treated with caution (Draeger et al. 2006). For instance, maghemisation of underwater basalts can lead to decrease of the primary natural remanent magnetization. In this case, the determination of the paleointensity of the geomagnetic field by the Telier method can give underestimated values, since the chemical magnetization is, as a rule, less than the thermo-remanent value obtained in the same field (Dunlop and Hale 1976; Carmichael 1977; Grommé et al. 1979).

Note, that the study of the oxidation of magnetite conducted not only on the natural materials, but on the artificial materials as well. For example, in the reference (Kechrakos et al. 2007) results of a study of the effect of maghemisation process on the magnetic characteristics and blocking temperature of a system of magnetite nanoparticles. Authors showed that increase of the proportion of the oxidized shell (maghemite) lead to decrease of the blocking temperature and have almost no effect on the spontaneous magnetization of a system of nanoparticles.

The oxidation mechanism is well described in the works of O'Reilly (1984) and Gallagher (1968), according to which the oxidation process beginning on the surface, where bivalent iron ions become trivalent. Further, as a result of diffusion processes, the bivalent ions leave vacancies and move from the center of nanoparticle towards the surface and form the oxidized layer. At low temperature (Ozdemir and Dunlop 2010) and the lowered diffusion threshold, the diffusion gradient abruptly breaks off in an isolated oxide layer. As a result, an oxidized shell is formed around a non-oxidized core, the so-called core-shell structure. However, despite the fact that core-shell structures have been studied for more than 50 years (Liu et al. 2004; Gallagher 1968; Ozdemir and Dunlop 2010; Banerjee et al. 1981), the effect of such a structure on the magnetic properties of partially oxidized nanoparticles remains insufficiently studied (Ge et al. 2014).

The aim of this work is to model the effect of the oxidation process on the hysteresis characteristics and the blocking temperature of a system of magnetite nanoparticles.

Model

We use the core-shell nanoparticle described in reference (Afremov and Ilyushin 2013):

1. we assume a uniformly magnetized ellipsoidal nanoparticle with volume $V = 4\pi QB^3/3$ and elongation Q (phase 1) containing uniformly magnetized ellipsoidal core with volume $v = \varepsilon V = 4\pi qb^3/3$ and elongation q (phase 2);
2. crystallographic anisotropy axes of both phases and major axes are parallel to each other and spontaneous magnetization vectors $I_S^{(1)}$ and $I_S^{(2)}$ are located in

same plane with the external magnetic field H parallel to axis Oz making angles $\vartheta^{(1)}$ and $\vartheta^{(2)}$, respectively;

In the absence of external magnetic field $(H = 0)$ and thermal fluctuations $(T = 0)$, nanoparticle will be in one of four states (Afremov and Ilyushin 2013):

- Magnetic moments of both phases are parallel and oriented along the axis Oz—*first state* «(↑↑)»;
- Magnetic moment of first phase is oriented along the axis Oz, and magnetic moment of second phase is oriented the opposite side—*second state* «(↑↓)»;
- Magnetic moment of both phases is antiparallel to the axis Oz—*third state* «(↓↓)»;
- Magnetic moment of second phase is oriented along the axis Oz, and magnetic moment of first phase is oriented the opposite side—*fourth state* «(↓↑)» (Fig. 15.1).

If temperature is not equal to zero then population (the probability of finding nanoparticle in one of four states) of four magnetic states n_i can be determined with the help of matrix exponential (Afremov et al. 2015):

$$\boldsymbol{N}(t) = \exp(\boldsymbol{\mathcal{W}}t) \cdot \boldsymbol{N}(0) + \int_0^t \exp(\boldsymbol{\mathcal{W}}(t-\tau))d\tau \cdot \boldsymbol{\mathcal{V}}, \tag{15.1}$$

where the matrix elements $\boldsymbol{\mathcal{W}}$, $\boldsymbol{\mathcal{V}}$, and vectors $\boldsymbol{N}(t)$, $\boldsymbol{N}(0)$ expressed through W_{ik} and $\boldsymbol{n}(t)$ respectively:

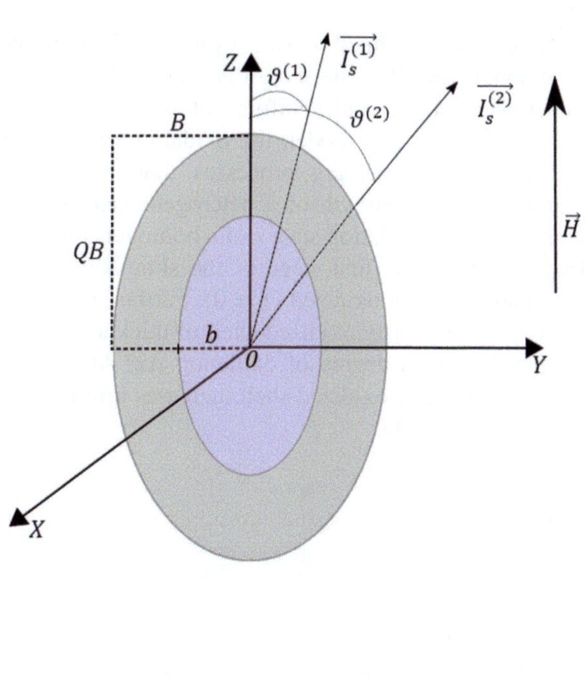

Fig. 15.1 The model of the core-shell nanoparticle

$$\mathcal{W}_{ik} = \begin{cases} -\sum\limits_{j \neq i, j=1}^{4} W_{ij} - W_{4i}, & i = k \\ W_{ki} - W_{4i}, & i \neq k \end{cases}, \quad \boldsymbol{\mathcal{V}} = \begin{pmatrix} W_{41} \\ W_{42} \\ W_{43} \end{pmatrix}, \quad \boldsymbol{N}(t) = \begin{pmatrix} n_1(t) \\ n_2(t) \\ n_3(t) \end{pmatrix},$$

$$\boldsymbol{N}(0) = \begin{pmatrix} n_{01} \\ n_{02} \\ n_{03} \end{pmatrix}. \tag{15.2}$$

Here $W_{ik} = f_0 \exp(-E_{ik}/k_B T)$ are the elements of a transition matrix from i-th equilibrium state to k-th state, $f_0 = 10^{10}\ \mathrm{s}^{-1}$ is the frequency factor, E_{ik} are potential barriers separating i-th and k-th states.

Equations (1) and (2) allows one to calculate magnetization of a system of nanoparticles:

$$\begin{aligned} I(\mathrm{t}) = \mathrm{c} \int \Big[\Big((1-\varepsilon) \mathrm{I}_{\mathrm{s}}^{(1)} + \varepsilon \mathrm{I}_{\mathrm{s}}^{(2)} \Big) (n_1(t,a) - n_3(t,a)) \\ + \Big((1-\varepsilon) \mathrm{I}_{\mathrm{s}}^{(1)} - \varepsilon \mathrm{I}_{\mathrm{s}}^{(2)} \Big) (n_2(t,a) - n_4(t,a)) \Big] F(a)\, da. \end{aligned} \tag{15.3}$$

where $c = N_0 V / V_0$ is the volume concentration of nanoparticles, $F(a)$ stands for the size distribution function of nanoparticles.

The assumption of a homogeneous distribution of the magnetic moments of the phases is due to the small size of magnetite nanoparticles compared to the width of the domain wall. In this paper we will study magnetite particles with a size less than 100 nm, while the width of the domain wall in magnetite is 2 times larger than this size for isometric particles and increases substantially with increasing elongation. For example, for $q = 1.5$, the width of the boundary is 450 nm (Afremov et al. 1998).

Study of the effect of single-phase oxidation on the magnetic properties of titanomagnetites has been carried out within the framework of the model of Scherbakov (1982) and Nikiforov et al. (2017).

According to this model, the oxidation of magnetite can be considered as two mutually complementary processes that ensure the formation of a core-shell nanoparticle: the growth of a heterogeneously oxidized shell and a simultaneous increase in the oxidation state z of a homogeneously oxidized core of a nanoparticle (Fig. 15.2). At the first stage of the simulation, it is assumed that the magnetite nanoparticle is not oxidized (z = 0). Further, two phases are formed in the particle, an inhomogeneously oxidized shell of thickness Δ, and a partially oxidized core of magnetite with a degree of oxidation Δz. At the k-th stage, the thickness of the inhomogeneously oxidized shell increases up to kΔ, and the degree of oxidation of the core up to kΔz.

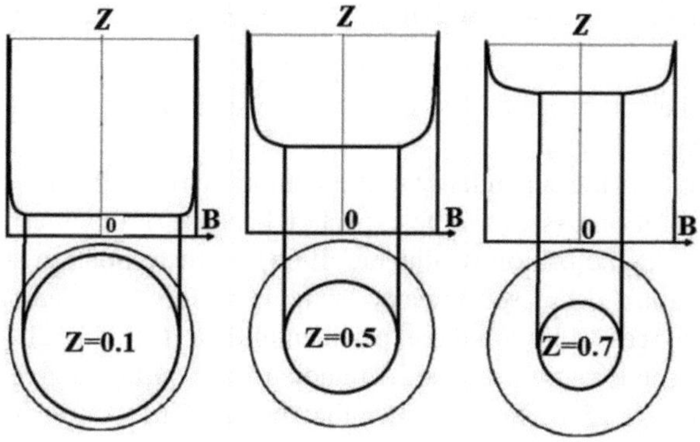

Fig. 15.2 Illustration for the oxidation model of magnetite nanoparticle. Distribution of the oxidation parameter z shown in the upper part of picture. Distribution is calculated within the framework of the model described in the reference (Nikiforov et al. 2017). Lower part of this picture is an illustration of core-shell nanoparticle with homogeneously oxidized core z_k and inhomogeneously oxidized shell ($z_k \leq z \leq 1$)

Selection of the Modeling Parameters

During the calculations of various types of remanent magnetization, it was assumed that as a result of oxidation of magnetite in a spherical nanoparticle, $a = 2B$ in size, a spherical magnetite core with radius b covered by a maghemite shell is formed. The size distribution function was taken from the distribution presented in the reference (Ge et al. 2014). In addition, we used the approximation of the experimentally established dependence of the spontaneous magnetization on the sizes a of the nanoparticles (Nikiforov et al. 2017) and the degree of oxidation z (Shcherbakov and Gribov 1986) was approximated by the following expressions:

$$I_s(a, z) = 480\left(1 - \frac{a_0}{a}\right)(1 - 0.23x)G, \quad a_0 = 4.07 \times 10^{-7}\text{cm}. \tag{15.4}$$

The dependence of the magnetic crystallographic anisotropy constant on the degree of oxidation was determined with the help of (Kechrakos et al. 2007)

$$K_A(a, z) = K_{ef}(a)\left(1 + 2.27x - 10.42x^2 + 6.54x^3\right), \tag{15.5}$$

and the dependence of the effective anisotropy constant on the size a of nanoparticles was determined by the well-known relation (Gubin et al. 2005).

$$K_{ef}(a) = K_V + \frac{K_S}{a}, \tag{15.6}$$

where, according to (Krupichka 1976; Perez et al. 2008) crystallographic anisotropy constants of magnetite are $K_V = -1.06 \times 10^5$ erg/cm^3, $K_S = 0.029$ erg/cm^2.

An Effect of Oxidation on the Hysteresis Characteristics

The magnetic characteristics of a system of core-shell nanoparticles were obtained using the hysteresis loops.

Figure 15.3 shows the dependence of the saturation magnetization of magnetite on the degree of oxidation. As was to be expected, because of the linear dependence of the spontaneous magnetization (Gallagher 1968) on the oxidation state z, the saturation magnetization decreases linearly with increasing z (Readman 1972). The obtained result agrees well with the experimental data (Ge et al. 2014). The dependence of the coercive field of magnetite nanoparticles on the degree of oxidation is presented on the Fig. 15.4a. It can be seen that the theoretical curve passes somewhat lower than the experimental curve. This is due to the fact that in the modeling we did not include particles with a size larger than 100 nm in the distribution function, assuming their number to be negligible. While a certain fraction of particles smaller than 100 nm can be in the superparamagnetic state, which led to lower theoretical values of not only the coercive field but also the ratio of the remanent saturation magnetization to the spontaneous saturation magnetization of magnetite (Fig. 15.4b).

Dependence of the Blocking Temperature on the Degree of Oxidation

The spectrum of the relaxation times is determined by the frequency matrix of the transitions from the i-th state to the k-th state: $W_{ik} = f_0 exp(-E_{ik}/k_B T)$. Since the eigenvalues of the transition matrix W_{ik} are the inverse times of the transition from one state to another one $1/\tau_k$, we will use the maximum of them, τ_{kmax}, to estimate the transition time to the equilibrium state, assuming that transitions with a shorter relaxation time have already been completed. Then, to estimate the blocking temperature, we can use $\tau_{kmax}(T_b) = t_{cal}$. Calculation time t_{cal} is 1 s.

When calculating the dependence of blocking temperature T_b of the system of $Fe_3O_4 - \gamma Fe_2O_3$ nanoparticles on the oxidation state, the experimental values

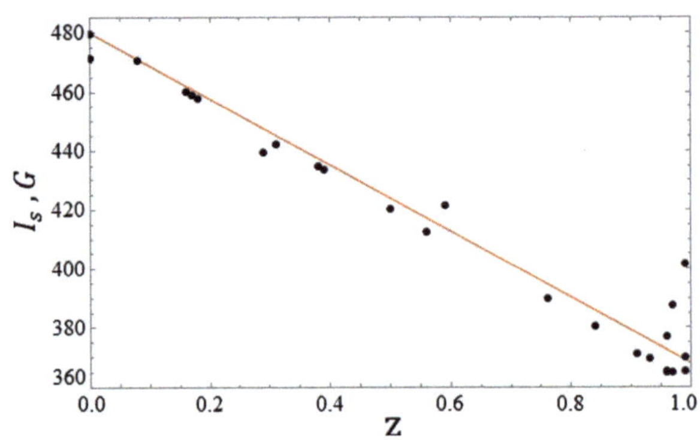

Fig. 15.3 Dependence of the saturation magnetization of magnetite I_s on the degree of oxidation z. The points indicate the experimental values of the saturation magnetization presented in the reference (Ge et al. 2014)

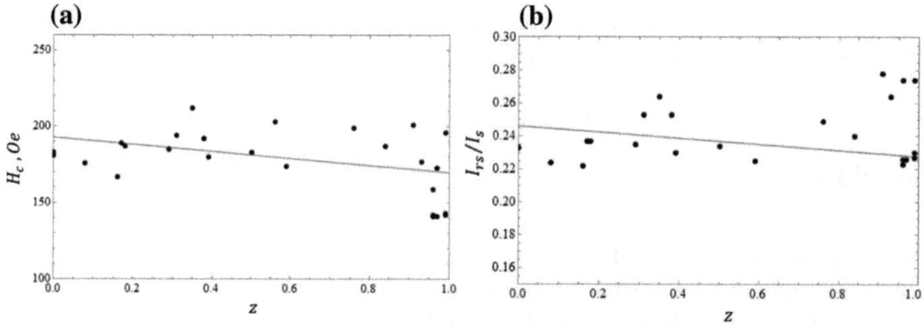

Fig. 15.4 Dependences of: **a** the coercive field H_c, **b** remanent saturation magnetization I_{rs} to saturation magnetization I_s ratio on the degree of oxidation z of magnetite. The points indicate the experimental values presented in the reference (Ge et al. 2014)

Table 15.1 Experimental values of the particle characteristics presented in the article (Kechrakos et al. 2007)

No.	Diameter of the core (nm)	Shell thickness (nm)	Magnetite fraction
1	5.0	2.0	0.184
2	6.4	1.9	0.261
3	7.8	1.1	0.493
4	5.6	2.2	0.230
5	4.4	2.6	0.106
6	18.0	1.1	0.712

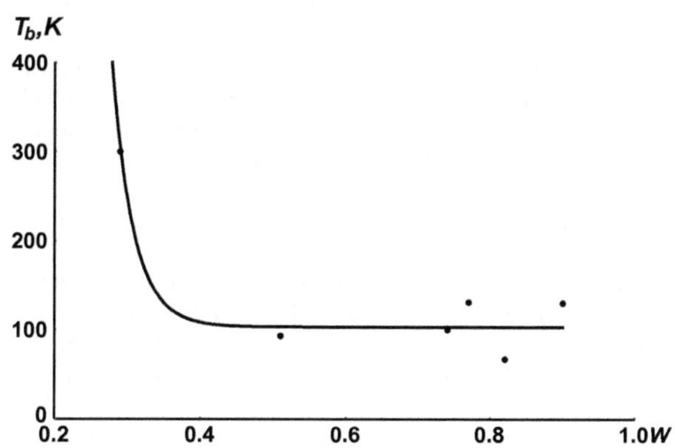

Fig. 15.5 Dependence of the blocking temperature T_b on the fraction of maghemite W for $Fe_3O_4 - \gamma Fe_2O_3$ nanoparticles. The points indicate the experimental values of T_b (Kechrakos et al. 2007)

(Kechrakos et al. 2007) of the particle characteristics presented in Table 15.1 were used.

The modeling results shown in Fig. 15.5 suggest that a decrease in the proportion of maghemite W leads to an increase in the blocking temperature T_b, which is associated with the growth of the magnetite core.

The results of the calculations agree qualitatively with the experimental values of the blocking temperature, given in (Kechrakos et al. 2007).

Conclusion

Modeling the effect of oxidation processes on the hysteresis characteristics of a system of magnetite nanoparticles carried out within the framework of a core-shell nanoparticle model (Afremov and Ilyushin 2013) showed that:

(a) The spontaneous saturation magnetization decreases with increasing degree of oxidation of nanoparticles,
(b) oxidation of magnetite slightly affects the coercive field H_c and the ratio I_{rs}/I_s. The results of the modeling are in good agreement with experimental data (Ge et al. 2014),
(c) an increase in the proportion of maghemite results in a decrease in the blocking temperature, which is confirmed by experimental results (Kechrakos et al. 2007).

References

Afremov L. L., Lamash B. E., Panov A. V. The metastable states of a pseudo-single-domain ferrimagnetic grain. Izvestiya. Physics of the Solid Earth. 1998. V. 34. № 11. P. 927–931.

Afremov L. L., Ilyushin I. G. // Effect of mechanical stress on magnetic states and hysteresis characteristics of a two-phase nanoparticles system, Journal of Nanomaterials, Volume 2013, 15p.

Afremov L. L., Il'yushin I. G., Anisimov S. V. Modeling the implications of chemical transformations for the magnetic properties of a system of titanomagnetite nanoparticles // Izvestiya, Physics of the Solid Earth. – 2015. – V. 51. – №. 5. P. 613–621.

Banerjee, S., J. King, and J. Marvin//A rapid method for magnetic granulometry with applications to environmental studies, Geophys. Res. Lett., 8(4), (1981), 333–336.

Carmichael C. // Phys. Earth Planet. Inter. 1977. 13. P. 332.

Carvallo C., Ozden Ozdemir, Dunlop J.D. // Geophys.J. Int. 2004. 156. P. 29.

Draeger U., Prevot M., Poidras T., Riisager J. // Geophys. J. Int. 2006. 166. P. 12.

Dunlop D., Hale C. // J. Geophys. Res. 1976. 81. P. 4166.

Gallagher, K. // Mechanism of oxidation of magnetite to gamma-Fe2O3, Nature, 217, (1968), 1118–1121.

Ge, K., W. Williams, Q. Liu, and Y. Yu // Effects of the core-shell structure on the magnetic properties of partially oxidized magnetite grains: Experimental and micromagnetic investigations, Geochem. Geophys. Geosyst., 15, (2014), 2021–2038, https://doi.org/10.1002/2014gc005265.

Gromm' e S., Mankinen E., Marshall M., Coe R. //J. Geophys. Res. 1979. 84. P. 3553.

Gubin S.P., Koksharov Yu.A., Khomutov G.B., Yurkov G.Yu., Magnetic nanoparticles: preparation, structure and properties. Russian Chem. Rev., 2005, 74, pp. 489–520.

Kechrakos D., Trohidou K. N., Vasilakaki M. Magnetic properties of dense nanoparticle arrays with core/shell morphology // Journal of Magnetism and Magnetic Materials. – 2007. – V. 316. – №. 2. P. e291–e294

Krupichka S., Physics of ferrites and related magnetic oxides. Moscow, 1976, 2, 504 p.
Liu, Q. S., S. K. Banerjee, M. J. Jackson, C. L. Deng, Y. X. Pan, and R. X. Zhu New insights into partial oxidation model of magnetites and thermal alteration of magnetic mineralogy of the Chinese loess in air, Geophys. J. Int., 158(2), (2004), 506–514.
Nikiforov V.N., Ignatenko A.N., Irkhin V. Yu., Size and surface effects on the magnetism of magnetite and maghemite nanoparticles. J. Exp. Theor. Phys., 2017, 124, pp. 304–310.
O'Reilly, W., Rock and Mineral Magnetism, 220 pp., Blackie Acad. and Prof, Glasgow, U. K (1984).
Ozdemir, O., and D. J.// Dunlop Hallmarks of maghemitization in low-temperature remanence cycling of partially oxidized magnetite nanoparticles, J. Geophys. Res., (2010), 115, B02101, https://doi.org/10.1029/2009jb006756.
Perez N., Guardia P., Roca A.G., Morales M.P., Serna C.J., Iglesias O., Bartolome F., Garcia L. M., Batlle X., Labarta A., Surface anisotropy broadening of the energy barrier distribution in magnetic nanoparticles. Nanotechnology, 2008, 19, 475704.
Prevot, M., A. Lecaille, and E. A. Mankinen. // Magnetic effects of maghemitization of oceanic crust, J. Geophys. Res., 86(B5), (1981) 4009–4020.
Readman P.W. and W. O'Reilly Magnetic Properties of Oxidized (Cation-Deficient) Titanomagnetites (Fe, Ti,□)3O4//J. Geomag. Geoelectr., 1972, Vol. 24, pp. 69–90.
Ryall P.J.C., Hall J. // Canad J. Earth. Sci. 1979. 16, Pt. 1. P. 496.
Shcherbakov V.P. The role of kinetics in the oxidation of titanomagnetite grains // Izv. AN SSSR. Ser. Physics of the Earth. 1982. № 5. P. 43–49.
Shcherbakov VP, Gribov SK Theory of oxidation of titanomagnetite grains with a diffusion coefficient that depends sharply on the degree of oxidation, Izv. AN SSSR, Physics of the Earth, 1986, No. 4, pp. 105–112.
Smith G., Banerjee S. // J. Geophys. Res. 1986. 91. P. 10337.
Stephenson A. // Geophys. R. J. astr. Soc. 1972. 29. P. 91.
Walderhaug H. // Geophys. J. Int. 1992. 111. P. 335.

Chapter 16
Low-Temperature Magnetic Properties and Magnetic Mineralogy of the Ropruchey Sill (Russian Karelia)

Andrei Kosterov, Elena S. Sergienko, Petr V. Kharitonskii, Svetlana Yu. Yanson and Irina A. Vasilieva

Abstract We report low-temperature magnetic properties of predominantly mafic rocks from the Mesoproterozoic Ropruchey sill, Karelia. Based on microscopic observations, the studied rocks can be classified into two groups, granodiorites and gabbro-diorites, with different magnetic mineralogy. In granodiorites, the only magnetic mineral is stoichiometric magnetite characterized by a Verwey transition temperature of 120–121 K. In gabbro-diorites, the magnetic mineralogy is also dominated by magnetite, which however contains 2–3% of Ti substitution atoms, lowering the Verwey transition temperature to 94–103 K. In addition, the gabbro-diorites appear to contain some amount of a more Ti-rich titanomagnetite, as well as another, yet unidentified magnetic phase, manifesting itself by a peculiar temperature and frequency-dependent behavior of magnetic susceptibility.

Keywords Low-temperature magnetometry · Magnetite · Verwey transition
Scanning electron microscopy · Microanalysis

Introduction

The study of the magnetic properties of rocks and their ferrimagnetic minerals at cryogenic temperatures has recently become an active research topic. From low low-temperature magnetic data it is possible to infer, in a non-destructive way, the magnetic mineralogy, domain state, and size of ferrimagnetic grains present in rocks. In combination with microscopic analysis to determine chemical composition and morphology of magnetic particles, low-temperature magnetometry helps to assess the nature and stability of natural remanent magnetization and its components.

However, it appears that low-temperature measurements of whole rocks are not always compatible with what is known about the behavior of sized single phase samples of the relevant magnetic minerals (cf. Moskowitz et al. 1998; Özdemir and

A. Kosterov (✉) · E. S. Sergienko · P. V. Kharitonskii · S. Yu.Yanson · I. A. Vasilieva
St. Petersburg State University, St. Petersburg, Russia
e-mail: a.kosterov@spbu.ru

D. Nurgaliev et al. (eds.), *Recent Advances in Rock Magnetism, Environmental Magnetism and Paleomagnetism*, Springer Geophysics,
https://doi.org/10.1007/978-3-319-90437-5_16

Dunlop 2003; Kosterov 2007; Engelmann et al. 2010; Church et al. 2011; Almeida et al. 2014). One possible reason is that interaction between e.g. magnetite and hemoilmenite phases may possibly alter the magnetic properties of both minerals compared with those of pure minerals. A possible approach to circumvent this is to study directly the magnetic properties of whole rocks combining magnetic methods with microscopy, X-ray, and microprobe characterization of the ferrimagnetic phases in this sample (e.g. Kosterov 2001a; Kosterov et al. 2009, 2018). The cited studies have shown that low-temperature magnetic properties of volcanic rocks exhibit an unexpectedly wide variability controlled primarily by the chemical composition of the relevant phases, but also by their grain size, style of exsolution, etc. It is therefore important to acquire more data on well characterized samples from different types of rocks. To contribute to this compilation, we here study a small collection of gabbro-diabases from the Mesoproterozoic Ropruchey layered sill, Karelia.

Samples and Methods

Ropruchey sill is a vast lopolith layered intrusion covering about 4500 km^2 and $\sim$170 m thick, which cuts the strata of the West Onega Trough considered to be of Vepsian age (Fedotova et al. 1999). The age of the sill is estimated at 1751 $\pm$ 3 Ma by U-Pb method on baddeleyite (Lubnina et al. 2012). Previous paleomagnetic studies (Damm et al. 1997; Fedotova et al. 1999) revealed the presence of a NRM component with high unblocking temperature, interpreted as primary. The paleomagnetic pole determined from this component (Φ = 39.1°N, Λ = 216.6°E, A95 = 6.7°) is considered as one of the key poles of the respective age for Baltica (Pisarevsky et al. 2014).

For the present study, we selected 16 samples approximately evenly distributed over the sill. For the initial assessment of their magnetic mineralogy, we measured the temperature dependence of low-field magnetic susceptibility using a KLY-3 susceptibility bridge equipped with a furnace. All measurements were carried out in argon atmosphere, and Curie temperatures were determined from the minima of the first derivatives of the heating curves, as suggested implicitly by Petrovský and Kapička (2006). Room temperature hysteresis loops in the maximum field of 1 T, and M_{rs} DC demagnetization curves (backfield curves) were traced using a PMC 3900 vibrating sample magnetometer. The same instrument was used to measure first-order reversal curves (FORC). The maximum field to measure FORCs varied from 150 to 250 mT, and field increments were chosen so as to equal to about 1/5 or less of the bulk coercive force. FORC distributions have been computed using the VARIFORC software (Egli 2013).

Low-temperature magnetic measurements included the following. Saturation isothermal residual magnetization (SIRM) have been acquired in a 5 T field at 2 K after cooling in zero (Zero Field Cooling, ZFC) and in a strong (5 T, Field Cooling, FC) magnetic field. These were subsequently demagnetized by warming to 300 K in zero field. The evolution of SIRM acquired at 300 K was also traced during a cooling-warming cycle in zero field. Temperature dependence of the low-field magnetic susceptibility was measured between 2 and 300 K at a frequency of 10 Hz. For selected samples, the temperature dependence of magnetic susceptibility was traced in limited temperature intervals as a function of frequency between 1 and 1000 Hz.

To characterise the morphology, structure, chemical composition, and alteration degree of magnetic minerals, nine representative samples were investigated with analytical scanning electron microscopy (SEM) using a system QUANTA 200 3D (FIA, the Netherlands) with an analytical complex Pegasus 4000 (EDAX, USA), and a tabletop microscope-microanalyzer TM 3000 (Hitachi, Japan). Qualitative and quantitative X-ray phase analysis was performed using a powder Bruker diffractometer D2 Phaser (Bruker AXS GmbH, Germany).

Results

Curie Temperatures and Room Temperature Hysteresis Properties

Characteristic temperature dependences of the magnetic susceptibility are shown in Fig. 16.1. Most curves exhibit relatively high reversibility. However, in some samples the shape of the cooling curve differs from that of the heating curve. In most cases, Hopkinson peaks, evident in heating curves, are considerably smoothed. The respective Curie temperatures as determined from the heating curves range from 550 to 585 °C indicating nearly stoichiometric magnetite as a major magnetic phase.

Typical hysteresis loops are plotted in Fig. 16.2, and hysteresis loop parameters are listed in Table 16.1. All samples except for sample 79 show relatively thin loops with coercive force ranging from 5 to 14 mT. Coercivity of remanence is also rather moderate, between 18 and 40 mT. Sample 79 is somewhat outstanding (Fig. 16.2e) having higher coercive force (17 mT) and especially much higher coercivity of remanence (58.7 mT) than the other samples. On the Day-Dunlop plot (Day et al. 1977; Dunlop 2002; Fig. 16.2f) the representative points fall somewhat below and to the left of SD-MD mixing trends. Again, sample 79 (shown by a triangle) is an exception yielding both relatively high M_{rs}/M_s and H_{cr}/H_c ratios, and plotting therefore on the other side of the mixing trend.

FORC diagrams for the selected samples are shown in Fig. 16.3. Their shape is quite common for pseudo single-domain magnetite bearing samples (*cf.* Roberts et al. 2014) with nearly circular or slightly elongated inner and onion-like outer contours. For most samples, the area of maximum distribution density is clearly separated from the $H_c = 0$ axis. However, in some samples, e.g. 104 (Fig. 16.3c), this is not the case. Expectedly, samples like 104 show also lower coercivity and M_{rs}/M_s ratio. Another exception is sample 79, which shows a diagram with the FORC distribution density extending both to low and high fields.

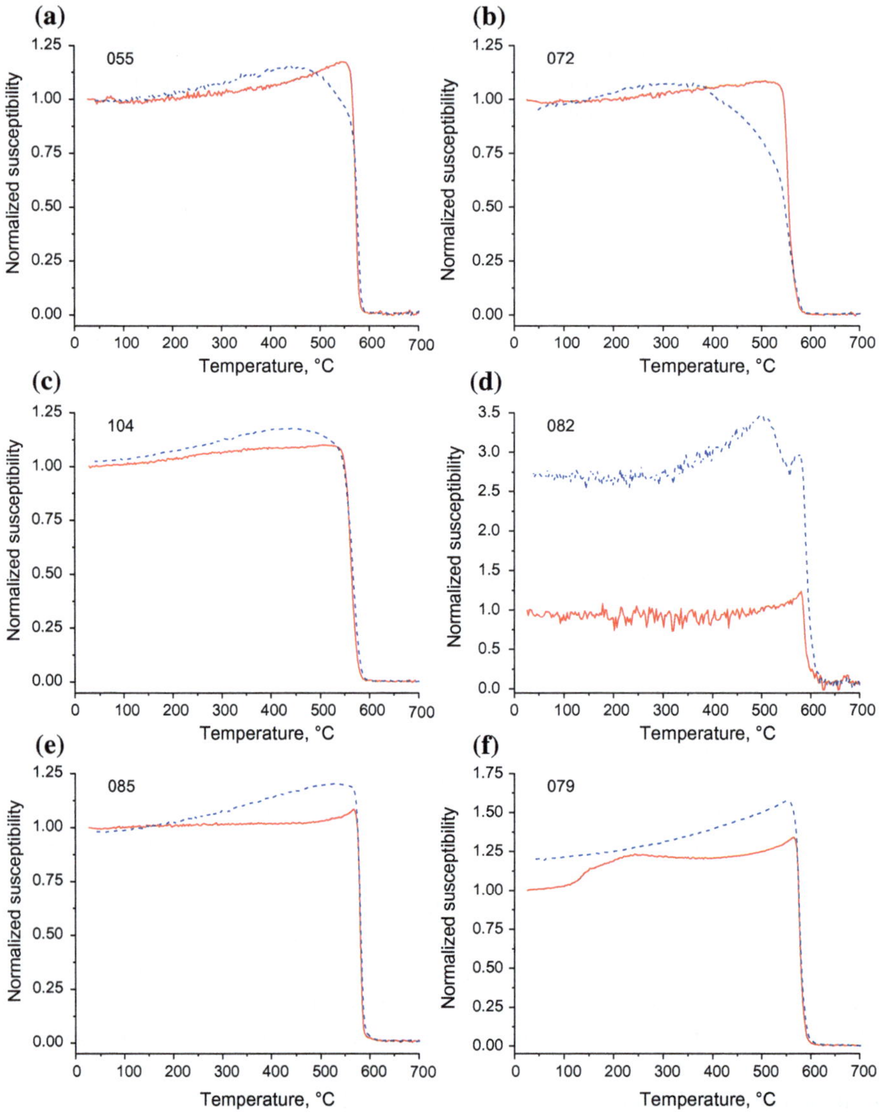

Fig. 16.1 Temperature dependence of the initial magnetic susceptibility between 20 and 700 °C. Heating and cooling curves are shown by solid and dashed lines, respectively

Low-Temperature Remanence

Plots of remanence versus temperature for representative samples are shown in Fig. 16.4. All samples exhibit a well pronounced Verwey transition. Furthermore, all the samples but one (79) are divided into two distinct groups (group I and group

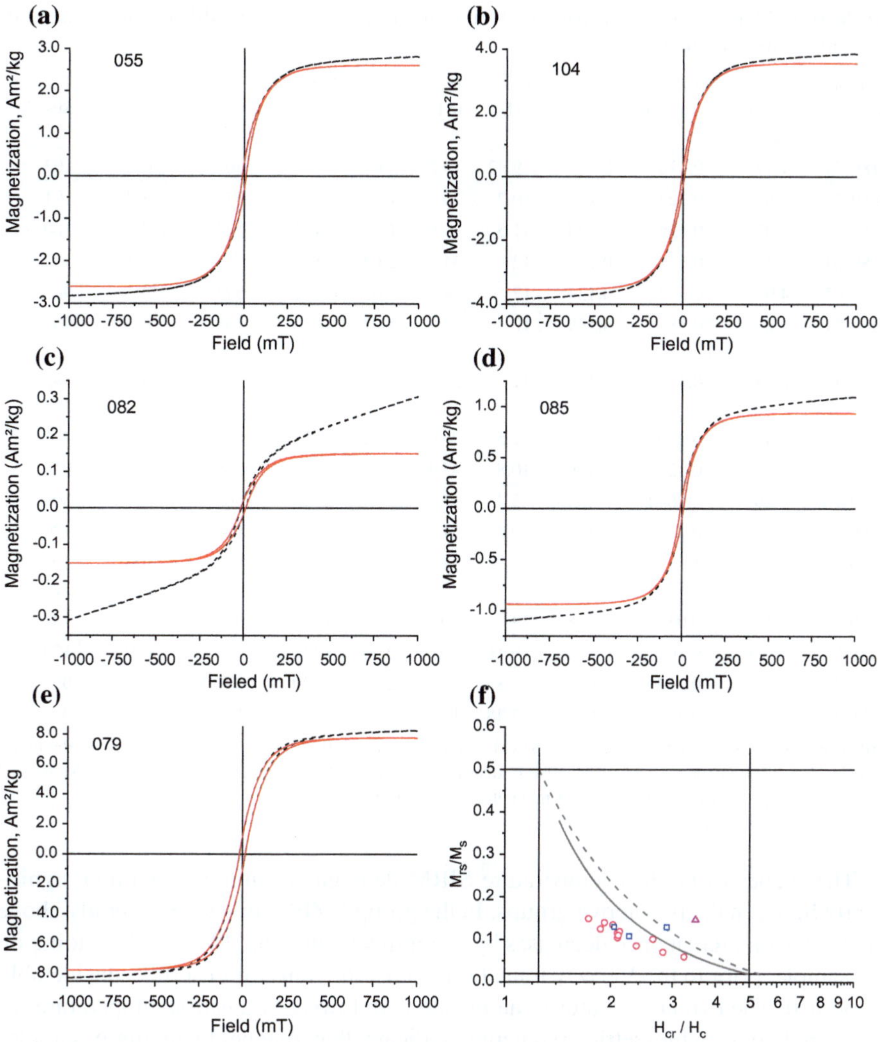

Fig. 16.2 **a–e** Hysteresis loops and **f** Day-Dunlop plot. Original loops are shown by dashed lines, loops corrected for high-field susceptibility—by solid lines. The lines in the Day-Dunlop plot show the characteristic trend from the SD to MD magnetic state (Dunlop 2002). Gabbro-diorites are shown by circles, granodiorites by squares, the unique sample 79 by a triangle

II thereafter) according to their transition temperature. In the first group (12 samples) Verwey transition temperatures range from 94 to 103 K. The second (3 samples) is characterized by significantly higher transition temperatures of 120–121 K. Finally, sample 79 appears unique in our collection, showing two clearly distinct Verwey transitions at 106 and 121 K, respectively.

Table 16.1 Magnetic hysteresis parameters, transition temperatures, and SIRM memory ratios for the samples used in the present study

Sample ID	M_s, Am^2/kg	M_{rs}, Am^2/kg	M_{rs}/M_s	H_c, mT	H_{cr}, mT	H_{cr}/H_c	Memory	T_C, °C	T_V Rem, K	T_V Sus, K
rr89-54	4.24	0.438	0.103	9.32	19.5	2.09	0.394	567	102	103
rr89-55	2.6	0.323	0.124	10.7	20	1.87	0.397	575	100	101
rr89-58	3.07	0.455	0.148	11.5	19.9	1.73	0.453	578	101	100
rr89-61	1.2	0.121	0.1	12	31.6	2.64	0.594	578	103	102
rr89-65	4.97	0.667	0.134	12	24.4	2.03	0.612	573	99	98
rr89-71	4.23	0.588	0.139	12	23	1.92	0.463	555; 574 (?)	97 (?)	96
rr89-72	3.88	0.445	0.115	11	22.9	2.09	0.478	554	101	88; 118
rr89-73	0.481	0.0567	0.118	12.5	26.6	2.12	0.528	550	99	–
rr89-75	2.93	0.32	0.109	10.8	22.7	2.1	0.378	568	95	94
rr89-76	1.09	0.0644	0.0593	5.4	17.5	3.24	0.328	565	98	100
rr89-77	0.777	0.1	0.129	10.5	21.4	2.05	0.427	585	120	120
rr89-79	7.74	1.13	0.146	16.7	58.7	3.5	0.749	577	106; 121	106; 120
rr89-82	0.15	0.0193	0.128	14	40.4	2.89	0.572	587	121	122
rr89-85	0.936	0.101	0.108	9.11	20.6	2.26	0.442	581	120	120
rr89-103	3.89	0.272	0.07	6.85	19.3	2.82	0.429	572	99	99
rr89-104	3.55	0.3	0.0845	7.48	17.7	2.37	0.385	562	94	90 ± 4

SIRM memory was calculated as a ratio of SIRM residue after the 300-2-300 K cycle in zero field to the initial SIRM; Verwey transition temperatures were determined from SIRM demagnetization curves (Rem) and temperature dependences of the in-phase magnetic susceptibility (Sus)

The shapes of the low-temperature SIRM demagnetization curves differ significantly between the two above groups. In the group I, ZFC curves lie generally above the FC curves, remanence decreases by a significant amount, 30–60% depending on the sample, prior to the Verwey transition, whilst the transition itself is considerably smoothed. The two latter features, alongside with lowered transition temperatures as compared to stoichiometric magnetite, indicate that magnetite in these samples contains 2–3% of substitution cations (in most cases, Ti, but also possibly Al and Mg) and/or vacancies. The fact that ZFC > FC implies a relatively large grain size, probably >10 μm (Kosterov 2001b, 2003; Brachfeld et al. 2002; Smirnov 2009). In contrast, the high transition temperature of the group II samples shows that their magnetite is very nearly stoichiometric, while the relation between ZFC and FC curves (FC $\geq$ ZFC) rather indicates a fine grain size, on the order of 1 μm or less.

Inspecting zero-field cycling of SIRMs acquired at 300 K (RT-SIRM for short, shown in Fig. 16.4 as dotted lines) yields an additional insight regarding the composition and grain size in our samples. This in particular applies to samples from group I which can be further classified by whether the cooling and the warming branches of a 300-2-300 K cycle diverge close to the Verwey transition temperature (sample 55, Fig. 16.4a) or some 20–30 K below it (sample 72,

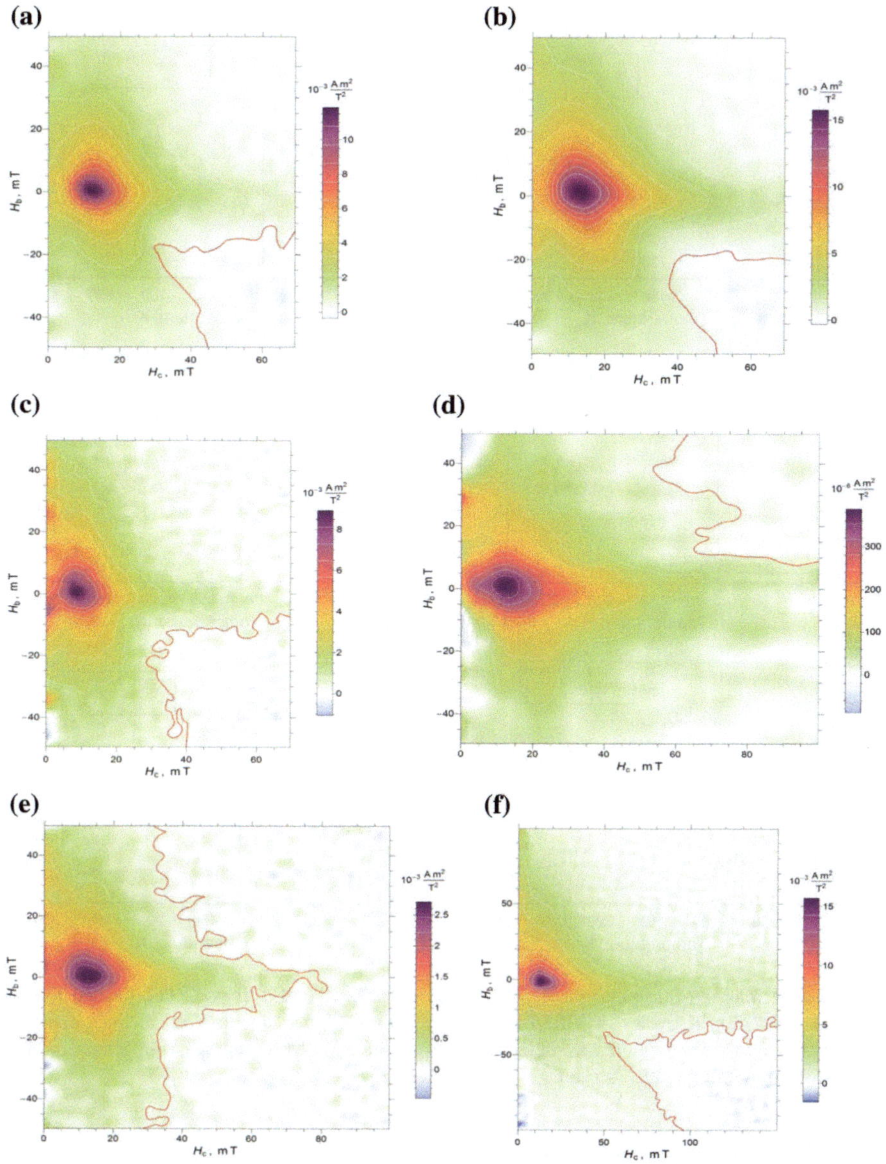

Fig. 16.3 FORC diagrams for representative samples. **a**—sample 55, **b**—sample 72, **c**—sample 104, **d**—sample 82, **e**—sample 85, **f**—sample 79. Red contours indicate the signal-to-noise ratio of 3

Fig. 16.4b). The latter might be due to a presence of a minor amount of a more Ti-rich phase not having the Verwey transition, in addition to the main phase with 2–3% of Ti substitution. This second phase would not be easily detectable by other methods.

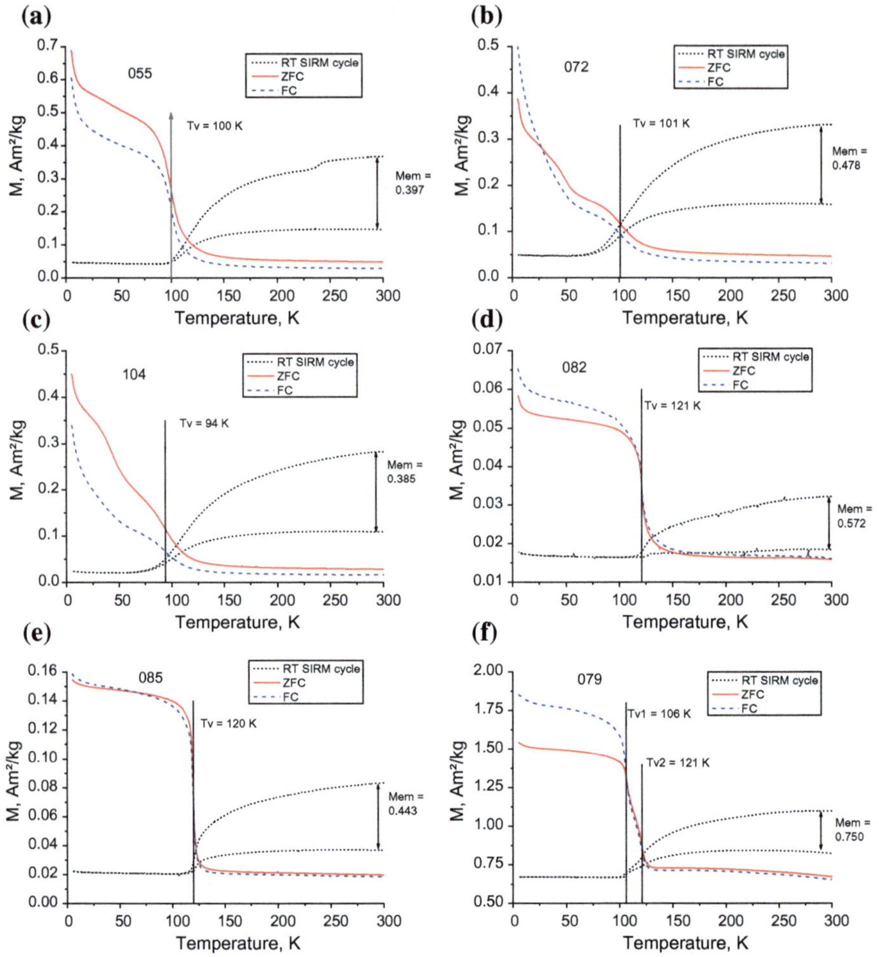

Fig. 16.4 Demagnetization curves of SIRM acquired at 2 K (solid lines—ZFC, dashed lines—FC) and 300-2-300 K cycles of SIRM acquired at 300 K (dotted lines). Vertical lines mark the Verwey transition temperatures as determined from the dSIRM/dT curves (not shown). Mem stands for the memory ratio (see text)

Low-Temperature Susceptibility

In Fig. 16.5 we plot the temperature dependence of the complex magnetic susceptibility in the 2–300 K range for the same six samples whose remanence is shown in Fig. 16.4. Again, group I and II are clearly separated also in the susceptibility data. Samples from the group I show rather gradual increase of the real part of susceptibility starting at quite low and highly variable temperature, between 30 and 70 K. The temperature, at which susceptibility increase begins, and the

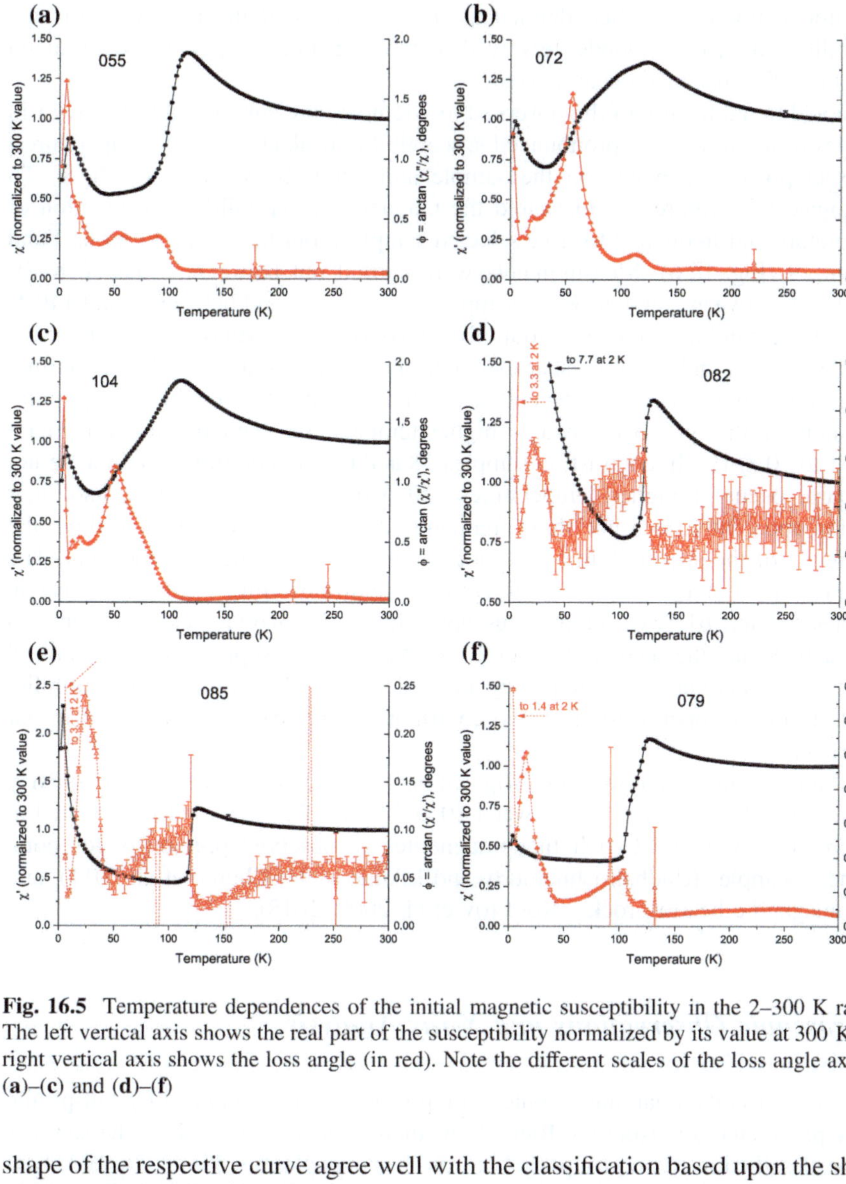

Fig. 16.5 Temperature dependences of the initial magnetic susceptibility in the 2–300 K range. The left vertical axis shows the real part of the susceptibility normalized by its value at 300 K; the right vertical axis shows the loss angle (in red). Note the different scales of the loss angle axes in (**a**)–(**c**) and (**d**)–(**f**)

shape of the respective curve agree well with the classification based upon the shape of RT-SIRM 300-2-300 K cycles. In the samples that contain a more Ti-rich titanomagnetite phase according to the shape of their RT-SIRM cycle, the increase of magnetic susceptibility starts already at about 30–35 K, while for samples not having the Ti-rich phase this temperature is considerably higher, between 65 and 80 K. Samples belonging to group II, as well as the unique sample 79, show a somewhat simpler behavior with very sharp susceptibility increase at the Verwey transition (two transitions in case of sample 79). In most samples the Verwey transition temperature, when defined as the maximum of the derivative of susceptibility curve, corresponds very well to the transition temperature determined from the LT remanence decay curve.

An additional feature of the observed low-temperature susceptibility behavior in samples from group I is a pronounced susceptibility peak at very low temperatures. Its exact position depends on the sample and varies between 4 and 20 K. To investigate this feature, we measured the magnetic susceptibility as a function of temperature and frequency for five selected samples, four belonging to group I and sample 79 (Fig. 16.6). Measurements were carried out between 2 and 50 K for sample 79, between 2 and 70 K for sample 55 (group Ia), and between 2 and 140 K for three remaining samples (group Ib). Two types of behavior are observed. Samples 55, 72, and 79 show sharp susceptibility peaks at about 6–8 K exhibiting an Arrhenius-type frequency dependence. The exact position of the susceptibility maximum shifts by 2–3 K towards higher temperatures with frequency increase from 1 to 1000 Hz. In contrast, in samples 58 and 61 susceptibility maxima are not as sharp, occur at higher temperatures (20 and 15 K, respectively), and their position apparently does not show frequency dependence. At the same time, the magnitude of the susceptibility peak does appear to be frequency-dependent. The imaginary part of the magnetic susceptibility indicates (Fig. 16.6b, d, f) that in samples 58 and 61 two different susceptibility peaks overlap, the first centered around 10 K and the second at about 20 K. We therefore suppose that the unusual shape of temperature and frequency dependence of the in-phase susceptibility results from the superposition of two contributions that cannot be resolved in our experiment.

All three samples that were investigated between 2 and 140 K show a frequency dependence of Arrhenius type between 40 and 80 K. This is most likely related to the presence of more Ti-rich titanomagnetites as observed previously for both, synthetic samples (Radhakrishnamurty and Likhite 1993; Church et al. 2011), and titanomagnetite bearing rocks (Kosterov et al. 2009, 2018).

Microscopic Observations and X-ray Analysis

Table 16.2 lists the quantitative phase compositions (wt%) obtained by full-profile X-ray phase analysis using the Rietveld method. Based on these data, the samples can be classified into two groups: gabbro-diorites (samples 55, 58, 72, 76, 103, 104) and granodiorites (samples 79, 82, 85). Among granodiorites, sample 79 is rather peculiar in that it contains a particularly high amount of mica. Samples 82 and 85 appear to have rather high amphibole content. Ore minerals detectable by the Rietveld method are represented by magnetite (2–5 wt%) and ilmenite (1–4 wt%).

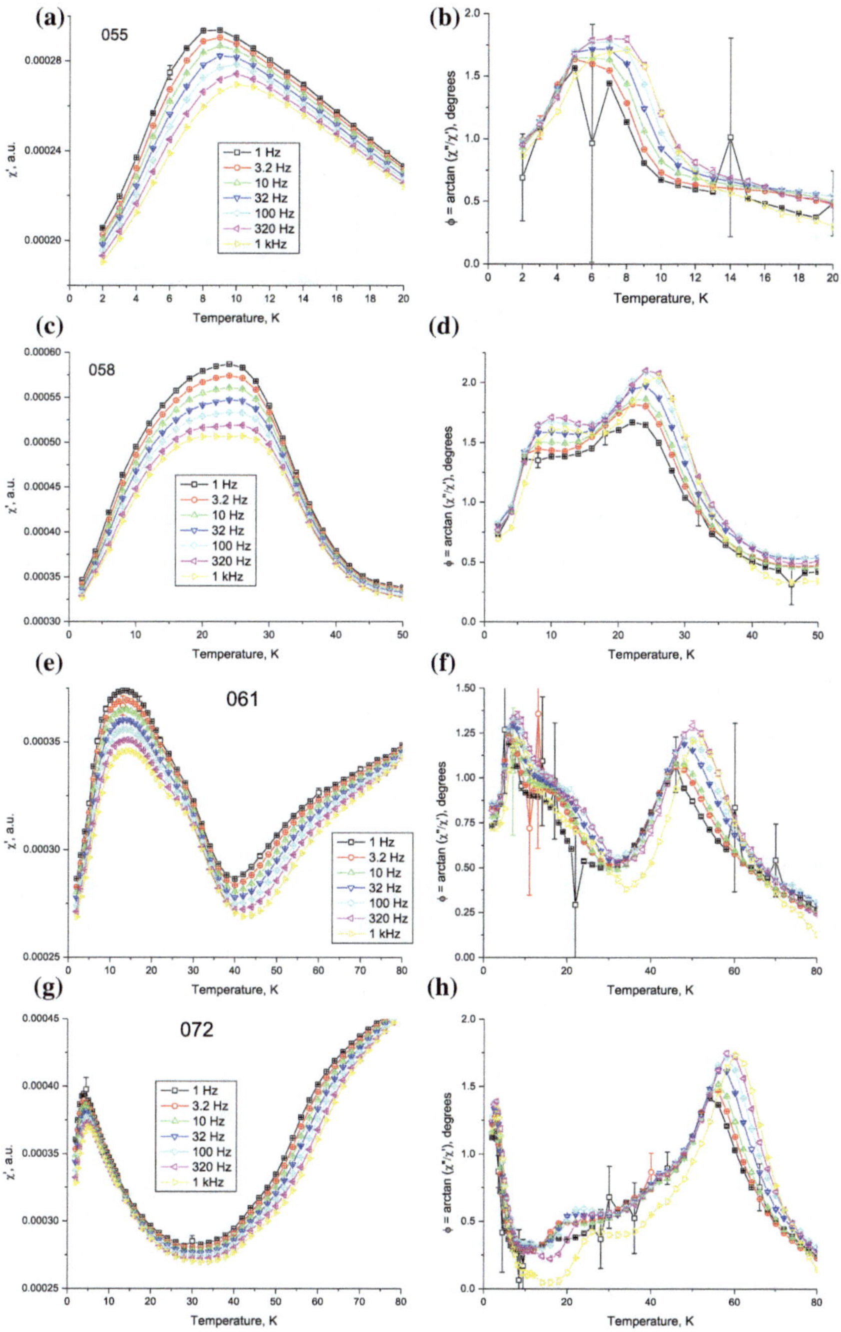

Fig. 16.6 Detailed plots of the temperature dependence of the initial magnetic susceptibility measured at different frequencies: **a**, **c**, **e**, **g**, **i**, real part of susceptibility; **b**, **d**, **f**, **h**, **j**, loss angle. Note the different temperature scales for each sample

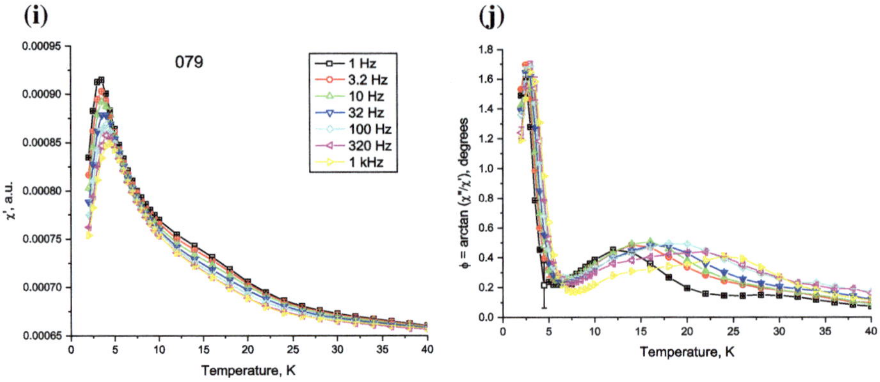

Fig. 16.6 (continued)

Table 16.2 Results of the X-ray phase analysis. $R_P = \frac{\Sigma\left|y_i^{obs}-y_i^{calc}\right|}{\sum y_i^{obs}}$ is the degree of convergence between calculated and experimental X-ray profiles. y_i is the intensity at each experimental point of an X-ray profile

Sample no.	55	58	72	76	79	82	85	103	104
Plagioclase andesine-labradorite composition	38	37	35	29	10	15	14	31	40
Pyroxene, diopside-augite composition	24	20	25	22	1	–	4	21	20
Pyroxene, enstatite-ferrosilite composition	6	6	8	3	–	2	3	5	5
Magnetite	3	3	2	4	2	–	–	5	3
Ilmenite	3	4	2	3	–	3	–	3	1
Quartz	9	7	5	7	24	16	20	8	7
Feldspars (albite and microcline)	5	11	12	11	17	25	26	6	6
Chlorite	3	4	2	2	12	6	6	6	2
Mica	3	4	2	5	26	3	2	3	5
Amphibole	6	3	6	12	1	29	24	12	11
Calcite	–	1	–	–	–	1	1	–	–
Talcum	–	–	<1	2	–	–	–	–	–
Epidote	–	–	–	–	7	–	–	–	–
R_p (%)	4.6	3.5	4.0	4.0	3.7	5.4	5.4	4.4	4.4

In SEM images of the ore fraction (Fig. 16.7), several characteristic morphological features of magnetic minerals are observed. Based on these the samples can again be classified into several types. Samples of gabbro-diorites (55, 72, 76, 103, 104) are characteristized by large grains of magnetite up to hundreds of microns in size, with two generations of exsolution structures. A fine second order mesh of ilmenite/hemoilmenite is observed, with cells up to 700–800 nm (Fig. 16.7a–f) and an overall trellis structure (Ramdohr 1980). In this group, specimen 58 appears somewhat peculiar, in that, in addition to the above structures, also fine submicron grains (200 nm–1.5 μm) with a near-wüstite composition were found (Fig. 16.7g, h). Besides, this sample is the only one where individual grains of pyrrhotite were found.

Samples 82 and 85 of the granodiorite group are characterized by skeletal magnetites, with grain sizes up to 500–600 μm. These grains also bear the signs of exsolution, forming a coarse lamellar (sample 82, Fig. 16.7i, j), or near-myrmekitic structure (sample 85, Fig. 16.7k, l). Again, sample 79 is clearly distinct from the others in terms of its magnetic mineralogy. Magnetite occurs here in two different generations and does not show any signs of oxidation or destruction. The first generation is represented by separate well-faceted crystals 3–10 μm in size (Fig. 16.7m). The second one occurs as dendritic structures developed exclusively in the quartz regions of the non-magnetic matrix (Fig. 16.7n).

Microprobe analysis of the ore fraction was performed on all samples, at 10 or more points. Structural and chemical formulae of minerals were calculated according to the oxygen method (Bulakh et al. 2014). The results are shown in a ternary Ti–Fe^{2+}–Fe^{3+} composition diagram constructed with the Triplot software (http://triplot.software.informer.com) (Fig. 16.8; Nagata 1961). The representative points form several clusters. For samples 82 and 85, the measurements are mainly concentrated at the magnetite-maghemite joint, though closer to the composition of magnetite. The few outliers however tend towards the rutile composition field suggesting at least a partial replacement of magnetite by rutile. The compositional peculiarities of samples 79 and 58 also manifest themselves sharply. For the sample 79, the representative points are located at near-magnetite composition, while for the sample 58 they are shifted towards wüstite, suggesting that the ferrimagnetic iron oxides contain little titanium and are strongly depleted in oxygen. Samples from the most numerous group of gabbro-diorites (Nos. 55, 58, 72, 76, 103, 104) form two diffuse linear compositional fields: "titanomagnetite" and "hemoilmenite". This reflects the presence of at least two generations of solid solution breakdown. Since the exsolution structure of the second generation is very fine, its characteristic size is much smaller than the dimensions of the X-ray spot, and therefore the average chemical composition of magnetite regions reflects the combined composition of magnetite and fine lamellae of (hemo)ilmenite.

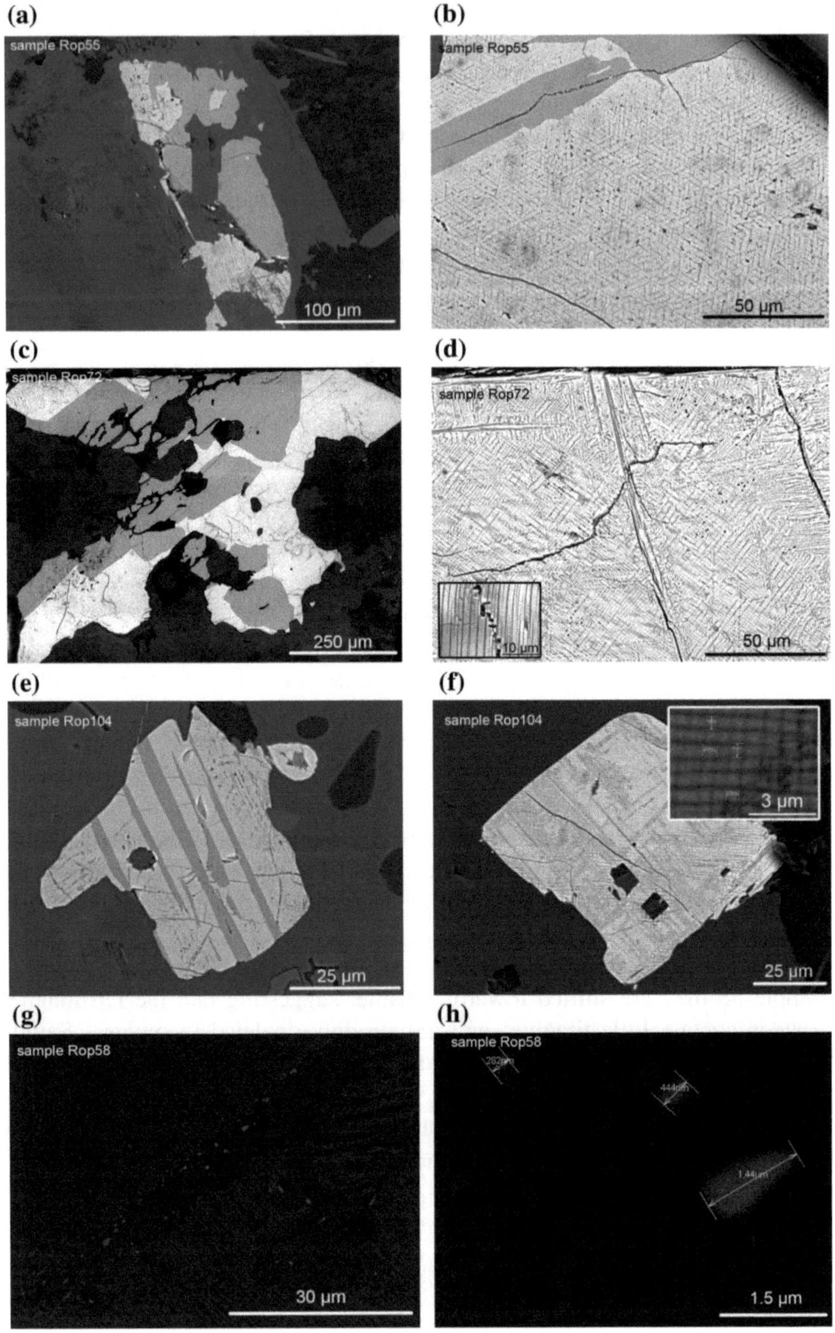

Fig. 16.7 Electron microscope images of the surface of polished thin sections. **a–h** gabbro-diorites, **i–l** granodiorites, **m**, **n** sample 79

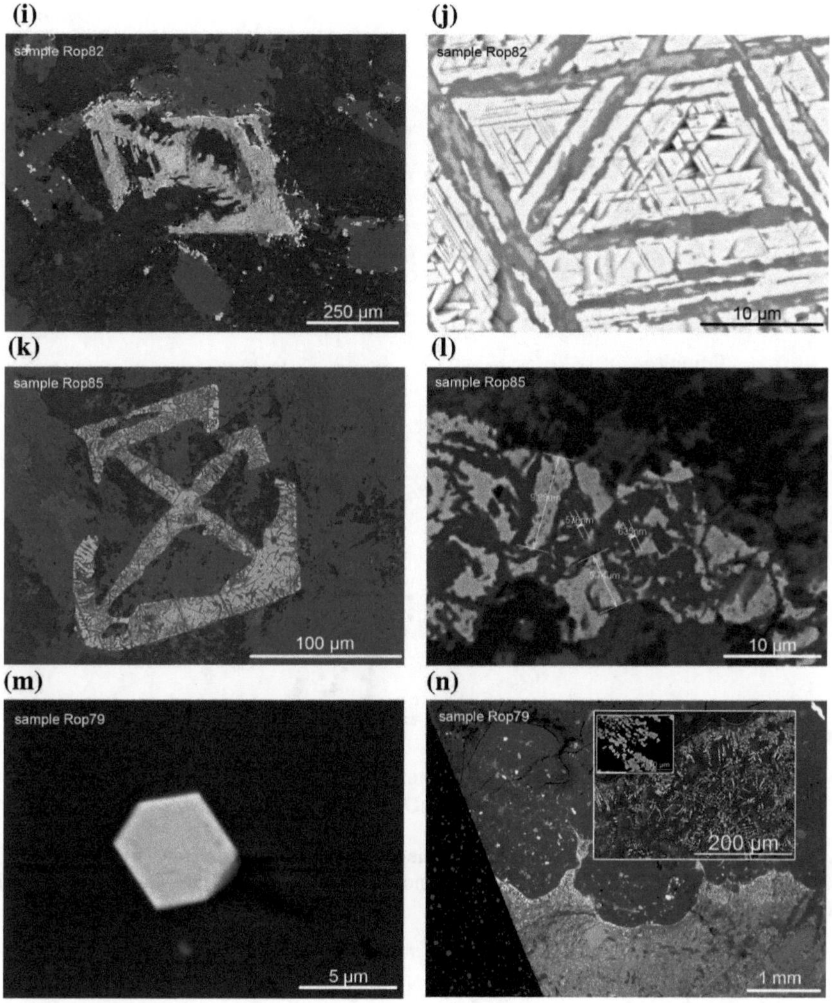

Fig. 16.7 (continued)

Discussion and Conclusions

The most notable feature of the Ropruchey sill samples is that their petrological and magnetic properties show excellent correlation. Indeed, samples with lowered Verwey transition temperatures all belong to the gabbro-diorite group according to SEM observations, while samples containing essentially stoichiometric magnetite are classified as granodiorites according to microscopic and microprobe data. The presence of two (and possibly more, considering the peculiar nature of sample 79) different groups of samples is probably not surprising bearing in mind the large size

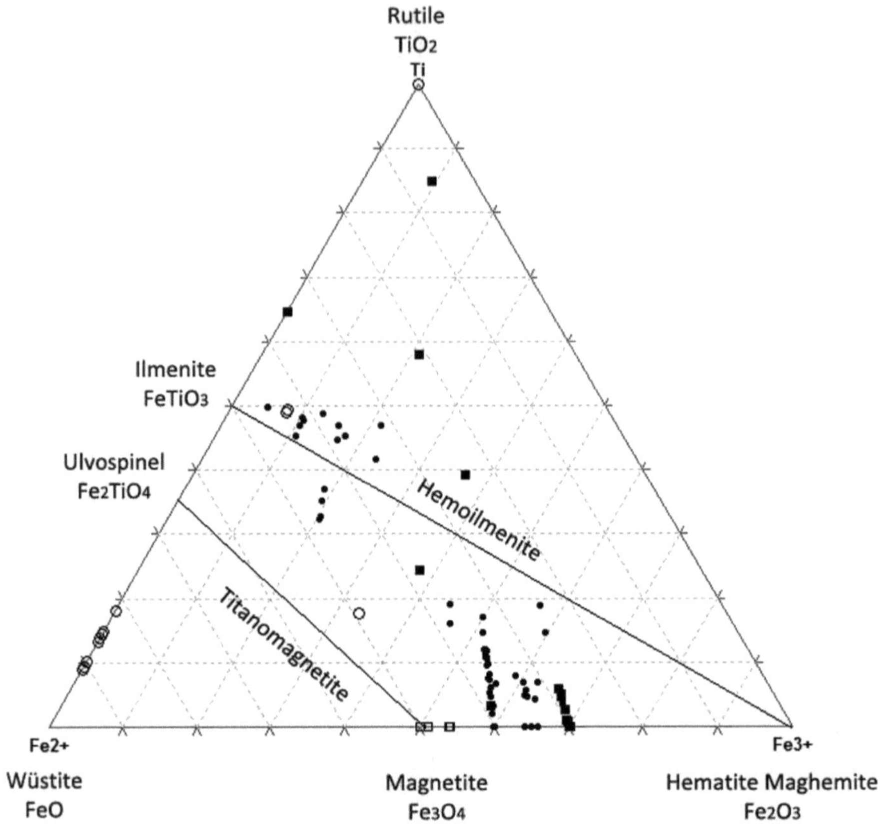

Fig. 16.8 Ternary Ti–Fe^{2+}–Fe^{3+} diagram of the ore mineral compositions according to SEM observations. ●—samples 55, 58, 72, 76, 103, and 104; ○—sample 58; ■—samples 82, 85; □—sample 79

and likely complex cooling history of the igneous body. On the other hand, low-temperature magnetometry appears to provide a relatively fast and simple, and at the same time sensitive method to classify the studied rocks into generic groups. For all except one of the examined Ropruchey sill samples the Verwey transition temperatures as determined from either remanence or magnetic susceptibility data are equal to within 2 K (Table 16.1). Therefore, LT susceptibility data alone are sufficient for at least a preliminary evaluation of the Verwey transition temperature, although comparison to LT remanence measurements is still strongly advised (*cf.* Kosterov 2001a).

The difference in the observed Verwey transition temperatures indicates a subtle but distinct difference in chemical composition between the (titano)magnetites found in gabbro-diorite, and those in granodiorite samples. In the former, Verwey transition temperatures in the 94–103 K range indicate that the major part of the titanomagnetite contains 2–3 atomic % of Ti (Kozłowski et al. 1996). Some samples from this group also seem to contain some amount of titanomagnetite with higher Ti content. This phase however must be relatively minor, because it does not manifest itself clearly in the high-temperature thermomagnetic data. Nevertheless, the characteristic frequency dependence of susceptibility in the 40–80 K range (Radhakrishnamurty and Likhite 1993; Church et al. 2011) allows with a high degree of certainty to infer its presence in the studied rocks. The granodiorite samples are remarkable in that they originate from nearly 2 Ga old rocks, and yet the magnetite in them appears to be essentially stoichiometric, as attested by its Verwey transition temperature of 120–121 K (Aragón et al. 1985; Kąkol and Honig 1989). The samples from both groups show little or no evidence for maghemitization. This supports the previous interpretation that the characteristic NRM component in the Ropruchey sill is primary (Damm et al. 1997; Fedotova et al. 1999).

Apart from the features associated with the Verwey transition, the gabbro-diorite samples from group I also show a rather complex magnetic behavior below ca. 30 K. An SIRM acquired at 2 K either shows an anomalously rapid decay below 15–25 K depending on the sample (Fig. 16.4a, c), or the ZFC and FC curves cross each other, as e.g. in sample 72 (Fig. 16.4b). Magnetic susceptibility also shows anomalous frequency-dependent peaks in this temperature range. In part, this behavior might be due to a more Ti-rich titanomagnetite (Kosterov et al. 2009, 2018). However, frequency-dependent susceptibility peaks occurring below 20 K are not typical for titanomagnetites, and hint to the presence of another, yet unidentified magnetic phase in these samples.

Acknowledgements The samples for this study were kindly donated by Boris Pisakin. Magnetic measurements were carried out at Kochi Core Center, Kochi University, Japan and supported by the JSPS long-term visiting fellowship to AK. Electron microscopy and X-ray studies have been carried out using the facilities of the St. Petersburg University Scientific Park. This paper has benefited from a review by Karl Fabian.

References

Almeida, T. P., A. R. Muxworthy, W. Williams, T. Kasama, and R. Dunin-Borkowski (2014), Magnetic characterization of synthetic titanomagnetites: Quantifying the recording fidelity of ideal synthetic analogs, *Geochemistry, Geophysics, Geosystems*, 15, 161–175.

Aragón, R., D. J. Buttrey, J. P. Shepherd, and J. M. Honig (1985), Influence of nonstoichiometry on the Verwey transition, *Phys. Rev. B*, 31, 430–436.

Brachfeld, S. A., S. K. Banerjee, Y. Guyodo, and G. D. Acton (2002), A 13200 year history of century to millennial-scale paleoenvironmental change magnetically recorded in the Palmer Deep, western Antarctic Peninsula, *Earth Planet. Sci. Lett.*, 194, 311–326.

Bulakh, A. G., Zolotarev, A. A., and Krivovichev V. G. (2014), *Structure, isomorphism, formulae, classification of minerals*. St. Petersburg, St. Petersburg University Publishers, 133 pp. (in Russian).

Church, N., J. M. Feinberg, and R. Harrison (2011), Low-temperature domain wall pinning in titanomagnetite: Quantitative modeling of multidomain first-order reversal curve diagrams and AC susceptibility, *Geochem. Geophys. Geosyst.*, 12, Q07Z27, https://doi.org/10.1029/2011gc003538.

Damm, V., T. S. Gendler, E. G. Gooskova, A. N. Khramov, M. Lewandowski, P. Nozharov, V. I. Pavlov, G. N. Petrova, S. A. Pisarevsky, and S. J. Sokolov (1997), Palaeomagnetic studies of Proterozoic rocks from the Lake Onega region, southern Fennoscandian Shield, *Geophys. J. Int.*, 129, 518–530.

Day, R., M. Fuller, and V. A. Schmidt (1977), Hysteresis properties of titanomagnetites: Grain-size and compositional dependence, *Phys. Earth Planet. Inter.*, 13, 260–267.

Dunlop, D. J. (2002), Theory and application of the Day plot (M_{rs}/M_s versus H_{cr}/H_c) 1. Theoretical curves and tests using titanomagnetite data, *J. Geophys. Res.*, 107, https://doi.org/10.1029/2001jb000486.

Egli, R. (2013), VARIFORC: An optimized protocol for calculating non-regular first-order reversal curve (FORC) diagrams, *Global and Planetary Change*, 110, Part C, 302–320.

Engelmann, R., A. Kontny, and D. Lattard (2010), Low-temperature magnetism of synthetic Fe-Ti oxide assemblages, *J. Geophys. Res.*, 115, B12107, https://doi.org/10.1029/2010JB000865.

Fedotova, M. A., A. N. Khramov, B. N. Pisakin, and A. A. Priyatkin (1999), Early Proterozoic palaeomagnetism: new results from the intrusives and related rocks of the Karelian, Belomorian and Kola provinces, eastern Fennoscandian Shield, *Geophys. J. Int.*, 137, 691–712.

Kąkol, Z., and J. M. Honig (1989), The variation of Verwey transition temperature with oxygen stoichiometry in magnetite, *Solid State Comm.*, 70, 967–969.

Kosterov, A. (2001a), Magnetic properties of subaerial basalts at low temperatures, *Earth Planets Space*, 53, 883–892.

Kosterov, A. (2001b), Magnetic hysteresis of pseudo-single-domain and multidomain magnetite below the Verwey transition, *Earth Planet. Sci. Lett.*, 186, 245–253.

Kosterov, A. (2003), Low-temperature magnetization and AC susceptibility of magnetite: effect of thermomagnetic history, *Geophys. J. Int.*, 154, 58–71.

Kosterov, A. (2007), Low-temperature magnetic properties, in *Encyclopedia of Geomagnetism and Paleomagnetism*, edited by D. Gubbins and E. Herrero-Bervera, Springer, Dortrecht, The Netherlands, 515–525.

Kosterov, A., G. Conte, A. Goguitchaichvili, and J. Urrutia-Fucugauchi (2009), Low-temperature magnetic properties of andesitic rocks from Popocatepetl stratovolcano, Mexico, *Earth Planets Space*, 61, 133–142.

Kosterov, A., E. S. Sergienko, P. V. Kharitonskii, and S. Yu. Yanson (2018), Low-temperature magnetic properties of basalts containing titanomagnetite with composition near $\sim$ TM30, *Isvestiya, Phys. Solid Earth*, 54, 134–149.

Kozłowski, A., Z. Kąkol, D. Kim, R. Zaleski, and J. M. Honig (1996), Heat capacity of $Fe_{3-\alpha}M_{\alpha}O_4$ (M = Zn, Ti, $0 \leq \alpha \leq 0.04$), *Phys. Rev. B*, 54, 12093–12098.

Lubnina, N. V., S. A. Pisarevsky, U. Söderlund, M. Nilsson, S. J. Sokolov, A. N. Khramov, A. G. Iosifidi, R. Ernst, M. A. Romanovskaya, and B. N. Pisakin (2012), New palaeomagnetic and geochronological data from the Ropruchey sill (Karelia, Russia): implications for late Palaeoproterozoic palaeogeography, in: *Supercontinent Symposium, Espoo 2012 Program and Abstracts*, 81–82.

Moskowitz, B. M., M. Jackson, and C. Kissel (1998), Low-temperature magnetic behavior of titanomagnetites, *Earth Planet. Sci. Lett.*, 157, 141–149.

Nagata, T. (1961), *Rock Magnetism*, Tokyo, Maruzen Co., 2nd ed., 350 pp.

Özdemir, Ö., and D. J. Dunlop (2003), Low-temperature behavior and memory of iron-rich titanomagnetites (Mt. Haruna, Japan and Mt. Pinatubo, Philippines), *Earth Planet. Sci. Lett.*, 216, 193–200.

Petrovský, E., and A. Kapička (2006), On determination of the Curie point from thermomagnetic curves, *J. Geophys. Res.*, 111, B12S27, https://doi.org/10.1029/2006jb004507.

Pisarevsky, S. A., S.-Å. Elming, L. J. Pesonen, and Z.-X. Li (2014), Mesoproterozoic paleogeography: Supercontinent and beyond, *Precambrian Res.*, 244, 207–225.
Radhakrishnamurty, C., and S. D. Likhite (1993), Frequency dependence of low-temperature susceptibility peak in some titanomagnetites, *Phys. Earth Planet. Inter.*, 76, 131–135.
Ramdohr, P. (1980), *The ore minerals and their intergrowths*, Pergamon Press., 2nd ed., 1207 pp.
Roberts, A. P., D. Heslop, X. Zhao, and C. R. Pike (2014), Understanding fine magnetic particle systems through use of first-order reversal curve diagrams, *Reviews of Geophysics*, 52, 557–602.
Smirnov, A. V. (2009), Grain size dependence of low-temperature remanent magnetization in natural and synthetic magnetite: Experimental study, *Earth Planets Space*, 61, 119–124.

Chapter 17
Preliminary Magnetic Investigation of Samples from Reference Permian-Triassic Sequence, Kemerovo Region, Russia

Dilyara Kuzina, Lina Kosareva, Ilmir Gilmetdinov, Radmir Aupov, Vladimir Silantiev, Vladimir Davydov, Irina Dogadina and Natalia Kuzmina

Abstract We performed magnetic investigations of samples from the Babyi Kamen' section in the left bank of the Tom' River, Kemerovo region, Russia (54° 23.079′N, 087° 32.105′E). This section is proposed as a reference for the Kuznetsk Basin. It has been studied since the 1930s and is widely described in the literature. The succession is formed by sandstone, siltstone, and claystone which contain a vast amount of tuffaceous material. Age of the samples is Permian-Triassic. The principal aim of our study is to find the stratigraphic position of the boundary between the two eras. Measurements of magnetic susceptibility (MS), natural remanent magnetization (NRM), hysteresis parameters, and induced magnetization versus temperature were carried out to determine samples magnetic mineralogy. According to the obtained data, the sequence is divided into two parts. In the lower one low values of the magnetic susceptibility, NRM are observed, remanence is carried by magnetite, and pyrite or siderite are also present. In the upper part higher values of magnetic parameters are observed, and magnetization carriers are magnetite and minor hematite.

Keywords Permian-triassic boundary · Magnetic properties · Magnetic susceptibility · Hysteresis parameters · Curie temperature

D. Kuzina (✉) · L. Kosareva · I. Gilmetdinov · R. Aupov · V. Silantiev
I. Dogadina · N. Kuzmina
Institute of Geology and Petroleum Technologies, Kazan Federal University,
Kazan, Russia
e-mail: di.kuzina@gmail.com

V. Davydov
Permian Research Institute, Boise State University, Boise, USA

D. Nurgaliev et al. (eds.), *Recent Advances in Rock Magnetism, Environmental Magnetism and Paleomagnetism*, Springer Geophysics,
https://doi.org/10.1007/978-3-319-90437-5_17

Introduction

The problem of catastrophic extinctions of biota is one of the most important and exceedingly debated problems, which attracts the attention of the broad science community. There are numerous hypotheses describing the causes of the Permian—Triassic (P-T) extinction, but none of them has been commonly accepted yet (Renne and Basu 1991; Retallack et al. 1998; Hallam and Wignall 1999; Heydari and Hassanzadeh 2003; Retallack and Jahren 2008; Grasby and Beauchamp 2009; Nastavko 2013). The majority of specialists tend to accept the model that proposes the Siberian trap volcanism and products of the volcanic explosions and traps interaction with surrounding rocks (coals, evaporates and sulfates) as the main source of the rapid growth of CO_2 (Knoll et al. 1996) and H_2S (Kump et al. 2005) amount in the atmosphere and in the oceans causing drastic climatic changes. The problem of the age of Siberian traps volcanism is thus the most critical for the clarification of the origin and tempo of the Permian-Triassic global extinction. If the main volcanic phase preceded the extinction, then the trap volcanism might be the main trigger of all the climatic, geochemical and biotic changes at the P-T boundary. Babyi Kamen' is the reference section for Kuznetsk Basin, because it is well exposed, but the position of the P-T boundary in this section is still under discussion. In terms of magnetostratigraphy, the boundary in the International Geologic Time Scale (IGTS) is constrained in Meishan section, South China, the same section where the GSSP of the boundary is established. However, some doubts do exist about the presence of primary paleomagnetic signals in these sections (Hounslow and Balabanov 2016). Magnetic and paleomagnetic investigations would help to constrain the position of the boundary which in turn will be helpful for understanding probable causes of Permian-Triassic extinction and particularly the role of Siberian traps in regards to the extinction. We report here new rock magnetic data from the Babyi Kamen' section obtained in the course of a pilot study preceding the planned detailed paleomagnetic research. This research is a part of the larger project to study the P-T boundary transition in the terrestrial and marine sections of Siberia.

Object of Investigations

Samples were collected from the Babyi Kamen' section in the left bank of the Tom' River, Kemerovo region, Russia (54° 23.079 'N, 087° 32.105 'E) (Fig. 17.1). This section is one of the best in the entire south Siberia and has been proposed as a reference of the P-T boundary in the Kuznetsk Basin (Kuzbass). The region is located in the northwestern part of the Altai-Sayan Folded Area. From the east, the Kuznetsk Basin is bounded by the Late Cenozoic mountain ranges of the Kuznetsk Alatau, and from the south by the Gornaya Shoria province, composed mainly of Paleozoic rocks. The Babyi Kamen' section has been studied since the 1930s and is

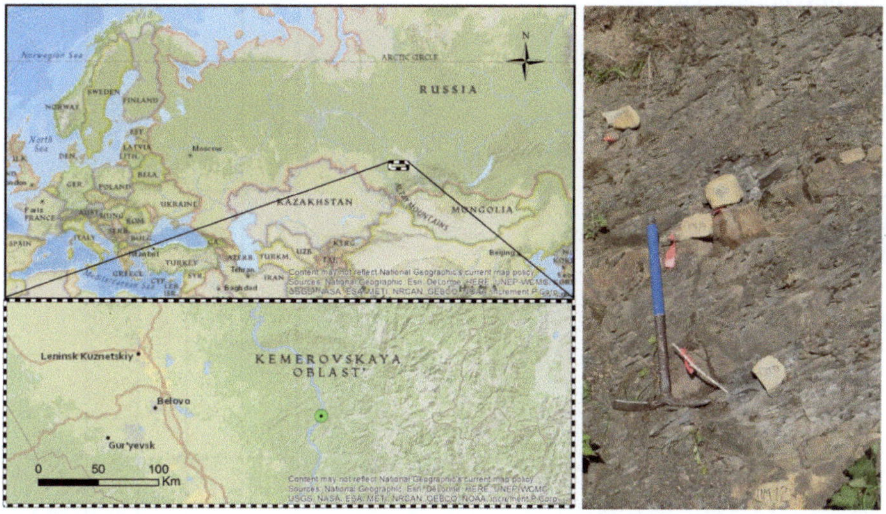

Fig. 17.1 Geographic location of the Babyi Kamen' section, sampling locality, and sampling site overview

well documented in the literature (Kazansky et al. 2005; Buslov et al. 2010; Nastavko 2013). The studied Permian—Triassic succession is composed of sandstone, siltstone, and claystone which contain a vast amount of tuffaceous material. The biostratigraphic boundary of the Lower Mal'tsevo and Tailugan suites which is traditionally drawn as the regional boundary between the Permian and Triassic systems in the Kuznetsk Basin, is defined by non-marine bivalves and conchostracans (Crustacea) (Papin and Chunikhin 2007; Buslov et al. 2010; Zharinova and Silantiev 2016). The lower Triassic Mal'tsevo Formation (T1 ml) in Babyi Kamen' section conformably overlaps the Upper Permian coal-bearing sediments (Betekhtina et al. 1986; Buslov et al. 2010).

Total of 77 oriented samples were collected over the thickness of 197 m. Samples were oriented by compass for paleomagnetic investigations. The sampling step was irregular. In the lowest 50 m of the studied interval it was on average 1.2 m, being denser in the beginning. At the end of the studied interval, samples were taken with a step of 10 m. The step of selection depended on the possibility of sampling and on changes in lithology. We tried as much as possible to take at least one sample from each designate bed. Thus, the samples represent different lithology rocks. For magnetic measurements, samples were ground into a powder in an agate mortar, to exclude contamination by magnetic particles. For NRM measurement and paleomagnetic investigations 1 inch diameter, 2 cm height cylinders were drilled in the laboratory (Fig. 17.2).

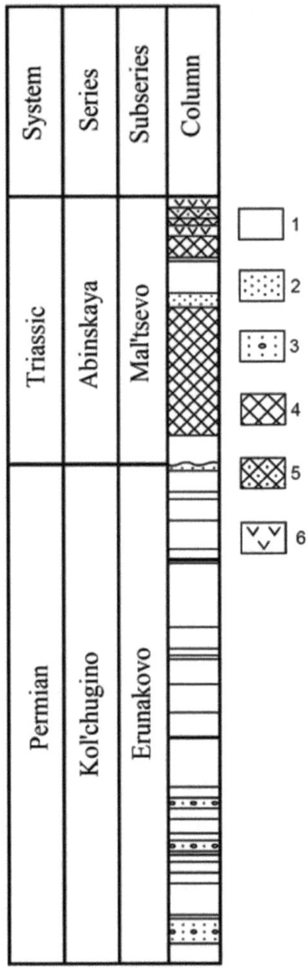

Fig. 17.2 Stratigraphic column of the Babyi Kamen' section modified after Papin and Chunikhin (2007). 1—siltstone and mudstone, 2—sandstone, 3—conglomerate, 4—tuff siltstone, 5—tuff mudstone, 6—basalt

Magnetic Measurements

Measurements of magnetic susceptibility, hysteresis parameters and induced magnetization versus temperature were carried out to determine samples magnetic mineralogy at the Laboratory of Paleoclimatology, Paleoecology, Paleomagnetism, Kazan Federal University. Magnetic susceptibility (MS) was measured using a Multi-function Kappabridge MFKA1-FA (AGICO, Czech Republic) at the 976 Hz frequency. All measured quantities were normalized by volume. Natural Remanent magnetization (NRM) was measured with a Cryogenic (SQUID) magnetometer 2G Enterprises (USA) located in a nonmagnetic room in the Laboratory of the Main Geomagnetic Field and Petromagnetism of the Institute of Physics of the Earth, RAS, Moscow.

To obtain the hysteresis parameters of the samples, we used the J_meter coercivity spectrometer (Iassonov et al. 1998; Nurgaliev and Yassonov 2009) which allows separate measurements of the remanent and induced magnetizations in magnetic fields up to 1.5 T at room temperature. The samples were magnetized from their natural state. The following parameters were deduced from the isothermal magnetization curves: saturation isothermal remanent magnetization (Jrs), saturation magnetization, excluding the effect of the paramagnetic component (Js), magnetization of the paramagnetic component in the field of 1.5 T (Jp), coercive force after removing the effect of the paramagnetic component (Bc), remanent coercivity (Bcr), and fields of maximum coercivity spectra for Jr acquisition (Bda) and demagnetization from saturated state (Bdb).

In order to determine the nature of magnetic susceptibility variation along the section we have separated the contributions of ferromagnetic and combined para- and diamagnetic components. Each component has a different origin related to changes in sedimentation processes. Dia-, paramagnetic components of the bulk susceptibility were determined by fitting a slope through the descending branch of the hysteresis loop in the interval between 1400 and 1500 mT. Ferromagnetic component was determined as the slope of the initial magnetization curve in low fields between 0 and 10 mT (Kosareva et al. 2015).

Thermomagnetic analysis is a major method for the diagnosis of ferromagnetic fraction composition in rocks. Differential thermomagnetic analysis (DTMA) of 30 collected samples uniformly distributed along the section was carried out using a Curie express balance (Burov and Yassonov 1986). Before measuring the dried sample was ground and weighed. Temperature dependences of induced magnetization up to 800 °C at a heating rate of 100 °C per minute in a constant magnetic field of 0.4 T were obtained. The curves of the first and second heating allow to evaluate possible mineralogical transformations in a sample.

Results

Analysis of 75 samples showed that the magnetic susceptibility values vary over a wide range of 6.06×10^{-5}–202×10^{-3} SI. NRM values also vary significantly from 0.259×10^{-3} to 0.457 A/m. Variations of magnetic susceptibility, NRM, and some of the magnetic parameters are shown in Fig. 17.3. Two zones with characteristic features on the magnetic parameters can be distinguished along the section. Lower part is remarkable by the low values of the MS, NRM, and the ferromagnetic component of the magnetic susceptibility. Upwards along the section, an increase in the magnetic parameters occurs, which is probably associated with an increase in the input of volcanic origin particles and weathering products of traps. Variations of the paramagnetic component are insignificant and are associated with the input of rock-forming minerals. This is due to the lithological composition of the rocks composing the section and past diagenetic changes. In the figure, the

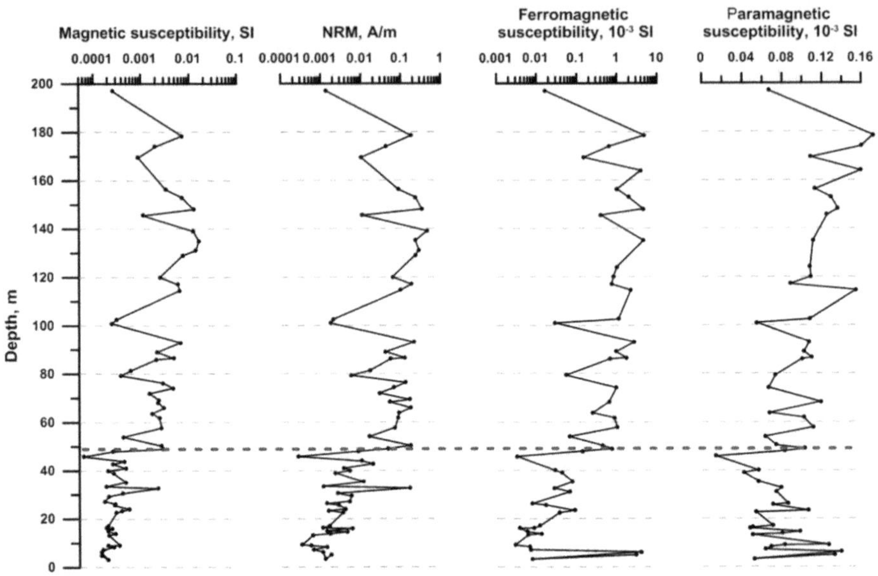

Fig. 17.3 Variations of the magnetic susceptibility, super/paramagnetic susceptibility, NRM along the section

boundary is marked with red dashed line. In future require additional lithological/ mineralogical and geochemical studies for more precise definition of the boundary.

The most common types of thermomagnetic curves are shown in Fig. 17.4. For the first group (below 48 m), curves as for sample 15, or sometimes as for sample 4 are usually observed. Growth of magnetization intensity during the first heating around 500 °C and following decrease at Curie temperature of 570 °C may indicate the presence of pyrite or siderite in the rock, which dissociate and form magnetite, or else burning of organic matter leading to reducing conditions in which magnetite is formed at the expense of iron-bearing clay minerals. The second group corresponds to the upper part of the section, above 48 m. Typical TMA diagrams are shown in the Fig. 17.4c, d. A characteristic signature of magnetite presence in the sample is a decrease of magnetization near 570–580 °C marking the Curie temperature of this mineral. Probably some hematite is also present in our samples, because magnetization is completely lost only near 640–660 °C. In the upper part of the section the main carrier of magnetization is magnetite.

Hysteresis parameters were obtained for most of the samples. Approximately 20% of the samples show an extremely low content of the ferromagnetic fraction. These samples were not used for further analysis of the magnetic properties. The Day-Dunlop diagram obtained from hysteresis parameters (Day et al. 1977; Dunlop 2002) allows one to draw conclusions about the presence in the sedimentary rocks of a mixture of multidomain grains, single domain particles (probably of biogenic origin) and possibly very fine superparamagnetic particles (Fig. 17.5).

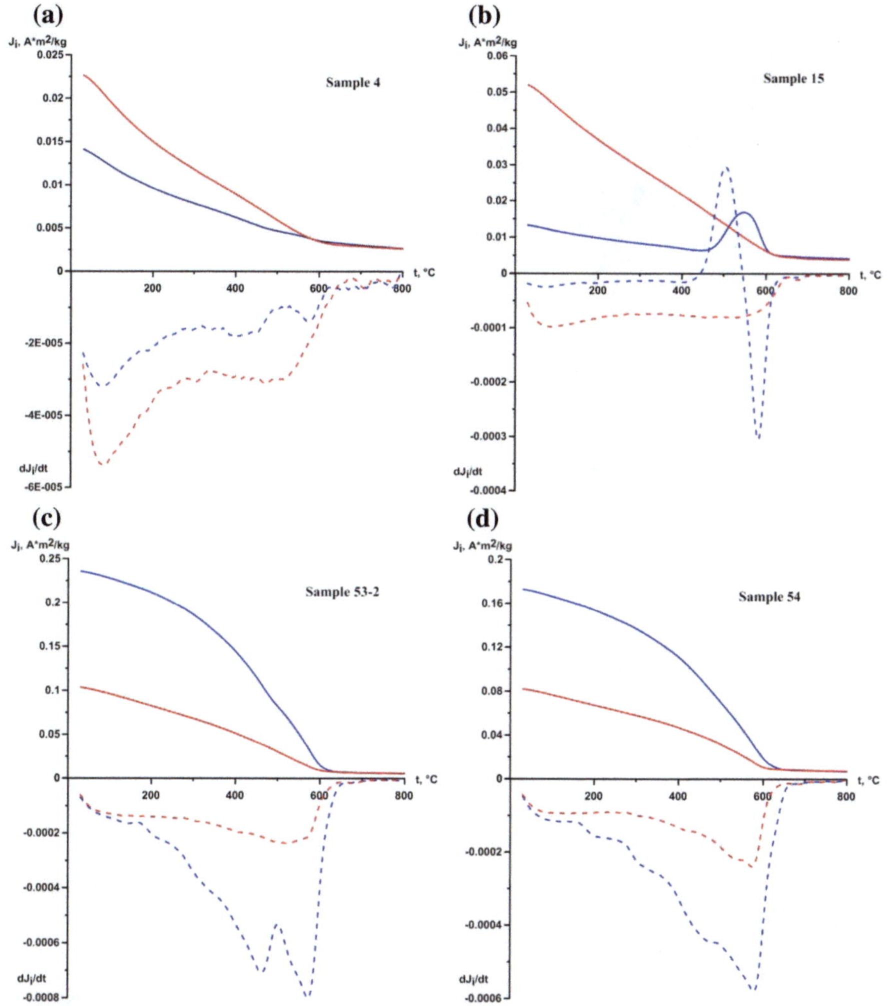

Fig. 17.4 Thermomagnetic curves—(**a**) and (**b**) lower part of the section, below 48 m; (**c**) and (**d**) upper part of the section. Continuous lines show integral curves, dashed lines—differential curves. First heating is shown in blue, second heating in red

According to the magnetic mineralogical composition and magnetic properties, samples can be divided into two groups. The first group outcrops in the lowest 48 m of the succession, and is represented largely by pyrite/siderite- and magnetite-bearing rocks with low values of magnetic susceptibility and NRM. Samples from the second group outcropping from 48 m upwards are characterized by presence of magnetite, and higher MS and NRM values. Using the magnetic mineralogy analysis, we provide a first-order estimation of the difference in rock mineralogy which can be associated with changes in sedimentation conditions. Next step of investigation will be the paleomagnetic analysis.

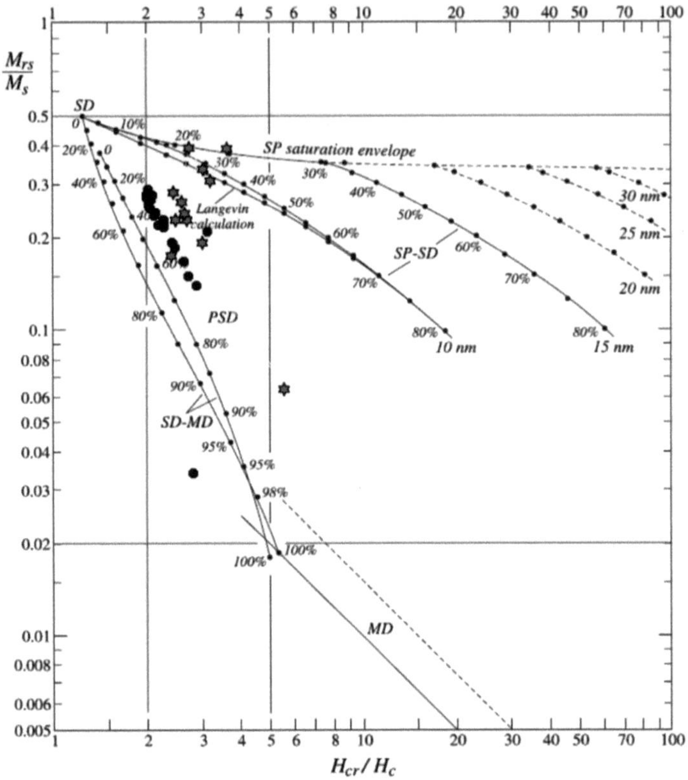

Fig. 17.5 Day-Dunlop plot of the samples. Grey dots show samples from the lower part of the section, black from the upper part

Conclusion

Magneto-mineralogical investigations of sedimentary rocks from the Babiy Kamen' section have been carried out. The results of thermomagnetic analysis show that in the lower part of the section, up to 48 m, pyrite or siderite and occasionally magnetite are present, whereas higher along the section magnetite is the main magnetization carrier. According to the variation of the magnetic susceptibility and NRM intensity the section is also clearly divided into two parts. The lower part is characterized by low values of magnetic susceptibility and NRM. The upper part of the section is characterized by higher values. The magnetic material in the section is represented by both single-domain and multidomain grains.

According to the data of magnetic studies, it is possible to designate the boundary for changing the magnetic parameters at 48 m from the bottom of studied interval. We may tentatively propose that the P–T boundary is near this level. This suggestion is based on the fact that magnetic properties of rocks should change during (and due to) the Permo-Triassic volcanism. To determine the exact position of the boundary, it is necessary to carry out additional studies, including paleomagnetic ones.

Acknowledgements Thanks to Andrei Kosterov, Vladimir Bakhmutov and an anonymous reviewer for very helpful comments to our article. This work was funded by the subsidy № 5.2192.2017/4.6 allocated to Kazan Federal University for the state assignment in the sphere of scientific activities and work was supported by the Ministry of Education and Science of the Russian Federation contract No. 14.Y26.31.0029 in the framework of the Resolution No. 220 of the Government of the Russian Federation.

References

Betekhtina, O.A., Mogucheva, N.K., Batyaeva, S.K., Kushnarev, M.P., 1986. Permian-triassic boundary in the type section of Mal'tsev Formation of Kuzbass: In: Yanshin, A.L; Dagis, A.S; Biostratigraphy of the Mezoic of Siberia and Far East of Russia. Transactions of Institute of Geology and Geophysics, v. 648, 31–38 (In Russian).

Burov B.V., Yassonov P.G. 1986. Kazan: Introduction to paleomagnetic analysis, Publishing house of KSU, Russia, p. 167.

Buslov M.M., Safonova I.Y., Fedoseev G.S., Reichow M.K., Davies K., Babin G.A., 2010. Permo-Triassic plume magmatism of the Kuznetsk Basin, Central Asia: geology, geochronology, and geochemistry. Russian Geology and Geophysics. vol. 51(9) pp. 1021–1036.

Day R., Fuller M., Schmidt V.A., 1977. Hysteresis properties of titanomagnetites: grain-size and compositional dependence, Physics of the Earth and Planetary Interiors, vol. 13, pp. 260–267.

Dunlop, D. J., 2002. Theory and application of the Day plot (M_{rs}/M_s versus H_{cr}/H_c), 1. Theoretical curves and tests using titanomagnetite data, J. Geophys. Res., 107(B3), https://doi.org/10.1029/2001jb000486.

Grasby, S.E., Beauchamp, B., 2009. Latest Permian to Early Triassic basin-to-shelf anoxia in the Sverdrup Basin, Arctic Canada. Chemical Geology 264, pp. 232–246.

Hallam, A., Wignall, P.B., 1999. Mass extinctions and sea-level changes. Earth-Science Reviews 48, pp. 217–250.

Heydari, E., Hassanzadeh, J., 2003. Deev Jahi Model of the Permian-triassic boundary mass extinction: a case for gas hydrates as the main cause of biological crisis on Earth. Sedimentary Geology 163, pp. 147–163.

Hounslow, M.W., Balabanov, Yu.P., 2016, A geomagnetic polarity timescale for the Permian, calibrated to stage boundaries. Geological Society Special Publication, V. 450 (1), pp. 61–103.

Iassonov P.G., Nourgaliev D.K., Bourov B.V., Heller F., 1998. A modernized coercivity spectrometer, Geologica Carpathica, Slovak Republic, vol. 49, issue 3, pp. 224–226.

Kazansky A.Yu, Metelkin D.V., Bragin V.Yu., Kungurtsev L.V., 2005. Paleomagnetism of the Permian-Triassic traps from the Kuznetsk Basin, Southern Siberia. Russian Geology and Geophysics, vol. 46(11), pp. 1089–1102.

Knoll, A. H., Bambach, R. K., Canfield, D. E., Grotzinger, J. P. 1996. Comparative earth history and the Late Permian mass extinction. Science, 273. pp. 452–457.

Kosareva L.R., Nourgaliev D.K., Kuzina D.M., Spassov S., Fattakhov A.V., 2015. Ferromagnetic, dia-/paramagnetic and superparamagnetic components of Aral sea sediments: significance for paleoenvironmental reconstruction, ARPN Journal of Earth Sciences, vol. 4, issue 1, pp. 1–6.

Kump, L. R., Pavlov, A., Arthur, M. A., 2005. Massive release of hydrogen sulfide to the surface ocean and atmosphere during intervals of oceanic anoxia. Geology, 33. pp. 397–400.

Nastavko A.V., 2013. Petrology of Permo-Triassic traps of the Kuznetsk Basin abstract of a thesis, Russia, p. 16.

Nurgaliev D.K., Yassonov P.G., Useful model "Coercitive spectrometer", patent 81805 registered in the State Register of Utility Models of the Russian Federation on March 27, 2009.

Papin Yu. S., Chunikhin S.A., 2007. Permian-triassic boundary in kuznetsk basin as a regional stratotype for west siberia. Lithosphere, №4. pp. 128–133.

Renne, P.R., Basu, A.R., 1991. Rapid Eruption of the Siberian Traps Flood Basalts at the Permo-triassic Boundary. Science, 253, pp. 176–179.

Retallack, G.J., Jahren, A.H., 2008. Methane release from igneous intrusion of coal during later Permian extraction events. Journal of Geology 116, pp. 1–20.

Retallack, G.J., Seyedolali, A., Krull, E.S., Holser, W.T., Ambers, C.P., Kyte, F.T., 1998. Search for evidence of impact at the Permian-triassic boundary in Antarctica and Australia. Geology 26 (11), pp. 979–982.

Zharinova, V.V., Silantiev, V.V., 2016. Conchostracans of the Maltsevo Formation from the Kuznetsk Coal Basin (Babiy Kamen section): Permian or Triassic? In: Paleostrat-2016. Annual meeting (scientific conference) of the Paleontological group of the Moscow Society of Naturalists and the Moscow branch of the Paleontological Society under the Russian Academy of Sciences (Moscow, 2016). Moscow, Borissiak Paleontological Institute Press, pp. 31–32.

Chapter 18
Magnetic Properties and Concentration of Heavy Metals in Soils of the Krasnyi Kut Town (Saratov, Russia)

M. V. Reshetnikov, A. S. Sheshnev, E. S. Sokolov and S. D. Shkodin

Abstract Anthropogenic activities in areas with high degree of urbanization leads to environmental pollution. Pollution of soils in cities threatens the health of the population. The main sources of pollution in urbanized territories are industrial enterprises and highway transportation.

Keywords Magnetic susceptibility · Soil · Heavy metals · Krasnyi Kut town

Introduction

Anthropogenic activities in areas with high degree of urbanization leads to environmental pollution. Pollution of soils in cities threatens the health of the population. The main sources of pollution in urbanized territories are industrial enterprises and highway transportation (Canbay et al. 2010).

The presence of heavy metals in soil serves as an accurate indicator of pollution). For soil pollution analysis chemical methods usually used consist of soil sampling, its preparation and analysis. Such methods are usually labor-intensive and expensive. In recent times, magnetic method of studying of soils is used and it is simple, cheap, informative and fast. Magnetic method makes it possible to study large number of samples in a short time.

Magnetic measurements are effective indicators when studying pollution caused by industrial and automobile emissions (Hoffmann et al. 1999; Kapicka et al. 1999; Lu et al. 2007). This method is based on the fact that most products of organic fuel combustion contain large amount of magnetic particles (Flanders 1994, 1999). Simultaneously, vehicles serve as sources of pollutants to the environment. It was discovered that environmental lead pollution occurs during combustion of gasoline (Chen et al. 1997). Besides this, roadside dust and soil polluted by various heavy metals enter the environment through transportation (Lu et al. 2005). Increase in

M. V. Reshetnikov (✉) · A. S. Sheshnev · E. S. Sokolov · S. D. Shkodin
Saratov State University, Saratov, Russia
e-mail: rmv85@list.ru

D. Nurgaliev et al. (eds.), *Recent Advances in Rock Magnetism, Environmental Magnetism and Paleomagnetism*, Springer Geophysics,
https://doi.org/10.1007/978-3-319-90437-5_18

magnetic susceptibility of soils has anthropogenic origin, for example it can be from industries and city dust containing magnetic particles (Lu et al. 2008; Blundell et al. 2009). Magnetic particles entering the environment from automobiles contain various heavy metals, for example Cu, Pb, and Zn (Beckwith et al. 1986) which are a part of paint coatings, fuel and tires (Blundell et al. 2009). Pollutants are transported through the atmosphere and get to the soil (Boyko et al. 2004) through atmospheric precipitation which serves as the main source of pollutants in the surface soil layer. Many researchers have recorded positive correlation between concentration of heavy metals and magnetic susceptibility of soils (Lecoanet et al. 2001; Xie et al. 2001; Hanesch et al. 2003; Jordanova et al. 2003). In recent times, magnetic mapping of soils is now used for the assessment of their anthropogenic state (Hay et al. 1997; Kapicka et al. 1999; Lecoanet et al. 2001; Lu et al. 2005). Lu et al. (Lu et al. 2008) found established a high correlation connection between magnetic susceptibility and the content of Cu and Zn in the soils near highways. Lu et al. (2008) suggested the use of magnetic susceptibility measurements *p* as a method for assessing pollution of soils by heavy metals (Cd and Pb). These data indicate that magnetic parameters (such as magnetic susceptibility and others) can be used as diagnostic methods of soil pollution by heavy metals.

The connection between magnetic susceptibility and concentration of heavy metals depends on various soil properties, and on selection. Magnetic properties depend on the size, concentration and type of magnetic minerals in the soil. Carbonate rocks are usually characterized by low magnetic susceptibility, but clays and sandstone show higher magnetic susceptibility. Boyko et al. (2004) believes that concentration of ferromagnetic particles in the soil is diluted with the availability of diamagnetic minerals, such as calcite, organic substance and others.

There isn't enough information on the interconnection between concentration of heavy metals and magnetic susceptibility in soils of cities, especially cities with a small amount of inhabitants (up to 50,000 people). In Russia there is a large number of such cities whose soils have never been analyzed for heavy metals and magnetic susceptibility. The purpose of our research was to determine the concentration of mobile forms of heavy metals and magnetic properties in soils of the settlement (called Krasnyi Kut town) in the territory of Saratov region, Russian Federation (Fig. 18.1). Also we attempted to establish an interconnection between magnetic properties of soils and concentration of mobile forms of heavy metals.

Materials and Methods

Research Area and Sampling

The research was conducted in the territory of Saratov region, Russian Federation. Saratov region is anagrarian and industrial region of Russia. In the territory there are companies for oil and gas production, transportation, storage and processing,

Fig. 18.1 Research territory

there are also steel manufacturing companies, electrical energy companies and others. In recent years, urbanization and industrialization of the territory of Saratov region have led to accumulation of heavy metals in soils of its large cities, for example, Saratov. The town Krasnyi Kut with a population of 14,773 persons (2016) was chosen as object of this research. The primary activity of the population is agriculture and small industrial production, there are no large industrial enterprises in the settlement's territory.

Parent rocks are generally presented by deposits of Neogene—Quarternary age. The main type of soil in this territory is chestnut carbonate soils of loamy structure. The sample area was fixed by GPS. The sampling scheme is represented in

Fig. 18.2 Sampling scheme for soil

Fig. 18.2. A total of 24 soil sample obtained using a plastic shovel were selected and put in plastic packages. Soil samples were quartered, dried and sifted through a sieve of 1 mm.

Magnetic Measurements and Chemical Analysis

In this research the set of magnetic parameters included magnetic susceptibility on low frequency (976 Hz, χLF), magnetic susceptibility on high frequency (3904 Hz, χHF) and also their frequency dependence (Frequency-dependent susceptibility) (χFD, %). Measurements were taken by means of a magnetic susceptibility measuring instrument MFK1-FB.

5 grams of the total sample was taken to evaluate the concentration of mobile forms of heavy metals. Extraction of mobile forms of heavy metals was carried out using 1 M HNO_3V solution during the 24 h. Concentration of Cu, Cd, Pb, Cr, Ni and Zn was determined on an atom- absorbing spectrophotometer. The pH and Eh values were also determined.

The actual concentration of each heavy metal was compared with its threshold limit value (TLV), while calculating its danger coefficient K_o using the formula for identifying the ecologically dangerous levels of concentration of heavy metals in a soil cover:

$$Ko = {}^{Ci}/_{MAC},$$

where

Ci—the content of a form of heavy metal in a sample, (mg/kg);

Maximum allowable concentration—the threshold limit value of the heavy metals form, [mg/kg].

Total coefficient of impurity of Zc was determined to enable us assess the extent of geochemical transformation of a soil cover by mobile forms of heavy metals using the formula:

$$Zc = \sum Ko_n - (n - 1), \text{where}$$

Zc total coefficient of impurity by heavy metals in test;
N quantity of the defined elements;
Ko coefficients of danger of the heavy metals defined in the sample.

When calculating Zc we used excess over the maximum allowable concentration (Ko), but not excess over the background (Ks), unlike recommendations.

Statistical processing of the experimental data was done using the application package Microsoft Excel 2010 (for Windows XP). The confidence limit was considered to be the difference in probable errors p 0.05 (95% a confidential interval) were considered reliable. Interrelations between chemical, magnetic indicators of the soil were estimated using Pearson's simple coefficients of correlation.

Results

Results in the Territory of the Town Krasnyi Kut

In the territory of the town Krasnyi Kut 24 soil samples were selected. Below are short results stating the concentration of heavy metals.

Mobile forms of chromium were determined in all the test samples concentration of chromium ranged from 1.45 to 33.78 mg/kg, with an average content of 3.97 mg/kg and at the maximum allowable concentration of 6.0 mg/kg. Changes in the coefficient of danger ranged from 0.24 to 5.63 with an average value of 0.66. Excess over maximum allowable concentration was recorded in only one sample №. 23 (5.63 maximum allowable concentrations).

Mobile forms of nickel were also recorded in all the test samples in concentrations from 4.56 to 28.07 mg/kg, with an average content of 9.26 mg/kg and at the maximum allowable concentration of 4.0 mg/kg. Changes in the coefficient of danger ranged from 1.14 to 7.02 with an average value of 2.31. Excess over maximum allowable concentrations were recorded in all the test samples with the maximum excess in sample №. 23 (7.02 maximum allowable concentrations).

Lead in its mobile form was found in all test samples in concentrations ranging from 8.60 to 61.89 mg/kg, with an average content of 27.95 mg/kg and a maximum allowable concentration of 6.0 mg/kg. Changes in the coefficient of danger ranged from 1.43 to 10.32 with an average value of 4.66. Excess over maximum allowable concentration were recorded in all test samples with the maximum excess in sample №. 10 (10.32 maximum allowable concentrations).

Mobile forms of copper were recorded in all the test samples in concentrations ranging from 5.46 to 325.39 mg/kg, with an average content of 29.41 mg/kg and a maximum allowable concentration of 3.0 mg/kg. Changes in the coefficient of danger ranged from 1.82 to 108.46 with an average value of 9.80. Excess over maximum allowable concentrations were recorded in all the test samples with the maximum excess in sample № 6 (108.46 maximum allowable concentrations). The geochemical order of mobile forms of heavy metals based on their concentrations in the territory of the city Krasnyi Kut was as follows: Zn > Cu > Pb > Ni > Cr > Cd.

While analyzing this laboratory research it was observed that the determined elements could be divided into two groups: the first group—nickel, lead, copper and zinc, in this group, excess over maximum allowable concentrations were practically recorded in all the test samples; the second group—chrome and cadmium, in which single excess over maximum allowable concentrations were recorded.

Calculation of the total coefficient of pollution showed us the following - in the explored territory, changes in Zc is ranged from 6.03 to 123.91 with an average of 21.43. 13 test samples (54.2%) belong to the category with admissible levels of pollution (Zc from 0 to 16), 7 test samples (29.1%) belong to the category with moderate dangerous levels of pollution (Zc from 16 to 32) and 4 test samples (16.7%) belong to the category with dangerous levels of pollution (Zc from 32 to

Table 18.1 The results of the statistical analysis

	Minimum	Maximum	Average	Standard deviation	Skewness	Kurtosis
Cr	1.45	33.78	3.97	6.71	22.9	4.76
Ni	4.56	28.07	9.26	4.48	14.37	3.45
Pb	8.6	61.89	27.95	16.16	−0.31	0.77
Zn	11.26	227.83	83.97	65.69	0.04	1.13
Cd	0.08	2.19	0.34	0.44	14.06	3.58
Cu	5.46	325.39	29.41	65.24	20.51	4.42
χLF	2.09×10^{-7}	4.14×10^{-6}	8.47×10^{-7}	7.85×10^{-7}	14.17	3.44
χHF	2.03×10^{-7}	4.17×10^{-6}	8.30×10^{-7}	7.94×10^{-7}	14.30	3.46
χFD	1.4	6.76	4.1	1.53	−0.89	−0.03

128). In normative documents the gradation used for Zc is calculated from the concentration coefficient (from excess over a background), we calculated the danger coefficient Zc (from excess over maximum allowable concentration).

Changes in the specific magnetic susceptibility of soils of Krasnyi Kut measured on low frequency (χLF) ranged from 2.09×10^{-7} to 4.14×10^{-6} m^3/kg with an average 8.47×10^{-7} m^3/kg. The magnetic perceptibility on high frequency (χHF) changes from 2.03×10^{-7} to 4.17×10^{-6} m^3/kg with an average 8.30×10^{-7} m^3/kg. The frequency dependence (Frequency-dependent susceptibility) (χFD) changes from 1.4 to 6.7%, with an average of 4.1% (Table 18.1).

The correlation analysis of analytical data for identification of possible paragenetic associations between petromagnetic and geochemical parameters which is presented in Table 18.2 was carried out.

It is visible from the Table 18.2 that in the explored territory two groups of geochemical paragenetic associations of elements were identified:

- the first group includes such associations as chromium-nickel, chromium-cadmium and nickel-cadmium with significant and high coefficients of correlation (0.86–0.93);
- the second group includes associations like lead-zinc, zinc-cadmium, lead-copper with significant, but not high coefficients of correlation (0.4–0.52).

Discussion

Relationships Between Heavy Metals and Magnetic Parameters in Soil

Analyzing data of the correlation analysis it is possible to assume that in the territory of the town Krasnyi Kut of compounds of mobile forms of heavy metals formed two paragenetic associations—chrome-nickel-cadmium and lead-zinc-

Table 18.2 Correlation between geochemical and petromagnetic characteristics in soils of the town KrasniyKut

	Cr	Ni	Pb	Zn	Cd	Cu	χ_{LF}	χ_{FD}	Zc
Cr	**1.00**	**0.93**	−0.15	0.13	**0.91**	−0.04	−0.15	0.11	0.06
Ni		**1.00**	−0.15	0.17	**0.86**	−0.03	−0.13	−0.01	0.08
Pb			**1.00**	**0.41**	−0.06	**0.52**	**0.55**	−0.26	**0.62**
Zn				**1.00**	**0.40**	0.10	0.08	−0.07	0.27
Cd					**1.00**	−0.03	−0.14	0.15	0.11
Cu						**1.00**	**0.90**	−0.27	**0.97**
χLF							**1.00**	**−0.40**	**0.87**
χFD								**1.00**	−0.27
Zc									**1.00**

n = 24; $p = 0.05$; r = 0.40
The significant correlation between geochemical and petromagnetic characteristics is marked as a bold text

copper. Thus, these elements have a uniform mineralogical origin inherited from parent rocks or a uniform source of technogenic origin. Confirmation of these assumptions demands more detailed mineralogical researches which aren't considered by us in this work. The correlation interrelation in couples χLF-Pb and χLF-Cu which indicates that oxides and hydroxides of iron in soils of the town Krasnyi Kut actively absorb cations of copper and lead is of special interest. Thus, we can claim that petromagnetic data can be possibly used for assessment of soil

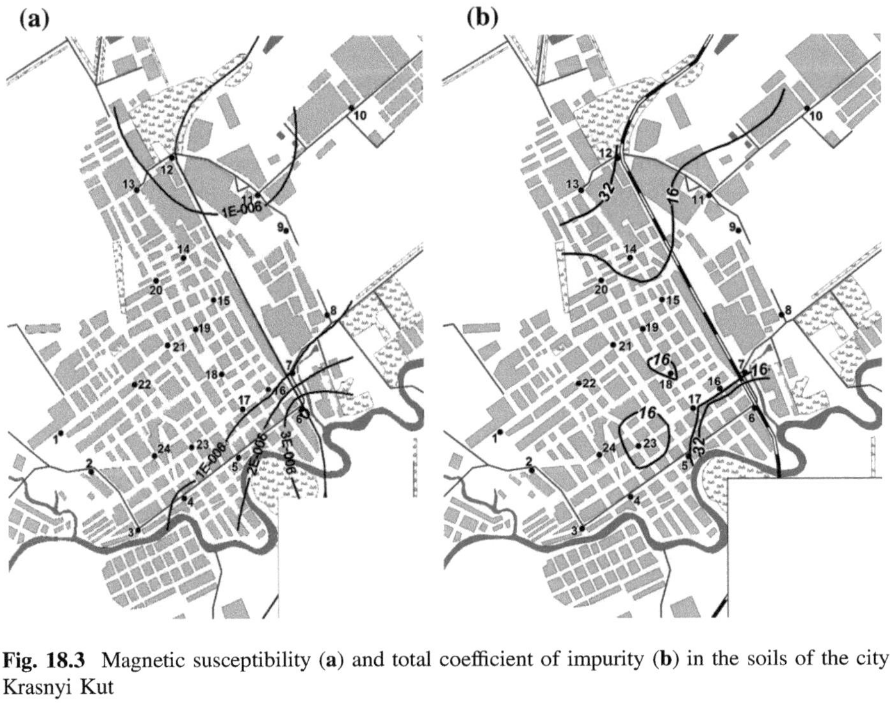

Fig. 18.3 Magnetic susceptibility (**a**) and total coefficient of impurity (**b**) in the soils of the city Krasnyi Kut

pollution by compounds of copper and lead of the town Krasnyi Kut. High correlation interrelations in couple χLF-Zc also demonstrates that in the studied samples there were close interrelations between magnetic susceptibility and total coefficient of pollution which allows us use petromagnetic methods for assessment of the ecological condition of the soil cover in the town Krasnyi Kut (Fig. 18.3).

Sources of Magnetic Particles in Soils

Anthropogenic pollution was the dominant factor influencing magnetic susceptibility of the soils. The role of magnetic susceptibility as an indicator of pollution was confirmed by its highly significant correlation with Pb, Cu and Z_c. There are three possible sources for the magnetic enhancement caused by anthropogenic magnetic particles: (1) emissions from vehicle exhaust and road surface; (2) emissions from fossil-fuel combustion processes: and (3) exotic materials (e.g., metallic fragments and building materials) incorporated into the soils.

Previous study had proven that multi-domain (MD) and stable single domain (SSD) size rangers rather than finer superparamagnetic (SP) grains dominated anthropogenic magnetic particles. In addition, χFD was used as a proxy of the concentrations of SP grains (Lu 2003). Table 18.2 showed that the χFD and magnetic susceptibility of the soils of this study changes inversely ($r = -0.4$, $P \leq 0.05$). This trend indicated that the anthropogenic particles (with fewer SP particles, and, therefore, a lower χFD) contained higher ferrimagnetic content than the natural background. Also, the low χFD indicated the dominance of magnetically soft (probably MD) ferrimagnetic particles.

Conclusions

Chemical analysis and magnetic measurement showed elevated Cu, Pb and Z_c levels together with a considerable enhancement in magnetic susceptibility in the soils. An unusually high level of magnetic susceptibility of the soils was attributed to the presence of magnetically soft MD ferrimagnetic particles, which were probably from automobile exhaust and other industrial activities. Therefore, the magnetic susceptibility of the topsoil proved to be a reliable indicator of anthropogenic pollutant input into the soil. Thus, magnetic measurements can provide valuable information on the distribution of contaminants in the soils and may be used as an efficient alternative technique for identifying and mapping heavy metal contamination due to human activities.The study was carried out at the expense of a grant from the Russian Science Foundation (project No. 17-77-10040).

References

Beckwith, P. R., Ellis, J. B., Revitt, D. M. and Oldfield, F. 1986. Heavy metal and magnetic relationships for urban source sediments. Phys. Earth Planet. Int. 42: 67–75.

Blundell, A., Hannam, J. A., Dearing, J. A. and Boyle, J. F.2009. Detecting atmospheric pollution in surface soils using magnetic measurements: A reappraisal using an England and Wales database. Environ. Pollut. 157: 2878–2890.

Boyko, T., Scholger, R., Stanjek, H. and MAGPROX Team. 2004. Topsoil magnetic susceptibility mapping as a tool for pollution monitoring: repeatability of in situ measurements. J. Appl. Geophys. 55: 249–259.

Canbay, M., Aydin, A. and Kurtulus, C. 2010. Magnetic susceptibility and heavy-metal contamination in top-soils along the Izmit Gulf coastal area and IZAYTAS (Turkey). J. Appl. Geophys. 70: 46–57.

Chen, T. B., Wong, J. W. C., Zhou, H. Y. and Wong, M.H. 1997. Assessment of trace metal distribution and contamination in surface soils of Hong Kong. Environ. Pollut. 96: 61–68.

Flanders, P. J. 1994. Collection, measurement and analysis of airborne magnetic particulates from pollution in the environment (invited). J. Appl. Phys. 75: 5931–5936.

Flanders, P. J. 1999. Identifying fly ash at a distance from fossil fuel power stations. Environ. Sci. Technol. 33: 528–532.

Hanesch, M., Scholger, R. and Rey, D. 2003. Mapping dust distribution around an industrial site by measuring magnetic parameters of tree leaves. Atmos. Environ. 37: 5125–5133.

Hay, K.L., Dearing, J.A., Baban, S.M.J. and Loveland, P. 1997. A preliminary attempt to identify atmospherically-derived pollution particles in English topsoils from magnetic susceptibility measurements. Phys. Chem. Earth. 22: 207–210.

Hoffmann, V., Knab, M. and Appel, E. 1999. Magnetic susceptibility mapping of roadside pollution. J. Geochem. Explor. 66: 313–326.

Jordanova, N. V., Jordanova, D. V., Veneva, L., Yorova, K. and Petrovsky, E. 2003. Magnetic response of soils and vegetation to heavy metal pollution a case study. Environ. Sci. Technol. 37: 4417–4424.

Kapicka, A., Petrovsk′y, E., Ustjak, S. and Machackova, K. 1999. Proxy mapping of fly-ash pollution of soils around a coal-burning power plant: a case study in the Czech Republic. J. Geochem. Explor. 66: 291–297.

Lecoanet, H., Leveque, F. and Ambrosi, J. P. 2001. Magnetic properties of salt-marsh soils contaminated by iron industry emissions (southeast France). J. Appl. Geophys. 48: 67–81.

Lu, S. G., Bai, S. Q. and Fu, L. X. 2008. Magnetic properties as indicators of Cu and Zn contamination in soils. Pedosphere. 18: 479–485.

Lu, S. G., Bai, S. Q. and Xue, Q. F. 2007. Magnetic properties as indicators of heavy metals pollution in urban topsoils: a case study from the city of Luoyang, China. Geophys. J. Int. 171: 568–580.

Lu, S. G., Bai, S. Q., Cai, J. B. and Xu, C. 2005. Magnetic properties and heavy metal contents of automobile emission particulates. J. Zhejiang Univ. Sci. B. 6: 731–735.

Lu, S. G. 2003. Chinese Soil Magnetism and Environment (in Chinese). Higher Education Press, Beijing. 240 p.

Xie, S., Dearing, J. A., Boyle, J. F., Bloemendal, J. and Morse, A. P. 2001. Association between magnetic properties and element concentrations of Liverpool street dust and its implications. J. Appl. Geophys. 48: 83–92.

Chapter 19
Rock-Magnetism and Granulometry of Late Palaeolithic Sites in the Tunka Rift Valley (Pribaikalie) as a Tool for Reconstruction of Sedimentary Environment

G. G. Matasova, A. A. Shchetnikov, I. A. Filinov, A. Yu. Kazansky, G. A. Vorobyeva, N. E. Berdnikova, E. O. Rogovskoy, E. A. Lipnina, I. M. Berdnikov and L. V. Lbova

Abstract Geological structure, rock-magnetic properties and grain size composition of Late Pleistocene-Holocene sediments containing archaeological horizons with cultural relics have been studied in the Tunka rift valley near Lake Baikal at two archaeological sites: "Tuyana" and "Zaktui". The closely neighboured sites are in a different geomorphological position: Tuyana is located on the valley slope, while Zaktui sits on the second terrace of Irkut river. The combined analysis of the granulometric and rock-magnetic parameters specifies the genetic deposition types and clarifies the general structure of both sections so that the sedimentation environment in the second half of the Late Pleistocene and Holocene could be reconstructed.

G. G. Matasova (✉)
Laboratory of Geodynamics and Palaeomagnetism, Trofimuk Institute of Petroleum Geology and Geophysics SB RAS, Novosibirsk, Russia
e-mail: MatasovaGG@ipgg.sbras.ru

A. A. Shchetnikov · I. A. Filinov
Institute of the Earth's Crust SB RAS, Irkutsk, Russia

A. A. Shchetnikov · I. A. Filinov · G. A. Vorobyeva
Vinogradov Institute of Geochemistry SB RAS, Irkutsk, Russia

A. Yu. Kazansky
Geological Department, Lomonosov Moscow State University, Moscow, Russia

A. Yu. Kazansky
Geological Institute of RAS, Moscow, Russia

N. E. Berdnikova · E. O. Rogovskoy · E. A. Lipnina · I. M. Berdnikov
Irkutsk State University, Irkutsk, Russia

N. E. Berdnikova · E. O. Rogovskoy · E. A. Lipnina · I. M. Berdnikov · L. V. Lbova
Institute of Archaeology, Ethnography SB RAS, Novosibirsk, Russia

D. Nurgaliev et al. (eds.), *Recent Advances in Rock Magnetism, Environmental Magnetism and Paleomagnetism*, Springer Geophysics,
https://doi.org/10.1007/978-3-319-90437-5_19

Introduction

The Tunka rift valley extends sublatitudinally over 200 km from the south-western termination of Lake Baikal to Lake Hovsgol (Fig. 19.1). It represents a deep tectonic depression between two large mountain systems: the alpine type Tunka range and the Khamar-Daban ridge. The rift valley comprises 6 basins divided by tectonic saddles. The basin floors are occupied by low and wide (up to 30 km) planes drained by the Irkut River—the main rift channel.

Archaeological studies of this area of Baikal region have been carried out for more than 100 years, but a systematic search for Palaeolithic locations was initiated in the early 1980s by G. I. Medvedev (Lbova et al. 2005). However, despite comprehensive studies of Quaternary deposits, which were carried out since 2010

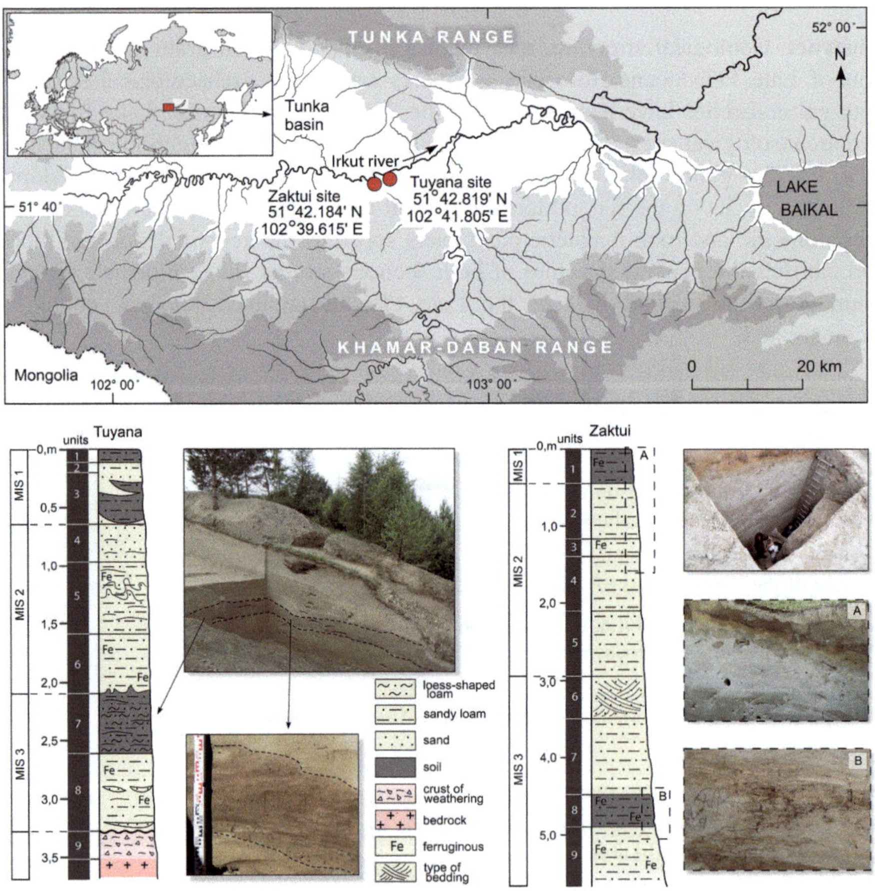

Fig. 19.1 Schematic map of the location of the Tuyana and Zaktui archaeological sites and the structure of the sections

(Shchetnikov et al. 2012; Kozyrev et al. 2014; Shchetnikov et al. 2015), a small number of archaeological sites (about 50) are known in Tunka valley up to now, and only ten of them have been identified as Palaeolithic. Most of those locations are based on lifted materials or singular objects found in the Pleistocene deposits (Berdnikova et al. 2014).

The relatively small number of archaeological sites of Pleistocene age in the Tunka region is mainly explained by the rifting environment during sedimentation, when subsidence and extension of the rift bottom was accompanied by continuous burial of older strata under young sediments and expansion of basin accumulations. In addition, the active riftogenic development in some areas of the Tunka valley is complicated by positive inverse and cyclical motions (Shchetnikov 2008; Ufimtsev et al. 2009). Such inverted sections of the Tunka valley, derived from the basin accumulation and subject to erosion crop out in distal segments of the basins at the foot of the Khamar-Daban ridge and on tectonic saddles between depressions (Shchetnikov 2017). Here, sedimentary complexes of Pleistocene age are exhumed, and the geomorphological position of the sections provides good prospects in the search and study of pre-Holocene geoarchaeological sites. Within one of these elevated areas, near the narrowing of the Irkut valley at the convergence with the Elovsky saddle, the greatest number of locations with Palaeolithic archaeological material was noted. During large-scale archaeological excavations in 2016, we studied two locations (Tuyana and Zaktui) and carried out rock-magnetic and grain-size analyses of the Quaternary sediments containing cultural relics. The main purpose of this study was to reconstruct the sedimentary environment of ancient human habitation.

Geological Position and Site Description

Tuyana site (51° 42′ 49.12″N, 102° 41′ 48.27″E) is located at the Khamar-Daban ridge piedmont on a gently inclined slope with reference marks of 15–35 m from the Irkut River shoreline (Fig. 19.1). It represents a unique Upper Palaeolithic archaeological site (Rogovskoy et al. 2017).

The section sediments excavated to a depth of 3.4 m comprise Late Pleistocene-Holocene sandy loams and loess-like loams. They are characterized by cryogenic deformation in the form of solifluction and redeposition of humic horizons of Middle Holocene soils. The soil humus has been dated at 5475 ± 15 14C BP (UCIAMS-185968) and 6300–6220 Cal. yr BP, respectively (Rogovskoy et al. 2017). The upper part of the Pleistocene sediments is presented by subbedded sandy loams which may be assigned to the Younger Dryas. Loess-like bedded loams deposited during MIS2 are located below. They are underlain by cryoturbated Late MIS 3 palaeosols lying on Early MIS 3 sediments.

Two series of 14C AMS dates have been obtained from bone remains: the first series from excavations of 2011 and 2013 cites 26000–36000 14C BP (30000–41000 Cal. yr BP) (Kozyrev et al. 2014); the second one from excavations from

2016 gives 27000 to 48000 14C BP (30000–48000 Cal. yr BP) (Rogovskoy et al. 2017). The significant scatter of absolute dates can be explained by redeposition and interfusion of sediments of different age which resulted from solifluction. Sedimentation processes in the section are characterized by multiple remobilizations. Redeposition traces are most strongly expressed in MIS3 sediments. Apparently, an intensive hill slide of slope sediments took place here during MIS2. Artifacts were first identified in cryoturbated palaeosol. Archaeological and faunal materials have been accumulated and preserved in soft sediments of small slope bumps, e.g. buried small ravines. Stone artifacts have been allocated provisionally as Late Palaeolithic Aurignacian industry with characteristic macro- and microforms.

Zaktui Section. (51° 42.184′N, 102° 39.615′E) is also situated at the Khamar-Daban ridge piedmont on the right bank of the Irkut River. Basin sediments here are involved in small amplitude inversion uplifts of the Khamar-Daban ridge piedmont tectonic structures and form gently waving tilted planes whose lower segments are characterized by accumulative or socle textures. The section, excavated to a depth of 5.5 m, is predominantly composed of loess-like sandy loams and sands in the upper part with traces of colluvial (slope wash) reworking. The lower part of the section is characterized by an increasing sand fraction. The sediments here represent floodplain and channel facies of alluvium. Crossbedded channel sands are exposed in the middle part.

A buried soil horizon of MIS3 age, strongly deformed by cryogenic processes, is fixed in the section base. Sediments of the upper part of the section are intensively ferruginated. Two microblades, made from flint and smoky quartz, have been found in a ravine slope in close proximity to the Zaktui site in the upper part of the layer (equivalent to layer 10) just below the lower palaeosol horizon. An elk bone (*Alces sp.*) from this layer has been dated with an age >50100 14C BP (OxA-25678) (Shchetnikov et al. 2015).

Methods

Geological field observations were accompanied by excavation of the sections, followed by an initial lithologic-stratigraphic description of the exposed sediments with the determination of their main diagnostic features on the basis of standard lithologic-facies analysis and the construction of a preliminary profile of the section. The lithological properties of the deposits were later specified by granulometric laboratory studies. Careful attention in field observations was paid to the description of the geomorphological position of the location, thereby characterizing the general features of the relief structure of the territory and the leading processes of local morpholithogenesis.

Samples were taken from the fresh outcrop surface at 10 cm intervals (5 cm intervals for some palaeosol horizons). Totally 42 samples were taken from the Tuyana section and 65 samples from the Zaktui section.

Granulometric studies were carried out using a Microtrac X100 particle size analyzer (INGG SB RAS). The results were divided into 60 size fractions (size from 704 to 0.026 μm) expressed in volume percentages for each sample. Particle aggregates were dispersed in an ultrasonic bath. For statistical analysis, the particle size fractions were classified into four groups following the scheme of Raukas (1981); sandy (>100 μm), coarse-silt (50–100 μm), fine-silt (10–50 μm) and clay (<10 μm). The following grain size parameters were calculated: mean grain size (D_{mean}), sorting, aggregation degree (degree of adhesion of particles), dynamic factor (F). D_{mean} was calculated as weighted average. The dynamic factor is the ratio of physical sand (the sum of sand and coarse-silt fractions) content to physical clay (the sum of fine-silty and clay fractions) content. This factor characterizes the dynamics of sedimentary environment. At $F > 1$, the saltation and creep process becomes the main sediment transportation mechanism with the source distance not exceeding 10 km. At $F < 1$, in contrast, suspension processes pre-dominate as well as post-depositional in situ sediment transformations such as pedogenesis.

Rock-magnetic measurements were carried out according to standard procedures (Evans and Heller 2003) with the help of a coercivity spectrometer [J-meter by Jasonov et al. (1998)]. The instrument measures both induced (J_i) and remanent (J_r) magnetization versus magnetic field in 0.5 mT intervals up to +700 mT and back to −700 mT. Determination of hysteresis parameters was made according to (Kosareva et al. 2015).

Initial susceptibility χ was calculated from the linear slope of the initial hysteresis curve at low fields between 0 and 10 mT. The paramagnetic contribution in the hysteresis loops was evaluated by subtracting the paramagnetic component of the bulk susceptibility (χ_p) which was determined by a linear fit through the descending branch of the hysteresis loops in the interval [+500; +700] mT. This correction gives also the values for J_{fer}—the ferromagnetic part of induced magnetization and J_p—the paramagnetic part of induced magnetization (both in the field of 700 mT). The paramagnetic contribution to total magnetization was estimated by the ratio J_p/J_i.

The superparamagnetic contribution ($J_{rm} - J_r$) was deduced from the loss of the J_r curve between 700 and 0 mT. The superparamagnetic susceptibility component (χ_{sp}) is the slope of the J_r loss. The ferromagnetic susceptibility part was calculated as $\chi_{fer} = \chi - \chi_p - \chi_{sp}$.

Coercive force (B_c) and remanent coercive force (B_{cr}) were calculated from the hysteresis loops. Thereby B_c was corrected for the paramagnetic contribution.

The magnetic "hardness" (S) was calculated as the ratio of the remanent magnetization in the reversed field of 300 mT to the saturation isothermal remanent magnetization in the field of 700 mT: $S = -J_r[-300]/J_r$. The "hard" part of saturation remanent magnetization (HIRM) was calculated as $HIRM = (J_r + J_r[-300])/2$.

Magnetic grain size and domain state were estimated plotting the ratios B_{cr}/B_c versus J_r/J_{fer} (Day et al. 1977). In addition, superparamagnetic particles are indicated by the χ_{fer}/J_{fer} ratio (Dekkers 2007).

Results and Discussion

Grain size analysis. According to the classification of Verzilin (1995), based on the ratio of the main granulometric fractions, the studied sections are composed mainly of sandy silt alternating with silty sand. According to the sedimentological classification of Sibirtsev modified by Kachinsky (1958), the main part of the deposits is represented by sandy loam and sand, but there are also layers of light loam in the Tuyana section. In the sandy silts the grain size distribution is unimodal in both sections with modes ranging from 62–74 (in Tuyana) to 52–88 μm (in Zaktui). In the silty sands the mode is shifted to more sandy fractions: 104–124 μm (in Tuyana) and 104–148 μm (in Zaktui) (Fig. 19.2). A weak additional peak of 598–704 μm appears in the lower part of the Zaktui section.

The mean grain size varies between 45–95 μm (Tuyana) and 60–150 μm (Zaktui) (Fig. 19.3). The grain size data help to correct the field description of the sections and to modify their structure (the boundaries of some layers and their thickness).

The Modern soil of both sections is characterized by similar grain size composition, a significant difference in granulometric parameters is observed only in the aggregation degree (46.5% in Tuyana and 30% in Zaktui). The layer that directly underlays the modern soil, differs considerably in composition: in Tuyana it is sandier. The average sand fraction here is 50% and the fine silt fraction is 12%, while the average sand fraction in Zaktui is 40%, the fine silt fraction is 21%. The coarse silt and clay fractions are approximately the same in the both sections, but the aggregation degree varies from 16% in Tuyana to 9% in Zaktui.

Further down the Tuyana section, grain size composition shows minor changes: the mean size is slightly decreasing (down to 45–50 μm) and the coarse-grained fractions decrease slightly in the buried soil (layer 5) where the aggregation degree increases to relatively high values (up to 40%). In the underlying layer, the mean size slightly increases (up to 58–62 μm), while the aggregation degree decreases (to 16%). All the Tuyana sediments are well sorted.

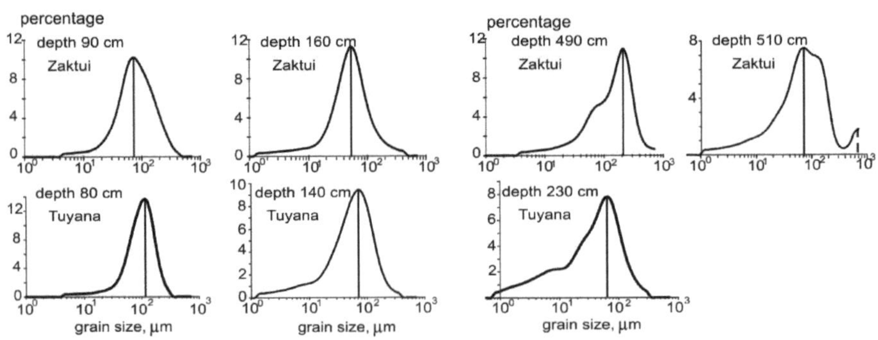

Fig. 19.2 Typical grain size distributions for different layers and lithologies

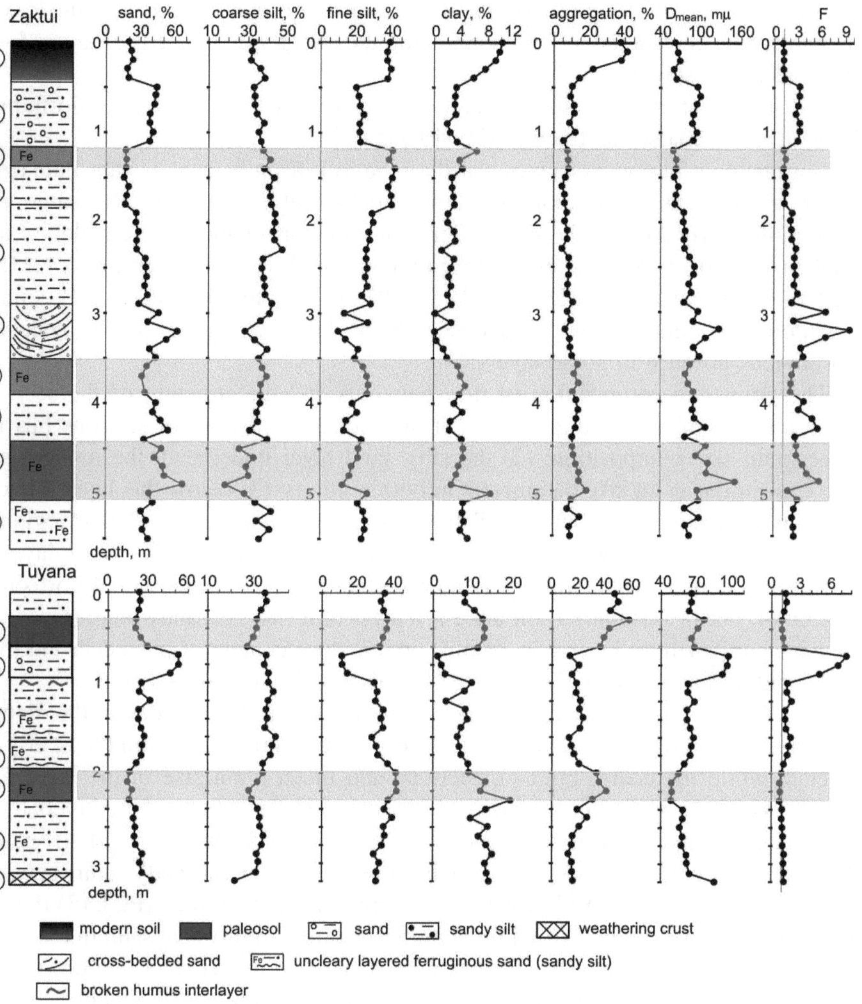

Fig. 19.3 Lithostratigraphy of the sections and their grain size composition. Numbers of the layers are in the circles on the left side of the column

The Zaktui section shows a quite different picture. First of all, the content of coarse-grained fractions and the mean grain size increase down the section. In the silt deposits of layer 6, the fine-grained fractions decreases sharply, and the clay fraction is reduced to zero, while the sand fraction and the average grain size increase. In the fossil pedocomplex (layer 9), the mean grain size and the content of sand fraction are even higher. At the same time, in those horizons the fine silt and clay fractions increase along with the aggregation degree, thus indicating pedogenic transformation of the sandy deposits. The deposits of layers 4 and 5 in the Zaktui section can be considered as well-sorted, and the underlying deposits of layers 6–10 are moderately sorted.

A distinct feature of the sediments in both sections is the relatively uniform high content of the coarse silt fraction (30–40%).

Correlation of the sections was based on the following features (1) absolute ages, which allowed us to attribute the lower palaeosol horizons to the same age interval-the early MIS3; (2) the stratigraphic position of the underdeveloped and implicit palaeosol, most likely corresponding to Meyendorff warming of MIS2 [12500–11850 years BC (Litt et al. 2007)]. before the Holocene; (3) similar granulometric composition of deposits from the upper parts of the sections; (4) obvious absence of some layers.

The grain size composition of the sediments and the structure of the sections show some similarities (Fig. 19.3): (1) the modern soil of both sections has the same grain size composition; (2) the silty sand layer underneath the modern soil also has similar grain size parameters in both sections; (3) below this layer one can trace an underdeveloped buried soil [the age of such soils in the Baikal region is about 13.3–15.1 kyr (Vorobyeva 2010; Vorobyova et al. 2016)]. This layer is poorly traced in Tuyana, while it is represented by an entire palaeosol horizon in Zaktui; (4) below this buried soil there is a layer of loess-like sandy loam (layer 3 in Tuyana, equivalent to layer 4 in Zaktui) with a close content of physical sand (59–62%) and almost the same D_{mean} size (64–65 μm) in the both sections.

The main differences between the sections begin further down the depths (Fig. 19.3): (1) layer 4 in Tuyana differs very little in grain-size from the overlying layer 3, while the coarse-grained fractions and mean grain size in the layer 5 at Zaktui essentially grow in comparison with layer 4 at Tuyana; (2) these layers (3, 4) in Tuyana are underlain by only one buried soil (layer 5), while in Zaktui section there are two palaeosol horizons which embrace a layer of sandy loam between them, so those layers (7–9) in Zaktui are forming a pedocomplex (PC); (3) this PC in Zaktui is capped by layer 6 with the highest content of physical sand in the entire section. This coarse-grained layer has no analogue in the Tuyana section; (4) the lowest layer in Zaktui under the buried soil (layer 10) differs significantly from the lowest layer under the buried soil in Tuyana (layer 6). The content of coarse-grained fractions in layer 10 is 1.5–2.5 times higher than in layer 6.

The results of the grain size analysis, geological observations and absolute dates allow modification of the structure of the sections and correlation between them (Fig. 19.4). The main differences in the sections appear below the layer 2. Since the traces of the buried soil in Tuyana are not developed into separate layers, it is suggested that layers 3, 4 in Zaktui correspond to layer 3 in Tuyana, and layer 5 in Zaktui corresponds to layer 4 in Tuyana. The layers 6–8 in Zaktui section have no analogues in the Tuyana section—those layers fall into hiatus. Layers 5 and 6 in Tuyana correspond to layers 9 and 10 in Zaktui (compare Figs. 19.3 and 19.4).

Fig. 19.4 Correlation between the Tuyana and the Zaktui section based on grain size, geological and 14C data

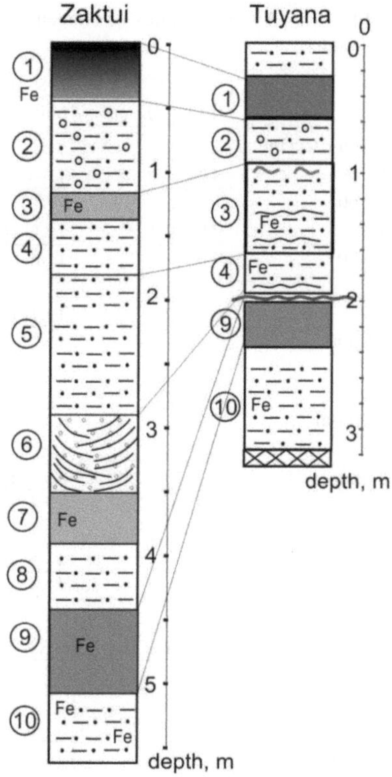

Rock-magnetic properties.

The following concentration-dependent rock-magnetic parameters were analyzed: specific magnetic low field susceptibility (χ), saturation magnetization (J_i), paramagnetic magnetization (J_p) and isothermal remanent saturation magnetization (J_r) and J_p/J_i ratio. The values of rock-magnetic parameters are summarized in Table 19.1; their behaviour versus depth in the sections is shown in Fig. 19.5.

Modern soil. Magnetic susceptibility χ in the modern soils from both sections varies in the same ranges with an average value 11–12 × 10^{-7} m^3kg^{-1}. The average J_r values in both sections are close to each other, although their limits of change vary considerably. The variation ranges of Ji differ significantly and do not overlap; the average J_i values in Zaktui are 1.5 times higher than in Tuyana. The paramagnetic contribution J_p/J_i in Tuyana runs up to 30%, compared with 17% in Zaktui. The behavior of the magnetic parameters indicates that in the modern soil of Zaktui the concentration of magnetic minerals is higher while the concentration of paramagnetic minerals is lower than in Tuyana.

Silty sand (layer 2 in both sections). The concentration of magnetic minerals in layer 2 from both sections is lower than in the modern soil. The values of the concentration-dependent parameters in both sections are close in magnitude, only J_r differs significantly: in Tuyana the average J_r value is almost 2 times higher than in

Table 19.1 Average values of concentration-dependent rock-magnetic parameters of the studied sediments

Layer №, depth (cm)	$\chi \times 10^{-7}$ (m^3kg^{-1})	$J_i \times 10^{-2}$ (Am^2kg^{-1})	$J_r \times 10^{-3}$ (Am^2kg^{-1})	$J_p \times 10^{-2}$ (Am^2kg^{-1})	J_p/J_i
Tuyana section					
1 30–60	8.2–14.2 10.9	13–17.3 14.7	10.9–15.5 12.8	3.6–4.3 4.0	0.25–0.3 0.27
2 60–90	8.8–10.3 9.6	16.4–18.4 17.3	12.7–14.8 14.0	2.9–3.7 3.2	0.16–0.21 0.18
3 90–160	6.8–10 8.2	13.7–17.7 15.6	10.1–12.9 11.3	3.9–4.6 4.2	0.23–0.31 0.27
4 160–190	6.6–6.9 6.7	13.4–13.6 13.5	8.5–9.1 8.8	4.2–4.3 4.25	0.31–0.32 0.315
5 (palaeosol) 190–230	5.7–7.5 6.4	12.7–14.7 13.4	8.6–9.7 9.0	4–4.3 4.2	0.28–0.34 0.32
6 230–310	6.6–8.4 7.6	13.5–16.3 14.8	8.6–12.6 11.3	3.2–4.1 3.8	0.22–0.3 0.26
Zaktui section					
1 0–40	11.7–13.7 12	21.2–22.6 22	12.3–14.7 13.3	3.5–3.8 3.6	0.15–0.17 0.16
2 40–110	8.1–9.7 8.8	15–18 16.3	7.3–9 8.4	2.3–3.0 2.7	0.15–0.17 0.16
3 (palaeosol) 110–140	9.4–9.9 9.6	18–19.1 18.6	10.9–11.5 11.1	3.1–3.8 3.4	0.17–0.20 0.18
4 140–180	9.3–9.9 9.5	17.9–18.7 18.2	11.5–11.9 11.7	3.4–3.9 3.6	0.18–0.22 0.20
5 180–290	6.8–11.6 9.6	13.1–20 17.3	7.9–10 8.8	2.7–3.3 3.0	0.14–0.25 0.18
6 290–350	7.7–16 11	13.7–28.6 19.5	6.4–10 8.2	1.5–3.3 2.3	0.05–0.18 0.13
7 (palaeosol) 350–390	6.4–8.1 7.4	13.4–15.9 14.8	7.4–7.9 7.6	2.7–3.2 2.9	0.17–0.24 0.20
8 390–440	5.9–9.2 7.5	11.8–16.5 13.9	6.2–7.7 7.1	2.2–2.6 2.4	0.15–0.22 0.18
9 (palaeosol) 440–510	4.6–8.5 6.5	8.9–15.4 12.3	5–8.1 6.7	1.8–2.9 2.4	0.14–0.24 0.20
10 510–550	7.2–7.9 7.6	13.8–15 14.5	7.7–8.3 8.0	2.7–3.0 2.9	0.19–0.22 0.20

Note In the numerator—minimum and maximum values, in the denominator—the average value of the parameter within the layer

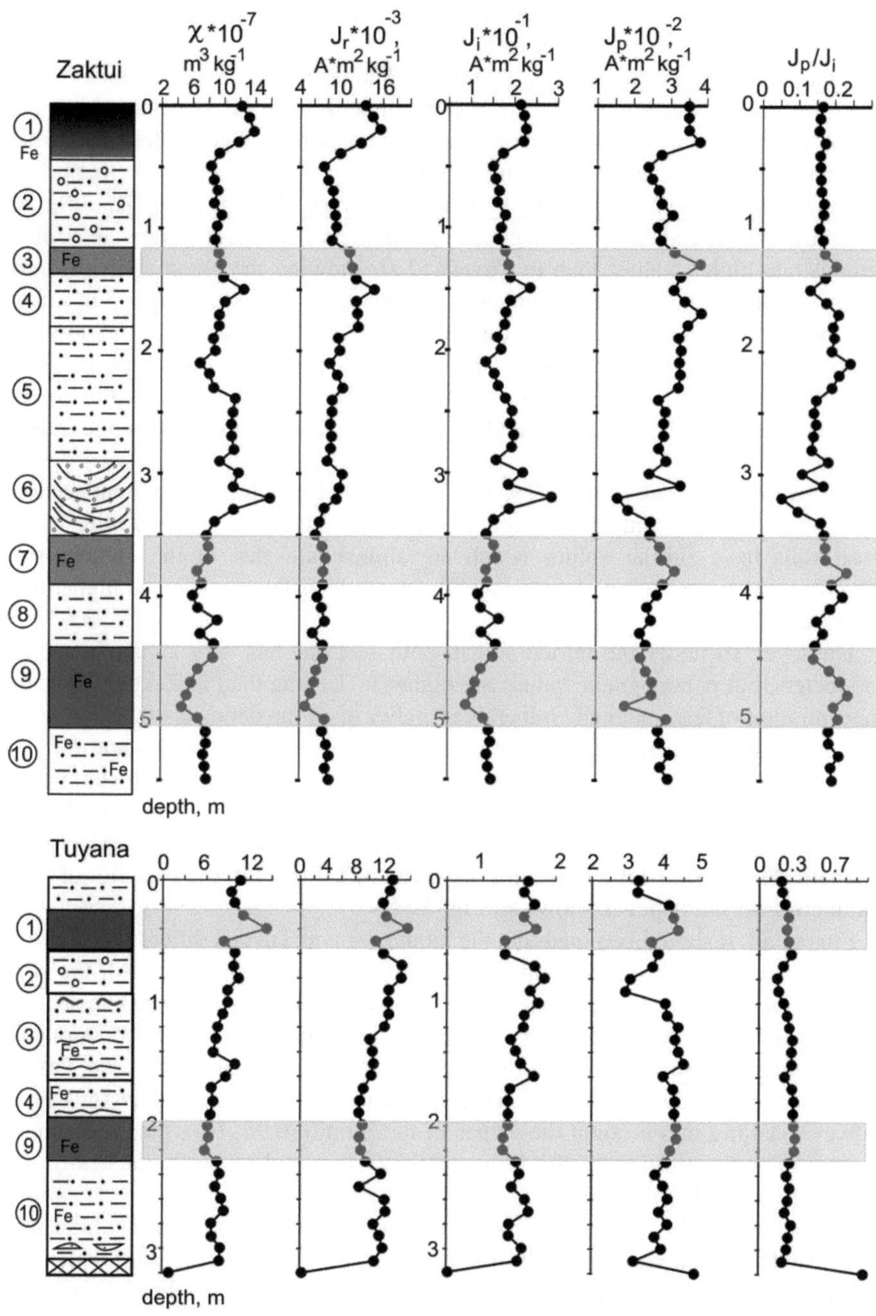

Fig. 19.5 Concentration-dependent rock-magnetic parameters as a function of profile depth

Zaktui, and the limits of J_r changes do not overlap. Both J_p and J_p/J_i values are somewhat higher in Tuyana than in Zaktui.

Undeveloped palaeosol (layer 3 in Zaktui). The layer 3 in Zaktui practically does not differ in the concentration of magnetic minerals from the deposits above and below, probably due to the lack of clearly expressed pedogenic features (small thickness, low degree of soil-forming processes).

Sandy silt (layers 3 and 4 in Tuyana; layers 4 and 5 in Zaktui). All concentration-dependent rock-magnetic parameters of these strata in Zaktui demonstrate higher values than in Tuyana. J_p/J_i showing the paramagnetic contribution on the contrary, is much higher in Tuyana than in Zaktui. Layer 5 in Zaktui may be divided into 2 sublayers (upper—relatively low magnetic and lower—relatively high magnetic), the boundary between them lies at the depth of 2.4 m. The highest values of the concentration-dependent parameters and the lowest paramagnetic contribution are found in the silty sand of layer 6 in the Zaktui section.

Buried pedocomplex (layers 7–9, see Figs. 19.4, 19.5). Layer 9 in both sections contains the lowest content of magnetic minerals. Magnetic susceptibility of the fossil soils have similar values which are almost half that of the modern soil. Moreover, the contribution of paramagnetic minerals in Tuyana is much higher than in Zaktui (Table 19.1).

The layer 10 under the buried soil in both sections has very similar magnetic parameters, but paramagnetic values are higher in Tuyana than in Zaktui. Thus, the concentration of ferromagnetic minerals is higher in all the deposits of Zaktui, while the contribution of paramagnetic minerals is somewhat higher in Tuyana.

Coercivity parameters. Coercivity data include the following parameters: coercive force (B_c), residual coercive force (B_{cr}), coercivity S-ratio, "hard" isothermal remanent magnetization (HIRM). The coercivity parameters characterizing the magnetic "hardness" of the sediments are given in Table 19.2; the behavior of these parameters versus depth is shown in Fig. 19.6.

First of all, it should be noted that the modern soil at Tuyana differs sharply from the other sediments in the section by its low residual coercivity B_{cr}. All the sediments at Tuyana have higher coercivity B_c than the sediments in the Zaktui section. The ranges of the mean values of B_c and B_{cr} in the sections (except the modern soil) do not intersect. According to the range of average values of HIRM, the Tuyana sediments also show hard magnetization about 4 times higher than in Zaktui. However, taking into account the values of the S ratio (0.98–1.0), the predominant carriers of magnetization in the sediments are supposed to be magnetically soft minerals (magnetite/maghemite).

On the base of the behavior of coercivity parameters the layer 5 in Zaktui can be divided into 2 sublayers, similarly it was done using the concentration-dependent parameters χ, J_i and J_r. Differences in the magnetic "hardness" of the sediments are explained by a very small admixture of high coercivity minerals (goethite/ hematite) in the lower part of the Zaktui section and in most of the layers of the Tuyana section (Fig. 19.6). The main reason of the differences in magnetic hardness,

Table 19.2 Coercivity parameters of the studied sediments

Layer №	B_c (mT)	B_{cr} (mT)	HIRM × 10^{-5} (Am^2kg^{-1})	B_{cr}/B_c	χ_{fer}/J_r, 10^{-4} m/A
Tuyana section					
1	6.7–7.3 / 7.1	24.3–29.4 / 26.9	0–5.8 / 1.4	3.5–3.9 / 3.7	7.1–8.8 / 8.1
2	8.5–9.6 / 9.0	38.3–40.4 / 39.3	2.3–1.4 / 1.9	4.2–4.6 / 4.4	6.2–6.8 / 6.5
3	6.6–9.3 / 8.2	35.9–40.8 / 39.0	0.4–7.3 / 4.7	4.3–5.7 / 4.8	5.7–8.7 / 6.5
4	8.1–8.4 / 8.2	39.4–40.6 / 40.1	0.8–5.1 / 3.1	4.8–5.0 / 4.9	6.5–7.7 / 7.2
9	7.9–8.7 / 8.5	40.2–41 / 40.5	1.3–9.4 / 5.1	4.6–5.2 / 4.8	5.7–7.6 / 7.2
10	7.5–10.6 / 9.3	39.1–45.2 / 43.2	3.6–8.0 / 6.5	4.3–5.2 / 4.7	5.3–6.8 / 6.1
Zaktui section					
1	6.3–6.5 / 6.4	32–38.7 / 35.1	0	4.9–6.8 / 5.6	9.0–9.3 / 9.0
2	5.1–5.8 / 5.4	36.4–38.1 / 37.5	0–0.4 / 0.08	6.4–7.6 / 7.0	9.5–10.1 / 10.0
3	6.5–7.1 / 6.7	36.3–37.3 / 36.9	0	5.3–5.8 / 5.6	7.4–8.6 / 8.0
4	6.4–7.4 / 7	35.8–35.9 / 35.9	0	4.8–5.6 / 5.2	7.9–9.0 / 8.4
5	4.3–6.8 / 5.3	32.3–38.2 / 35.5	0–0.2 / 0.03	5.5–7.9 / 6.8	10.9–13.3 / 12.4
6	3.0–5.6 / 4.5	30.4–37 / 34.7	0	6.5–10 / 7.9	11.2–17.3 / 13.0
7	4.9–6.4 / 5.5	36.8–39.7 / 38	0–1.5 / 0.03	6.2–7.6 / 7	7.9–10.4 / 9.2
8	4.8–6.4 / 5.6	36.3–39.8 / 38	0–2.8 / 1.1	6.3–7.6 / 6.9	8.4–11.4 / 10.1
9	4.9–6.6 / 6.0	35.9–41.1 / 39.1	0–1.9 / 0.08	6.1–7.4 / 6.6	8.4–10.9 / 9.2
10	5.6–6.7 / 6.0	38.2–40.2 / 39	0–2.2 / 1.4	6.0–6.8 / 6.5	8.2–9.5 / 9.0

Note In the numerator—minimum and maximum values, in the denominator—the average value of the parameter within the layer

however, lies in the different predominant sizes of magnetic grains constituting the magnetic fraction.

Grain size dependent rock-magnetic parameters include the ratios of various magnetic parameters—J_r/ J_{fer}; B_{cr}/ B_c; χ_{fer}/ J_r (Table 19.2), and indicators of superparamagnetic grains χ_{fer}/ J_{fer}; (J_{rm}-J_r) and χ_{sp} (Fig. 19.7). The first two ratios clearly describe the structure of the Zaktui section, indicating an increase in the

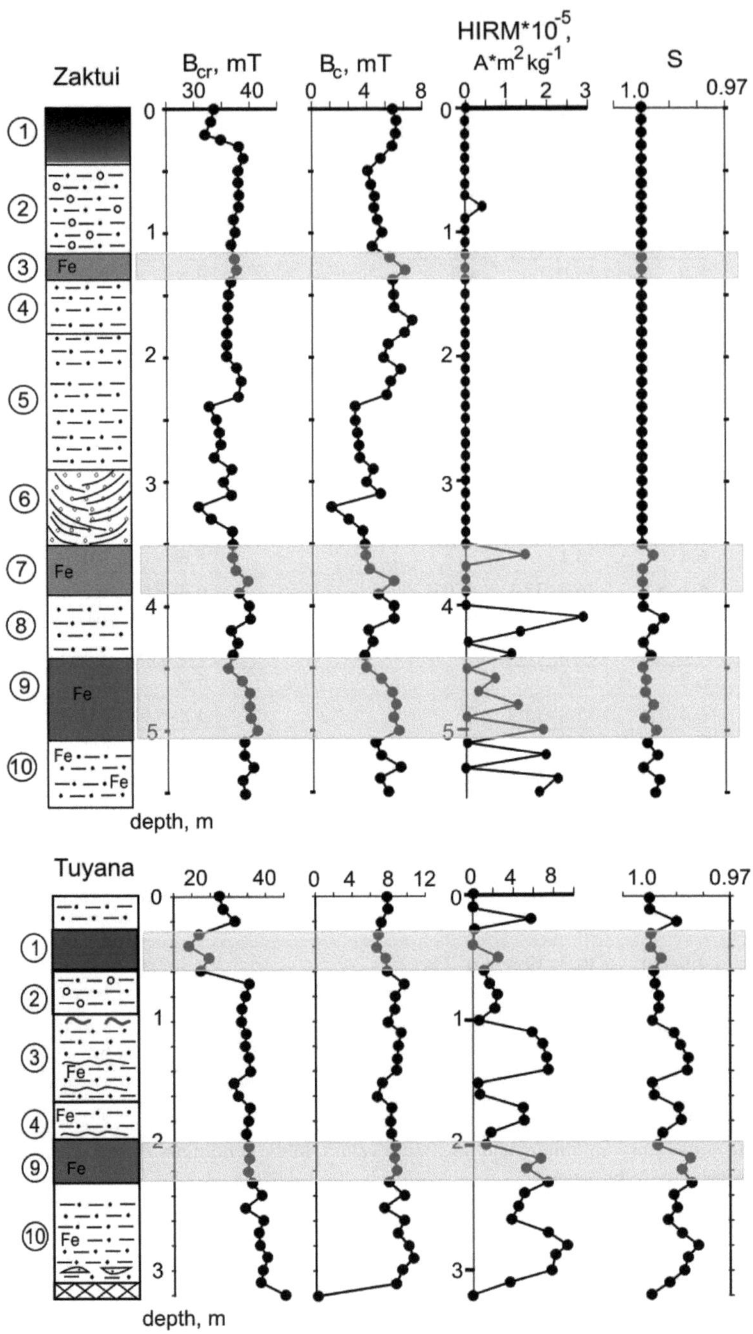

Fig. 19.6 Coercivity parameters as a function of depth at Zaktui and Tuyana

magnetic grain size in the sandy layers and a decrease in the buried soils. Those ratios closely correlate with each other (correlation coefficient $\boldsymbol{r}$ = 0.87) but are practically unrelated to the third ratio (χ_{fer}/J_{fer}) (Fig. 19.7).

The situation in Tuyana is different: the second and the third ratios here are related closely ($\boldsymbol{r}$ = 0.74), but they do not correlate with the first parameter (B_{cr}/B_c). The change limits of B_{cr}/B_c and χ_{fer}/J_r ratios indicate a smaller grain size in all sediments from Tuyana, with the ranges of χ_{fer}/J_r mean values in both sections not overlapping.

The indicators of the presence of superparamagnetic (SP) grains in both sections, J_{rm}-J_r and χ_{sp}, are closely correlated. The correlation coefficient $\boldsymbol{r}$ in Tuyana equals 0.97 and in Zaktui it is 0.94. The SP parameters, however, correlate only in Tuyana with those indicating the effective magnetic grains size, i.e. for (J_{rm}-J_r) $\boldsymbol{r} = 0.74 - 0.77$; for χ_{sp} $\boldsymbol{r} = 0.57 - 0.60$. In Zaktui such correlation is absent. The absolute magnitudes of the SP parameters are similar in both sections.

The domain state of the magnetic grains—Day plot (Day et al. 1977)—shows an elongated data distribution which mainly occupies the field typical for multidomain (MD) grains. Only the upper part of the Tuyana sediments may contain up to 10% pseudosingle-domain (PSD) grains. The magnetic grain size distribution in Zaktui is much wider with the grains generally larger (all grains in the MD field) (Fig. 19.8).

Thus, the magnetic fraction of the sediments from both sections consists mainly of magnetically soft minerals–MD grains of magnetite, possibly oxidized superficially to maghemite.

The differences between the content of magnetic minerals in the studied sections are as follows:

(1) A small amount of high coercivity minerals is present in the sediments at Tuyana, which increases down the section and reaches a maximum content in the layers 9–10. The sediments of the upper part of the section (layers 1–6) at Zaktui, however, contain only low coercivity magnetic minerals, but there is a very tiny admixture of high coercivity minerals in its lower part, but less than in Tuyana (cf. Fig. 19.6).

(2) All grain size dependent rock-magnetic parameters in the Tuyana sediments demonstrate smaller size of magnetic grains and smaller scatter of these sizes than in the Zaktui sediments. The content of PSD magnetite grains in the modern soil and the underlying silt in Tuyana comes up to 10%.

Reconstruction of Sedimentation Environment and Habitat Conditions of the Ancient Humans in the Tunka Depression

A combined analysis of the geological, grain size and rock-magnetic data on the Late Pleistocene deposits from archeological sites Tuyana and Zaktui which contain the products of the vital activity of ancient man, offers the possibility to reconstruct

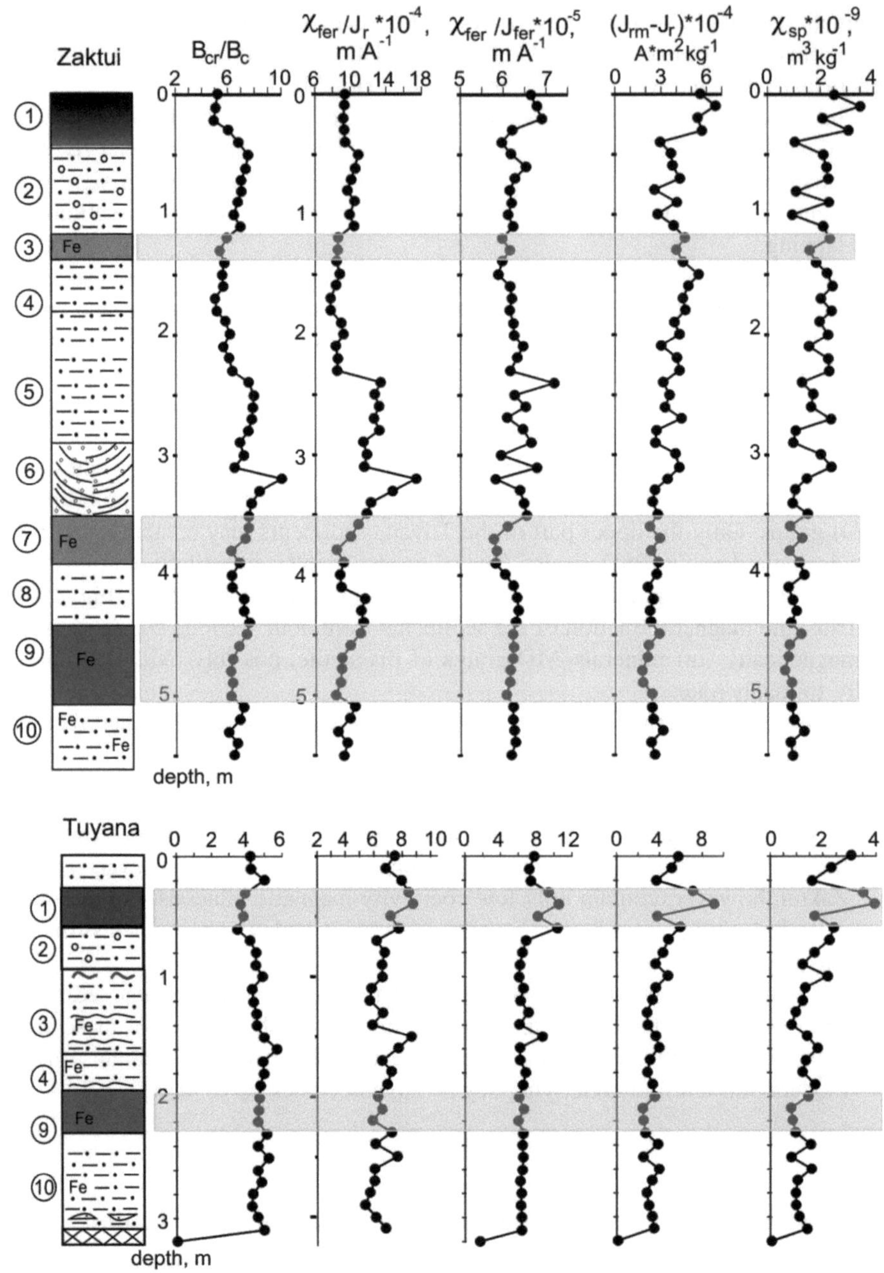

Fig. 19.7 Grain size dependent magnetic parameters at Zaktui and Tuyana

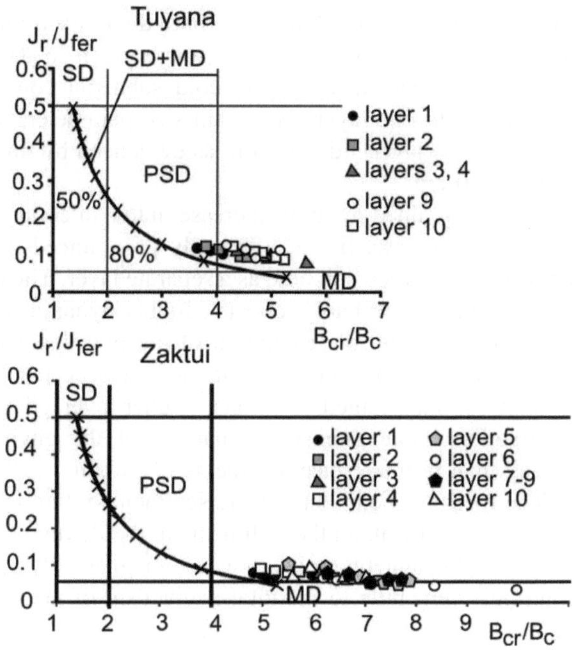

Fig. 19.8 Domain state of the magnetic grains at Tuyana and Zaktui on the "Day plot" Day et al. (1977). SD + MD mixture curve after Dunlop (2002)

sedimentation conditions in these sections of the Tunka Depression with a high degree of probability. First of all it should be noted that grain size similarity is observed only for the upper parts of the sections, i.e. the uppermost 2 m in Tuyana and the uppermost 3 m in Zaktui. The most probable age of this part of the sections may be assigned to MIS2-MIS1 [cf. Shchetnikov et al. (2015)]. Structure, grain size composition and rock-magnetic properties in the lower parts of the sections at Tuyana and Zaktui are quite different despite their geographic proximity. These differences are ascribed to the different geomorphological position of the sections.

The Zaktui terrace section. The buried PC was formed on top of sandy sediments in a dynamically unstable environment, but during a relatively warm period (possibly early MIS3). The upper layer of the silty sand is a fluvial deposit with fine-grained fractions washed out. It was formed in the floodplain part of the Irkut River under a variable sedimentation regime. Grain size analysis suggests the main input of sand-sized material for the lower part of the section (layers 6–10) from local sources. The upper layers (5, 4) were formed in cold subaerial conditions with a gradual decrease in wind intensity and a reduction in dynamics of the sedimentation environment. These processes were accompanied by a gradual warming that ended with the period of deposition of the buried soil of layer 3. Most likely, the layers 4, 5 can be integrated into one horizon with a gradual change in sedimentation conditions. The silty sands of layer 2 have a colluvial (slope wash) origin. Those deposits are depleted in fine-grained fractions.

The Tuyana slope (colluvial) section. The entire sequence of sediments here was formed in a subaerial environment. The layer underneath the buried soil represents a typical loess-like sandy loam. The buried soil itself on the base of the proposed age can be compared with the lower soil horizon in the Zaktui section (MIS3). The layers (4, 3) were deposited in cold subaerial conditions; probably they can be integrated into one layer. The grain size parameters suggest a gradual warming at the end of the layer 3 deposition, as evidenced by traces of buried soil at the depth of 100–110 cm.

The fine-grained fractions increase in this interval while aggregation and average grain size decrease, but insufficiently pronounced differences do not allow us to single out the buried soil here as a separate layer. The overlying layer 2 is composed of silty sand and probably reflects a highly dynamic sedimentation of material from local sources. With the exception of the two upper layers with similar grain size composition, the Tuyana colluvial section is generally characterized by a higher content of fine-grained fractions, higher average grain size, higher degree of aggregation, more intensive dynamics of the environment and better sediment sorting than in the terrace sediments of Zaktui.

The rock-magnetic data of the sediments refine and detail the structure of both sections. In general, all the sediments in the Zaktui section are more magnetic, have coarser ferromagnetic grains, they are magnetically soft and paramagnetic minerals contribute very little to the total magnetism of the deposits.

Conclusions

We have studied geological structure, rock-magnetic properties and grain size composition of Late Pleistocene-Holocene sediments containing archaeological horizons with cultural relics at the two archaeological sites Tuyana and Zaktui.

The sites are located 3 km from each other, but have a different geomorphological position: Tuyana is located on the valley slope, while Zaktui is on the second terrace above the floodplain. The studied strata are represented by interbedded subaerial and alluvial deposits and fossil soils. Grain size analysis attributes the sediments to sandy silt with a unimodal fraction distribution (mode 52–88 μm) alternating with silty sands (mode 104–148 μm); only in Zaktui the silty sand has a bimodal distribution with a secondary mode of 598–704 μm. Rock-magnetic studies show that least magnetic are fossil soils, most magnetic are the modern soils and subaerial deposits. In colluvial sediments there are interlayers with both high and low values of magnetic parameters. Multidomain magnetite is the main carrier of magnetization in all cases. Despite the geographic proximity of the sections, their structure, granulometric composition and magnetic properties of their sediments are different due to their geomorphical location. A combined analysis of the granulometric and the rock-magnetic parameters has specified the genetic types of deposits and has clarified the general structure of the sections so that the sedimentation environment in the second half of the Late Pleistocene and Holocene could be reconstructed.

Even small differences in climatic conditions (increase/ decrease of temperature by a few degrees, fluctuations in humidity by several percent, changing wind force, etc.) are of utmost significance for human life. This study shows that such small climatic fluctuations can be well documented in the grain size composition and magnetic properties of loess-like deposits.

Combination of rock-magnetic and granulometric parameters suggests that the culture-containing horizons of the Tuyana section were deposited under more favorable (mild) climatic conditions than those of the Zaktui section.

At Zaktui archaeological site, the sediments of layers 6–10 are not very promising for the preservation of artifacts in situ. Due to the genesis of sediments in this part of the section, any archaeological material most likely has been redeposited. The sediments comprising the upper part of the section more likely may contain traces of ancient human activity, especially inside and near layer 3 in Zaktui section. The slope deposits in the Tuyana section and in other parts of the Tunka depression with similar geomorphological position are generally more promising for archaeological findings.

Acknowledgements We are deeply grateful to the reviewer for taking the time to provide quite valuable comments and suggestions.

The study was supported by RFBR (No. 16-05-00586) and RNF (No. 16-17-10079, 14C dating, field studies) grants. The archaeological part of the work was carried out within the framework of the state task of the Ministry of Education and Science (project No. 33.2015.2017/4.6) "Palaeoenvironment, Extreme Natural Conditions and Fossil Cultures of Baikal Siberia in the Anthropogene". This study has also been carried out following the plans of the scientific research of the Geological Institute of RAS (for AYK, project No. 0135-2016-0009).

References

Berdnikova N.E., Berdnikov I.M., Rogovskoy E.O., Timoshchenko A.A., Ulanov I.V., Sokolova N.B., Popov A.A., Kogai S.A. New archaeological objects in the Tunka basin (Baikal rift zone) // Eurasia in the Cenozoic. Stratigraphy, Paleoecology, Culture. 2014. Issue 3. P. 19–41. (in Russian).

Day R., Fuller M., Schmidt V.A. Hysteresis properties of titanomagnetites: Grain-size and compositional dependence // Phys. Earth Planet. Inter. - 1977. - Vol. 13. - P. 260–267.

Dekkers M.J. Magnetic proxy parameters. In Gubbins, David; Herrero-Bervera, Emilio. Encyclopedia of geomagnetism and palaeomagnetism. Dordrecht: Springer. - 2007. pp. 525–535.

Dunlop, D.J. Theory and application of the Day plot (M_{rs}/M_s versus H_{cr}/H_c) // J. Geophys. Res. Solid Earth. - 2002. - Vol. 107. - Iss. B3. - P. 2046–2067.

Evans M.E., Heller F. Environmental Magnetism. - New York : Academic Press, 2003. - 299 p.

Jasonov P.G., Nurgaliev D.K., Burov B.V., Heller F. A modernized coercivity spectrometer // Geologica Carpathica. - 1998, Vol. 49, N 3. - P. 224–226.

Kachinsky N.A. Mechanical and microaggregate composition of soils, methods of its study. Izd-vo AN SSSR, Moscow. – 1958. 258 p. (in Russian).

Kosareva L.R., Nourgaliev D.K., Kuzina D.M., Spassov S., Fattakhov A.V. Ferromagnetic, dia-/paramagnetic and superparamagnetic components of Aral Sea sediments: significance for paleoenvironmental reconstruction // ARPN Journal of Earth Sciences, - 2015. - Vol. 4. - No. 1. -P. 1–6.

Kozyrev A., Shchetnikov A., Klement'ev A., Filinov I.A., Fedorenko A., White D. The early Upper Palaeolithic of the Tunka rift valley, Lake Baikal region, Siberia // Quaternary International. – 2014. – Vol. 348. – P. 4–13.

Lbova L.V., Lipnina E.A., Medvedev G.I., Novoseltseva V.M., Postnov A.V., Fedorenko A.B. Preliminary archaeological zoning of the Eastern Sayan, problems and prospects for searching Stone Age sites // Problems of archeology, ethnology, anthropology of Siberia and adjacent territories. - 2005. - Vol. 11, part 1. - P. 150–156 (in Russian).

Litt T., Behre K.-E., Meyer K.-D., Stephan H.-J., Wansa S. Stratigraphische Begriffe für das Quartär des norddeutschen Vereisungsgebietes // Eiszeitalter und Gegenwart (Quaternary Science Journal). 2007. Vol. 56 (1/2). 7–65.

Raukas A.V. Classification of detrital rocks and sediments by granulometric composition. Academy of Sciences of the Estonian Soviet Republic, Institute of Geology, 1981. 24 p. (in Russian).

Rogovskoy E.O., Berdnikova N.E., Lipnina E.A., Vorobyova G.A., Berdnikov I.M., Kuznetsov A.M., Lokhov D.N., Shchetnikov A.A. Palaeolithic features of the Tunka Valley // V (XXI) All-Russian Archaeological Congress. - Barnaul: The ASU Publishing House, 2017. - P. 871–872 (in Russian).

Shchetnikov, A.A.. The manifestation of a Gobi mechanism of orogenesis in the Baikal rift system (using the Tunka rift as an example). Geography and Natural Resources. – 2008. Vol. 29, - 226–229.

Shchetnikov A.A., Khenzykhenova F.I., Klement'ev A.M., Simakova A.N., Semenei E.Y., Filinov I.A. Changes of environments and climate during the Late Pleistocene and Holocene reconstructed from aeolian and colluvial deposits of the Zaktui site (Tunka rift valley, Baikal region) // Quaternary International. – 2015. – Vol. 355 - 80-89.

Shchetnikov A.A., White D., Filinov I.A., Rutter N. Late Quaternary geology of the Tunka rift basin (Lake Baikal region) // Journal of Asian Earth Sciences. – 2012. – Vol. 46 - 195–208.

Shchetnikov A.A. Morphotectonic inversion in the Tunka rift basin (southwestern Baikal region) // Russian Geology and Geophysics. - 2017. Vol. 58 (7) - 778–786.

Ufimtsev G.F., Shchetnikov A.A., Filinov, I.A. Inversions in the contemporary geodynamics of the Baikal rift system // Russian Geology and Geophysics. – 2009. Vol. 50 (7) -618-627.

Verzilin N.N. On the classification of sedimentary rocks in lithologic-palaeogeographic investigations // Russian Geology and Geophysics. - 1995, Vol. 36 (11) – 131–141.

Vorobyeva G.A. Soil as a record of natural events of the Baikal region: problems of evolution and classification of soils. - Irkutsk: Publishing House of Irkutsk State University, 2010. – 205 p. (in Russian).

Vorobyova G.A., Berdnikova N.E., Lipnina E.A., Rogovskoy E.O., Berdnikov I.M. Palaeosols of the Sartan age in the cross-sections of geoarcheological objects of the Baikal region // Soil as a connecting link in the functioning of natural and anthropogenically transformed ecosystems. - Irkutsk: Publishing House of Irkutsk State Univ., 2016. - 233-239 (in Russian).

Chapter 20
Application of Methods of Rocks Magnetism for Determination of Criteria of Perspectiveness of Ore-Formation

J. H. Minasyan

Abstract The study of rock magnetism is of great scientific and practical interest for solving a wide range of geological and geophysical problems (Konstantinov in The structures of the gold ore fields in Transcaucasian ragion and the main regularities of their location. Izv. Academy of Sciences of the USSR, 1973; Sargsyan in Role of enclosing rocks in the metasomatism and zonality of its products on the example of the gold ore deposits. Nedra, Moscow, 1972; Sholpo in Use of rock magnetism for solving geological problems. Nedra, Leningrad, p. 182, 1977). In solving this problem an important stage is the determination of the composition and the conditions for the formation of minerals of rocks. This article poses the problem as a result of the application of magnetic methods for studying the rocks of the Zod deposit (Armenia) in order to clarify the conditions for the formation of the enclosing rocks of ore formations and to determine the characteristic criteria for promising ore-bearing formations.

Keywords Rocks · Magnetic mineral · Pyroxenite · Serpentinite
Peridotit

Research Area

The Zod deposit is located in a very interesting and geologically difficult area—in the zone of the ophiolite belt. Although the area was studied by various geological and geophysical methods, systematic studies of the magnetic properties and composition of the magnetic minerals of rocks common in the field were not carried out. Five rock complexes participate in the structure of the Zod ore field: metamorphic formations of the Paleozoic to the Mesozoic, volcanic-sedimentary formations of the ophiolite

J. H. Minasyan (✉)
Institute of Geophysics and Engineering Seismology after
A. Nazarov NAS, Yerevan, Republic of Armenia
e-mail: julyaminas@yandex.ru

D. Nurgaliev et al. (eds.), *Recent Advances in Rock Magnetism, Environmental Magnetism and Paleomagnetism*, Springer Geophysics,
https://doi.org/10.1007/978-3-319-90437-5_20

association of the Lower Senonian, terrigenous- sedimentary rocks of the Upper Senonian, Miocene volcanics and neoplasms of the Quaternary system. The most ancient breeds of the region, which were serpentinized in the Upper Cretaceous, are peridotites. It is believed that as a result of the introduction of gabbroids into peridotites, weakened contact zone forms, to which the main metasomatic processes are confined (Konstantinov 1973; Sargsyan 1972). Before approaching the study of the magnetic properties of rocks directly from ore zones, it is necessary to clearly visualize the composition of the magnetic minerals and the magnetic properties of the rocks representing the "background", on which the areas recycled by ore metasomatic processes are distinguished (Minasyan and Minasyan 2016). Such a "background" is serpentinites (the product of earlier changes in peridotites), pyroxenites and gabbroids, widespread on the deposit. Sampling was carried out from three sites, to varying degrees subjected to changes. The first part is represented by the least altered species—serpentinites, pyroxenites and gabbroids. The second part is the continuation of the first, where talc-bearing serpentinites and talc-carbonate rocks appear. The third part is characterized with superimposed subsequent changes and is represented by quartz-talc-carbonates and carbonate rocks. Dike of quartz liparites, which have been significantly altered by ore processes, which can also carry gold mineralization, are also common in the field. But the study of the magnetic properties of these rocks was not within the scope of this article.

Methods of Study

To solve this problem, the determination of the crystallization temperature of ferromagnetic minerals of rocks widespread in the field was of great interest. On the study of individual samples from the Zod deposit, it was known that magnetite and pyrrhotite occur in the rocks of the deposit (Konstantinov 1973; Sargsyan 1972). However, there were no definite data on whether there are other ferromagnetic minerals, what kind of rocks they are related to, at what stages of the change and especially at what temperatures they crystallized.

In order to determine the secondary heating of rocks, there are interesting samples, in the ferromagnetic fraction of which pyrrhotite is present. Many studies have been devoted to the study of changes in pyrrhotite under heating, a brief survey of which is contained in (Brodskaya 1980; Brodskaya and Gendler 1979; Popov et al. 1982; Sholpo 1977). The results of these studies gave grounds to assume that it is possible to determine the temperatures of secondary heating of rocks by the nature of the changes in pyrrhotite during heating.

The main methods of studying the collection were: determination of Is(t) and Irs (t) changes under continuous heating and saturation parameters Is(H) and H′cs for successive heating. The Is(t) curves were taken in a magnetic field, and Irs(t) in a zero field. Demagnetization curves were also compiled for the variable magnetic field Iri(h), In(h), Irs(h), Irt(h), H′cs, and coercive spectra. Since the behavior of these parameters for the main ferromagnets (magnetite, titanomagnetite, hematite, maghemite, pyrrhotite, etc.) is known, it was assumed that the solution of the inverse problem is also possible. It is believed that the rock containing pyrrhotite underwent reheating if: 1. the sample under investigation has a $Is(t)/Is_0(t)$ curve with a "platform" in the temperature region 210–260 °C, and the curve $Irs(t)/Irs_0(t)$ has an inflection of about 325 °C (Curie point of pyrrhotite) changes in saturation parameters are not characteristic for pyrrhotite; 2. the sample has a smooth curve Is $(t)/Is_0(t)$, the Curie point near 325 °C is clearly determined from the $Is(t)/Is_0(t)$ and $Irs(t)/Is_0(t)$ curves, and the saturation parameter changes for successive heats are characteristic of pyrrhotite (Brodskaya 1980; Brodskaya and Gendler 1979). If the destructive field does not increase or decrease at temperatures above 450 °C, then hexagonal pyrrhotite predominates in the sample. If the destructive field increases substantially, then pyrrhotite is monoclinic. If the rock is heated to temperatures of 550–600 °C, the curves for changing the saturation parameters, which are similar in both the case of hexagonal and monoclinic pyrrhotite, will indicate the reheating.

Research Results

Figure 20.1 shows the characteristic thermomagnetic curves for the rocks of the first part. Table 20.1 shows the main petromagnetic characteristics of the rocks studied.

Of ferromagnetic minerals, serpentinites and pyroxenites contain mainly magnetite, the Curie point of which is located in the temperature range 575–625 °C (probably cation deficient magnetite). However, serpentinites and pyroxenites clearly differ in magnetic properties. Magnetite in serpentinites was oxidized, as indicated by the presence of maghemite and hematite in the rock. The saturation magnetization Is (H) of serpentinites turned out to be substantially larger than that of pyroxenites.

The destructive saturation field of H′cs, on the contrary, in serpentinites and gabbros is considerably less than in pyroxenites. Almost all samples of serpentinites contain maghemite along with magnetite, and hematite is also found in some samples, which are a product of oxidation of magnetite. In pyroxenites, the ferromagnetic fraction is represented only by magnetite, which has not undergone oxidation. All this suggests that pyroxenites were formed later serpentinites.

The section of the second part studied is a continuation to the north-east of the first section. Figure 20.2 shows the characteristic thermocouples, and in Table 20.2 the main petromagnetic characteristics of the samples of the second part are shown. The main part of the collection consists of talc-carbonate rocks, representing a metasomatic zone with a thickness of ~15 m. In the lower part of the section, pyroxenite, similar to the pyroxenite of the first section, occurs. The Curie point of these samples is located in the interval 575–625–650 °C. In rocks there is magnetite, which in almost all samples is oxidized.

A feature of the samples of this section is the presence of a large amount of maghemite, which is clearly established by the changes in the parameters Is/Is_0, Irs/Irs_0 and H′cs after heating. A sharp increase in the Curie point of the samples of the

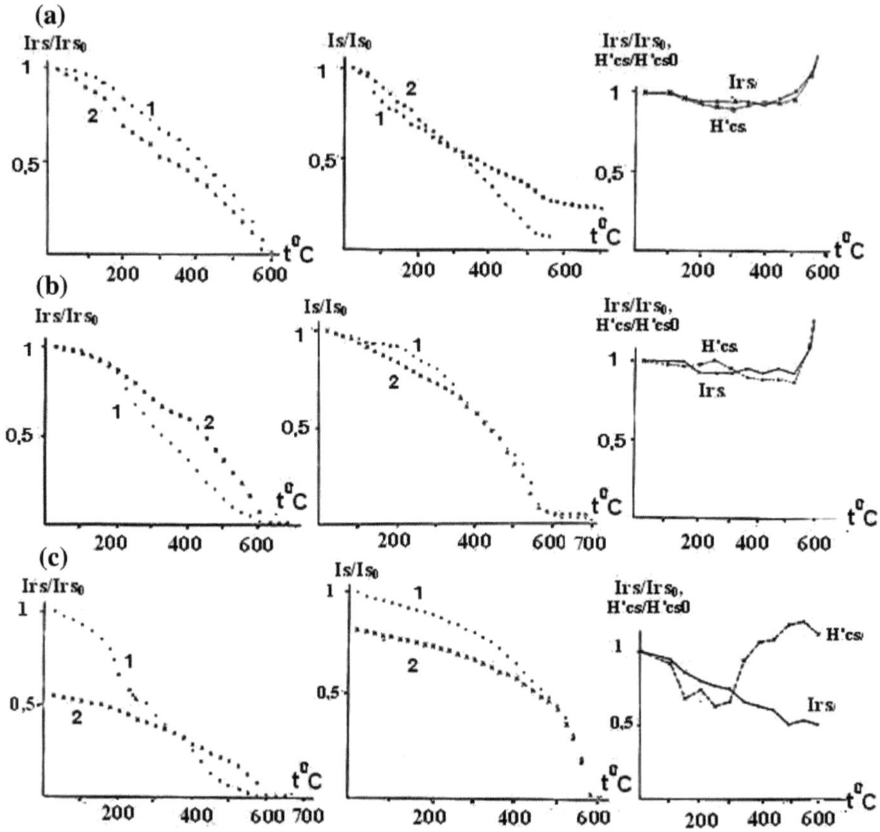

Fig. 20.1 Characteristic thermomagnetic curves of the rocks of the first part. **a** gabbros; **b** pyroxenites; **c** serpentinites. 1 primary heating; 2 secondary heating

central part of the section was noted, compared with samples from the periphery of the section, from 575–600° to 675 °C. In some samples, there is a component with a Curie point of 675 °C, which corresponds to hematite. This gives grounds to state that in addition to oxidized magnetite and maghemite, hematite is present here, the amount of which increases in the center of the section. This means that the central part of the section is the most altered, from where the hydrothermal solutions came, which led to serpentinization and talc-bearing of the rocks in the region. According to geological data, the rocks of this region less altered and relict breeds. The results of thermomagnetic studies showed the opposite. Maghemite in the rocks of this section is a later low-temperature mineral. The studied section of the third part is located at a great distance from the ore zones. The difference of this part is that the rocks of this part were subjected to a weak effect of ore processes.

Here, mainly carbonate-talc-bearing serpentinites, quartz-talc-carbonate and quartz-carbonate rocks are common. Thermomagnetic studies of the rocks of this

Table 20.1 The petromagnetic characteristics of the samples of the first part (group)

Rocks	$\frac{Ist}{Is_0}$	$\frac{Irst}{Irs_0}$	$H^{/}_{cs}$	Composition of ferromagnetic fraction	Temperature of formation I_n (°)
Serpentinite	0.77	0.56	410	Magnetite Maghemite	≥ 575
Serpentinite	1.07	0.85	325	Magnetite Hematite Maghemite	625
Serpentinite	0.79	0.76	215	Magnetite Maghemite Hematite	≤ 575
Serpentinite	0.85	0.74	192	Magnetite Maghemite	575
Pyroxenite	0.54	0.40	355	Magnetite Maghemite Hematite	>575
Pyroxenite	0.89	0.8	680	Magnetite	600
Pyroxenite	1.2	1.3	780	Magnetite	≥ 575
Gabbro	3.5	1.5	380	Maghemite	600

$\frac{Ist}{Is_0}$—Isothermal saturation magnetization versus temperature / Isothermal saturation magnetization at roomtemperature
$\frac{Irst}{Irs_0}$—Isothermal remanent saturation magnetization versus temperature/ Isothermal remanent saturationmagnetization at room temperature
$H^{/}_{cs}$—Coercitive force
I_n—Natural remanent magnetization

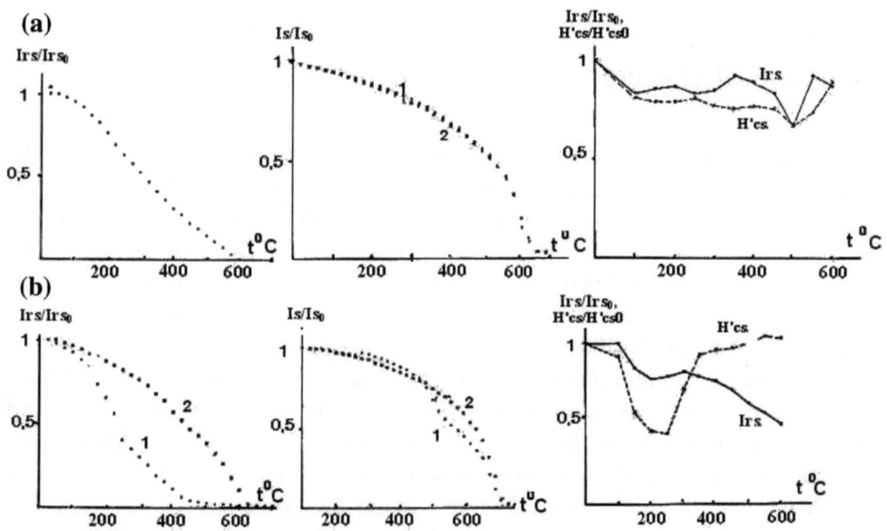

Fig. 20.2 Characteristic thermomagnetic curves of the rocks of the second part. **a** carbonated pyroxenite; **b** talc-carbonate rock; 1 primary heating; 2 secondary heating

Table 20.2 The petromagnetic characteristics of the samples of the second part (group)

Rocks	$\frac{Irst}{Irs_0}$	$\frac{Ist}{Is_0}$	$H^{/}_{cs}$	Composition of ferromagnetic fraction	Temperature of formation I_n (°)
Pyroxenite	1.07	0.98	280	Magnetite	600
Talc-carbonate	0.4	0.32	–	Magnetite	575
Talc-carbonate	0.52	0.72	350	Magnetite	620
Talc-carbonate	0.44	0.38	360	Magnetite Hematite	675
Talc-carbonate	0.59	0.53	–	Magnetite	625
Talc-carbonate	0.47	0.32	–	Magnetite	650
Talc-carbonate	0.64	0.66	–	Magnetite Hematite	675
Serpentine-talc	0.9	3.6	385	Magnetite Hematite	275

$\frac{Ist}{Is_0}$—Isothermal saturation magnetization versus temperature / Isothermal saturation magnetization at roomtemperature
$\frac{Irst}{Irs_0}$—Isothermal remanent saturation magnetization versus temperature/ Isothermal remanent saturationmagnetization at room temperature
$H^{/}_{cs}$—Coercitive force
I_n—Natural remanent magnetization

section (Fig. 20.3) showed that the main ferromagnetic for mentioned rocks them is cation deficient magnetite. In individual samples, magnetite is oxidized to maghemite, and almost everywhere to hematite.

In some samples, a small amount of warmed λ-pyrrhotite and pyrite was detected. Since they are found in quartz-carbonate rocks, this may mean that pyrrhotite and pyrite crystallized only at the ore stage.

Discussion of Results

Based on the study of the magnetic properties of rocks remote from ore zones, but to varying degrees, we can say that the serpentinites, pyroxenites and gabbroids, which represent the "background" by which later processes of talc-bearing and carbonatization have passed, contain mainly magnetite. In these rocks, apart from magnetite, maghemite and hematite are contained, which are products of oxidation of magnetite. The crystallization temperature of magnetite for pyroxenites is >575 °C, and for serpentinites 400–500 °C. This suggests that ferromagnetics in these rocks were formed under different conditions and at different times. In all likelihood pyroxenites are later formations than serpentinites. Later, to the ore stage, metasomatic processes led to the formation of talc-carbonate rocks. Quartz-talc-carbonate and quartz-carbonate rocks were formed at the ore stage of metasomatism. According to the composition and magnetic properties of ferromagnetic minerals, the "background" rocks and rocks altered by ore processes vary significantly. In these rocks, the content of magnetite decreases (sometimes completely

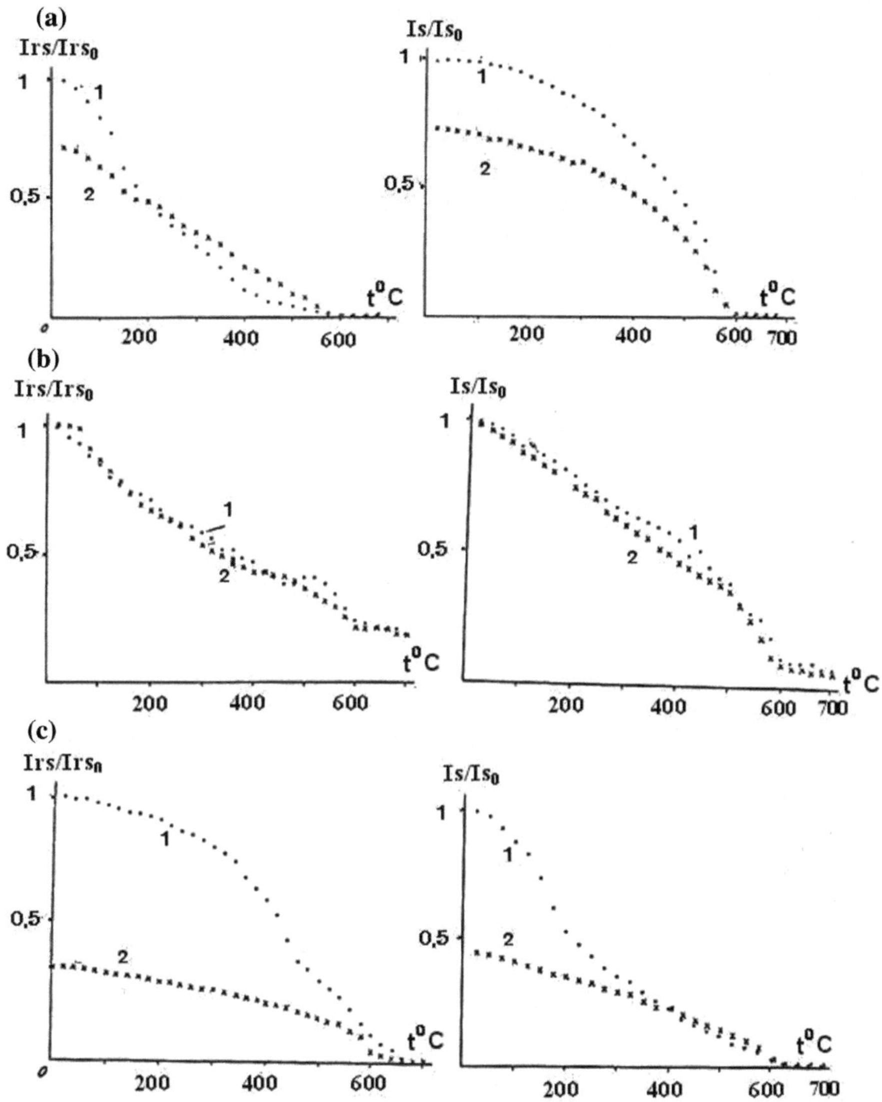

Fig. 20.3 Characteristic thermomagnetic curves of the third part. **a** carbonate-talc-bearing serpentinite; **b** quartz-talc carbonate rock; **c** talc-carbonate rock; 1 primary heating; 2 secondary heating

absent) and iron sulphides appear, mainly pyrrhotite and pyrite. Talc-carbonate rocks (metasomatites up to the ore stage) on sites not affected by ore processes contain only magnetite. Pyrrhotite and pyrite appear near the ore bodies. A small amount of pyrrhotite in talc-carbonate rocks and near the gold ore core indicates that it has undergone a second heating. Consequently, it can be assumed that pyrrhotite crystallized at one of the first stages of ore formation. Hence, the temperature of its reheating must characterize the maximum temperature of the processes following its crystallization. A later stage of ore formation crystallized mainly pyrite, which is also found in enclosing rocks. It should be noted that there are many of them especially in ore bodies. This means that the main amount of pyrite is released at the last fading stages of metasomatism.

We can conclude that the results suggest the possibility of applying the method of reheating (as more promising) in other fields of Armenia.

References

Brodskaya S. Yu. The pyrrhotite as a geothermometer for rock reheating. Izv. Academy of Sciences of the USSR, Physics of the Earth No. 3, 1980.

Brodskaya S. Yu. Gendler T.S. On the nature of changes in the λ-type pyrrhotite in γ-transition area. Izv. Academy of Sciences of the USSR, Physics of the Earth No. 12, 1979.

Konstantinov M.M. The structures of the gold ore fields in Transcaucasian ragion and the main regularities of their location. Izv. Academy of Sciences of the USSR, ser. Geological, No. 5, 1973.

Minasyan J.H. Minasyan J.H. Petromagnetic mapping of the territory of Armenia. Coll. of articles "Investigation of magnetic properties of rocks", Voronezh, 2016, pp. 240–244.

Sargsyan G.A. Role of enclosing rocks in the metasomatism and zonality of its products on the example of the gold ore deposits. Coll. of articles, Nedra, Moscow, 1972.

Popov L.V. Bazilev B.A. Scherbakov V.P. Temperature range for magnetization of oceanic spinel peridotites. Coll. articles "Investigation of magnetic properties of rocks", Vladivostok, 1982, pp. 81–90.

Sholpo L.E. Use of rock magnetism for solving geological problems. Leningrad, "Nedra" 1977, 182 p.

Chapter 21
Rock Magnetic Properties of Pleistocene Tephras from the Polovinka Section of the Central Kamchatka Depression

A. N. Naumov, S. S. Burnatny, P. S. Minyuk and A. G. Zubov

Abstract Tephra deposits are important stratigraphic markers, particularly those originating from large explosive eruptions, such as documented from Kamchatka volcanoes. This study is focused on the magnetic characteristics of the Late Pleistocene tephras from the Polovinka outcrop. The aim is to detect possible peculiarities in the rock magnetic properties of different tephras, in order to evaluate their suitability for tephrochronological and correlation purposes. The Polovinka section is located on the right bank of the Kamchatka River in the Central Kamchatka depression (158° 55.670′E, 54° 54.654′N). For this study, twenty tephra layers from Middle to Late Pleistocene sediments were studied. The geochemical analyses indicate that the tephras are composed of basaltic-andesite, andesite, dacite and rhyolite. The magnetic mineral fraction consists of titanomagnetites with variable titanium content and therefore different Curie points (Tc). Titanium-rich titanomagnetites with an average titanium content between 20.42 and 25.23% show Curie points of 50–65 °C. In contrast, middle- and poor-Ti titanomagnetites have Curie points between 335 and 480 °C. Titanomagnetites with Tc = 455–515 °C were thermally unstable during heating in argon, while titanomagnetites with a Tc between 335 and 480 °C display thermal instability during heating in air. The magnetic properties of tephra depend on the granulometry. The values of coercive force (Bc), remanent coercive force (Bcr) and the ratio of saturation remanence/saturation magnetization (Jrs/Js) show a directed increase with decreasing grain size, while Bcr/Bc decreases, indicating a different domain state of the magnetic particles in the grain size fractions. The values of magnetic susceptibility (MS) and Js decrease with decreasing size of the fractions for the most tephra samples. The Curie temperatures of magnetic minerals do not depend on the size of the

A. N. Naumov (✉) · S. S. Burnatny · P. S. Minyuk
North East Interdisciplinary Scientific Research Institute n.a.
N.A.Shilo of FEB RAS, Magadan, Russia
e-mail: naumov@neisri.ru

A. G. Zubov
Institute of Volcanology and Seismology of FEB RAS,
Petropavlovsk-Kamchatski, Russia

D. Nurgaliev et al. (eds.), *Recent Advances in Rock Magnetism, Environmental Magnetism and Paleomagnetism*, Springer Geophysics,
https://doi.org/10.1007/978-3-319-90437-5_21

granulometric fractions. Various magnetic mineralogical and geochemical characteristics of some interlayers of tephra allow its use as stratigraphic markers.

Keywords Kamchatka · Tephra · Pleistocene · Titanomagnetite
Geochemistry · Rock magnetism

Introduction

The Kamchatka Peninsula is one of the most active volcanic regions in the world. The volcanoes of Kamchatka are assigned to three main volcanic belts (Fig. 21.1b): Sredinny Range, Eastern volcanic front (EVF) and Central Kamchatka depression (CKD) (Portnyagin et al. 2005; Ponomareva et al. 2007; Bindeman et al. 2010).

Historically active volcanoes are located only at the Eastern volcanic front and in the CKD. Up to 37 large volcanic centers have been active during the Holocene (Ponomareva et al. 2007). Periods of strong Holocene volcanic activity occurred within the time intervals of 7500–7800 and 1300–1800 ^{14}C years BP (Braitseva et al. 1995). Most Late Pleistocene calderas were formed 30,000–40,000 years ago (Braitseva et al. 1995; Ponomareva et al. 2007; Bazanova et al. 2016). The explosive eruptions have resulted in widespread dispersal of tephra-fall deposits.

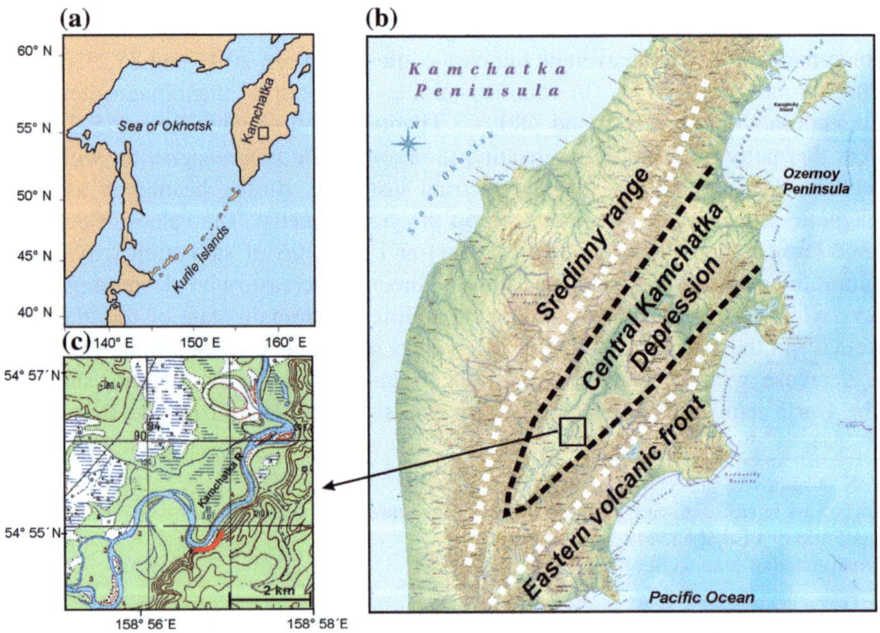

Fig. 21.1 Schematic map showing study area (**a**, **b**) and location of Polovinka exposure, which is marked by the thick red line (**c**)

Tephra deposits are important stratigraphic markers, particularly those originating from large explosive eruptions, such as been documented from Kamchatka volcanoes. These types of eruptions are characterized by a rapid ash dispersal, which often covers large areas. They form key horizons and can be used for precise dating of past environmental events and climatic changes (e.g., Ponomareva et al. 2004, 2010; Kyle et al. 2011; Dirksen et al. 2011; Bazanova et al. 2016; Derkachev et al. 2016). Furthermore, tephrochronological studies in lake and marine environments have been used for constraining age-depth models. However, before using tephra layers for chronology and correlation it is first necessary to identify them. A variety of methods have been employed for identification, including geochemical, petrological, magnetic, mineralogical, and optical proxies (e.g., Robertson 1993; Lowe et al. 2015; Ponomareva et al. 2015).

Pleistocene tephra layers often occur in the sedimentary sequences of the CKD (Braitseva et al. 1968, 2005; Kuprina 1970; Kiryanov et al. 1981). This study is focused on the magnetic characteristics of the Late Pleistocene tephra from the Polovinka outcrop. The aim is to detect possible peculiarities in the rock magnetic properties of different tephras in order to evaluate their suitability for tephrochronological and correlation purposes.

Study Area

The studied area is located in the Central Kamchatka depression. The CKD is a huge tectonic structure falling more or less along a northeast direction and ranges from the source of the Kamchatka river to the Ozernoy Peninsula (Fig. 21.1b).

The depression is located between the Sredinny Range and the Eastern volcanic front and is filled with Neogene and Quaternary sediments (Braitseva et al. 1968; Kuprina 1970). Voluminous basaltic-intermediate active volcanoes of Kliuchevskoi group, Shiveluch, and Bezymyanny are located in the CKD (Portnyagin et al. 2005; Bindeman et al. 2010). The magmas in the Sredinny Range are mostly K_2O-rich and contain biotite, while the majority of EVF rocks are K_2O-poorer (Volynets 1994).

The oldest sedimentary deposits are exposed at Generalka uplift in the central part of the CKD (Braitseva et al. 1968).

The studied Polovinka section is located in southern part of Generalka uplift on the right bank of the Kamchatka River, 20 km downstream from the village of Kirganik, in the Milkovskiy district of Kamchatka (158° 55.670′E, 54° 54.654′N, Fig. 21.1c). It is a steep outcrop with a height of about 80 m.

Lithology

There are five lithological units (from bottom to top), whose names refer to those introduced by (Braitseva et al. 1968).

Unit 1 (Lacustrine Middle Pleistocene Blue clay). This unit is represented by bluish-gray, horizontally laminated clays in the lower part and homogeneous clays in the upper part of the sequence. Interbedded layers of sand and white ash (up to 5 cm thick, tephra 17) as well as small pebbles have been noted in the unit. The unit thickness is about 4.0 m.

Unit 2 (Middle Pleistocene alluvium). Unit 2 is composed of cross-bedded, greenish-gray sands, with interlayers of gravel and pebbles and rarely lenses of sandy silt. Brown crusts of iron oxyhydroxides are observed within the sediments. Abundant fossil tree trunks and branches occur in the lower part of the sequence. The thickness of the unit is 24 m.

Unit 3 (Middle-Pleistocene moraine). Unit 3 consists of cross-bedded, pebble layers with a sandy clay aggregate. The material inside the cross-bedded series is unsorted; cobbles are typical. Foreign lenses of pebbles and sand blocks were noted in the lower part of the unit. These blocks are covered with layers of sand and silt, indicating the occurrence and direction of flow during the accumulation of the sediments. The transition between the second and third unit is gradual. The unit thickness is about 18 m.

Unit 4 (Late Pleistocene lacustrine-fluvial). This unit is characterized by interbedded bluish-gray clays, silt and brown pebbles, and by sand. A tephra layer (tephra 18) (5 cm thick) and ice-wedge casts occur 40 cm from the bottom. The thickness of Unit 4 is 2.7 m.

Unit 5 (Late Pleistocene polygenetic). Unit 5 was studied in 6 sequences. Sediments are mainly gray to olive-gray, sandy silt and fine-grained sands, the alternation of which gives horizontal stratification that is especially clear in the lower part (5 m) of the unit. Remnants of decomposed brown-colored organic matter were also present. The sediments are ferruginated locally and in different layers. A layer of humic silt (2.5 m thick) occurs in the middle part of Unit 5. The color of the sediments is dark gray in the lower part of the unit but becomes lighter up-section. Organic-rich (e.g., peat, branches) layers of dark brown color provided material for radiocarbon dating. Numerous cryogenic disturbances are characteristic for Unit 5 with a few ice-wedge casts noted in the unit. The thickness of Unit 5 is about 35 m.

Interbeds of tephra of various colors and granulometry distinguish the Polovinka section. In most cases, the ash layers are not of constant thickness, and occur in the form of lenses and pockets of various thicknesses (up to 15–20 cm). There are twenty tephra layers, to which were assigned numbers (Figs. 21.2 and 21.3). Most of the tephra layers are located in the upper and lower parts of Unit 5. Tephra 17 is from Middle Pleistocene Unit 1 and consists of volcanic glass (97%) and plagioclase (3%) (Kiryanov 1981). The colors of the tephras in Unit 5 and 4 vary from white to gray to black, and from fine-grained to coarse-grained (Fig. 21.3). Tephra 12 is the most clearly seen ash in the exposure. This layer is disturbed everywhere by permafrost processes. Numerous lenses (few mm thick) of tephra are found throughout Unit 5.

Fig. 21.2 Polovinka exposure showing lithological units and tephra layers (doted lines and numbers)

Method

The Polovinka section was sampled in detail for several multidisciplinary studies. In this work we present rock magnetic, mineralogical and geochemical data of the tephra. About 500 g of tephra samples were collected from all the visible tephra layers and were stored in the plastic bags.

The magnetic low-field susceptibility (MS) of air-dried, weighed sampled was measured on a MFK1-FA Kappabridge (AGICO Ltd., Brno, Czech Republic) with sensitivity 3×10^{-8} SI. Each sample was measured three times and the mean is reported here. The analytical results are expressed in units of 10^{-6} m^3/kg.

Hysteresis loops were measured for all weighed samples using an automatic coercivity spectrometer (Burov et al. 1986; Jasonov et al. 1998) to estimate the bulk magnetic grain size through a Day plot (Day et al. 1977; Dunlop 2002a, b), i.e. Jrs/Js versus Bcr/Bc. The coercivity spectrometer has sensitivity of 1×10^{-8} Am^2 and 1×10^{-6} Am^2, for the remanent and induced magnetic moment, Jr and Ji respectively. The maximum specimen volume is 1.92 cm^3.

Saturation magnetization, and coercive force were determined after correction for the high-field slope which in our case resulted mainly from paramagnetic contributions.

The paramagnetic susceptibility (MSp) was obtained from the high-field slopes of the hysteresis curves. In order to determine the Curie temperature of magnetic minerals, the temperature-dependent susceptibility curves from bulk tephra samples

Fig. 21.3 Tephra of Polovinka section. The size of each square is 2 cm

and a few magnetic extracts were measured in an argon and in air atmosphere using a MFK1-FA Kappabridge (AGICO Ltd., Brno, Czech Republic) with CS-3 system. The heating and cooling rates were 10–12 °C min^{-1}. Samples were heated continuously from room temperature to 700 °C and cooled back to room temperature. The Curie temperature was determined using the Hopkinson peaks and the temperature dependence of inverse susceptibility values (Lattard et al. 2006; Petrovský and Kapička 2006; Fabian et al. 2013).

The low-field variation of magnetic susceptibility was investigated on all tephra samples using a MFK1-FA Kappabridge in magnetic fields ranging from 2 to 700 A/m.

Major and trace elements of the tephra were analyzed with a S4 Pioneer X-ray fluorescence spectrometer (Bruker, Germany) and a VRA-30 XRF spectrometer (Germany), respectively. Elemental compositions were determined using the fundamental parameters method (Borkhodoev 2002).

The chemical composition of the polished grains obtained from the magnetic extract was determined with a Camebax microprobe with energy dispersive spectroscopy INCA EMERGY Oxford. Magnetic minerals were extracted from bulk sediments using a hand-held rare earth magnet housed in a plastic sheath.

Result

Geochemistry

According to the geochemical data, the Polovinka tephras are of intermediate and acidic composition and plot in basaltic-andesite (2 tephra), andesite (4 tephras), dacite (7 tephras) and rhyolite (6 tephras) fields in a total alkali-silica (TAS) diagram (Le Maitre et al. 2002; Fig. 21.4). Tephras 10, 12, and 13 are enriched in Fe_2O_3, TiO_2, MgO and CaO (Table 21.1), whereas tephras 2, 16 and 20 are depleted in these elements.

Strong positive correlations occur between Fe_2O_3 and TiO_2 (r = 0.95), Fe_2O_3 and CaO (r = 0.94), and Fe_2O_3 and MgO (r = 0.86). Conversely, strong negative correlations were noted for SiO_2 and Fe_2O_3 (r = −0.97), SiO_2 and TiO_2 (r = −0.93), and SiO_2 and CaO (r = −0.97). K_2O and Al_2O_3 show a negative correlation (r = −0.59). Positive correlations also occur between Al_2O_3 and P_2O_5 (r = 0.60), Al_2O_3 and MnO (r = 0.62), and P_2O_5 and MnO (r = 0.48).

Magnetic Properties

Magnetic properties of the tephras are inhomogeneous: MS = 0.6 − 14.7 × 10^{-6} m^3/kg; Js = 0.06 − 1.38 Am^2/kg, Jrs = 0.005 − 0.31 Am^2/kg, Bc = 1.2 − 36.4 mT, and Bcr = 14.6 − 80.7 mT. Highest values of MS and Js occur in tephras of basaltic-andesite compositions (tephras 9 and 10), dacite composition (tephra 9) and rhyolite composition (tephra 14). Black tephras 10 and 14 are characterized by the highest value of Jrs (Fig. 21.5). Tephras 6, 10 and 18 show high values of coercive force (Bc) and remanence coercivity (Bcr) as well as high Jrs/Js ratios.

Hysteresis loops provide information about the domain state of ferrimagnetic minerals (e.g. Dunlop and Özdemir 1997).

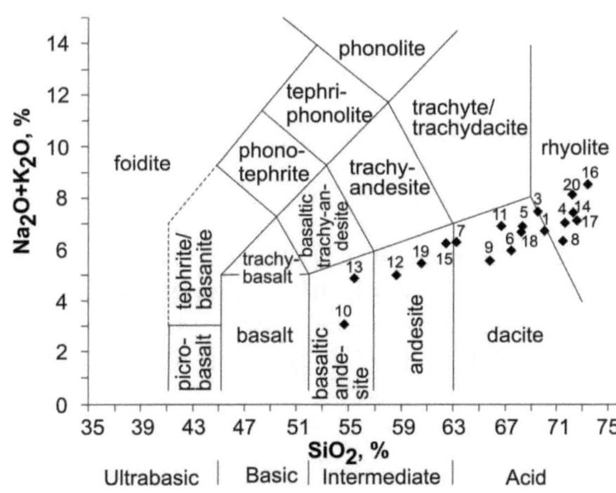

Fig. 21.4 Diagram of total alkali and silica (Le Maitre et al. 2002) for tephras from the Polovinka exposure

Table 21.1 Content (weight %) of major elements for tephra

Tephra	SiO_2	TiO_2	Al_2O_3	Fe_2O_3	MnO	MgO	CaO	Na_2O	K_2O	P_2O_5	Total
1	70.03	0.47	15.56	3.08	0.09	1.19	2.70	4.10	2.63	0.15	100.00
2	75.12	0.27	13.90	1.81	0.06	0.42	1.86	4.50	2.03	0.04	100.01
3	69.56	0.42	15.83	2.49	0.07	1.11	2.90	4.04	3.40	0.16	100.01
4	71.70	0.49	14.74	2.66	0.09	0.79	2.40	4.44	2.57	0.12	100.00
5	68.32	0.26	17.80	2.38	0.09	0.78	3.39	4.80	2.06	0.12	100.00
6	67.46	0.41	15.61	3.87	0.09	2.51	3.91	4.15	1.78	0.20	99.99
7	63.31	0.63	17.02	5.68	0.13	1.80	5.05	4.54	1.73	0.10	99.98
8	71.53	0.43	14.56	3.13	0.12	0.68	3.15	4.80	1.50	0.10	100.00
9	65.85	0.44	16.03	5.03	0.07	2.87	3.98	3.73	1.82	0.19	100.01
10	54.70	1.03	17.45	10.82	0.18	4.52	8.09	2.45	0.62	0.15	100.01
11	66.72	0.61	16.27	4.57	0.12	1.52	3.10	4.52	2.36	0.20	99.99
12	58.69	0.85	16.84	8.30	0.12	3.73	6.32	3.42	1.56	0.17	100.00
13	55.51	1.26	17.43	10.82	0.25	3.15	6.50	3.95	0.91	0.23	100.01
14	72.32	0.31	14.63	2.26	0.08	0.91	2.05	4.26	3.13	0.06	100.01
15	62.54	0.78	16.95	5.91	0.10	2.26	5.02	3.95	2.25	0.24	100.00
16	73.40	0.27	14.39	1.41	0.07	0.51	1.40	4.12	4.37	0.05	99.99
17	72.60	0.39	14.34	2.29	0.08	0.64	2.47	4.06	3.03	0.11	100.01
18	68.24	0.49	17.32	3.15	0.09	1.26	2.68	4.29	2.36	0.13	100.01
19	60.64	0.88	18.47	5.95	0.15	2.21	6.12	4.16	1.27	0.17	100.02
20	72.21	0.31	14.99	1.84	0.07	0.64	1.79	3.86	4.23	0.06	100.00

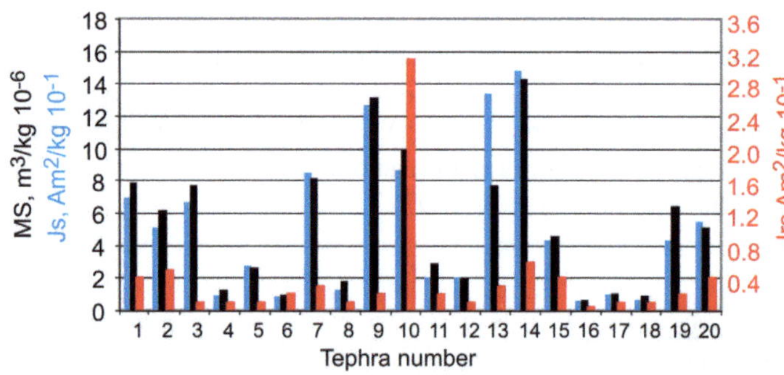

Fig. 21.5 Magnetic susceptibility (MS), saturation magnetization (Js), and saturation remanence (Jrs) of the Polovinka tephras

Tephra samples display different coercivity and paramagnetic contributions. Tephras 6, 10, and 18, which have high ratios of Jrs/Js, show the highest coercivity and paramagnetic components: Bc = 14.24 − 36.35 (22.38 on average) mT, Bcr = 35.72 − 80.68 (54.29) mT, and MSp = 0.0701 − 0.4690 (0.1950) × 10^{-6} m^3/kg.

Tephras 15, 16 and 17, the oldest of the Polovinka tephras, are characterized by lower values of these parameters: Bc = 6.16 − 7.02 (6.47) mT, Bcr = 33.67 − 48.62 (39.67) mT, and MSp = 0.0283 − 0.1251 (0.0633) × 10^{-6} m^3/kg.

The main group of samples (except samples 6, 10, 18, 15, 16, 17) displays the lowest coercivity and paramagnetic components: Bc = 1.17–8.98 (4.28) mT, Bcr = 14.6–44.23 (27.82) mT, and MSp = 0.0117–0.1010 (0.048) × 10^{-6} m^3/kg (Fig. 21.6, samples 1, 7, 5, 20, 3).

The modified Day plot (Day et al. 1977; Dunlop 2002a, b; Fig. 21.7) includes hysteresis data for all tephra samples. The tephras plot between the SD + MD and SP + SD mixing curves but closer to and paralleling the SD + MD mixing curve. Jrs/Js is contained between 0.01 and 0.12, while Bcr/Bc ranges from 4.45 to 12.52 for the main point-cluster. Samples 3 and 9 show the only truly MD values (Jrs/Js = 0.01, Bcr/Bc = 12.18–12.52). The black tephra 10 has points located in the upper part of the diagram on the SP + MD mixing curve with a SP volume fraction of 20%. Points for the fine-grained samples of tephra 6 and 18 are displayed close to the SD + MD mixing curves with Jrs/Js = 0.20–0.22, and Bcr/Bc = 2.51–2.81.

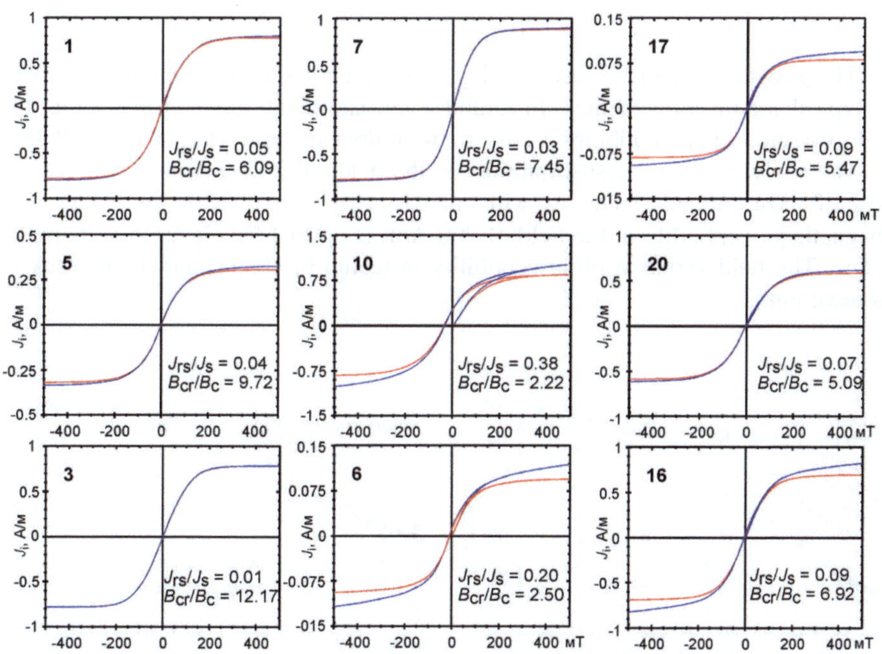

Fig. 21.6 Hysteresis loops for representative samples. The blue and red curves represent before and after the correction for the paramagnetic component. Sample numbers (depths) are marked in bold font. Jrs/Js and Bcr/Bc ratios are counted after correction for the paramagnetic component

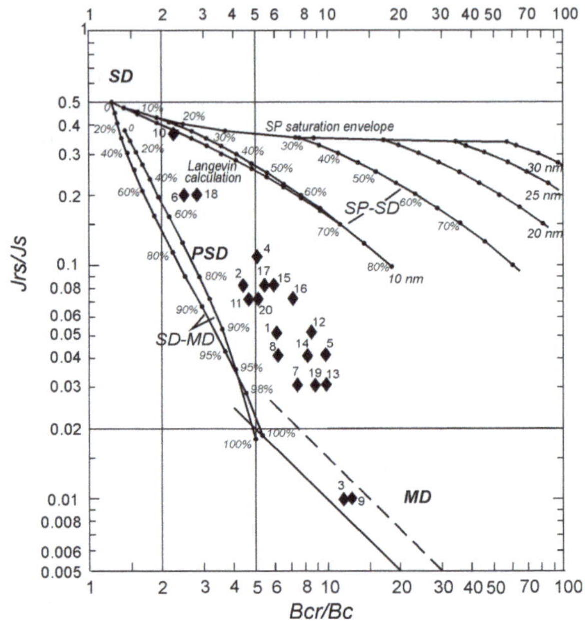

Fig. 21.7 Hysteresis ratio for tephra samples compared to SD + MD and SP + SD mixing curves (Dunlop 2002a, b)

The growth of magnetic susceptibility with an increasing magnetic field in rocks is typical for titanomagnetites. In multidomain and single domain magnetite, the field variation of susceptibility is very weak or does not exist (Hrouda et al. 2006).

The study of magnetic susceptibility in the fields of 2–700 A/m shows that the MS of most samples increases with increasing magnetic field (Fig. 21.8). The magnetic susceptibility in the field of 700 A/m is up to 17% higher than its initial value. The field variation of susceptibility in tephra 6, 10, 11, and 18 is weak to nonexistent.

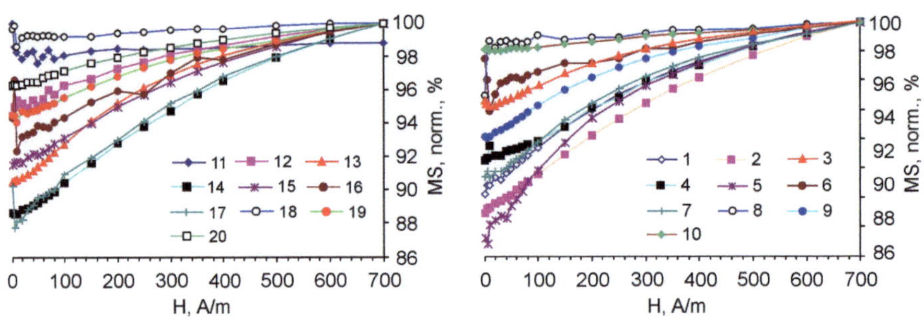

Fig. 21.8 Magnetic susceptibility of tephra versus field variation. MS is normalized to highest value of MS

Temperature Dependent Magnetic Susceptibility

Temperature dependent magnetic susceptibility provides information about Curie point of the magnetic minerals present in the sample (Fig. 21.9). It is also highly sensitive to mineralogical changes during thermal treatment. One notes that the mineralogical alterations of the Polovinka tephras during heating in argon and air atmosphere are different (Fig. 21.9, sample 14).

The behavior of the tephras during heating in argon can be divided into three general groups.

1. Tephra with a single mineral phase belongs to the first group (tephras 10, 18). The heating curves are characterized by a sharp decrease in susceptibility at 560–575 °C. This behavior indicates that magnetite is a major contributor to the susceptibility. The cooling curves display the same Curie points of magnetite. Magnetic susceptibility after a heating cycle is slightly increased. On the heating and cooling curves of tephra 6, the Curie point of magnetite is indistinct. The decline in susceptibility on heating is not complete until 700 °C, possible indicating the presence of hematite.
2. The heating and cooling curves of the second tephra group are indicative of 3–4 magnetic phases. The heating curves display a clear susceptibility drop between 50 and 65 °C (Fig. 21.9, samples 1, 2, 5, 9, 12, 17 M). This first phase, referred to as phase A, is thermally stable and is visible in the cooling curves. Phase B shows a decrease in magnetic susceptibility at 467 °C (tephra 1), 395 °C (tephra 2), 400 °C (tephra 5), 475 °C (tephras 9), and 480 °C (tephras 12). This magnetic phase also is stable during thermal treatments. Curie points of phase B are clearly seen in the cooling curves. Phase C displays a decrease in susceptibility between 470 and 515 °C. On the cooling curves, this phase is not evident, whereas phase B is more clearly visible. Phase D shows drops in susceptibility at 560–600 °C, marking the presence of magnetite. On the heating run for tephra 5 and 9, the magnetite phase is indistinct.
3. The third group of tephra includes all magnetic phases mentioned above except phase A, which has the lowest Curie point. The heating curves show that the thermally stable phase B has Curie points between 335 and 470 °C. Phase C is thermally unstable. Heating curves show the decrease of MS in the range between 455 and 515 °C and then a shift to lower temperatures during cooling. On the heating and cooling curves, the Curie point of magnetite is clearly visible. On the reversible curves of the second run (in argon), only Phase B and magnetite are visible (Fig. 21.9, sample 14).

The thermal stability of magnetic phases during heating in air is different than in argon (Fig. 21.9, sample 14). On the heating curves, Curie points of phases B and C are clearly visible. Magnetic phase B is thermally unstable. This phase is not seen on the cooling curves, whereas phase C is more clearly visible. Magnetic susceptibility increased after the heating run. The heating and cooling curves of the second run are nearly reversible.

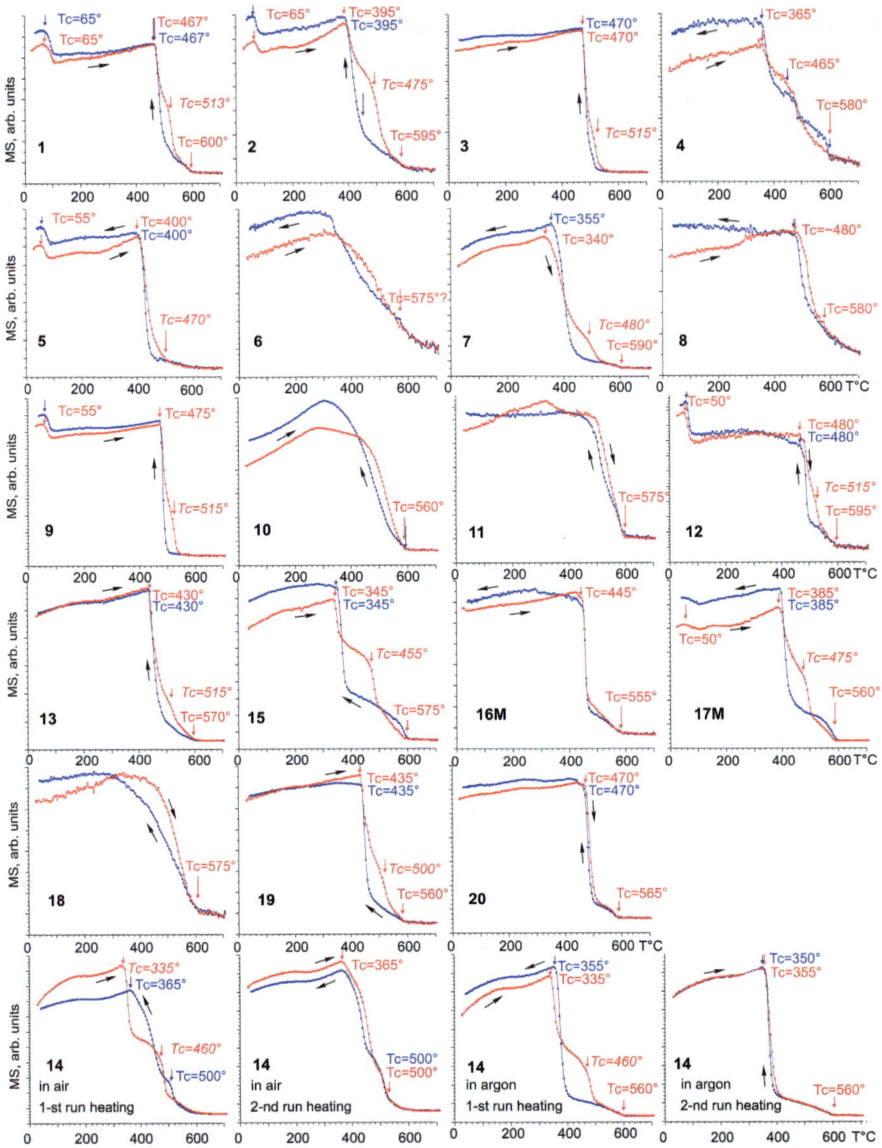

Fig. 21.9 Magnetic susceptibility versus temperature for the Polovinka tephra. Black arrows indicate the heating and cooling runs. Tephra sample numbers are marked by bold font. Curie points are indicated by red and blue color arrows

Microprobe Analysis and Energy Dispersive Spectroscopy

The elemental composition of polished grains from the magnetic extract of tephra indicated three phases of titanomagnetites. The average content of iron and titanium in the main phase are 65.02 and 4.43%, respectively. Minimum titanium content is less than 1% (Table 21.3). These Ti-poor titanomagnetites have significant Al (up to 1.92% on average), Mg (up to 2.2% on average), Mn (up to 0.88%), and vanadium (up to 0.51% on average) (Table 21.3; Fig. 21.10h). Some grains included an admixture of Si, P and Ca.

The second phase includes T-rich titanomagnetite with iron and titanium having average concentrations of 44.08 and 23.26%, respectively (bold font in Table 21.3). In addition to the thermomagnetic data, the grains characterizing this phase were found in tephras 10, 11, 16, and 19. Mg and Mn are the main impurities in the T-rich titanomagnetites, while V, Si, P, and Ca were rarely found (Table 21.3; Fig. 21.10b).

The third phase of titanomagnetites (cursive font in Table 21.3) was revealed in tephra 1 (2 analyses), 10 (4 analyses), and 11 (2 analyses). The average content of iron and titanium are 47.32 and 13.55%, respectively. All grains of this phase contain Al, Mg, and Mn.

Back scattered electron images indicate no evidence of high temperature oxidation (Fig. 21.10). Composition of the grains is almost homogenous, which is indicated by their elemental mapping (Fig. 21.10c–f, I–l). Some grains of titanomagnetite either have other mineral phases or are not included in any magnetic matrix.

Discussion

Tephra layers of the Polovinka section are characterized by different magnetic properties, and their chemical and mineral composition.

Geochemical control of the magnetic properties

According to the geochemical data, the tephras are basaltic-andesite, andesite, dacite and rhyolite in composition. Tephra of intermediate composition (tephras 10, 12, and 13) are located only in the lower part of the Late Pleistocene Unit 5. These tephras are enriched in Fe_2O_3, TiO_2, MgO and CaO, depleted in SiO_2 and are important stratigraphic markers. Our single radiocarbon sample was obtained two meters below the tephra 9 (Fig. 21.1). Fine plant detritus from this sample yielded an age of 39640 ± 1230 year BP (Lab. No MAG-2172). Organic material for dating are very rare in the sediments of Unit 5. The radiocarbon age of the samples from the Late Pleistocene Unit 4 was reported as nonfinite (Braitseva et al. 2005).

Tephra 3, 14, 16 and 17 and 20 enriched in K_2O (3.13–4.37%) suggesting that a source of tephra was located in the Sredinny Range (Braitseva et al. 2005).

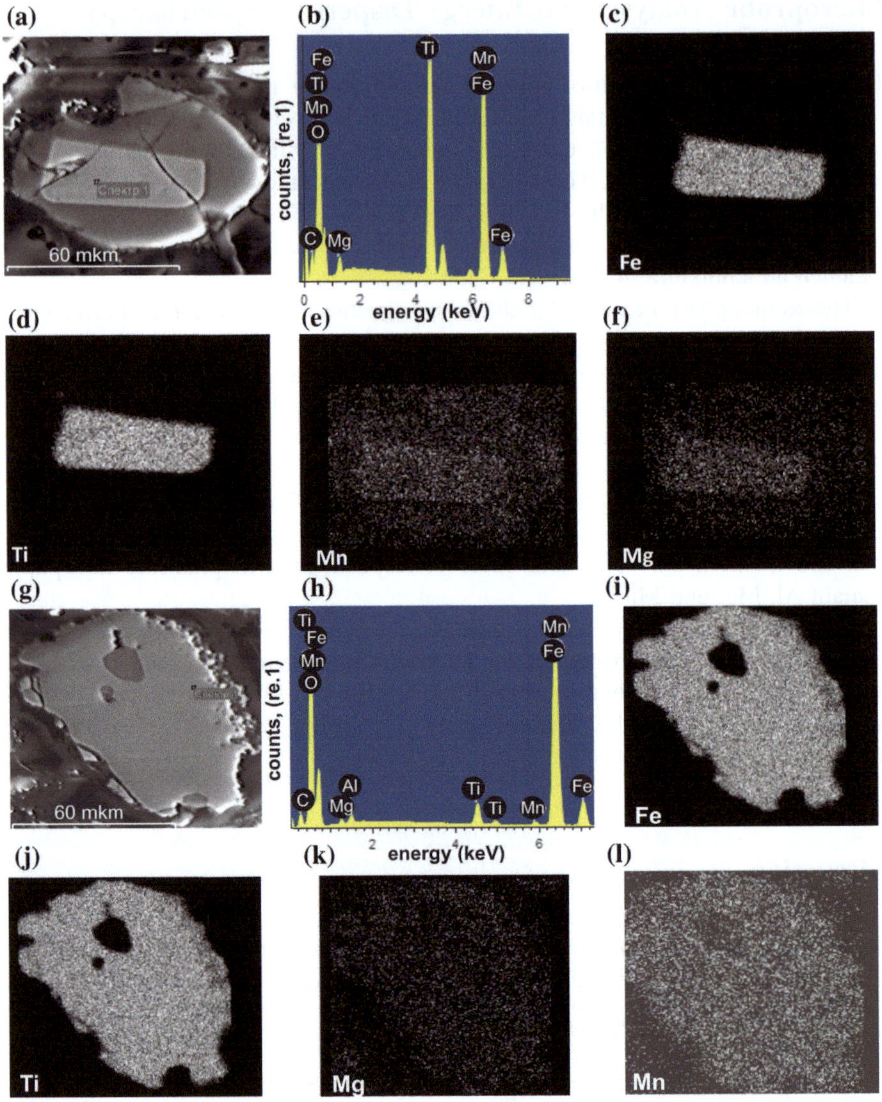

Fig. 21.10 Back scattered electron images (**a**, **g**), energy dispersive spectroscopy spectra (**b**, **h**) and elemental mapping (**c**, **d**, **e**, **f**, **I**, **j**, **k**, **l**) of Ti-rich titanomagnetite (upper set of images) and Ti-poor titanomagnetite (lower set of images)

The total content of SiO_2 in tephras varies from 54.7 to 75.12%. Negative correlations were noted for SiO_2 and MS ($r = -0.31$), SiO_2 and Js ($r = -0.19$), SiO_2 and Jrs ($r = -0.19$), SiO_2 and Bc ($r = -0.32$), SiO_2 and Bcr ($r = -0.27$). Positive correlation occur between Fe_2O_3 and MS ($r = 0.37$), Fe_2O_3 and Js ($r = 0.22$), Fe_2O_3

and Jrs (r = 0.51), Fe_2O_3 and Bc (r = 0.36), Fe_2O_3 and Bcr (r = 0.32). Same coefficients correlation observed between TiO_2 and magnetic parameters.

Granulometrical control of the magnetic properties

The tephras have inhomogeneous magnetic properties. Magnetic susceptibility and Js are higher in tephras 9, 10, 13, and 14. The maximum value of Jrs occurs in tephra 10. Relatively high Ms, Js and Jrs characterize the tephras (1, 2, 3, and 20) from the top of the section.

The oldest tephras (15, 16, and 17), located in Middle Pleistocene Unit 3 and Late Pleistocene Unit 4, display low values of MS, Js and Jrs but relatively high values of Bc, Bcr. These characteristics are caused by weathering. Reddish and brownish Fe-oxyhydroxide crusts are typical for all sediments of the Polovinka section. On the other hand, the highest values of Bc, Bcr are in tephras 6, 10, and 18. These tephras plot in the upper part of Day diagram (Fig. 21.7).

During the eruption, the ejecta are transported and dispersed by the wind. The particle size distribution of ash particles depends on the distance to the eruption source. Proximal tephra is coarser then tephra from the distal location (e.g., Kyle et al. 2011; Ponomareva et al. 2015). It can be therefore assumed that the magnetic properties of tephra will depend on the granulometry. To test this, the samples of tephra were separated by dry sieving into four size fractions: >100, 100–63, 63–40 and <40 μm which magnetic properties were studied.

Grain size of the tephras 12 and 14 is >100 μm, while grain size tephras 2, 3, 6, and 7 is <100 μm. The abundance >100 μm grain size fraction is 86–95% in the tephras 9, 10, 13, and 19 (Fig. 21.11). These tephras are intermediate composition and show the high values of MS, Js and Fe_2O_3 (Figs. 21.4 and 21.5; Table 21.1). The <40 μm grain size fraction is dominate in the acid tephras 4, 8, 16, 17 and 20 (Fig. 21.4). These tephras is characterized by low values of the MS, Js, Jrs, and Fe_2O_3.

The values of Bc, Bcr and Jrs/Js show a directed increase with decreasing fraction, while Bcr/Bc decreases indicating a different domain state of the magnetic particles in the fractions. The values of MS and Js decrease with decreasing size of the fractions for the most tephra (Fig. 21.12, tephra 4, 8, 16, 17, 20). The fractions of <40 μm predominate in these tephras, therefore, at distal locations, as a result of granulometric differentiation of the tephra during transport, tephra will be less magnetic.

The MS and Jrs of some samples (5, 8, 10, and 13) are increased in fine fractions with maximum values for fraction 100–63 μm (Fig. 21.12). These tephra can be more magnetic in remote areas.

The values of Bcr, Bc, Jrs/Js of the basaltic-andesite tephra 10 display decrease with decreasing fraction while Bcr/Bc increase in fine fractions. Magnetic minerals in the tephra 10 are included in larger paramagnetic grains. This tephra shows the highest paramagnetic susceptibility (Table 21.2).

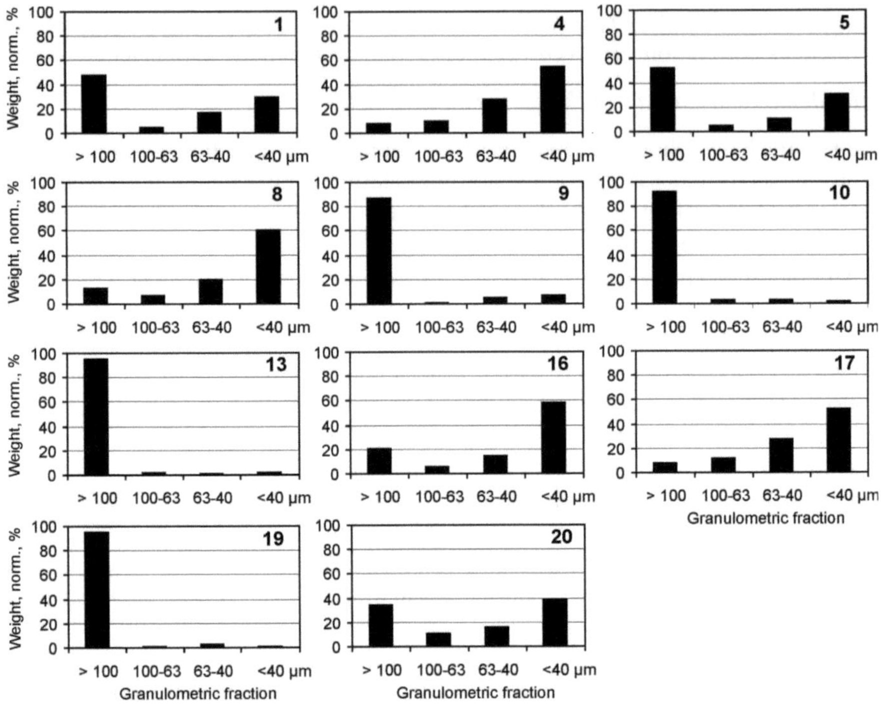

Fig. 21.11 Histogram of grain size fractions of tephra. Weight of fractions is normalized to bulk weight. Tephra sample numbers are marked by bold font

The obtained data indicates the different magnetic properties of tephra from proximal and distal locations. Thermomagnetic study indicates the magnetic mineralogical data are more consistent and identical in different fractions.

Magnetic mineralogy

Titanomagnetite with variable titanium content is the main magnetic mineral type of the tephras. It can be subdivided into three different phases.

A Ti-rich titanomagnetite (phase A) which contains 23.26% titanium, and 44.08% iron and shows Curie points between 50 and 65 °C. Mg and Mn are the main impurities in the Ti-rich titanomagnetite. The Ti-rich titanomagnetite phase is almost stable under the thermal treatments. The magnetic extract from tephra 9, which contains titanomagnetite with Tc = 55 °C was placed in a container with boiling water to separate the Ti-rich titanomagnetite. Then, by magnet the magnetic fraction 1 was extracted from boiling water. The fraction 2 that remained in the container (titanomagnetite phase with a low Curie point) was studied as well as the extracted magnetic fraction 1. Two fractions were heated to 160 °C (Fig. 21.13). The first fraction consists entirely of titanomagnetite with a low Curie point (Fig. 21.13a). Heating and cooling curves of both fractions are irreversible; cooling

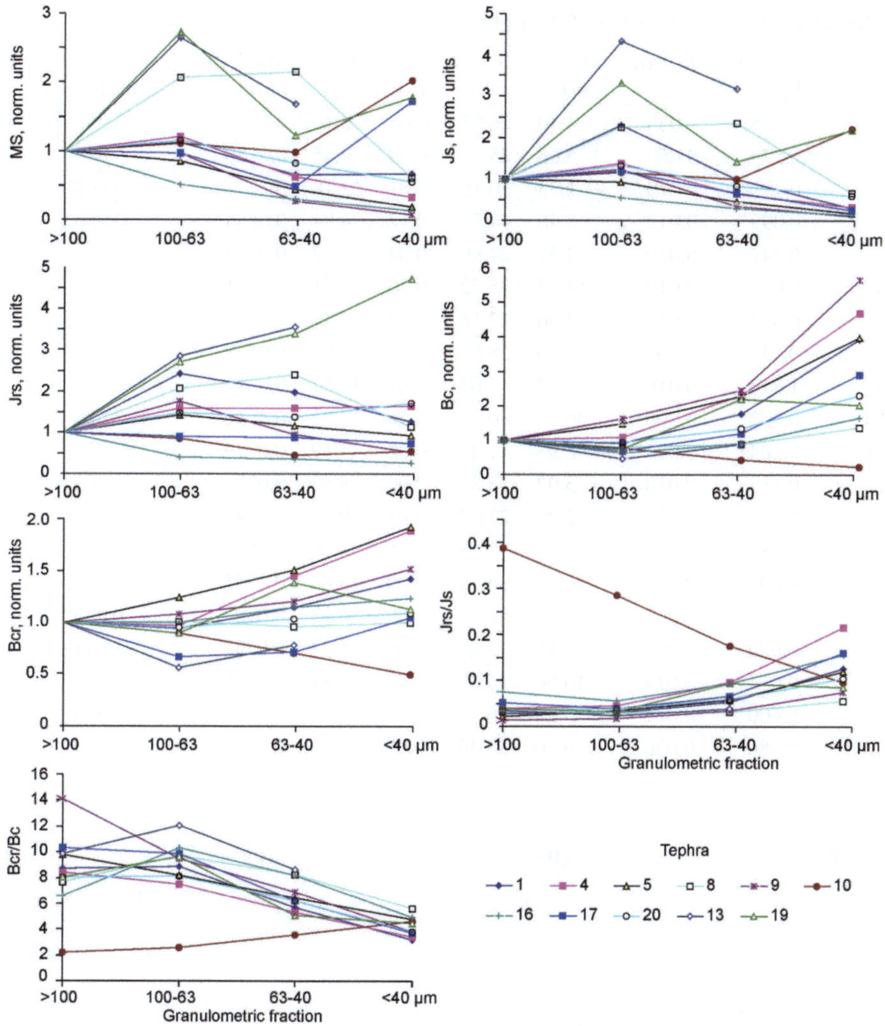

Fig. 21.12 Magnetic parameters of tephras versus granulometry. MS, Js, Jrs, Bcr and Bc are normalized by the values of MS, Js, Jrs, Bcr and Bc of granulometric fraction >100 μm

curves show a shift of the Curie point to 60 °C. The presence of Ti-rich titanomagnetite with a low Curie point in both fractions (Fig. 21.13a, b) suggests that titanium-rich titanomagnetite is not the product of a separate eruption. Ti-rich titanomagnetites were found in Late Pleistocene (Unit 5) tephras 1, 2, 5, 9, 12 and 17 and Middle Pleistocene Unit 1 using thermomagnetic experiments and additionally in tephras 10, 11, 16, 19 by microprobe analysis.

Table 21.2 Magnetic properties of the tephras

Tephra	Js, Am^2/kg	Jrs, Am^2/kg	Bc, mT	Bcr, mT	Jrs/Js	Bcr/Bc	MSp, 10^{-6} m^3/kg	MS, 10^{-6} m^3/kg
1	0.775	0.038	4.40	26.83	0.05	6.10	0.0139	6.912
2	0.578	0.050	5.68	25.28	0.09	4.45	0.0750	5.047
3	0.773	0.010	1.33	16.19	0.01	12.18	0	6.691
4	0.099	0.012	8.96	44.23	0.12	4.94	0.0554	0.966
5	0.242	0.010	2.57	24.99	0.04	9.72	0.0356	2.751
6	0.087	0.017	14.24	35.72	0.20	2.51	0.0701	0.836
7	0.802	0.025	2.30	17.16	0.03	7.46	0.0292	8.481
8	0.156	0.007	4.66	28.08	0.05	6.02	0.0577	1.296
9	1.313	0.016	1.17	14.63	0.01	12.52	0	12.610
10	0.802	0.306	36.35	80.68	0.38	2.22	0.4690	8.683
11	0.264	0.022	8.98	41.23	0.08	4.59	0.0779	2.035
12	0.181	0.010	3.97	33.83	0.06	8.52	0.0233	2.003
13	0.738	0.025	2.80	27.42	0.03	9.80	0.0840	13.360
14	1.383	0.062	3.38	27.58	0.05	8.16	0.0858	14.760
15	0.405	0.037	6.25	36.74	0.09	5.88	0.1251	4.307
16	0.053	0.005	7.02	48.62	0.09	6.93	0.0283	0.593
17	0.085	0.008	6.16	33.67	0.09	5.47	0.0365	1.011
18	0.064	0.014	16.56	46.49	0.22	2.81	0.0466	0.679
19	0.600	0.021	3.22	28.53	0.04	8.85	0.1005	4.384
20	0.486	0.036	6.56	33.46	0.08	5.10	0.0584	5.479

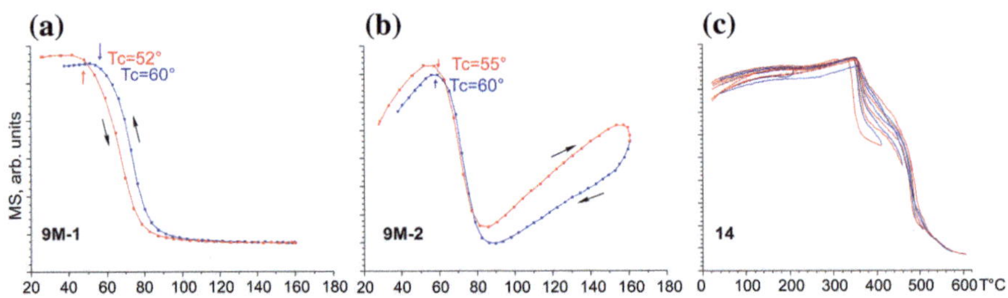

Fig. 21.13 Magnetic susceptibility versus temperature curves for separated magnetic fraction 1 (**a**) and 2 (**b**) of tephra 9 (see text) and changes of magnetic susceptibility during a succession of heating cycles of tephra 14 (**c**)

A Ti-poor titanomagnetite (phase B and C), which is characteristic for most tephras, with an average content of iron and titanium are 65.02 and 4.43%, respectively. These titanomagnetites were heated in argon and air and display irreversible curves. Heating curves in argon and air are similar but cooling curves are completely different indicating the different mineralogical changes during heating.

We suppose that the Ti-poor titanomagnetites consist of both oxidized (maghemitized) and non oxidized phases. It is known that maghemitization increases the Curie temperature of titanomagnetite (e.g., Nishitani and Kono 1983; Dunlop and Özdemir 1997). Tephras 3, 4, 6, 7, 8, 12, 13, 14, 15 and 18 consist of one mineral phases according to the elemental composition (Table 21.3). On the other hand, the heating curves (MS-T) are distinct and indicate two phases (Fig. 21.9). During heating in air, oxidation of titanomagnetites occurs, and the temperature Curie phase B (lower Tc) is shifted to the area of high temperatures. Tc gradually increased with increasing heating cycles (Fig. 21.9, tephra 14, Fig. 21.13c). It is possible that an inversion of titanomaghemites occurred simultaneously as the magnetic susceptibility is decreased after the heating run. The composition of the inversion products depends on the content of titanium and the level of oxidation of the titanomaghemite, but the products are titanomagnetite or magnetite with one or more rhombohedral phases (Özdemir 1987).

Heating in argon displays instability in phase C (higher Tc). The Curie temperature in this phase is shifted to the Curie temperature of phase B, indicating titanomagnetite formation from reduction of titanomaghemite. MS following the

Table 21.3 Microprobe analyses of magnetic extracts from tephra, wt%

Tephra	Fe	Ti	Al	Mg	Mn	Si	V	P	Ca	O
1	**43.7 (7)**	**23.52 (7)**		**1.29 (7)**	**0.82 (7)**					**37.78 (7)**
	38.92 (2)	13.86 (2)	4.57 (1)	0.56 (2)	0.66 (2)				0.27 (1)	49.82 (2)
	50.16 (20)	2.26 (20)	1.92 (20)	2.20 (17)	0.66 (17)	1.39 (1)	0.24 (9)		0.79 (6)	39.66 (20)
2	**43.74 (4)**	**22.89 (4)**	**0.41 (2)**	**1.73 (4)**	**1.51 (4)**	**0.74 (1)**				**38.01 (4)**
	64.79 (16)	5.40 (16)	1.08 (16)	1.83 (16)	0.45 (15)		0.38 (16)			34.91 (16)
3	68.52 (28)	2.78 (28)	0.91 (28)	0.58 (28)	0.88 (28)	0.31 (1)	0.22 (5)			32.24 (28)
4	64.53 (26)	5.83 (26)	1.02 (26)	0.99 (26)	0.58 (26)	0.81 (2)	0.31 (23)	1.09 (1)	1.61 (1)	32.54 (26)
5	**44.48 (4)**	**23.9 (4)**		**1.52 (4)**	**0.61 (4)**					**38.4 (4)**
	66.62 (25)	4.21 (25)	1.03 (25)	1.10 (25)	0.49 (25)		0.35 (23)			32.67 (25)
6	63.51 (26	6.43 (26)	1.15 (26)	0.90 (25)	0.35 (26)	1.17 (2)	0.29 (15)	0.61 (1)	1.28 (1)	32.59 (26)
7	66.00 (30)	5.87 (30)	1.02 (29)	0.96 (28)	0.51 (30)		0.34 (24)			33.00 (30)
8	68.27 (26)	2.66 (26)	1.11 (25)	0.58 (26)	0.33 (14)	0.29 (2)	0.30 (25)			32.45 (26)

(continued)

Table 21.3 (continued)

Tephra	Fe	Ti	Al	Mg	Mn	Si	V	P	Ca	O
9	**44.77 (2)**	**23.08 (2)**		**0.93 (2)**	**1.06 (2)**					**34.75 (2)**
	68.24 (25)	3.72 (25)	0.92 (25)	0.66 (25)	0.77 (25)		0.25 (9)	0.41 (2)		32.32 (25)
10	**41.67 (2)**	**25.23 (2)**	**0.21 (1)**	**1.23 (2)**	**0.88 (2)**		**0.46 (1)**			**37.19 (2)**
	55.72 (4)	13.23 (4)	0.72 (4)	0.71 (3)	0.46 (4)	0.33 (1)	0.62 (3)			35.25 (4)
	67.26 (16)	3.27 (14)	1.14 (14)	0.75 (11)	0.64 (11)	0.61 (1)	0.53 (10)		0.52 (1)	31.96 (16)
11	**47.17 (2)**	**20.42 (2)**		**0.53 (2)**	**0.85 (2)**					**37.48 (2)**
	60.64 (2)	8.60 (2)	1.21 (2)	1.09 (2)	0.44 (2)		0.50 (2)			33.32 (2)
	67.23 (24)	2.21 (24)	1.03 (24)	0.66 (22)	0.39 (15)	1.17 (2)	0.31 (24)		1.34 (2)	32.50 (24)
12	66.48 (29)	4.33 (29)	0.88 (29)	0.74 (29)	0.60 (29)	0.43 (1)	0.25 (17)	1.55 (1)	1.32 (2)	32.58 (29)
13	66.53 (30)	4.62 (30)	0.89 (30)	0.83 (30)	0.59 (30)	1.33 (1)	0.29 (16)		0.9 (2)	33.69 (30)
14	63.01 (29)	5.60 (29)	1.87 (29)	1.88 (29)	0.29 (22)		0.50 (29)	0.28 (1)	0.4 (1)	33.91 (29)
15	63.58 (30)	5.59 (30)	1.84 (30)	1.87 (30)	0.29 (26)	0.2 (1)	0.51 (30)			33.89 (30)
16	66.33 (15)	3.78 (15)	0.96 (15)	0.85 (15)	0.85 (14)		0.38 (6)			32.48 (15)
	43.05 (12)	**23.87 (12)**		**1.16 (12)**	**1.26 (12)**					**37.69 (12)**
17	65.37 (18)	4.93 (18)	1.00 (18)	0.97 (18)	0.57 (18)		0.36 (17)			33.07 (18)
	44.35 (12)	**23.05 (12)**	**0.18 (1)**	**1.45 (12)**	**0.53 (12)**			**0.18 (1)**		**37.60 (12)**
18	61.61 (4)	6.69 (4)	1.22 (4)	1.87 (4)	0.38 (4)		0.52 (4)			34.14 (4)
19	64.78 (23)	3.76 (22)	1.16 (22)	0.93 (23)	0.87 (22)		0.32 (6)			33.24 (23)
	43.77 (6)	**23.40 (6)**		**0.98 (6)**	**1.34 (6)**					**38.17 (6)**
20	68.53 (29)	2.86 (29)	0.74 (29)	0.51 (29)	1.09 (29)	0.22 (1)	0.23 (10)	0.32 (1)	0.57 (1)	32.57 (29)

In brackets are number of analysis. Different fonts mark the different magnetic phases (see text) heating cycle is increased. Differences of the MS-T curves during heating of titanomaghemites in air and argon has been shown, for example, on weathering of basaltic material of Antarctic (Chevrier et al. 2006).

The third phase of titanomagnetite with contents of titanium 47.32%, and iron 13.55% was found in a few grains of tephras 1, 10, 11. Because of low content these phase no visible on MS-T curves (Fig. 21.9).

Conclusion

Tephra layers from Middle and Late Pleistocene sediments of the Polovinka section have different geochemical and mineralogical composition and magnetic properties. The tephras are of basaltic-andesitic, andesitic, dacitic and rhyolitic compositions.

Ti-rich titanomagnetites with average titanium contents between 20.42 and 25.23% show a Curie point between of 50 and 65 °C. The Curie points of middle-and poor-Ti titanomagnetites range between 335 and 480 °C. Titanomagnetites with Tc = 455–515 °C are thermally unstable during heating in argon while titanomagnetites with a Tc range between 335 and 480 °C displays thermal instability during heating in air. The magnetic mineralogical data are identical in the different grain size fractions of the tephras.

Various magnetic mineralogy, magnetic properties and geochemistry can be used to discriminate between different tephras and thus aiding the detection of stratigraphic markers in other depositional realms.

Acknowledgements These studies were funded by the Far East Branch Russian Academy of Sciences (15-I-2-067).

References

Bazanova, L.I., Melekestsev, I.V., Ponomareva, V.V., Dirksen, O.V., and Dirksen, V.G. Late Pleistocene and Holocene volcanic catastrophes in Kamchatka and Kuril Islands. Part. 1. Types and classes of catastrophic eruptions as the leading components of volcanic catastrophism. Journal of Volcanology and Seismology. 2016. Vol. 10 (3). P. 151–169.

Bindeman, I.N., Leonov, V.L., Izbekov P.E., Ponomareva, V.V., Watts, K.E., Shipley, N.K., Perepelov, A.B., Bazanova, L.I., Jicha, B.R., Singer, B.S., Schmitt, A.K., Portnyagin, M.V., and Chen, C.H. Large-volume silicic volcanism in Kamchatka: Ar-Ar, U-Pb ages and geochemical characteristics of major pre-Holocene caldera-forming eruptions. J. Volcanol. Geotherm. Res. 2010. Vol. 189(1–2). P. 57–80.

Borkhodoev, V.Ya. Accuracy of the fundamental parameter method for x-ray fluorescence analysis of rocks, X-Ray Spectrom. 2002. Vol. 31. P. 209–218.

Braitseva, O.A., Melekestsev, I.V., and Sulerzhitskii, L.D. New data on the Pleistocene deposits in the Central Kamchatka depression. Stratigraphy and Geological correlation. 2005. Vol. 13 (1). P. 106–115 (in Russian).

Braitseva, O.A., Melekestsev, I.V., Evteeva, I.S., and Lupilina, E.G. Stratigraphy of the Quaternary deposits and glaciations of Kamchatka. Moscow: Nauka Press, 1968, 226 pp.

Braitseva, O.A., Melekestsev, I.V., Ponomareva, V.V., and Sulerzhitsky, L.D. The ages of calderas, large explosive craters and active volcanoes in the Kuril–Kamchatka region, Russia. Bull. Volcanol. 1995. Vol. 57. P. 383–402.

Burov, B.V., Nurgaliev, D.K., and Yasonov, P.G., Paleomagnitnyi analiz (Paleomagnetic Analysis), Kazan: KGU, 1986. (In Russian).

Chevrier, V., Mathé, P.-E., Rochette, P., and Gunnlaugsson, H.P. Magnetic study of an Antarctic weathering profile on basalt: Implications for recent weathering on Mars. Earth and Planetary Science Letters 244 (2006) 501–514.

Day, R., Fuller, M., and Schmidt, V.A. Hysteresis Properties of Titanomagnetites: Grain_Size and Compositional Dependence. Phys. Earth Planet. Inter. 1977. Vol. 13. P. 260–267.

Derkachev, A.N., Nikolaeva, N.A., Gorbarenko, S.A., Portnyagin, M.V., Ponomareva, V.V., Nürnberg, D., Sakamoto, T., Iijima, K., Liu, Y., Shi, X., Lv H., and Wang, K. Tephra layers in the Quaternary deposits of the Sea of Okhotsk: Distribution, composition, age and volcanic sources. Quaternary International. 2016. Vol. 425. P. 248–272.

Dirksen, O., Van den Bogaard. C., Danhara, T., and Diekmann, B. Tephrochronological investigation at Dvuh-yurtochnoe lake area, Kamchatka: Numerous landslides and lake tsunami, and their environmental impacts. Quaternary International. 2011. Vol. 246. P. 298–311.

Dunlop, D., and Özdemir, Ö. Rock Magnetism: Fundamentals and Frontiers, Cambridge: Cambridge University Press, 1997.

Dunlop, D.J. Theory and application of the Day plot (*M*rs/*M*s versus *H*cr/*H*c) 1. Theoretical curves and tests using titanomagnetite data. J. Geophys. Res. 2002a. Vol. 107. P. 56–60.

Dunlop, D.J. Theory and application of the Day plot (*M*rs/*M*s versus *H*cr/*H*c) 2. Application to data for rocks, sediments, and soils. J. Geophys. Res. 2002b. Vol. 107. No. B3, https://doi.org/10.1029/2001jb000487.

Fabian, K., Shcherbakov, V.P., and McEnroe, S.A. Measuring the Curie temperature. Geochem. Geophys. Geosyst. 2013. Vol. 14. P. 947–961.

Hrouda, F., Chlupáčová, M., and Mrázová, S., Low-Field Variation of Magnetic Susceptibility as a Tool for Magnetic Mineralogy of Rocks. Phys. Earth Planet. Inter. 2006. Vol. 154. P. 323–336.

Jasonov, P.G., Nourgaliev, D.K., Burov, B.V., and Heller, F. A modernized coercivity spectrometer, Geologica Carpathica. 1998. Vol. 49. P. 224—225.

Kiryanov V.Yu. On Possibility of correlationof Ash horizons in Pleistocene deposits of the Central Kamchatka depression. Volcanology and Seismology. 1981. Vol. 6. P. 30–38. (in Russian).

Kuprina N.P. A stratigraphy and sedimentation history of Pleistocene deposits of Central Kamchatka. Moscow: Nauka Press, 1970. 149 pp.

Kyle, P.R., Ponomareva, V.V., Schluep, R.R. Geochemical characterization of marker tephra layers from major Holocene eruptions, Kamchatka Peninsula, Russia. International Geology Review. 2011. Vol. 53(9). P. 1059–1097.

Lattard, D., Engelmann, R., Kontny A. and Sauerzapf, U. 2006. Curie temperatures of synthetic titanomagnetites in the Fe-Ti-O system. Reassessment of some methodological and crystal chemical effects. J. Geophys. Res. 2006. Vol. 111, B12S28.

Le Maitre, R.W., Streckeisen, A., Zanettin, B., Le Bas, M.J., Bonin, B., Bateman, P., Bellieni, G., Dudek, A., Efremova, S., Keller, J., Lamere, J., Sabine, P.A., Schmid, R., Sorensen, H., and Wool, A.R. Igneous rocks. A Classification and Glossary of Terms, in: Recommendation of the International Union of Geological Science Subcommission on the systematics of Igneous rocks, 2nd Edn., 2002. Cambridge University Press, 254 pp.

Lowe, J.J., Ramsey, C.B., Housley, R.A., Lane, C.S., and Tomlinson, E.L., RESET Team. RESET Associates. The RESET project: constructing a European tephra lattice for refined synchronisation of environmental and archaeological events during the last c. 100 ka. Quaternary Science Reviews. 2015. Vol. 118. P. 1–17.

Nishitani, T., and Kono, M. Curie temperature and lattice constant of oxidized titanomagnetite. Geophys. J. R. Astron. Soc. 1983. Vol. 76. P. 585–600.

Özdemir, Ö. Inversion of titanomaghemites. Phys. Earth Planet. Int. 1987. Vol. 46 (1–3). P. 184–196.

Petrovský, E. and Kapička, A. On determination of the Curie point from thermomagnetic curves. J. Geophys. Res. 2006. Vol. 111. B12S27, https://doi.org/10.1029/2006jb004507.

Ponomareva, V., Portnyagin M., and Davies S.M. Tephra without Borders: Far-Reaching Clues into Past Explosive Eruptions. Front. Earth Sci. 2015. https://doi.org/10.3389/feart.2015.00083.

Ponomareva, V.V., Kyle, P.R., Melekestsev, I.V., Rinkleff, P.G., Dirksen, O.V., Sulerzhitsky, L. D., Zaretskaia, N.E., and Rourke, R. The 7600 (14C) year BP Kurile Lake caldera-forming eruption, Kamchatka, Russia: stratigraphy and field relationships. Journal of Volcanology and Geothermal Research. 2004. Vol. 136. (3–4). P. 199–222.

Ponomareva, V.V., Kyle, P.R., Pevzner, M.M., Sulerzhitsky, L.D., and Hartman, M. Holocene eruptive history of Shiveluch volcano. Kamchatka Peninsula. In: Eichelberger J., Izbekov P., Ruppert N., Lees J., Gordeev E. (Eds). Volcanism and Subduction: The Kamchatka Region. AGU Geophysical Monograph Series. 2007. Vol. 172. P. 263–282.

Ponomareva, V.V., Melekestsev, I.V., Bazanova L.I., Bindeman, I.N., Leonov, V.L., and Sulerzhitsky, L.D. Volcanic catastrophes on Kamchatka during Middle Pleistocene and Holocene. In: Extreme environmental hazards and catastrophes. Moscow: IPSE, 2010. P. 219–237.

Portnyagin, M., Hoernle, K., Avdeiko, G., Hauff, F., Werner, R., Bindeman, I., Uspensky, V., and Garbe-Schönberg D. Transition from arc to oceanic magmatism at the Kamchatka-Aleutian junction. Geology. 2005. Vol. 33. No. 1. P. 25–28.

Robertson, D. J. Discrimination of Tephra Using Rockmagnetic Characteristics. J. Geomag. Geoelectr. 1993. Vol. 45. P. 167–178.

Volynets, O.N. Geochemical types, petrology and genesis of Late Cenozoic volcanic rocks from the Kurile–Kamchatka island–arc system. Int. Geol. Rev. 1994. Vol. 36 (4). P. 373–405.

Chapter 22
Basalts of the Zhom-Bolok Lava River as a Possible Sources of Metallic Iron in Sediments of Local Lakes: Thermomagnetic and Microprobe Justification

D. M. Pechersky, A. Yu. Kazansky, A. Kozlovsky, G. P. Markov, A. A. Shchetnikov and V. A. Tselmovich

Abstract The results of thermomagnetic and microprobe studies of basalts of the Zhom-Bolok lava river (East Sayan) are given in the article. On the example of the Zhom-Bolok basalt lava river the process of settling of metallic iron particles in a basaltic lava has been demonstrated. Some of examples show the distribution of metallic iron particles during their settling within basaltic lava flows. It has been shown that the main source of metallic iron in lake sediments is the volcanic rocks of the Zhom-Bolok basaltic lava river, detrital material being created through erosion and weathering. Out of numerous studied objects Zhom-Bolok region is a single example of significant enrichment of sediments with metallic iron of terrestrial origin in the immediate vicinity of large eruptions of basaltic lava.

D. M. Pechersky (✉) · G. P. Markov · V. A. Tselmovich
Institute of Earth Physics RAS, Moscow, Russia
e-mail: diamarmp@gmail.com

D. M. Pechersky
Institute of Geology and Petroleum Technologies, Kazan Federal University, Kazan, Russia

A. Yu. Kazansky
Moscow State University, Moscow, Russia

A. Yu. Kazansky
Geological Institute of RAS, Moscow, Russia

A. Kozlovsky
IGEM RAS, Moscow, Russia

A. A. Shchetnikov
Institute of the Earth's Crust SB RAS, Irkutsk, Russia

A. A. Shchetnikov
Institute of Geochemistry SB RAS, Irkutsk, Russia

D. Nurgaliev et al. (eds.), *Recent Advances in Rock Magnetism, Environmental Magnetism and Paleomagnetism*, Springer Geophysics,
https://doi.org/10.1007/978-3-319-90437-5_22

Accordingly the halo of scattering of metallic iron particles from the centers of volcanic eruptions and/or the fall of large meteorites and other cosmic bodies to Earth is very limited.

Keywords Thermomagnetic analysis · Microprobe analysis · Sediments
Basalts · Native iron · Magnetite

Introduction

We found particles of metallic iron in the lake sediments of the Zhom-Bolok region, East Sayan, in all the samples studied. Their age varies from recent to 10–15 thousand years ago, the iron concentration is above 0.001% (Pechersky et al. 2018). In numerous previously studied sections of sediments in different regions, of different ages and different origins, the vast majority of metallic iron particles discovered everywhere have an extraterrestrial origin and are associated mainly with interplanetary dust depositing on the Earth surface (Pechersky 2015). The metallic iron of clearly terrestrial origin is presents only in the young lacustrine sediments from the Zhom-Bolok region. This follows, first of all, from the histogram of concentrations of iron particles, similar to the histogram of the concentrations of magnetite, a mineral of undoubtedly terrestrial origin, and sharply differing from the bimodal histogram of iron concentrations in all other previously studied sediments of the Earth, which is primarily due to the presence of a distinct "zero" group samples, i.e. samples where no particles of metallic iron were found (Fig. 22.1).

The terrestrial origin of the metallic iron in the lake sediments of the Zhom-Bolok region naturally led to an idea of the connection of this iron with volcanic activity, since most of the lakes are in close proximity to the Zhom-Bolok lava river. The highest concentration of native iron is observed in the sediments of Lake Khara-Nur, which is in direct contact with the Zhom-Bolok lava river (Fig. 22.2a). Lake Hikushka is located ~800 m from the edge of the Zhom-Bolok lava river, and the concentration of iron in its sediments is much lower than in those of the Khara-Nur Lake (Fig. 22.2b). The lowest concentration of native iron is observed in the sediments of Lake Tuhuren-Nur (Fig. 22.2c), located behind the ridge ~3 km both from the border of the Zhom-Bolok lava river and from volcanoes.

However, direct communication requires very frequent eruptions of volcanoes during the accumulation of lake sediments, because iron particles are present in significant amounts in all samples of lake sediments. It remains to relate it to the weathering products of the volcanic rocks of the region. In this article we attempt to answer these questions.

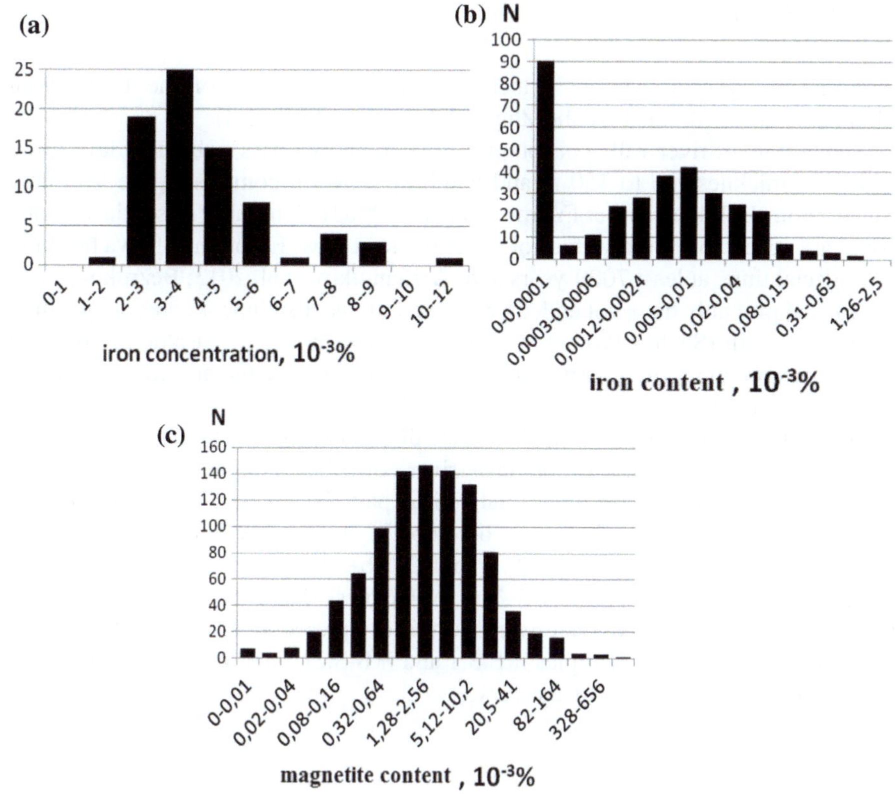

Fig. 22.1 Concentration histograms of native iron (**a**, **b**) and magnetite (**c**) in the sediments. Generalization of TMA results for three sediment sections from Zhom-Bolok region (**a**) and previously studied sediments in all sections (**b**, **c**) (Pechersky et al. 2018)

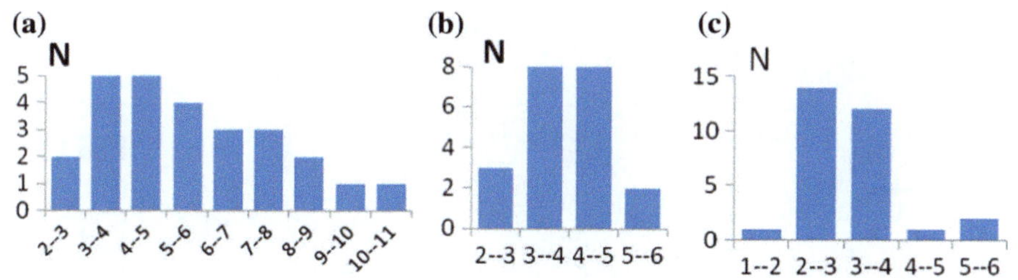

Fig. 22.2 The concentration of native iron (10^{-3}%) in the sediments of the lakes Khara-Nur (**a**), Hikushka (**b**) and Tuhuren-Nur (**c**) (Pechersky et al. 2018)

Research Area

Zhom-Bolok volcanic region is located in the eastern part of the East Sayan Mountains and is drained by the Zhom-Bolok river (Fig. 22.3).

The bottom of river valley is completely overlain by series of basaltic lava flows with total thickness up to 150 m and width up to 4 km comprising the lava field called Zhom-Bolok lava river (Yarmolyuk et al. 2003). A number of scoria volcanic cones are raised above the surface of the upper lava flow. Eruptions of lava began in post-glacial time, at least 7000 years ago (Arzhannikov et al. 2013; Bezrukova et al. 2016), and probably have not ended yet. Descriptions of sediment cores are given in accordance with (Shchetnikov et al. 2016) and descriptions of basaltic samples are given in accordance with (Yarmolyuk et al. 2003). To solve the problem, a series of basalt samples was selected: from the volcanoes Treschinny, Kropotkin and Ostanets, including volcanic bombs, from the cinder cone and from a series of basalt flows of the Zhom-Bolok lava river, distributed along its entire length (Fig. 22.3). To study the settling of native iron particles inside a lava flow, samples of basalts were collected from two outcrops in the Uherik dell (Figs. 22.3 and 22.4a) and Waterfall area (Figs. 22.3 and 22.4b). The section in Uheric dell is represented by a single lava flow 5–7 m in thickness. Here the frontal part of lava flow is cut by the canyon of Oka river. Basaltic samples were taken from the base of the flow near the contact (samples ЖБ-3.1 and ЖБ-3.2); middle (sample ЖБ-2) and upper (sample ЖБ-1) parts of the flow.

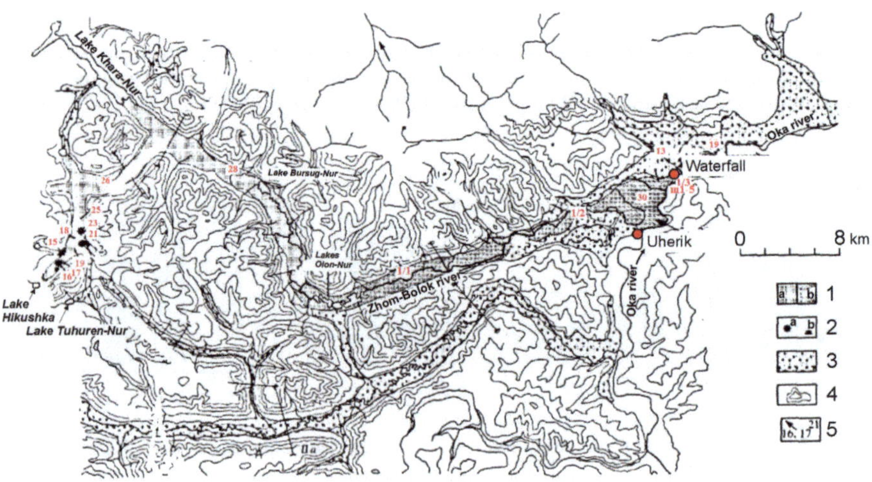

Fig. 22.3 Scheme of the structure of the Zhom-Bolok lava river (Yarmolyuk et al. 2003). 1—series of flows, 2—cinder volcanic cones; 3—moraine deposits; 4—relief contours; 5—sampling points and samples numbers (Yarmolyuk et al. 2003); samples 1/1, 1/2 and 1/3—were collectedby A. Kozlovsky, Щ1–Щ5—were collected by A. A. Shchetnikov. Sampling sites Waterfall and Uherik (sampling points of ЖБ)—see text

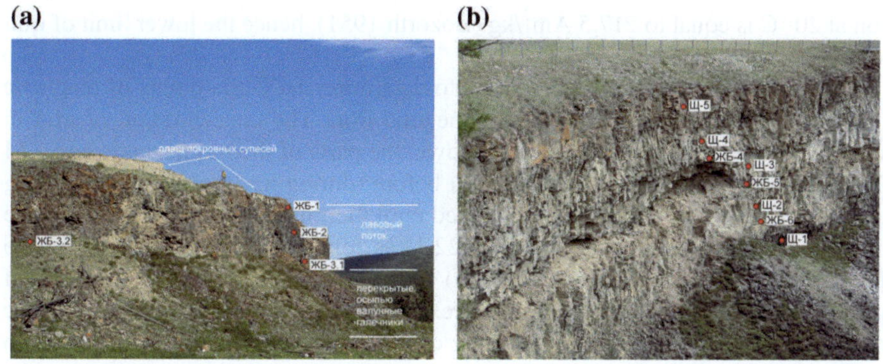

Fig. 22.4 Places of basalts samples from the outcrops of the Uherik dell (Fig. 22.4a) and near the waterfall (Fig. 22.4b)

The section in Waterfall area is located near the 14 m waterfall at the confluence of Zhom-Bolok River with Oka River (Figs. 22.3 and 22.4b). The section consists of five interbedded basaltic flows. The full thickness of the first (lowest) flow is buried under a talus and the visible thickness of the flow is about 2 m. Sample Щ–1 was collected at 1 m below the surface of the first flow. The second flow has a thickness of 2.5–3 m. The sample Щ-2 was collected approximately 1 m below the roof of the flow and sample ЖБ-6 was taken from its bottom. The third flow has a thickness of 0.7–1.0 m. Two samples Щ-3 and ЖБ-5 were collected near the bottom of this flow. The fourth flow has a thickness of 4–5 m. One sample Щ-4 was collected approximately 2 m below the surface of the flow and one sample ЖБ-4 was collected from its bottom. The uppermost fifth flow has a thickness of 3–4 m; one sample Щ-5 was taken 1 m below its surface.

Research Methods

The studies included thermomagnetic analysis (TMA), electron microscopy, optical and microprobe (MPA) analyzes. TMA was conducted in the Laboratory of geomagnetism and petromagnetism IPE RAS using thermo-vibrating sample magnetometer designed by N. Anosov and Yu. Vinogradov. TMA measurements were carried out by G. Markov and A. Titov. The sensitivity of the instrument to the magnetic moment is 3×10^{-8} Am^2, basalt sample weighting 1.5–2 g. For TMA, continuous measurements of the samples magnetic moment are performed during heating in a DC magnetic field of 1 T from the room temperature to 800 °C and subsequent cooling to the room temperature. All measurements were conducted in air with the heating speed ~60°/min. The sensitivity of the instrument to the specific magnetization is 3×10^{-5} Am^2/kg. Specific saturation magnetization for iron at 20 °C is equal to 217.5 Am^2/kg (Bozorth 1951), hence the lower limit of iron detection should be $\sim 1.4 \times 10^{-5}$ wt%.

Thermomagnetic curves M (T) allow to determine the Curie points of magnetic minerals in samples, including that of metallic iron. The concentration of iron in samples was determined by TMA as follows: the parts of thermomagnetic curve M (T) above hematite Curie temperature but before (during heating) and after (during heating) of iron Curie point were continued until their intersection with the ordinate axis, i.e. to the room temperature (Fig. 22.1) as follows. The first (heating) curve (parabola below the Curie point of iron) is the extension of the M (T) curve of native iron with a given Curie point; it excludes all contributions to of magnetic minerals with Curie points less than that of iron. The second curve (cooling from the Curie point of iron) is a paramagnetic curve (hyperbola) which is parallel to the paramagnetic part of the cooling curve M(T), because there are no other magnetic minerals above the Curie point of iron. The difference between those two curves on the ordinate axis is the value of the specific saturation magnetization of the iron (M_s) in the sample, because thermomagnetic measurements were carried out in the saturation magnetic field for iron. The ratio M_s to the specific saturation magnetization of iron (217.5 Am^2/kg) will give the iron content in the sample. Examples of TMA curves are in Fig. 22.5.

The accuracy of the Curie point determination is $\sim 5°$. The accuracy of the described concentration determination is low, but it is sufficient for variations in the iron concentration in the studied sediments by orders of magnitude, and the deficiencies in the accuracy of the TMA are compensated by the simplicity of the method, and of sample preparation and by the speed of the data acquisition.

Microprobe analysis (MPA) and other microscopic studies were performed by V. A. Tselmovich at Borok Geophysical Observatory of the Institute of Physics of the Earth, RAS. `Samples for MPA were crushed and dispersed in an ultrasonic cleaner bath. Magnetic fraction extracted by a permanent magnet was applied to a

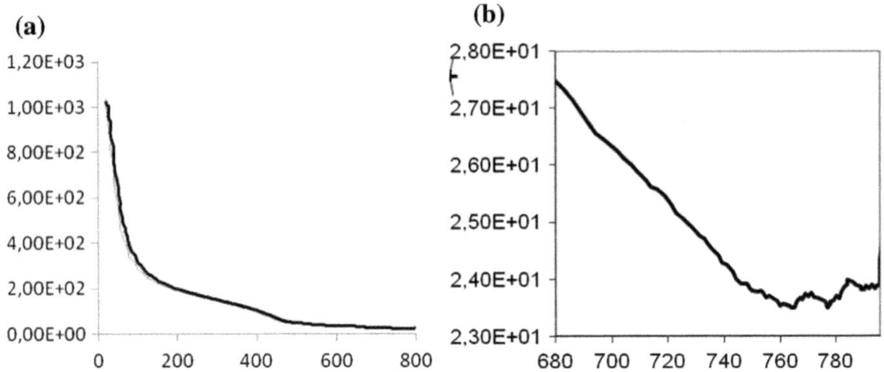

Fig. 22.5 Example of TMA curve (sample OK-1/23). Figure 22.5a shows titanomagnetite with Tc = 80 °C ($\sim$70% of sample magnetization), Fig. 22.5b shows iron with T_c = 743 °C and T_c = 760 °C

double-sided conductive carbon adhesive tape and rolled up with a glass rod so that the surface of the particles was oriented parallel to the surface of the sample table. Such approach allowed to analyze the particles without polishing. MPA was performed at the voltage of 20 kV and the current of 0.2 nA, the probe beam size was ~0.2 μm, the size of the analyzed area was 2 μm. In most cases, MPA and TMA results are consistent, however some discrepancies exists due to the fact that TMA is carried out on whole piece of rock while MPA—uses the enriched magnetic fraction from a much larger sediment sample.

Results of Thermomagnetic Analysis

The results of TMA are summarized in Table 22.1.

All studied basalts are fresh and homogeneous. This can be seen from the uniform values of the specific magnetization M_{20} (Table 22.1; Fig. 22.6a) and the paramagnetic magnetization measured at 800 °C (M_{800}), which reflects the total iron content in the rock (Table 22.1; Fig. 22.6b), and, especially, from permanent presence of a homogeneous primary magmatic titanomagnetite with T_c = 70–80 °C (Table 22.1; Figs. 22.6c and 22.7). The homogeneity of titanomagnetite attests to homogeneous T-fO_2 conditions in the magma chamber, which does not changes both before and during the eruption of the melt and inside the lava flow. The composition of titanomagnetite corresponds to the depth of the chamber about 70 km (65–75 km) (Pechersky et al. 1975; Pechersky and Didenko 1995).

As a result of secondary single-phase oxidation of titanomagnetite, its Curie point increases to 130–230 °C (Table 22.1; Fig. 22.7), and as a result of heterophase oxidation of titanomagnetite, a mineral similar to magnetite with T_c ~550 °C and even hematite with T_c = 665–680 °C appears.

Note to Table 22.1: N—sample number; km—distance from the place of sampling to the center of eruptions, km; M_{20} is the specific magnetization of the sample, measured before its heating, 10^{-3} Am^2/kg; M_{800}—specific magnetization of the sample measured at 800 °C, $10^{-3} Am^2/kg$; Mt is the specific magnetization of the sample, measured at room temperature after its heating to 800 °C, 10^{-3} Am^2/kg; T_c is the Curie point of titanomagnetite (TM), magnetite (MT), hematite (HM), and iron; S_{TM} is the concentration of titanomagnetite with Tc < 100 °C, 10^{-3}%; CFe is the concentration of iron, 10^{-3}%. For each sample, the top row is the Curie points, determined on the heating curve M (T), the bottom row is the Curie point determined on the cooling curve M (T). ? - Curie point was not exactly determined.

The total magnetization of the sample (from <5 to 60%, Table 22.1) is attributable to magnetite. The concentration of homogeneous titanomagnetite and the degree of its oxidation do not depend on the distance from eruption center. During laboratory heating to 800 °C, titanomagnetite undergoes a heterophase oxidation with neoformation of magnetite, which results in a noticeable increase magnetization by a factor of 2–6 and respectively in negative correlation (R = −0.63) between the Curie point of titanomagnetite and M_t/M_{20} ratio. Regardless of the

Table 22.1 Magnetic properties of Zhom-Bolok basalts

N	Rock	km	M_{20}	M_{800}	Mт/M_{20}	TcTM	TcMT	TcHM	TcFe	Cтм	CFe
жмб-1/1	Lava flow, top,	36	245.4	28.4	3.56	75(55)			755 775	4	0.019
							560				
жмб-1/2	Lava flow, top,	52	228.4	23.9	2.09	125(40)	540			3.6	0
							565	665			
жмб-1/3	Lava flow, top,	65	134.8	24.5	4.66	75(61)	560		760?	4.1	0.005
							560				
ок-1/9	Lava flow, top,	68	343.8	22.5	1.37	230(73)	520		730? 765	6.3	0.015
							540				
ок-1/13	Lava flow, top,	63	431.2	24.3	1.94	80(62)	?		730? 765	10.6	0.014
							560				
ок-1/15	Lava flow, top,	0	197.2	23	3.77	70(58)	555		770?	5.7	0.007
	volc. Treshchin.						565				
ок-1/16	Bomb	0	480.9	22.3	4.03	70(70.5)	500		760	16.8	0.023
	volc. Kropotkin						550				
ок-1/17	Lava flow	65	?	24.8			470?		762? 785?		0.03
							540				
ок-1/18	Cinder cone	2	4010.9	25.4	0.76		505	690?	745?775?	44.5	0.008
							450	670			
ок-1/19	Lava flow, top,	1	266.7	26.3	6.42	90(73)	580		735?	7.5	0.008
							580	675?			
ок-1/21	Bomb	0	2013.9	25.6	1.19	170(21)	510	670	730	14.2	**0.46**
							490	680			
ок-1/23	volc. Ostanets	1	788.5	21.3	2.18	80(71)	460 550	670	745 760	27.6	**0.35**
							550	670			

(continued)

Table 22.1 (continued)

N	Rock	km	M_{20}	M_{800}	$M_{т}/M_{20}$	TcTM	TcMT	TcHM	TcFe	$C_{тм}$	CFe
ок-1/25	Lava flow, top,		372.9	23.2	2.59	80(37)	450 550			6.9	0
							555				
ок-1/26	Lava flow, top,	8	1491.7	25.9	1.05	115(19.7)	470? 550		725?	9.8	0.013
							540				
ок-1/28	Lava flow, top,	12	587.9	25.2	1.85	90(47)	450 545	680?	760 780	13.8	0.011
							515?? 560	675			
ок-1/30	Lava flow, top,	59	230.4	26.8	4	125(35)	550?		755??	3.2	0
							570		750??		
щ1	Lava flow, top,	65	219.2	25.7	?	100(71)	560(1.2)			6.2	0
	Waterfall						560				
щ2	Lava flow, top,	65	210.9	24.2	6.3	120(65)	550(1)			5.5	0
	Waterfall						550				
щ3	Lava flow, bottom	65	283.7	26.1	2.9	70(69)	560?(1)		740 770	10	0.02
	Waterfall						560				
щ4	Lava flow, middle	65	445.8	21	1.97	150(76)	550(5)	680?	765	9.7	0.01
	Waterfall						560	680?			
щ5	Lava flow, bottom	65	776.5	25.5	1.25	150(34)	530(58)	670?	728	7.5	0.014
	Waterfall						560	675?			
жб-1	Lava flow, top,	60	380.3	34.2	2.17	60(27) 135(20)			765?	5.1	0.05
	Uherik						550				
жб-2	Lava flow, middle	60	1530	38	0.97		560		725 755 790?	0	0.2
	Uherik						560				
жб-3.1	Llow, bottom, clinker	60	1048.7	41.8	1.24	70(4)??	535	685	720? 775?	1.2	0.02
	Uherik						550	685			

(continued)

Table 22.1 (continued)

N	Rock	km	M_{20}	M_{800}	Mт/M_{20}	TcTM	TcMT	TcHM	TcFe	Cтм	CFe
жб-3.1a	Flow, bottom, clinker	60	241.4	29.8	3.46	80(50)	580?	685	750 765	3.45	0.15
	Uherik						580	680			
жб-3.2	Lava flow, bottom	60	306.3	32.1	3.07	80(50)	560		745 785	4.4	0.18
	Uherik						580				
жб-4	Flow, bottom, clinker	65	2475.6	28.4	0.96		530		740 755	0	0.11
	Waterfall						570				
жб-4a	Lava flow, bottom	65	2496.3	11.9	0.85		490		740 760	0	0.24
	Waterfall						540				
жб-5	Lava flow, bottom	65	166.6	33.4	5.31	80(38)	560?	680?	710? 755?	1.8	0.12
	Waterfall						570	685			
жб-6	Lava flow, bottom	65	635.9	31.5	2.02		540	675	735? 770?	0	0.1
	Waterfall						560		760?		

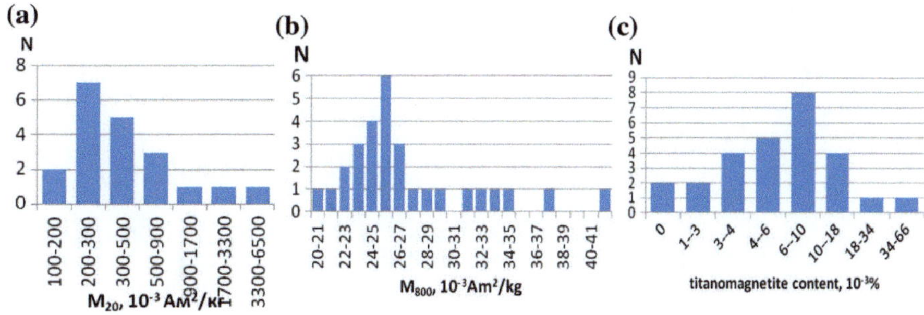

Fig. 22.6 M_{20} **a** and M_{800} **b** values and the concentration of titanomagnetite **c** in basalts

Fig. 22.7 Curie points of titanomagnetite in basalts

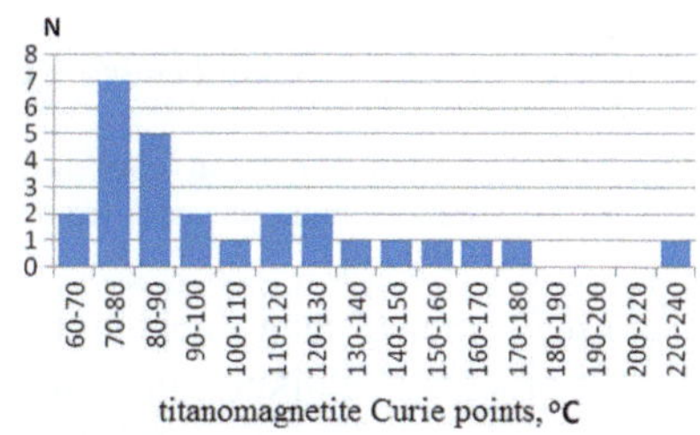

degree of oxidation of basalts (this is seen from titanomagnetite behavior), basalts contain metallic iron from zero to $\leq 0.5 \times 10^{-3}\%$ (Table 22.1; Fig. 22.8), which is noticeably less than in the lake sediments of the region (Fig. 22.2). We emphasize the bimodal distribution of iron concentrations (Fig. 22.8): the first mode ($0.01{-}0.02 \times 10^{-3}\%$) refers to samples from the upper parts (tops) of the flows, the second mode ($0.08{-}0.16 \times 10^{-3}\%$)—to samples from their bottom parts. Maximal iron concentrations are typical for to samples from the centers of volcanic eruptions and then sharply decreases along the flow (Table 22.1; Fig. 22.9).

This can be explained by the fact that most of the samples were taken from the upper parts of the flow which are depleted in heavy iron particles due to the simple sinking of those particles towards the bottom of the lava flow during its displacement from the center of the eruption. A second possible mechanism of iron removal

Fig. 22.8 Concentration of iron in basalts

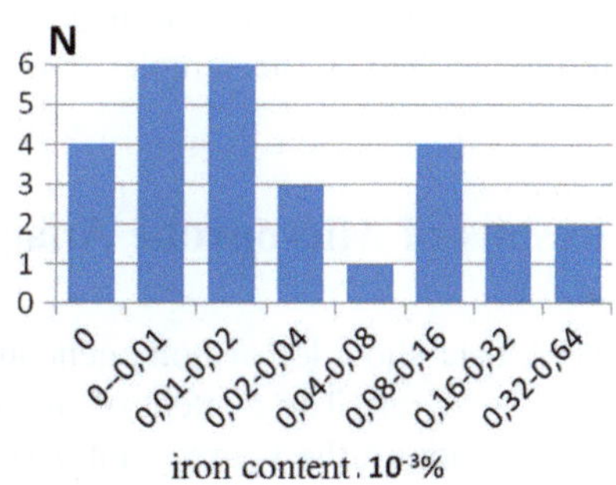

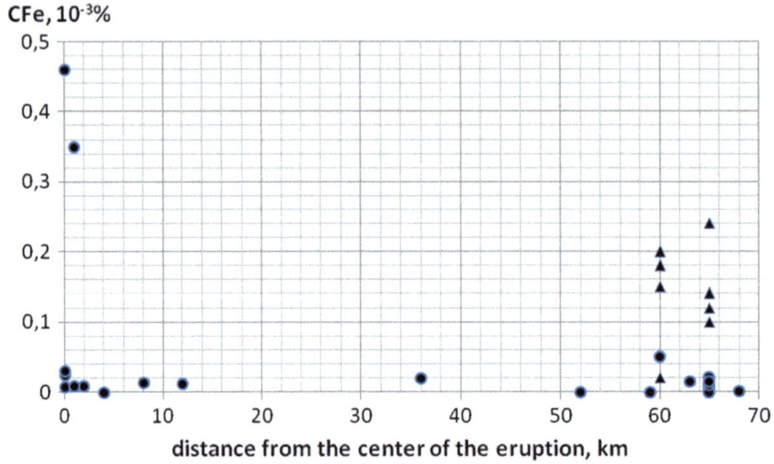

Fig. 22.9 Distribution of iron concentration (10^{-3}%) along the Zhom-Bolok lava river (see Fig. 22.3). Triangles mark samples from the bottom of the flows nearby Zhom-Bolok mouth

from the center of lava flows is that iron particles are oxidized and destroyed. But such a process is unlikely for samples where titanomagnetite with T_c < 100 °C not stable under oxidizing conditions is retained. The heterophase oxidation of titanomagnetite is more common for scoria. For example, the sample OK-1/18 (Table 22.1) from scoria cone is devoid of titanomagnetite with T_c < 100 °C, while a mineral close to magnetite is present, and an appreciable rehomogenization of titanomagnetites resulted from heating of the sample to 800 °C. This is manifested in a decrease of the Curie temperature to 450 °C, and a noticeable drop in the magnetization to about 70% of the initial value. The concentration of metallic iron is significantly reduced in scoria samples, down to its complete absence. The low concentration of iron in basalts, and the predominance in them of homogeneous primary titanomagnetite with $T_c \leq 100$ °C, absent in sediments and, vice versa, the presence of "pure" magnetite in sediments, absent in basalts, indicates that basalts from the lava flows of the region are probably not the only sources of metallic iron in the sediments. In the geological structure of the region Paleozoic granitoids, gabbros and gabbro-norites are widely distributed. They may be additional sources of metallic iron in lake sediments.

Curie temperatures of iron (i.e., iron composition) range from 712 to 780 °C, more than half of the samples containing a nearly pure iron (Fig. 22.10).

Results of Microprobe Analysis

MPA data show lot of homogeneous titanomagnetite grains with a size from ~5 m to ~60 μm. The content of titanium varies from 8 to 19%, 15–16% prevail. It corresponds to the Curie point 70–80 °C. Less common is ilmenite. Only single

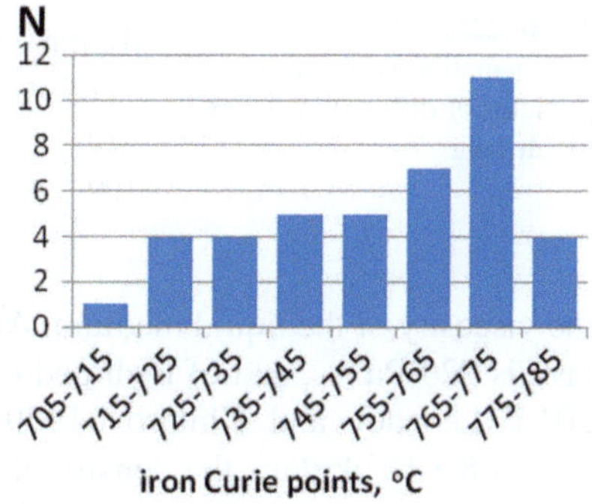

Fig. 22.10 Curie points of iron in basalts

small particles of native iron were found in basalts, which does not contradict TMA data on a very low iron concentration 0–0.0007% (Table 22.1). As in lake sediments, various alloys are found in basalts, and more often than in sediments, chromium iron (10.7–17.7% Cr, grain size 3–15 μm), alloys of the more complex composition of FeCrNi, FeNiSn; CuZn, CuZnPb. As in the sediments (Pechersky et al. 2018), in the basalts there are particles of nickel. A similar set of exotic metal particles in sediments and basalts, especially the great similarity of nickel particles (Pechersky et al. 2018), unusual both for basalts and for sediments, suggests that basalts are sources of such particles in lake sediments of the Zhom-Bolok region.

Discussion

Let us consider the behavior of metallic iron particles in basalts starting from the moment they appear in the melt. Metallic iron in the form of rare small particles appears in the basalt melt at a high temperature likely before it is outpouring onto the surface in the form of lava. This can be seen, for example, from the relatively high content of metallic iron in volcanic bombs up to $0.5 \times 10^{-3}\%$ (Table 22.1; Fig. 22.9), reflecting the state of the melt at the time of eruption. Melt poured on the Earth's surface is characterized by high mobility, reduced viscosity, which is determined by its high temperature [>1200 °C according to the study of melt inclusions (Yarmolyuk et al. 2003)].

In a liquid basalt lava, heavy metal iron particles should be sinking. Let us try to determine the rate of sinking of iron particles for a simple case of settling a spherical iron particle in a laminar flow of liquid basaltic lava by the Stokes formula:

$$V_{oc} = gd_p^2(\rho_p - \rho_c)/18\mu_c \quad (1)$$

where V_{oc} is the particle settling rate, g is the acceleration of gravity, d_p is the particle diameter, ρ_p, ρ_c are the particle and liquid density, respectively, and μ_c is

Table 22.2 The rates of sedimentation of the iron particles of different sizes in a basalt lava

$d_ч$, μm	V_{oc}, mm/hour
1	~0.0001
10	~0.01
100	~1

the viscosity of the liquid medium. At a temperature of 1200 °C, the viscosity of dry lava is 220 Pa • s, and of hydrated one—165–60 Pa • s (Balkhanov and Bashkuev 2015; Lebedev and Khitarov 1979; Persikov 1984); We assume: g = 10 m/s^2, $\rho_p = 7.8 \cdot 10^3$ kg/m^3, the density of the lava is $\rho_c = 2.6 \cdot 10^3$ kg/m^3, the viscosity of the lava is $\mu_c = 100$ Pa s. Substituting these values into formula (1), we obtain the immersion rates of iron particles of different sizes in basaltic lava (Table 22.2).

The mode of the size of iron particles in sediments, and in magmatic rocks is 10–20 μm, and the upper limit of the size is 100 μm, the particle settling rate of 100 μm particles is ~1 mm/hour (Table 22.2). An iron particles will settle at such a rate from the surface of the flow to a depth of 0.5 m in 21 days. How far will a lava float away? For example, lava of the Bilyukai volcano (eruption of 1938) and Kilauea (eruption in 1954) moved near the center of the eruption at a velocity of 5–6 m/min; lava of the Great Fissure Tolbachinsky eruption (1975–1976)—2 m/min (Fedotov 1984). Far from the center of the eruption, the lava velocity is likely to be less than 1 m/min, for example, the flow of the Bilyukai volcano more than 1 km from the center of the eruption moved at 0.6 m/min. In the latter case, iron particles will dive to 0.5 m at about 20 km from the eruption center. It follows that in the majority of samples collected near the surface of the Zhom-Bolok lava river iron particles should not be detected as seen from Fig. 22.9. At the end of the Zhom-Bolok lava river (~70 km from the eruption center), an appreciable fraction of iron particles should be at the bottom of the flow. This can be seen in the Waterfall sampling point at the confluence of Zhom-Bolok and Oka rivers, where 5 flows are exposed. In the sample Щ-1 from the first (lower) flow iron was not detected. Also it was not detected in the sample Щ-2, taken 1 m from the surface of the second flow. The iron content of 0.12×10^{-3}% was found in the sample ЖБ-6 taken from the bottom of the second flow. The sample Щ-3 was taken near the base of the third flow, and the iron content in it was 0.02×10^{-3}%. The sample of ЖБ-5 was taken from the bottom of the third flow, and has the iron content 0.14×10^{-3}%. The sample Щ-4 was selected approximately 2 m from the top of the fourth flow, the iron content in it was 0.01×10^{-3}%, the sample of ЖБ-4 was taken from the bottom of the fourth flow, and the iron content in it was 0.24×10^{-3}%. The sample Щ-5 was taken from a depth of about 1 m from the top of the fifth flow, the iron content in it is 0.014×10^{-3}%. In the outcrop of the lava flow on the Uherik dell (Fig. 22.2), the sample of ЖБ-1 basalt from the surface of the flow contains 0.05×10^{-3}% of iron; while in the middle part of the same flow, in the sample ЖБ-2, the iron content is 0.2×10^{-3}%. At the basal part of the lava flow on contact with boulder layer in the sample ЖБ-3.1 iron content was

Table 22.3 Distribution of iron concentrations in basaltic flows of the Zhom-Bolok lava River

	Min	Max	Average
Top of the flow	0	$0.05 \times 10^{-3}\%$	$0.007 \times 10^{-3}\%$ (17 samples)
Lower part of the flow	0.02	$0.24 \times 10^{-3}\%$	$0.12 \times 10^{-3}\%$ (10 samples)

$0.15 \times 10^{-3}\%$. In 50 m up the valley, the sample of ЖБ-3.2 was also selected from the basalt flow base, the iron content in it was $0.18 \times 10^{-3\%}$.

The summary results of iron content in the lava flows are present it the Table 22.3. The main conclusions from the Table 22.3 are as follows: the relative accumulation of iron particles take place in the lower part of lava flows as a result of settling of iron particles in lavas.

Thus, the obtained TMA results do not contradict the calculations.

How did the iron particles arrive into the studied lakes? In order to verify a possibility of air transport of terrestrial iron particles by volcanic ash, we have checked the presence of traces of volcanic material in previously studied sites. Only three such sites were found.

(1) Miocene volcanoclastic turbidites of ~160 m thickness (ODP hole 386). 22 particles of native iron were found here. It is less than in neighboring terrigenous sediments (33 particles), and in terrigenous-carbonate deposits of the Miocene (ODP hole 391A) with no volcanic material, where 50 native iron particles were found (Pechersky et al. 2013).
(2) Permo-Triassic sediments from Meyshan section (China). Above and below the Permian—Triassic boundary there are interlayers of volcanic ash, but particles of native iron were not found in the ash layer, although they are found in the surrounding terrigenous sedimentary rocks (Korchagin 2013).
(3) Permo-Triassic tuffs between the flows of Siberian Traps. The native iron was not found (Pechersky 2015).

Consequently, the above way of a broad global spread of native iron and its enrichment in sediments is highly improbable. Obviously, native iron particles during the air transport should be rapidly deposited near the volcanic vent.

Estimation of a distance of air transport of magnetic particles with the ash plume from the volcanic eruption was performed by Evans (1999) on an example of the 1980 eruption of Mount St. Helens. The concentration of magnetic particles was estimated from the value of magnetic susceptibility. It was found that the magnetic content decreases exponentially with distance from the source: by a factor of 10 at a distance of 450 km, 100 at a 880 km and 1000 at a 1300 km. The content of native iron in the Earth's volcanic rocks and in volcanic ash is less than tens of ppm (Karpov 2014), and therefore, the native iron in the sediments may be detected by TMA no farther than a few hundred kilometers from the source. This explanation is quite consistent with a unique accumulation of native iron in lake sediments located only in the immediate vicinity of active volcanoes of Zhom-Bolok region.

Obviously, this conclusion applies not only to the volcanic sources of native iron, but also to a similar situation in the case of large impacts of space bodies.

A significant part of volcanic material must to arrive to the bottom sediments of lakes from the slopes of the catchment basin after ash-falls. The latter fact, combined with the multiphase nature of volcanic eruptions with short breaks, throughout the post-glacial period (Shchetnikov et al. 2018), could provide a virtually continuous and abundant supply of tephra to the lake sedimentation reservoirs. The constant presence of native iron in the sediments can be explained by supply from weathering products of volcanic rocks. In the last case the area of detectable concentartion of iron particles would be even smaller than in the case of air transport. This conclusion casts doubt on the possibility of tracing the metallic iron halos from large impact events in a global scale.

Conclusions

Analysis of TMA and MPA results allowed us to solve a variety of problems:

Firstly, on the example of the Zhom-Bolok basalt lava river the process of settling of metallic iron particles in basaltic lava has been demonstrated;

Secondly, it has been shown that the main source of metallic iron in lake sediments in East Sayan Zhom-Bolok area is the volcanic rocks of the Zhom-Bolok basaltic lava river, main part of detrital material being created through erosion and weathering.

Thirdly, it has been established that the halo of scattering of metallic iron particles from the centers of volcanic eruptions and/or the fall of large meteorites and other cosmic bodies to Earth is very limited. This follows from a vivid example: out of numerous studied objects Zhom-Bolok region is a single example of significant enrichment of sediments with metallic iron of terrestrial origin in the immediate vicinity of large eruptions of basaltic lava. Consequently, it is unfeasible to find traces of halos of the global iron spread resulting from the fall of large cosmic bodies, as well as from large volcanic eruptions.

Acknowledgments We are extremely grateful to the reviewer A. Kosterov for a detailed in-depth review and careful editing of the English text.

The geological part of the research was supported by RFBR grant 15-05-01811 and RSF grant 16-17-10079, microprobe—RFBR 16-05-00703a, thermomagnetic measurements—with the support of the RF Government grant No. 14.Z50.31.0017. This study has also been carried out following the plans of the scientific research of the Geological Institute of RAS (for AYK, project no 0135-2016-0009).

References

Arzhannikov S.G., Ivanov A.V., Arzhannikova A.V.,, Demonterova E.I., Orlova L.A., Pigati J.. Stages of formation of Zhom-Bolok lava fields (Eastern Sayan) / Geodynamic Evolution of the Lithosphere of the Central Asian Mobile belt. Irkutsk: IZK SO RAN, 2013.-P.22–24.(in Russian).

Balkhanov V.C, Yu.B. Bashkuyev. Lava eruption on Flat Tolbachik (Kamchatka-2012) // Journal of Technical Physics, 2015, Volume 85, no. 8, p. 145–146.

Bezrukova E.V., Shchetnikov A.A., Kuzmin M.I., Sharova O.G., Kulagina N.V., Letunova P.P., Ivanov E.V., Kraynov M.A., Kerber E.V., Filinov I.A., Levina O.V. First Data on the Environment and Climate Change within the Zhom-Bolok Volcanic Field (Eastern Sayan Mountains) in the Middle–Late Holocene // Doklady Earth Science. 2016. Vol. 468, No. 3. P. 527–531.

Bozorth R.M. Ferromagnetism. D.Van Nostrand Comp. Inc.. Toronto-New York-London. 1951. 784 pp.

Evans M.E. Magnetoclimatology: a test of the wind-vigour model using 1980 Mount St. Hellens ash. // Earth. Planet. Sci. Lett. 1999. V.172. No 3–4. P.255–259.

Fedotov S.A. The large fissure Tolbachik eruption. Kamchatka 1975–1976. Moscow: Nauka, 1984. 638 p.

Karpov G.A. Native metals in volcanic ashes // Science in Russia. 2014. №4 (202). P.19–28. (in Russian).

Korchagin O.A. Fossil micrometeorites, microtektites and microcyistites: research methodology, classification and impact-stratigraphic scale. In: Proc. The stratigraphy at the beginning of the XXI century - trends and new ideas. Essays on Russian regional geology. M .: Geolkart-GEOS. 2013. Iss.6. P 112–142 (in Russian).

Lebedev E.B., Khitarov N.I. Physical properties of igneous melts. Moscow: Nauka, 1979. 200 p.

Pechersky D.M. Distribution of particles of native iron and Fe-Ni alloys on planets / / ed. Saarbrücken: Palmarium Academic Publishing. 2015. 56 pp.

Pechersky D.M., A.Yu. Kazansky, G.P. Markov, A.A.Tselmovich, A.Shchetnikov. The rare phenomenon of accumulation in the lake sediments of particles of metallic iron of terrestrial origin: the Jombolok volcanic region, the East Sayan. // Physics of the Earth, 2018, №1. FROM.

Pechersky D.M., Bagin V.I.. Brodsky S.Yu., Sharonova Z.V. Magnetism and the conditions for the formation of igneous rocks. L.: Nauka, 1975. 288 pp. (in Russian).

Pechersky D.M., Didenko A.N. Paleoasianic ocean. Izd. UIPE, 1995, 296 pp.

Pechersky, D. M., Gil'manova, D. M., Markov, G. P., Murdmaa, I. O., Nurgaliev, D. K., Tsel'movich, V. A. and Sharonova Z. V. Native Iron and Other Magnetic Minerals in the Sediments of the Northwestern Atlantic: Thermomagnetic and Microprobe Evidence. *Izvestiya, Physics of the Solid Earth,* 2013, 49, 426–447.

Persikov E.S. Viscosity of igneous melts. M .: Nauka, 1984. 160 p.

Shchetnikov A.A., Bezrukova E.V., Filinov I.A., Ivanov E.V., Kerber E.V.. Lake morpholitogenesis in the "Valley of Volcanoes" (Zhom-Bolok lava field, Eastern Sayan) *// Geography and natural resources, 2016, No. 3, P. 37–48.*

Shchetnikov AA, Bezrukova EV, Kerber EV, Belozerova O.Yu., Kuzmin MI, Ivanov EV, Krainov MA, Filinov IA The first results of tephrochronological studies of Late Pleistocene-Holocene volcanic eruptions in the valley of the Zhom-Bolok River (Eastern Sayan) // DAN, 2018 (in press).

Yarmolyuk V.V., Nikiforov A.V., Ivanov V.G. Structure, composition, sources and mechanism of valley outpouring of Zhom-Bolok lava flows (Holocene, South Baikal volcanic region) // Volcanology and seismology, 2003, No. 5, p. 41–59.

Part III
Magnetostratigraphy

Chapter 23
Magnetostratigraphy of the Reference Sections of the Cisuralian Series (Permian System)

Yu. P. Balabanov, R. Kh. Sungatullin, G. M. Sungatullina, L. R. Kosareva, M. S. Glukhov, P. G. Yakunina, A. O. Zhernenkov, V. V. Antonenko and A. A. Churbanov

Abstract For the first time, paleomagnetic and rock-magnetic studies of the Dal'ny Tulkas and Mechetlino sections located in the Southern part of the Cisuralian foredeep have been performed and proposed as international standards for the global correlation (GSSP) of the lower boundaries of the Artinskian and Kungurian Stages. It was determined that the rocks of the sections are weakly magnetic (J_n—0.13–6.08 × 10^{-3} A/m, æ—−0.6–21.8 × 10^{-5} SI units). The increase of the magnetic characteristics occurs in the following direction: limestone → carbonate concretions → calcareous argillites → argillites → sandstones. There were detected no inversions of the magnetic pole in the frontier sections of the Kiaman reversed polarity hyperchron. Paleomagnetic and rock-magnetic data were compared with the results of paleontological studies, which allowed a paleofacial reconstruction of depositional environment in the Cisuralian Epoch. It is confirmed that paleomagnetic data can be an important additional element when considering sections as reference objects of the Global Boundary Stratotype Section and Point (GSSP).

Keywords Permian · Cisuralian · Paleomagnetism · Lithology Geochemistry · Magnetostratigraphy · Reference sections · GSSP Dal'ny tulkas · Mechetlino · Cisuralian foredeep

Yu. P. Balabanov (✉) · R. Kh. Sungatullin · G. M. Sungatullina · L. R. Kosareva M. S. Glukhov · P. G. Yakunina · A. O. Zhernenkov · V. V. Antonenko A. A. Churbanov
Kazan (Volga Region) Federal University, Kremlyovskaya St, 18, Kazan, Russia 420008
e-mail: balabanov-geo@mail.ru

D. Nurgaliev et al. (eds.), *Recent Advances in Rock Magnetism, Environmental Magnetism and Paleomagnetism*, Springer Geophysics,
https://doi.org/10.1007/978-3-319-90437-5_23

Introduction

In 2015, the XVIII International Congress on Carboniferous and Permian Systems was held in Kazan (Southern Urals 2015). The congress was followed by a geological excursion to the sections of the Lower Permian located on the western slope of the Southern Urals in the territory of the Republic of Bashkortostan. The participants got acquainted with the sections—potential candidates for the international standards of the lower boundaries of the Cisuralian Series of the Permian System of the International Chronostratigraphic Chart (ICS)—Mechetlino and Dal'ny Tulkas (Fig. 23.1). The latter section (Gafuri area of Bashkortostan) represents the boundary deposits of the Sakmarian and Artinskian Stages, the Mechetlino section (Salavatsky area) of the Artinskian and Kungurian Stages respectively. According to Russian and foreign stratigraphers (Chuvashov et al. 2011, 2013; Chernykh et al. 2012; Kotlyar et al. 2016), the above sections have good chances of recognizing them as reference objects for global correlation in the status of the so-called "golden nails" of (the Global Boundary Stratotype Section and Point—GSSP) International Stratigraphic Chart.

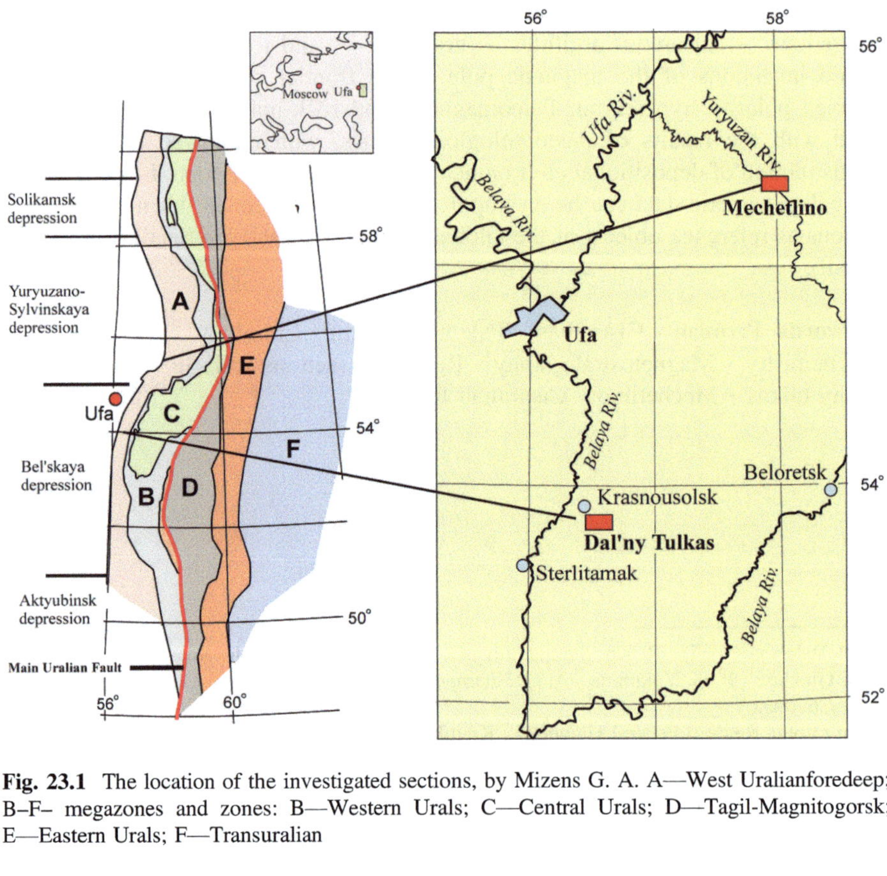

Fig. 23.1 The location of the investigated sections, by Mizens G. A. A—West Uralianforedeep; B–F– megazones and zones: B—Western Urals; C—Central Urals; D—Tagil-Magnitogorsk; E—Eastern Urals; F—Transuralian

During the excursion it was revealed that the Uralian sections do not fully meet the modern requirements of the International Stratigraphic Commission for stratotypes of the lower stage boundaries as GSSP objects. They require substantial additional studies, including the uncovering of additional intervals of the section, the conduct of geochemical and isotopic studies, the obtaining of U-Pb zircon ages, and the additional justification of the boundaries by various groups of fauna and flora. Furthermore, no paleomagnetic and rock-magnetic studies have been carried out on these sections, which are an integral part of the correlation of standard stratigraphic objects, and are also of great importance for studying paleogeography and tectonic reconstructions (Golovanova et al. 2017). In order to conduct an integrated exploration of the Mechetlino and Dal'ny Tulkas sections, the additional stripping of rocks was done in 2016 (Fig. 23.2), followed by detailed sampling for various types of analyzes.

Fig. 23.2 Dal'ny Tulkas (at the top) and Mechetlino (at the bottom) sections. Left side—during International Congress, 2015, right—after stripping in 2016

General Characteristics of Objects

The investigated sections are located in the axial part of the Cisuralian Foredeep (see Fig. 23.1) and belong to the classical flysch formation (Misens 1997; Tevelev et al. 2016). They are characterized by the absence of depositional breaks and they are studied in detail in a biostratigraphic respect (Chuvashov et al. 1990; Chuvashov and Chernykh 2011; Southern Urals 2015).

The Dal'ny Tulkas section is exposed in a roadside quarry on the right bank of the same named river—the left tributary of the Usolka River flowing into the Belaya River. The section is represented by a series of terrigenous-carbonate rocks (calcareous argillites, loamy slightly dolomitized limestones) of the Sakmarian and Artinskian Stages (Fig. 23.3). The layers of carbonate and carbonate-argillaceous rocks are separated by thin (up to the first centimeters) interlayers and films of argillites. The interlayers of organogenic detritus limestones are found in the Artinskian part of the section. The lenses and lenticular interlayers of limestone concretions are present at different levels. Numerous plant remains are found throughout the section, scales and whole specimens of fish are common. There are many radiolarians in the carbonate-argillaceous rocks. Organogenic-detrital limestones contain fragments of crinoid segments, brachiopod shells, bryozoans, foraminifera and shells of cephalopods in some places. The thickness of the considered interval of the section is 13.8 m. The 8 m belong to the Sakmarian Stage, and 5.8 m to the Artinskian. This section is considered as a GSSP reference object of the lower boundary of the Artinskian Stage.

The Mechetlino section (see Fig. 23.1) is located on the right valley slope of the Yuryuzan river and it is a candidate for the global stratotype of the lower boundary (GSSP) of the Kungurian Stage of International Stratigraphic Chart (Chuvashov and Chernykh 2011. The carbonate-terrigenous deposits of the upper part of the Artinskian and lower parts of the Kungurian Stage (see Fig. 23.3) with a total thickness of 10 m are exposed in a small quarry here. Therefore, upon the recommendation of the participants of the International Congress on Carboniferous and Permian, the Artinskian sediments were stripped down the slope (see Fig. 23.2). The thickness of the exposed interval reached 16.4 m (including 11.8 m in the Artinskian Stage, and 4.6 m in the Kungurian Stage). The section is composed of turbidites, lying among clayey, carbonate-argillaceous and sandy-argillaceous rocks. Sandstones are from fine—to coarse grained quartz, greywackes by composition. The organic remains (foraminifera, fragments of algae, crinoids, brachiopods, bryozoans) are rather common, there are many plant residues. Carbonate layers are represented by organogenic detritus limestones with diverse remains of fauna and flora: fragments of bryozoans, crinoids, needles of sea urchins, fragments of brachiopod shells, foraminifera, algae, calamites, sponge spicules, there are conodonts, ammonoids, fishes.

Further studies were conducted in 2016 after uncovering of new intervals in the Dal'ny Tulkas and Mechetlino sections (see Fig. 23.2) (Kotlyar et al. 2016). This article presents the first data on paleomagnetic and rock-magnetic studies of

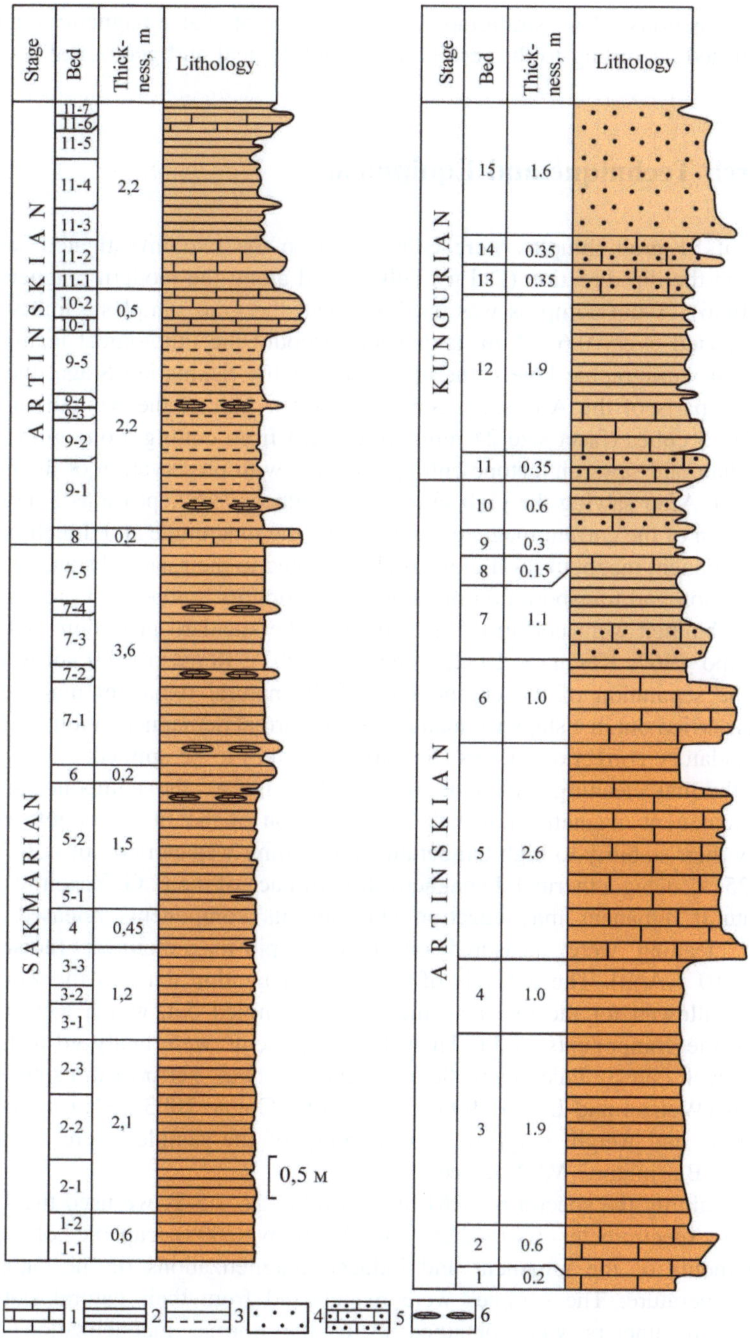

Fig. 23.3 Lithological columns of the Dal'ny Tulkas (left) and Mechetlino (right) sections. 1—limestones, 2—argillites, 3—calcareous argillites, 4—sandstones, 5—calcareous sandstones, 6—carbonate concretions

reference sections. The stratigraphic referencing of paleomagnetic data was accomplished according to the results of paleontological and geological research.

Research Technique and Equipment

A total of 51 grab samples were collected from the two investigated sections: oriented in the dip direction (Dal'ny Tulkas) and along the modern magnetic field (Mechetlino). A dip compass was used to orient the grab samples. Grab samples were collected every 0.6–1.5 m, taking into account the lithological features and represent 4 stratigraphic levels: the upper part of the Sakmarian Stage, the lower and upper parts of the Artinskian Stage, the lower part of the Kungurian Stage. From 3 to 5 cubes (facet size 24 mm) were taken from cutting a one grab sample using a diamond saws on a stone-cutting machine with preservation of the primary orientation. After sawing the grab samples, the cubes were kept for 1–2 months in the direction of the current magnetic field so that the magnitude and direction of the natural remanent magnetization of the rocks returned to their natural state, disrupted after selection and transportation of samples. Laboratory studies and processing of the data obtained were performed according to the standard procedure (Khramov and Sholpo 1967; Khramov 1982; Zijderveld 1967). Remasoft 3.0 software was used. The separation of the components of the natural remanent magnetization (Jn) was carried out in a stepwise manner using thermal demagnetization techniques with mandatory correction factors (for dip and strike) to be applied.

The thermal cleaning technique was used to isolate the components of the natural remanent magnetization with the separation of the Jn vector into components by their stability to high temperatures. Cleaning was carried out in 6–7 steps up to 325 °C using a thermal demagnetization furnace of the P. G. Yasonov design. The natural remanent magnetization (Jn) and its components released during thermal cleaning were measured on a JR-4 spin-magnetometer (noise level 0.005×10^{-3} A/m). The analysis of the demagnetization data on the Zijderveld diagrams allowed for the linear segments to be singled out, which were used to calculate the components of Jn. Then the components were analyzed at the collection level (sites). Paleomagnetic data analysis was performed using special programs (Watson and Enkin 1993; Enkin 1994; Cogne 2003). Measurements of the volume and weight magnetic susceptibility of the samples were carried out using the "Bartington" MS2 device.

For obtaining the spectra of IRM magnetization to 1.5 T, we used the J_meter coercivity spectrometer (Nurgaliev and Yasonov 2009) capable of separate measurements of the remanent and induced magnetizations of the samples at room temperature. The samples were magnetized from their natural state. The following parameters were obtained using the normal magnetization curves: normal remanent saturation magnetization (Jrs), saturation magnetization, excluding the effect of the paramagnetic component (Js), the magnetization of the paramagnetic component in the field of 0.5 T (Jp), the coercivity force after removing the effect of the paramagnetic component (Bc), the remanent coercivity force (Bcr), the position of maximum coercivity spectra of NRM (Bda), the position of maximum coercivity spectra of NRM demagnetization from saturated state (Bdb).

In order to determine the nature of changes of magnetic susceptibility we have identified the contribution of the different components—ferromagnetic ($æ_f$), paramagnetic and diamagnetic ($æ_p$), as well as superparamagnetic ($æ_{sp}$). Each component has a different nature of origin related to sedimentation processes changes. Methods of determining the contribution of paramagnetic component to the total signal by inductive normal magnetization curve, determining ferromagnetic component by inductive normal magnetization curve minus paramagnetic component, determining the superparamagnetic component by Jr curve and its total value were applied to all samples of core column (Kosareva et al. 2015).

Thermomagnetic analysis is a main type of diagnosis of ferromagnetic fraction's composition in rocks. For conducting the thermomagnetic studies samples were taken from different depths of the cross sections. Using the Curie express-balance (Burov et al. 1986), we measured the temperature dependence of induced magnetization at a heating rate of 100 °C per minute up to 800 °C. The curves were measured at intensity of the magnetic field 0.2 T. The curves of the first and second heating enabled us to estimate the mineralogical transformations in a sample. The second curve was measured on the same sample after its cooling to room temperature.

To study the internal structure of magnetite microspheres, the X-ray computed tomography method was used. This was accomplished using X-ray microtomograph Phoenix V | tome | X S 240, equipped with two X-ray tubes: a microfocus tube with a maximum accelerating voltage of 240 kV/320 W and a nanofocus tube with a maximum accelerating voltage of 180 kV/15 W. The CT scanning was conducted at an accelerating voltage of 70 kV and a current of 170 mA. The scanning resolution was 0.9 μm.

The Results of Paleomagnetic Research

Paleomagnetic studies were carried out in the rocks of the boundary (Sakmarian-Artinskian and Artinskian-Kungurian) intervals. A total of 34 grab samples from the Dal'ny Tulkas section and 17 grab samples from the Mechetlino section (Tables 23.1 and 23.2) were studied. The 99 and 52 cubes were produced respectively from these grab samples.

Rock magnetic properties. The studied Sakmarian-Artinskian and Artinskian-Kungurian boundary sediments have a relatively small value of Jn. In the Dal'ny Tulkas section, this parameter varies from 0.40 to 5.10×10^{-3} A/m (Table 23.1; Fig. 23.4) with an average value of 2.1×10^{-3} A/m. In the Mechetlino section, it varies from 0.13 to 6.08×10^{-3} A/m (Table 23.2; Fig. 23.5) with an average value of 1.0×10^{-3} A/m. The magnetic susceptibility (æ) of rocks of the Dal'ny Tulkas

Table 23.1 Magnetic characteristics of the Sakmarian-Artinskian deposits of the Dal'ny Tulkas section

Rocks	N	n	æ, 10^{-5} SI units			J_n, 10^{-3} A/m		
			Min	Max	x	Min	Max	x
Argillites	2	23	13.00	18.90	16.12	1.70	4.72	2.81
Calcareous argillites	9	67	5.50	13.60	10.36	1.05	3.42	2.23
Carbonate concretions	2	7	2.00	5.00	3.69	0.50	0.83	0.62
Limestones	2	7	−0.60	1.80	0.49	0.40	5.10	1.25

Note N is the number of layers; n is the number of cubes, x is the average value

Table 23.2 Magnetic characteristics of Artinskian-Kungurian deposits of the Mechetlino section

Rocks	N	n	æ, 10^{-5} SI units			J_n, 10^{-3} A/m		
			Min	Max	x	Min	Max	x
Calcareous sandstones	6	24	12.90	21.80	18.14	0.50	2.62	1.40
Argillites	1	5	11.90	16.60	14.34	0.21	1.16	0.72
Calcareous argillites	4	13	4.10	10.10	6.06	0.13	6.08	1.20
Limestones	4	10	−0.30	3.30	1.48	0.23	0.76	0.44

Note N is the number of layers; n is the number of cubes, x is the average value

section varies from -0.6 to 18.9×10^{-5} SI units (Table 23.1; Fig. 23.4) with an average value of 9.71×10^{-5} SI units. In the Mechetlino section æ varies from -0.3 to 21.8×10^{-5} SI units (Table 23.2; Fig. 23.5) with an average value of 11.3×10^{-5} SI units. Thus, all the studied rocks are weakly magnetic.

A positive correlation relationship is established, with the exception of single samples, between the values of Jn and æ (Figs. 23.6, 23.7, 23.8 and 23.9). It is most clearly demonstarted in the Dal'ny Tulkas section, where the paired correlation coefficient between the magnetic parameters is 0.64.

Magneto-mineralogical investigations. In order to determine the carriers of magnetization and the nature of magnetic fraction formation, a set of magneto-mineralogical studies was carried out.

Analysis of Day diagram (Dunlop 2002) reveals homogeneity of the ferromagnetic fraction of cross section Mechetlino (Fig. 23.10). The magnetic material is represented by pseudo- single-domain and multidomain material with a noticeable influence of superparamagnetism, and to a lesser extent by single-domain grains. This suggests that the main carriers of sediment magnetization are terrigenous minerals.

Differential thermomagnetic analysis (DTMA) of the samples of Mechetlino section (Fig. 23.11) leads to conclusion that the content of ferromagnets in the studied rocks is low. As an example, the results for 2 samples are given. The sample 2-M (layer 2) characterizes limestones with higher-than-usual value of æ_pairs. According to the DTMA data, the sample contains paramagnetic iron compounds. The sample 3502-3 M (layer 3) is represented by argillites with low values of

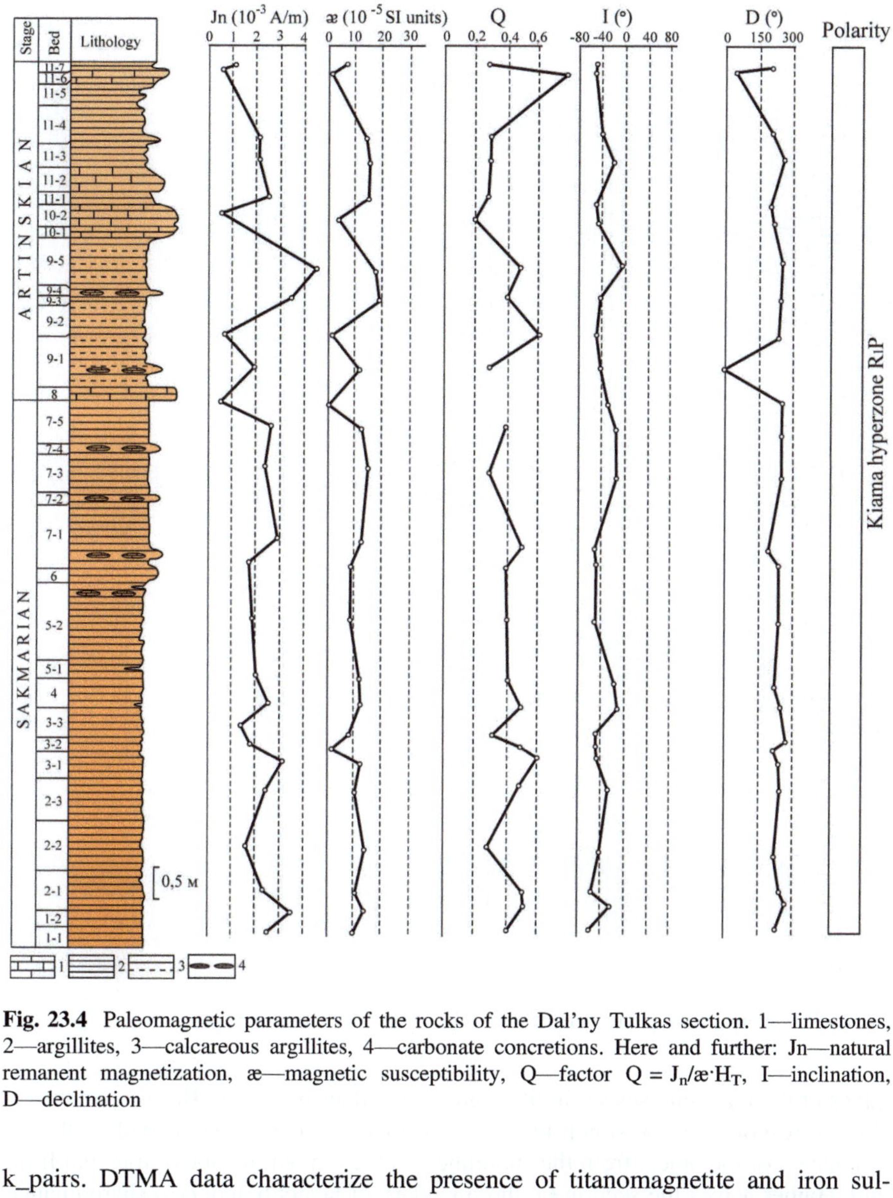

Fig. 23.4 Paleomagnetic parameters of the rocks of the Dal'ny Tulkas section. 1—limestones, 2—argillites, 3—calcareous argillites, 4—carbonate concretions. Here and further: Jn—natural remanent magnetization, æ—magnetic susceptibility, Q—factor Q = J_n/æ·H_T, I—inclination, D—declination

k_pairs. DTMA data characterize the presence of titanomagnetite and iron sulphides (pyrite) in the sample. In addition, a magnetite microsphere has been found in layer 12 (Fig. 23.12), the origin of which we associate with the deposition of cosmic dust (Sungatullin et al. 2017) or impact events.

The analysis of variations in the values of magnetic susceptibility and coercive parameters in Mechetlino section (Fig. 23.13) indicates a good correlation between lithological, paleontological and rock-magnetic parameters. By these parameters, the zone of the deep sea basin is clearly distinguished, characterized by decreased

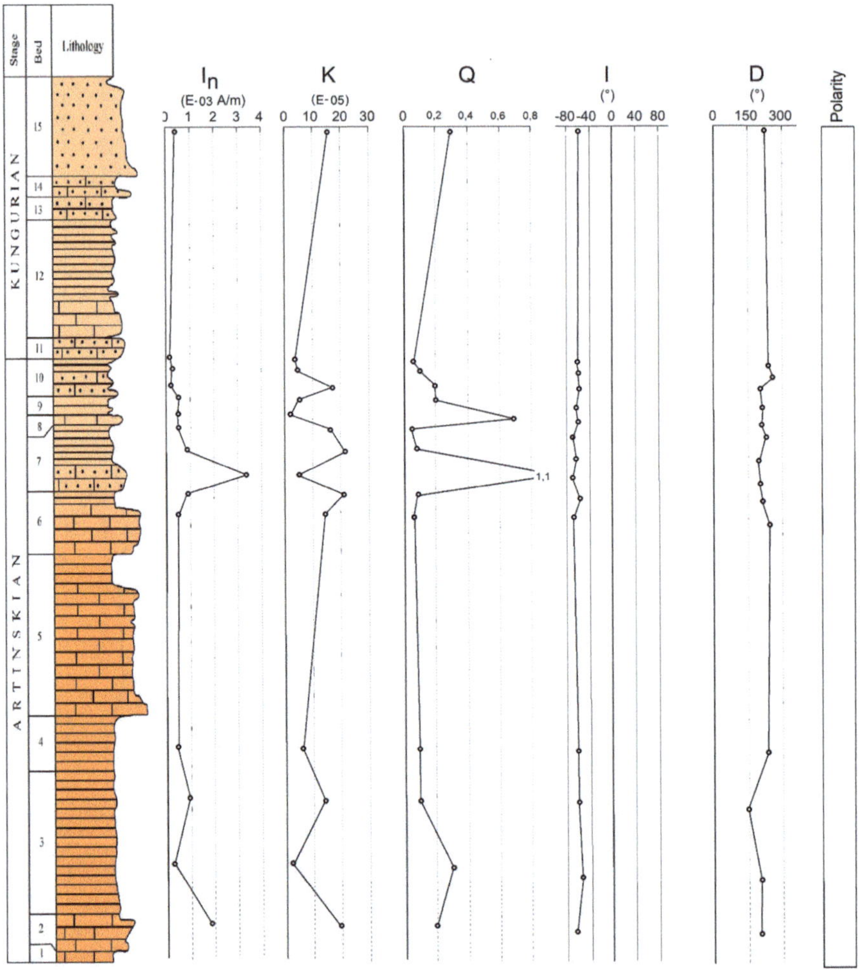

Fig. 23.5 Paleomagnetic parameters of the rocks of the Mechetlino section

values of the magnetic susceptibility, enhanced values of Bc and Bcr/Bc parameter. This corresponds to a small input of magnetic minerals associated with the remoteness of the object from the shoreline and the limited income of material from land. Higher across the section (at the boundary of layers 6 and 7), a sharp change in the magnetic parameters is recorded, which is apparently due to the beginning shallowing of the sea basin and an increased input of terrigenous material with magnetic minerals. Various components of the magnetic susceptibility, calculated using coercive spectrometry, also indicate a transfer from deep-sea to a more shallow-water sedimentary environment (Fig. 23.14). The paramagnetic component ($æ_p$) characterizes the clay input from the shore. The behavior of the ferromagnetic and superparamagnetic components in the upper part of the section indicates an

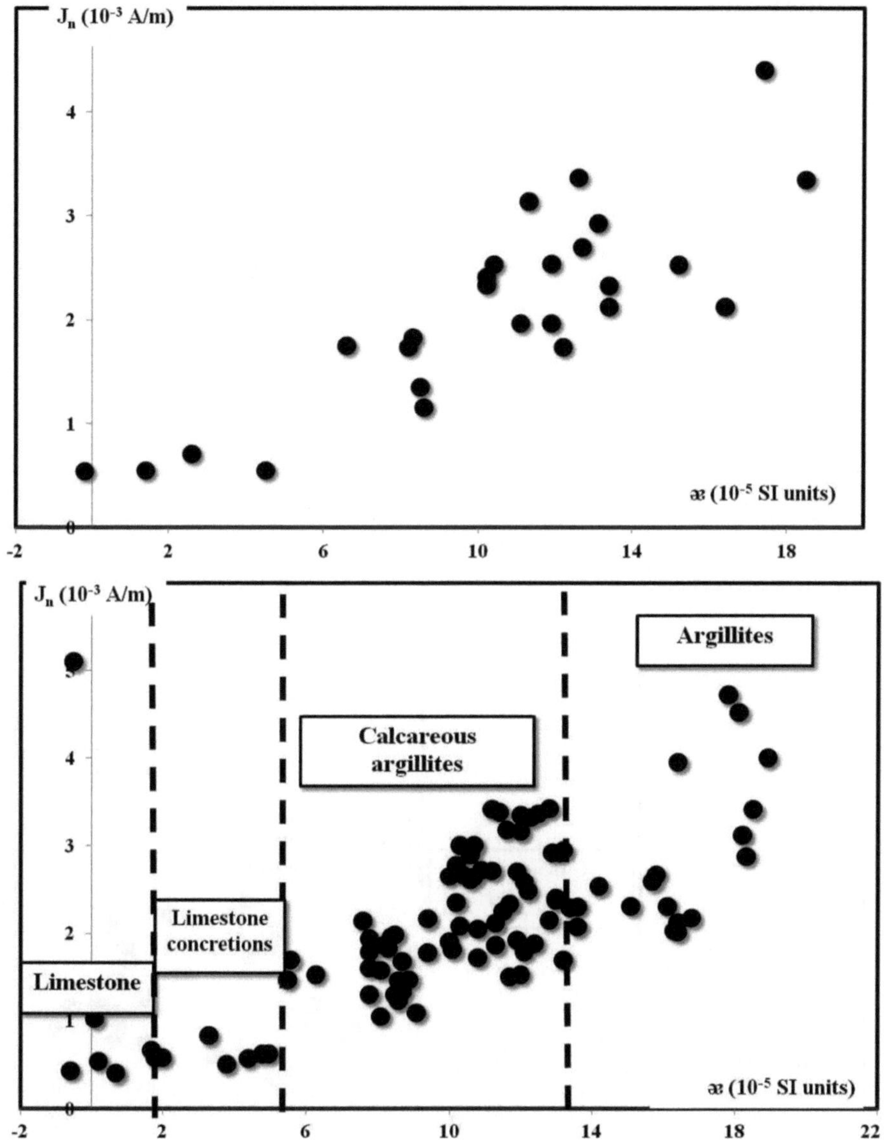

Fig. 23.6 Magnetic susceptibility and remnant magnetization. Dal'ny Tulkas section. At the top —averaged data for 27 samples, at the bottom—ordinary data for 99 cubes

increase in the amount of the biogenic magnetic component in this interval. This may be due to the climate warming and development of biota. However, the study of this section requires additional research by both magnetic and other paleoecological (geochemical, faunistic, palynological) methods.

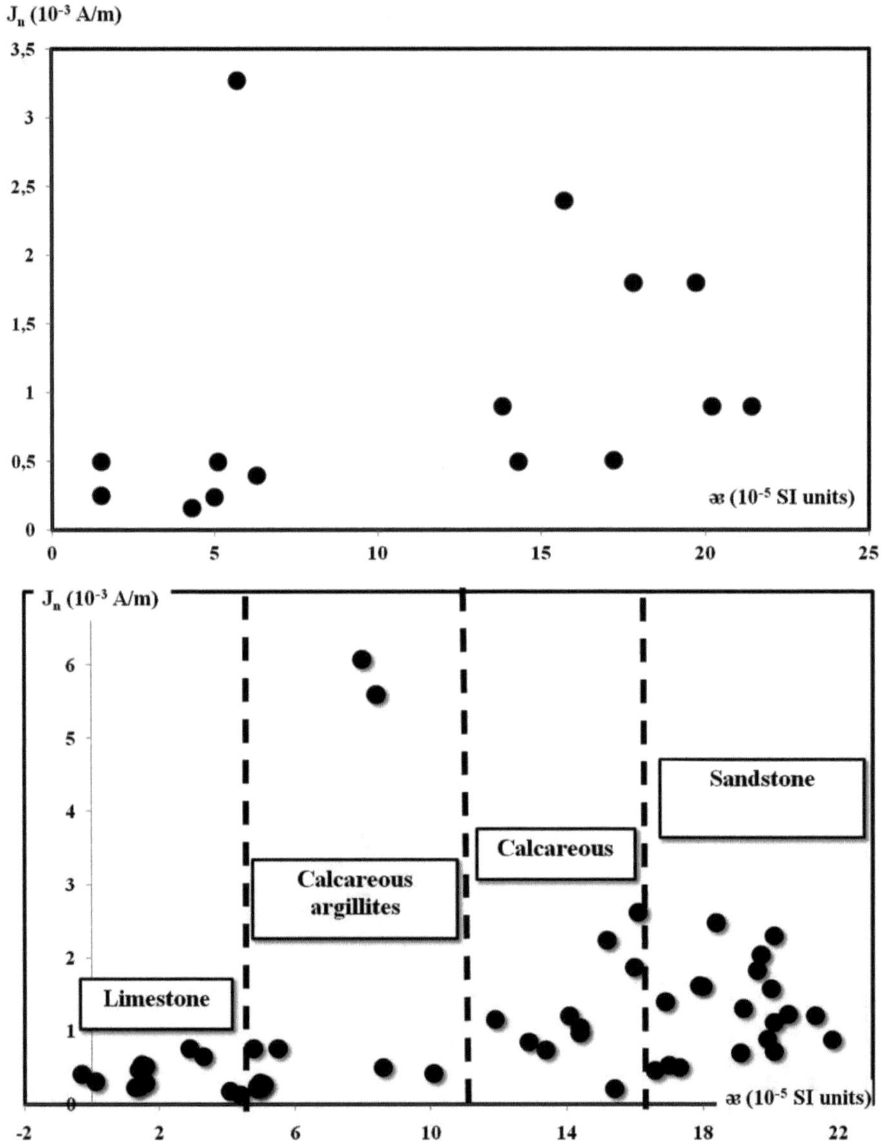

Fig. 23.7 Magnetic susceptibility and remnant magnetization. The Mechetlino section. At the top—averaged data for 15 samples, at the bottom—ordinary data for 52 cubes

Despite the low magnetization of the samples of the Mechetlino section, magnetic investigations have made it possible to single out the stages of changes in sedimentary environment at the boundary of Artinskian and Kungurian Stages. The stage of deep-sea sedimentary environment corresponds to the Artinskian age (layers 1–6). The shallow-water conditions existed during Kungurian age, when

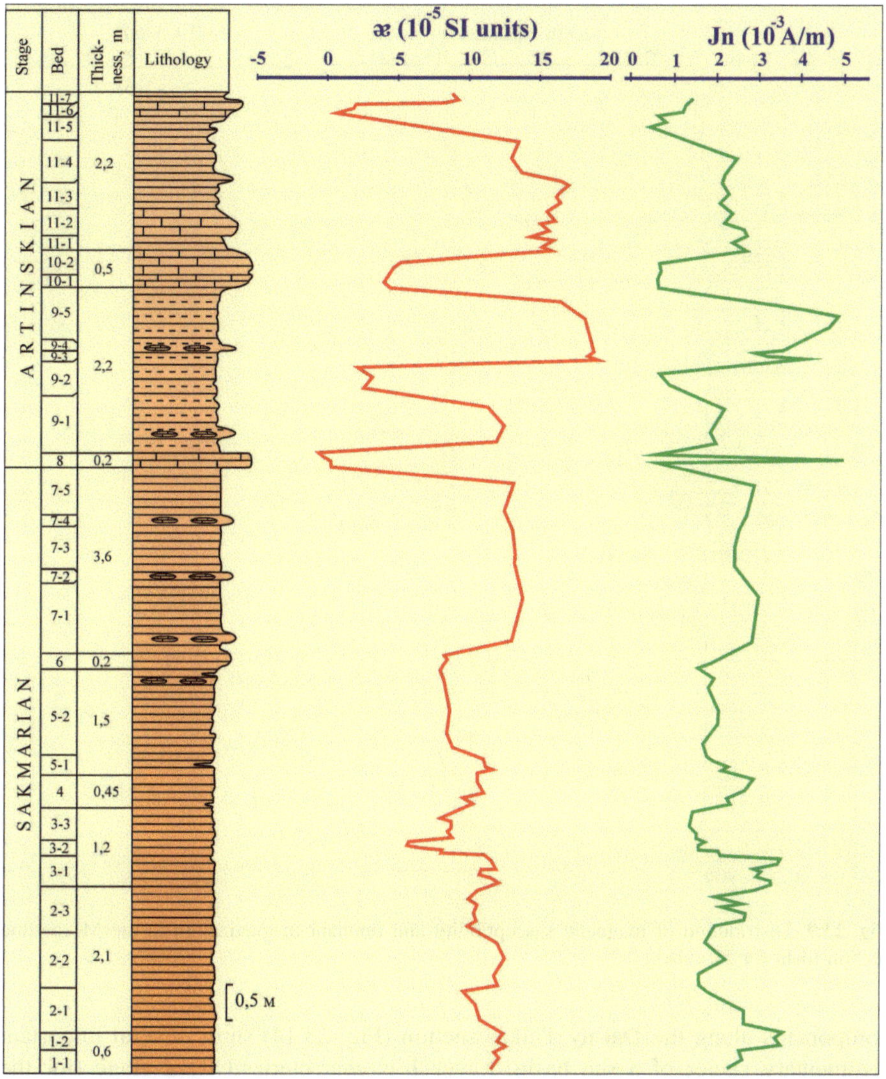

Fig. 23.8 Distribution of magnetic susceptibility and remnant magnetization of the Dal'ny Tulkas section (data for 99 cubes)

aridization of the climate began. Layer 15 is marked by an increase of the ferromagnetic and superparamagnetic material, which is most likely due to an increase in the population of magnetotactic bacteria (Kirschvink and Gould 1981), which are sensitive to the slightest changes in the environment.

The Day plot for the Dal'ny Tulkas section (Fig. 23.10) is generally identical to the Mechetlino section. Hysteresis parameter variations of the Dal'ny Tulkas section are shown in Fig. 23.13. The variations of the magnetic susceptibility

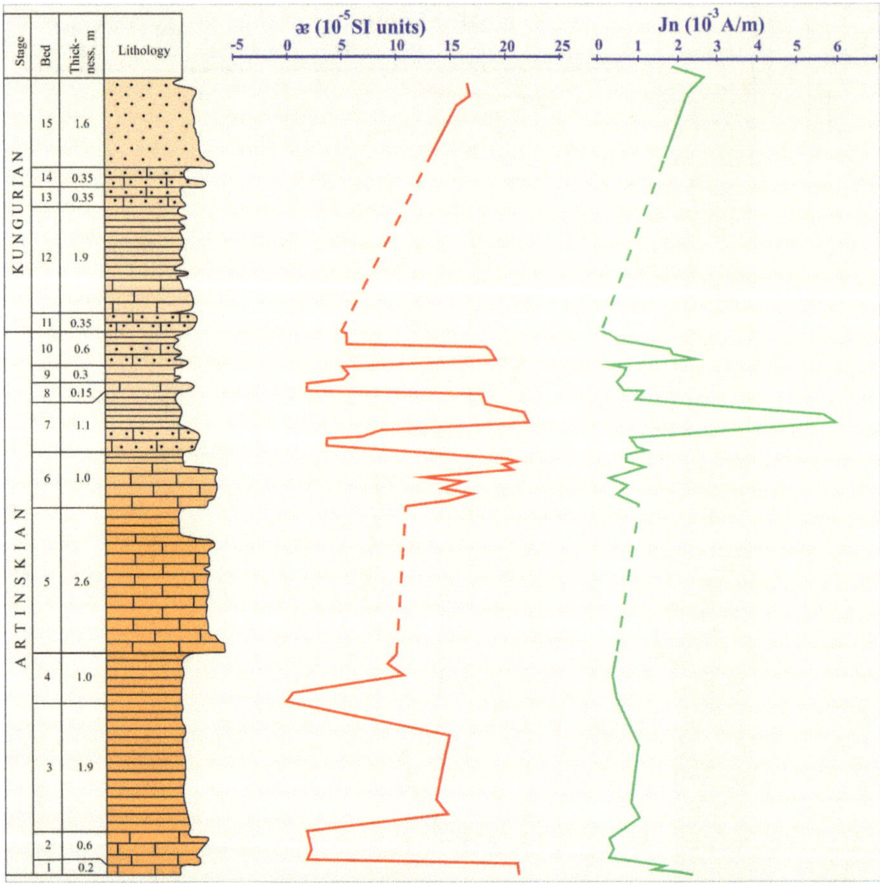

Fig. 23.9 Distribution of magnetic susceptibility and remnant magnetization in the Mechetlino section (data for 52 cubes)

components along the Dal'ny Tulkas section (Fig. 23.14) show several important sedimentary stages of a sea basin. Layer 1 is characterized by average (for the section in total) values of magnetic susceptibility and rock-magnetic parameters, which indicates the stability of sedimentary processes in coastal-marine conditions. Layer 2 shows an increase in the ferromagnetic and superparamagnetic components, which is probably due to the presence of significant amount of biogenic magnetic component. Apparently, the climatic conditions for development of the colony of magnetotactic bacteria and preservation of magnetofossils were optimal. Layer 4 is characterized by a slight increase in æ, which indicates an increase in the input of clastic matter. The coastal zone or shallow warm creeks, estuaries were preserved in the area of the Dal'ny Tulkas section. The massive accumulation of radiolarians indicates a profuse silica input into the sea basin due to active volcanic activity. The same conditions are also characteristic for the layers 5 and 6.

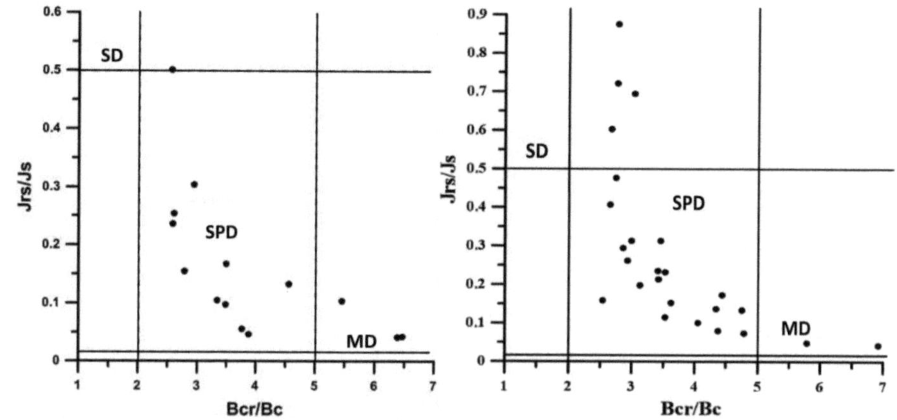

Fig. 23.10 Day plot. At the left—Mechetlino section, at the right—Dal'ny Tulkas section Particles: SD—single domain, PSD—pseudo-single domain, MD—multidomain

Layers 7 and 8 are characterized by lower values of æ and a paramagnetic component, which indicates a decreased input of terrigenous material and an increase in the sea depth. However, here it is necessary to single out an interlayer 7-2. It is defined by the increase in the value of æ and the paramagnetic component, as well as by the large content of multidomain particles, which is most likely due to a significant amount of clastic material input. The number of plants decreased. Layer 7 is marked by the appearance of conodonts, which is associated with a deepening (transgression) of the basin and/or a decrease in supply of terrigenous material from the land. For the layer 9 there is an increase in æ and a paramagnetic component, indicating a marine regression. A single-domain fraction predominates in the interlayers 9-3 and 9-4, most likely of biogenic origin, which is fixed by an increased content of the ferromagnetic fraction and enhanced values of the coercive force. In the layer 11-2, the value of the paramagnetic component decreases, which may be due to the input of a diamagnetic fraction. In general, the layers 9-11 show a tendency for a gradual increase in the magnetic susceptibility, which, in turn, corresponds to a decrease in sea level.

The direction and origins of the natural remanent magnetization of rocks. According to the pattern of distribution of the Jn vectors, the rocks of the Dal'ny Tulkas section differ from the formations of the Mechetlino section (Fig. 23.15). In the first section, the Jn vectors predominate, forming elongated regions on the stereogram characterizing the rocks to a different degree of remagnetization (metastable). We think that the elongated form of the set of directions is not related to the inclination shallowing due to the predominant presence of the superparamagnetic fraction of titanomagnetite in the studied samples (Fig. 23.10), the size of which is known to be less than 0.03 μm (Kirschvink and Gould 1981). The structure of titanomagnetite is cubic, which determines the isometric shape of the grains of this mineral. Therefore, taking into account also marine, i.e. more quiet in

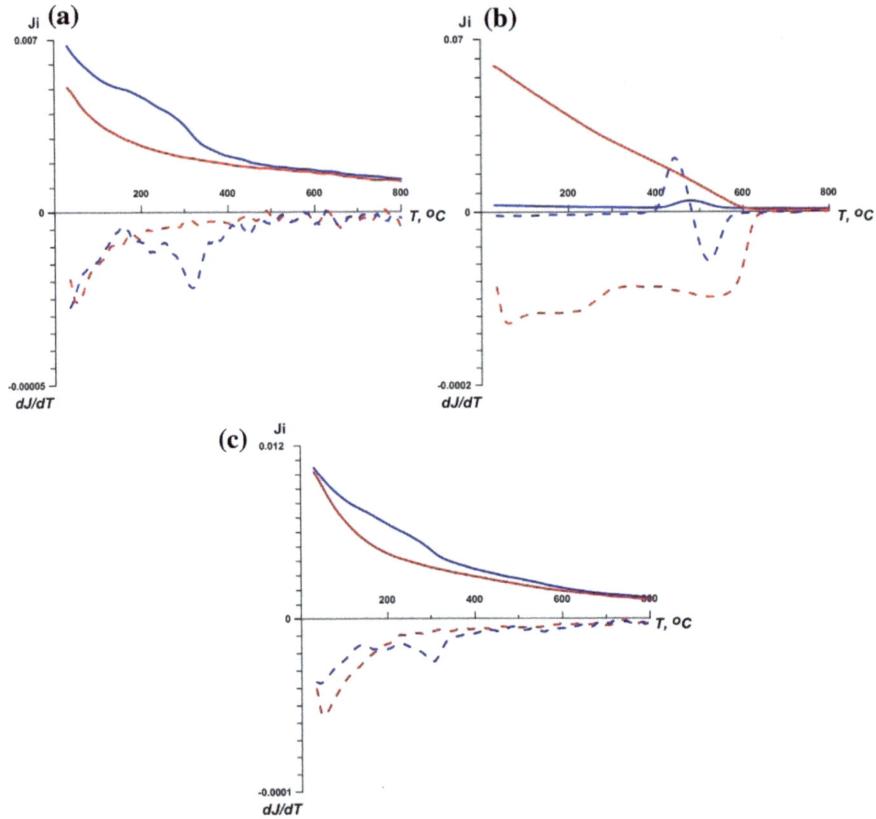

Fig. 23.11 The TMA diagrams in the field of 0.4 T. Integral (at the top) and differential (at the bottom) curves of the first (blue) and second (red) heating. Samples from the Dal'ny Tulkas section: **a** sample 5-1; **b** sample 7-4 and **c** sample 11-3

comparison with the continental depositional conditions of a studied rocks, the coefficient of inclination flattening, in our case will be close to unity.

In the second section, the vectors are grouped around the direction of the primary negative magnetization. The accuracy concentration of the J_n vectors increases substantially after the samples are thermally cleaned to temperatures of 275–300 °C, and the projections of these vectors on the sphere are grouped around the direction of the primary magnetization. The absence of secondary alteration of rocks after their formation is a proof of the primary nature of the Jn's distinguished directions. The characteristic component of magnetization has orientational, and the secondary component—a viscous nature.

After heating to 200–250 °C, an unstable component of magnetization is destroyed, the directions of which are chaotic. Most probably it is the sum of viscous

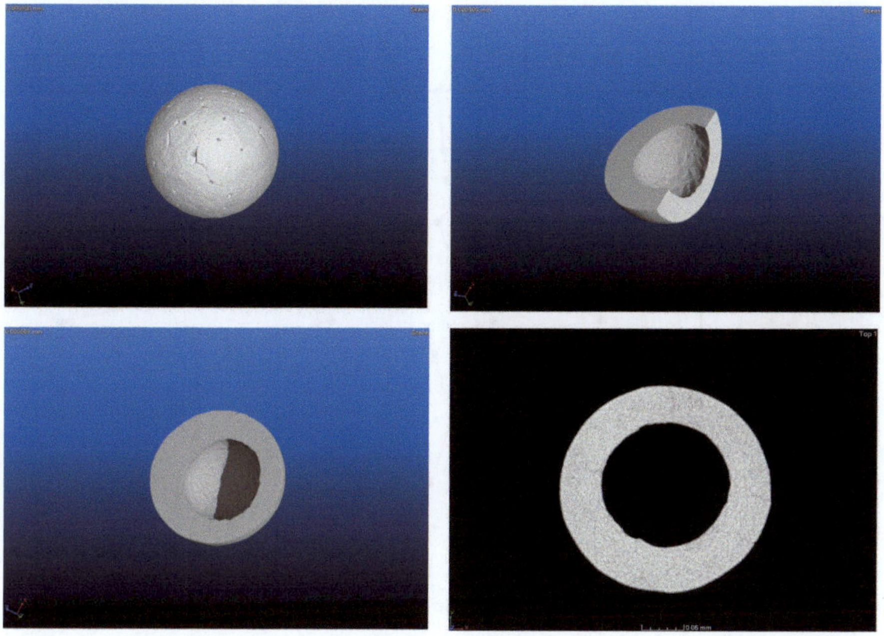

Fig. 23.12 Tomographic (2D and 3D) images of the magnetite microsphere. Mechetlino section. Layer 12. Diameter of microsphere—300 μ

and laboratory magnetizations. In some samples, when heated above 250 °C, other components are not released, however, in the greater part of the collection, the medium-temperature component (MTC) of the magnetization is reliably released in the range of 300–350 °C, often constituting more than 90% of the remanent magnetization. It should be noted that all the studied samples has negative MTC (Figs. 23.4 and 23.5).

Minerals-carriers of magnetization of sedimentary rocks were determined using DTMA, which was performed by measuring induced magnetization on a self-recording torsion balance (Burov and Yasonov 1979). The three diagrams of successive heating up to 800 °C were obtained for the samples. The curves of the differential thermomagnetic dJ/dT analysis are shown in Fig. 23.13. The DTMA curves of sedimentary rocks have a U-shaped form and represent not the Curie points of individual ferromagnetic minerals, but a spectrum of blocking temperatures corresponding to magnetization carriers of a certain dimension. The absence of Curie points in sedimentary rocks is caused by the presence of ferromagnetic mineral grains of different sizes. An analysis of these diagrams shows that the magnetic phase with a Curie temperature of 200–350 °C on the differential curve of the first heating and the 'knee' of the thermomagnetic curve in almost the same

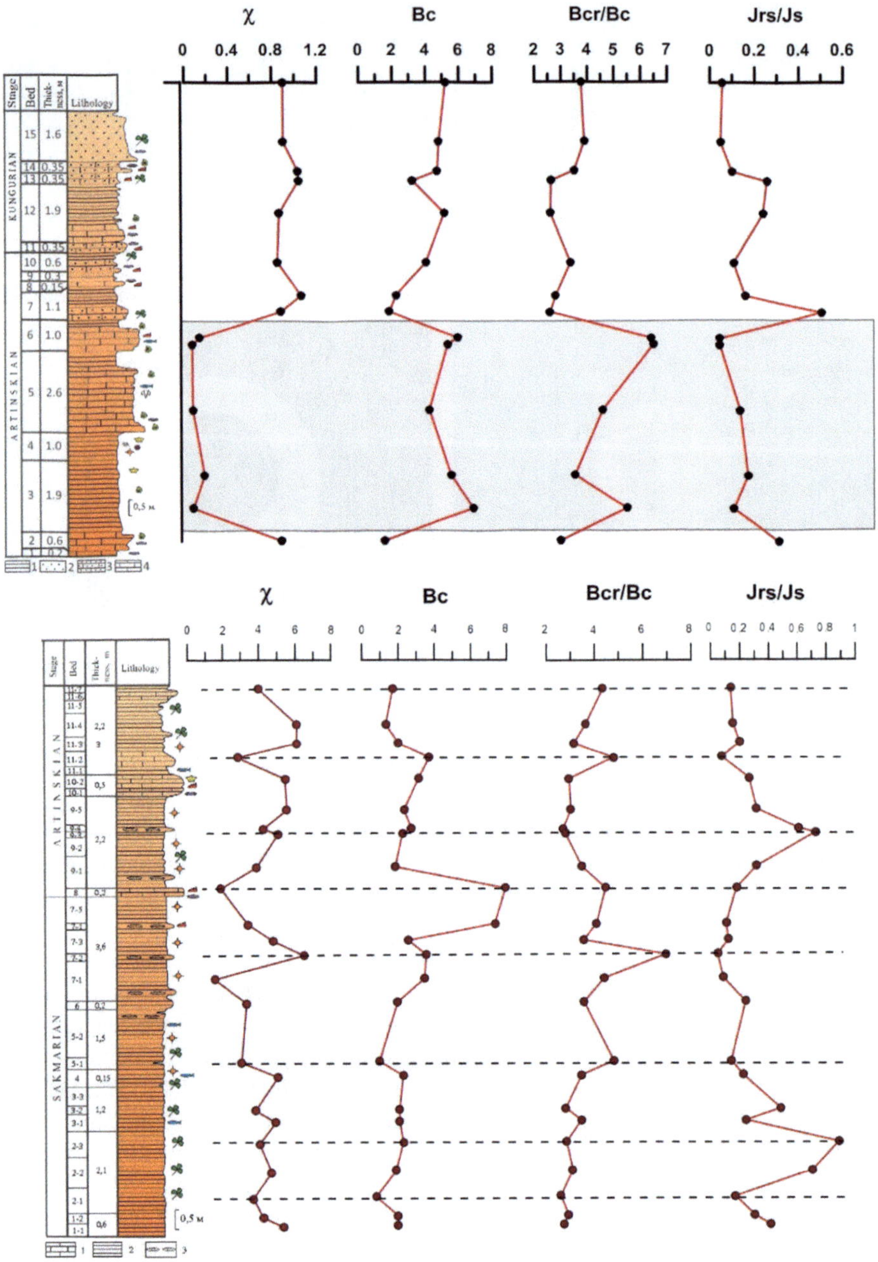

Fig. 23.13 Scheme of comparison of lithological, paleontological data and hysteresis parameters. At the top—Mechetlino section, at the bottom—Dal'ny Tulkas section. Gray filling—deep water basin

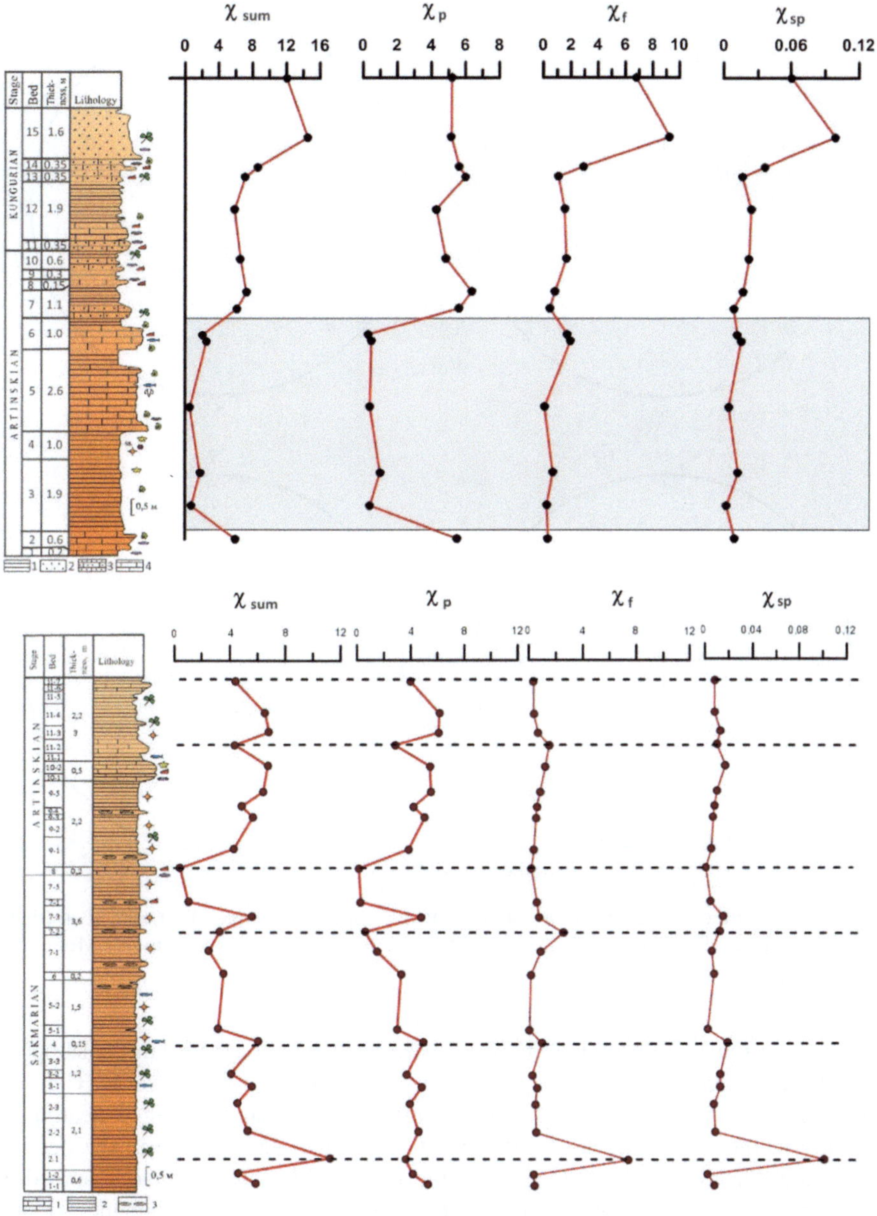

Fig. 23.14 Scheme of comparison of lithological, paleontological data and components of magnetic susceptibility calculated using coercive spectrometry. At the top—Mechetlino section, at the bottom—Dal'ny Tulkas section. Gray filling—deep water basin

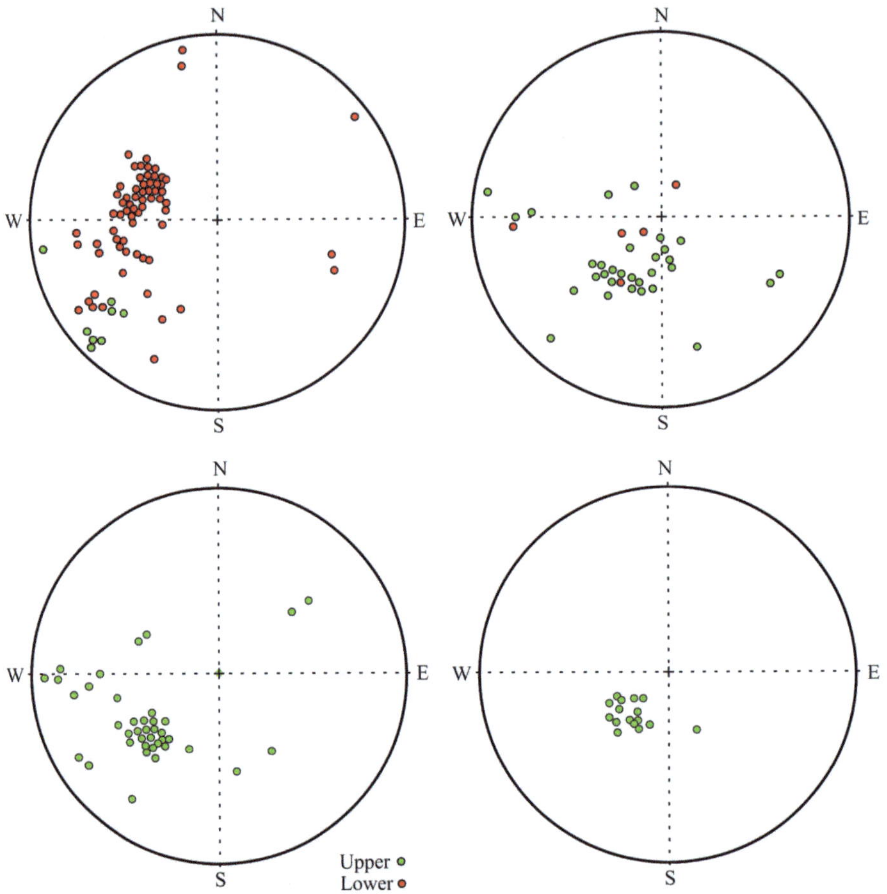

Fig. 23.15 Distribution of directions of the NRM components before (left) and after (right) temperature magnetic cleaning. At the top—Dal'ny Tulkas section, at the bottom—Mechetlino section

temperature range are explained by the presence of titanomagnetite in the samples. Based on the results of thermal demagnetization, the carrier of the MTC in all samples is titanomagnetite. The optimum mode of thermal cleaning is heating of samples to the temperatures of 175–250 °C for 2.5 h. An increase in the heating temperature leads to a sharp decrease in the values of J_n and to a chaotic spread of the remanent magnetization vectors, which apparently is a consequence of the manifestation of the false signal (paramagnetic noise) (Fig. 23.16-3). As you may see in Fig. 23.16-2, about 15–20% of the primary magnetization remains in the

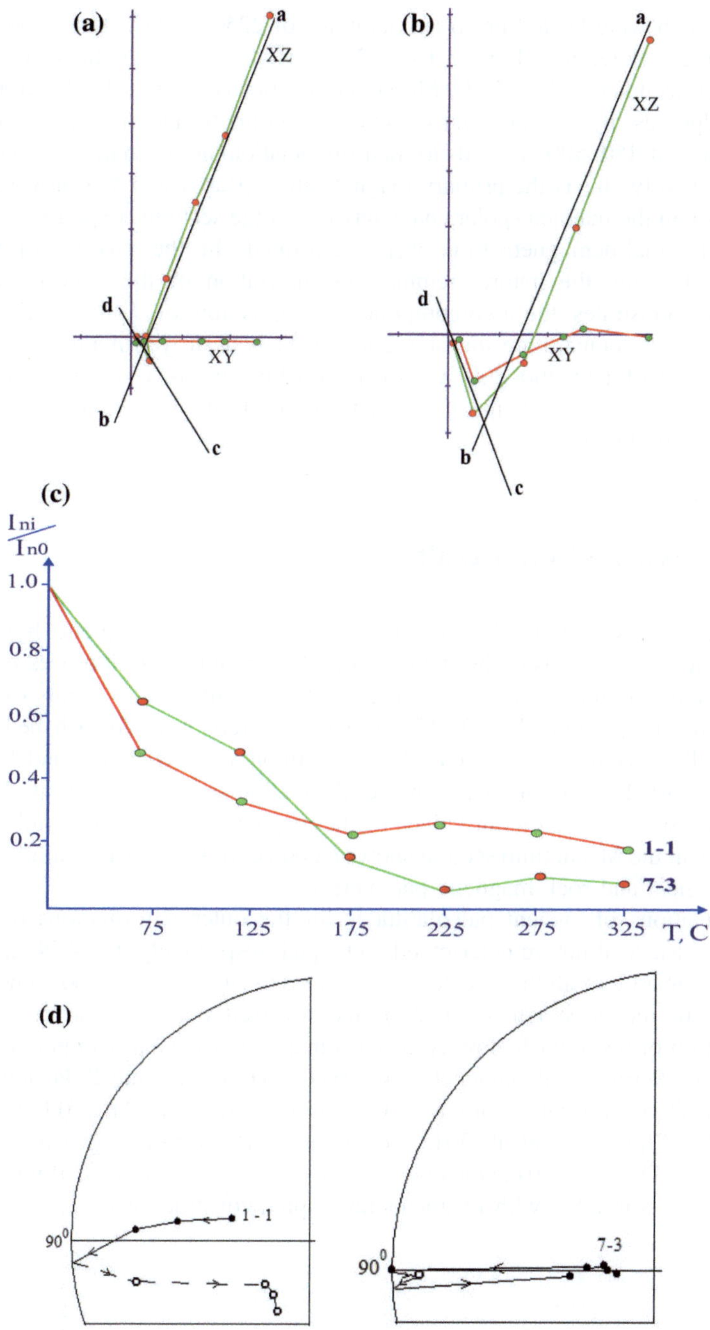

Fig. 23.16 Diagrams of Zijderveld. Dal'ny Tulkas section. **16-1** a) sample 1-1, b) sample 7-3; a-b direction of viscous component of magnetization; c-d direction of characteristic component of magnetization; **16-2** the combined figure; **16-3** stereographic projections of magnetization vectors, sample. 1-1 (left), 7-3- (right)

samples which were heated up to temperatures of 225–275 °C. As can be seen in the Zijderveld diagrams (Fig. 23.16-1, 23.16-3), the yield to the characteristic component occurs at 225–275 °C. Moreover, as practice shows, the disseminations of iron sulphides- pyrite—are very common in marine formations, heating them to temperatures of 450–500 °C leads to their dissociation and formation of magnetite, which negatively affects the primary magnetization (Burov and Yasonov 1979). In order to obtain the magneto-polar characteristics of the sections, experiments on the stepwise thermal-demagnetization were performed. In the process of thermal magnetic cleaning, the natural remnant magnetization of the majority of rock samples demonstrates the two-component composition of J_n (Figs. 23.15 and 23.16). The low-temperature unstable component is destroyed mainly in the temperature range of 100–200–300 °C. Most probably, this is a viscous component. The characteristic component of magnetization of the reverse polarity begins releasing from 200 °C.

The Discussion of the Results

310 million years ago the Kiaman reversed polarity paleomagnetic hyperchron began, which corresponds to the stabilization of internal terrestrial processes during the subsequent around 60 million years. Practically only single events of normal polarity are known here (Fig. 23.17). The U-Pb zircon ages for ash beds in the Dal'ny Tulkas section determined 288–291 million years (Schmitz and Davydov 2012) and 280–283 million years for the Mechetlino section (Ogg et al. 2016).

An analysis of variations in values of the magnetic susceptibility and coercive parameters in the Mechetlino section shows a good correlation between lithological, paleontological and rock-magnetic parameters.

The positions of ancient paleolatitudes for the outcrop formations of Dal'ny Tulkas and Mechetlino are determined and equal, respectively, to 34°N and 46°N. A common feature of all the selected components of Jn is the correspondence of the directions to the geomagnetic field of the reversed polarity. The criterion for identification of distinguished reversed paleomagnetic zone and its reference to the international Permian paleomagnetic time scales (Hounslow and Balabanov 2016; Ogg et al. 2016) are paleontological (Chuvashov and Chernykh 2011; Chernykh et al. 2012; Chuvashov et al. 2013; Southern Urals 2015) and geochronological (Schmitz and Davydov 2012) data on the Dal'ny Tyulkas and Mechetlino sections, which do not contradict with magnetostratigraphic constructions.

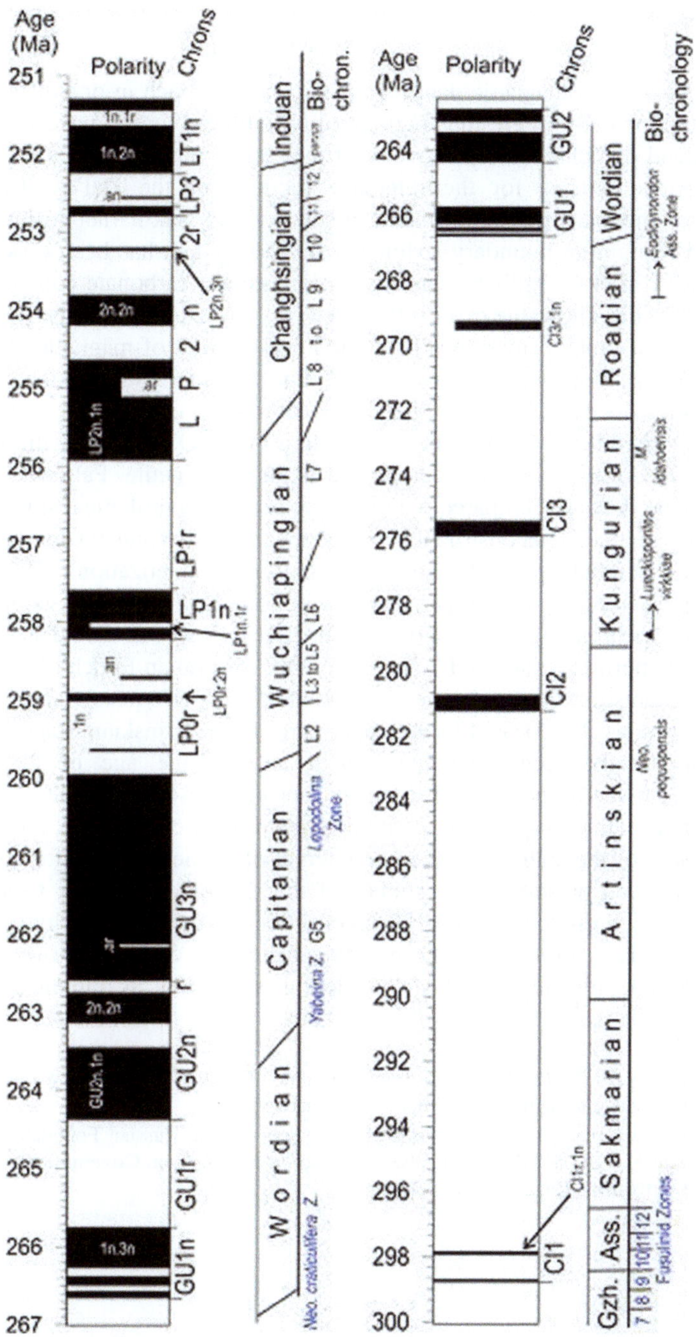

Fig. 23.17 The composite section of the geomagnetic polarity of the Permian system (Hounslow and Balabanov 2016) and the studied intervals (red filling)

Conclusion

The scarcity of paleomagnetic data for the Cisuralian Epoch in one of the world's most stratigraphically interesting regions (the Southern Urals) make the material important and demonstrates the novelty of the results obtained. As a result of the first performed studies for the reference sections of the Dal'ny Tulkas and Mechetlino, the paleomagnetic characteristics of the Sakmarian-Artinskian and Artinskian-Kungurian boundary sediments are received. It has been observed that all the studied rocks (argillites, sandstones, limestones, carbonate concretions) are weakly magnetic. The value of Jn is $0.13–6.08 \times 10^{-3}$ A/m, and æ is from −0.6 to 21.8×10^{-5} SI units. Limestones have the lowest values of magnetic characteristics, including negative values of æ. Further increase in the magnetic characteristics occurs in the direction: carbonate concretions → calcareous argillites → argillites → sandstones. According to the DTMA data, the main carrier of the medium-temperature component is titanomagnetite. Paleomagnetic and petromagnetic data are compared with the results of paleontological studies, which allowed a paleofacial reconstruction of sedimentary environment for the studied sections in Cisuralian epoch. No horizons of normal magnetization (N-horizons) in Kiaman reversed monopolar hyperzone R were found, which confirms the modern world magnetochronological scale (Ogg et al. 2016). However, the interval between chrons of the normal polarity CI 2 and CI 3 in the Cisuralian Epoch of the Permian System is interesting from paleomagnetic point of view (Hounslow and Balabanov 2016). This interval is located in the upper part of the Artinskian Stage and in the lower part of the Kungurian Stage with an absolute age of 282–275 Ma (Fig. 23.17). It should be noted that the CI 3 event is absent from the International Paleomagnetic Time Scale of the Permian System (Ogg et al. 2016), and therefore the expansion of the study area and age interval of studies in the future will allow for more detailed paleomagnetic studies of the Cisuralian sections to search for inversions and episodes of direct magnetization in the Kiaman hyperchron as global correlatives. From here, the paleomagnetic data can become an important magnetostratigraphic element when considering the sections as reference objects of GSSP—"golden nails".

Acknowledgements *Credits* The authors are very grateful to A. V. Latyshev, whose comments have made it possible to significantly improve the manuscript.

The research was carried out with the financial support of the Russian Foundation for Basic Research (grant No. 16-05-00306a) and also was supported by Russian Government's Program of Competitive Growth of Kazan Federal University.

References

Balabanov U. P., Sungatullin R. H., Sungatullina G. M. Paleomagnetic and petromagnetic characteristics of the reference sections Cisuralian Series Permian System // International conference on paleomagnetism and rock magnetism (Kazan, 2–7 October 2017). Book of abstracts / Compilers: L. A. Fattakhova, D. M. Kuzina. Kazan: Kazan University Press, 2017. P. 9.

Beck M. E. On the shape of paleomagnetic data sets // Journal of Geophysical Research. 1999. V. 104, No. B11. P. 25427–25441.

Burov B.V., Nurgaliev D.K., Yasonov P.G. Paleomagnetic analysis. Kazan: Kazan University Press, 1986. P. 167.

Burov B.V., Yasonov P.G. Introduction to differential thermomagnetic analysis. Kazan: Kazan University Press, 1979. P. 159.

Chernykh V. V., Chuvashuv B. I., Davydov V. I., Schmitz M. D. Mechetlino section: a candidate for the Global Stratototype Section and Point (GSSP) of the Kungurian stage (Cisuralian, Lower Permian) // Permophiles. 2012. V. 56. P. 21–34.

Chuvashov B. I., Chernykh V. V., Shen S., Henderson C. M. Proposal for the Global Stratotype Section and Point (GSSP) for the base-Artinskian Stage (Lower Permian) // Permophiles. 2013. № 58. P. 26–34.

Chuvashov B. I., Chernykh V. V.. The Mechetlino section (Southern Urals) - potential limitotype of the lower boundary of the Kungurian stage // Dokl. AN. 2011. Vol. 441. No. 5. P. 657–660.

Chuvashov B. I., Dyupina G. V., Misens G. A., Chernykh V. V. *Opornye razrezy verkhnego karbona i nizhnei permi zapadnogo sklona Urala i Priural'ya* (Upper Carbonaceous and Lower Permian Reference Sections of the Western Slope of the Urals and the Cis-Urals). Sverdlovsk: UrO AN USSR, 1990. 402 p.

Cogne J. P. PaleoMac: a Macintosh application for treating paleomagnetic data and making plate reconstructions // Geochem. Geophys. Geosyst., 2003. V. 4(1). P. 1007, https://doi.org/10.1029/2001gc000227.

Dunlop D. J. Theory and application of the Day plot (Mrs./Ms. versus Hcr/Hc). 1. Theoretical curves and tests using magnetite data // J. Geophys. Res. Solid Earth. 2002. V. 107, no B3. P. EPM 4-1-EPM 4-22.

Enkin R. J. A computer program package for analysis and presentation of paleomagnetic data. Pacific Geoscience Centre, Geological Survey of Canada. 1994. P. 16.

Golovanova I. V., Danukalov K. N., Kadyrov A. F, Khidiyatov M. M., Salmanova R. Yu., Shakurov R. K., Levashova N. M., Bazhenov M. L. Paleomagnetism of sedimentary strata and the origin of the structures in the western slope of the South Urals// Institute of Physics of the Earth. 2017. No. 2. P. 148–156.

Hounslow M. W. & Balabanov Yu. P. A geomagnetic polarity timescale for the Permian, calibrated to stage boundaries // Geological Society of London. 2016. V. 12 / http://www.sp.lyellcollection.org.

Iassonov P. G., Nourgaliev D. K., Bourov B. V., Heller F. A modernized coercivity spectrometer // Geologica Carpathica. 1998. V. 49, No 3. P. 224–226.

Khramov A. N., Sholpo L. E. Paleomagnetism. Principles, methods and geological applications of paleomagnetology. Leningrad: Nedra, 1967. 252 p.

Khramov A. N. (Eds.). Paleomagnetology. Leningrad, 1982. 312 p.

Kirschvink J. L., Gould J. L. Biogenic magnetite as a basis for magnetic field detection in animals // BioSystems. 1981. V. 13. P. 181–201.

Kosareva L. R., Nourgaliev D. K., Kuzina D. M., Spassov S., Fattakhov A. V. Ferromagnetic, dia-/paramagnetic and superparamagnetic components of Aral sea sediments: significance for paleoenvironmental reconstruction // ARPN Journal of Earth Sciences. 2015. V. 4, no. 1. P. 1–6.

Kotlyar G. V., Chernykh V. V., Mizens G. A., Sungatullin R. Kh., Sungatullina G. M., Kutygin R. V., Filimonova T. V., Davydov V. I., Nurgalieva N. G., Balabanov Yu. P., Linkina L. I., Gareev B. I., Batalin G. A. Multidisciplinary study of the Dal'ny Tulkas section, Southern Urals // Kazan Golovkinsky Stratigraphic Meeting – 2017 and Fourth All-Russian Conference «Upper Palaeozoic of Russia» . Upper Palaeozoic Earth systems: high-precision biostratigraphy, geochronology and petroleum resources. Abstract volume. Kazan, September, 19–23, 2017 / D. K. Nurgaliev, V. V. Silantiev (Eds.). Kazan: Kazan University Press, 2017. P. 99–100.

Kotlyar G. V., Chernykh V. V., Mizens G. A., Sungatullin R. Kh., Sungatullina G. M., Kutygin R. V., Filimonova T. V., Davydov V. I., Nurgalieva N. G., Balabanov Yu. P., Linkina L. I., Gareev B. I., Batalin G. A. Multidisciplinary study of the Mechetlino section, Southern Urals // Kazan Golovkinsky Stratigraphic Meeting – 2017 and Fourth All-Russian Conference «Upper Palaeozoic of Russia» . Upper Palaeozoic Earth systems: high-precision biostratigraphy, geochronology and petroleum resources. Abstract volume. Kazan, September, 19–23, 2017 / D. K. Nurgaliev, V. V. Silantiev (Eds.). Kazan: Kazan University Press, 2017. P. 101–102.

Kotlyar G., Sungatullina G., Sungatullin R. GSSPs for the Permian Cisuralian Series stages // Permophiles. 2016. No. 63. P. 32–37.

Mizens G. A. Upper paleozoic flysch of the Western Urals. Ekaterinburg: UrO RAN, 1997. P.230.

Nurgaliev D. K., Yasonov P. G. Coercive spectrometer. The patent of the Russian Federation for utility model No. 81805.-2009. Bulletin of FIPS No. 9.

Ogg J. G., Ogg G. M., and Gradstein F. M. A Concise Geological Time Scale 2016. Elsevier, 2016. 234 p.

Schmitz M. D., and Davydov V. I. Quantitative radiometric and biostratigraphic calibration of the Pennsylvanian–Early Permian (Cisuralian) time scale and pan-Euramerican chronostratigraphic correlation // Geological Society of America Bulletin. 2012. V. 124, No. 3/4. P. 549–577.

Southern Urals. Deep water successions of the Carboniferous and Permian. A Field Guidebook of XVIII International Congress on Carboniferous and Permian. Pre-Congress A3 Trip, August, 6–10, 2015 / V. V. Chernykh, B. I. Chuvashov, V. I. Davydov, C. M. Henderson, S. Shen, M. D. Schmitz, G. M. Sungatullina, R. Kh. Sungatullin, J. E. Barrick, O. P. Shilovsky. Kazan: Tatarstan Academy of Sciences Press, 2015. 88 p.

Sungatullin R. Kh., Sungatullina G. M., Zakirov M. I., Tsel'movich V. A., Glukhov M. S., Bakhtin A. I., Osin Yu. N., Vorobiev V. V. Cosmic microspheres in the Carboniferous deposits of the Usolka section (Urals foredeep) // Geology and geophysics. 2017. Vol. 58. No. 1. P. 74–85.

Tevelev A. V., Prudnikov I. A., Hotylev A. O., Baraboshkin E. E., Popov S. S. Structural paragenesis of flysch complexes in the central part of the Pre-Urals foreland basin // Vestnik MGU. Series 4 Geology. 2016. No.3. P. 11–16.

Watson G. S., Enkin R. J. The fold test in paleomagnetism as a parameter estimation problem // Geophys. Res. Lett. 1993. Vol. 20. P. 2135–2137.

Zijderveld J. D. A. A. C. demagnetization of rocks: analysis of results // Methods in paleomagnetism. New York, Elsevier Science, 1967. P. 254–286.

Chapter 24
General Magnetostratigraphic Scale: Present Status and Outlook of Development

A. Yu. Guzhikov

Abstract Geomagnetic Polarity Time Scale (GPTS) and Generalized Magnetostratigraphic scale (GMSS) which is in use in Russia synthesize the paleomagnetic data of the world. Ideally, there should be only one global scale of geomagnetic polarity (GPTS or GMSS, or a new scale constructed through their synthesis). All other paleomagnetic scales may be of local or regional (possibly, sub-global) character. But since the GPTS chrons are referenced to the International Stratigraphic Scale (ISS), and the GMSS magnetozones—to the General Stratigraphic Scale (GSS), the necessity of equiponderant coexistence of the GPTS and the GMSS will persist until all the correlation problems between the two scales are solved or until these are merged together. Practical use of a domestic version of paleomagnetic scale for geologic investigations requires construction of an updated and verified GMSS version—with regard to recent magnetostratigraphic data. Simultaneously with the process of data verification, procedural problems should be solved which may arise while further development of the General Magnetostratigraphic Scale: to continue improving the GMSS hierarchic structure on the quantitative basis, to take into account the diachronism of biostratigraphic boundaries for integration of a paleomagnetic scale with detailed biostratigraphic units, to develop theoretical grounds for the use of paleomagnetic criteria for identifying the boundaries of common stratigraphic units.

Keywords Magnetostratigraphy · Geological Time Scale · Geomagnetic Polarity Time Scale · General Magnetostratigraphic Scale · Magnetic chron

Geomagnetic Polarity Time Scale (GPTS) and Generalized Magnetostratigraphic scale (GMSS) which is in use in Russia synthesize the paleomagnetic data of the world. Ideally, there should be only one global scale of geomagnetic polarity (GPTS or GMSS, or a new scale constructed through their synthesis). All other paleomagnetic scales may be of local or regional (possibly, sub-global) character.

A. Yu.Guzhikov (✉)
Saratov State University, Saratov, Russia
e-mail: aguzhikov@yandex.ru

D. Nurgaliev et al. (eds.), *Recent Advances in Rock Magnetism, Environmental Magnetism and Paleomagnetism*, Springer Geophysics,
https://doi.org/10.1007/978-3-319-90437-5_24

But since the GPTS chrons are referenced to the International Stratigraphic Scale (ISS), and the GMSS magnetozones—to the General Stratigraphic Scale (GSS), the necessity of equiponderant coexistence of the GPTS and the GMSS will persist until all the correlation problems between the two scales are solved or until these are merged together. Practical use of a domestic version of paleomagnetic scale for geologic investigations requires construction of an updated and verified GMSS version—with regard to recent magnetostratigraphic data. Simultaneously with the process of data verification, procedural problems should be solved which may arise while further development of the General Magnetostratigraphic Scale: to continue improving the GMSS hierarchic structure on the quantitative basis, to take into account the diachronism of biostratigraphic boundaries for integration of a paleomagnetic scale with detailed biostratigraphic units, to develop theoretical grounds for the use of paleomagnetic criteria for identifying the boundaries of common stratigraphic units.

Classification of the geomagnetic polarity scales by their construction methods into magnetostratigraphic (magnetozones are referenced to relative age), magnetochronological (magnetozones are referenced to absolute age) and anomaly scales (the magnetozone sequence is identified with spatial distribution of linear magnetic anomalies—LMA) is of mere historical interest nowadays. The present-day scales—the Geomagnetic Polarity Time Scale (GPTS), representing a part of the Geological Time Scale (GTS) (Ogg et al. 2016), and the General Magnetostratigraphic Scale (GMSS) (Khramov and Shkatova 2000)—synthesize the global paleomagnetic data by means of all the three methods. In both scales, the magnetic-polarity structures of the intervals from the Bajocian to the Quaternary are based on the results of the reference section studies and on the LMA data, while those from the Cambrian to the Aalenian—exclusively on magnetostratigraphic data. There are still no generally accepted paleomagnetic scales for the Precambrian.

Ideally, there should be only one global scale of geomagnetic polarity (GPTS or GMSS, or a new scale constructed through their synthesis). All other paleomagnetic scales may be of local or regional (possibly, sub-global) character. But since the GPTS chrons are referenced to the International Stratigraphic Scale (ISS), and the GMSS magnetozones—to the General Stratigraphic Scale (GSS), the necessity of equiponderant coexistence of the GPTS and the GMSS will persist until all the correlation problems between the two scales are solved or until they are merged together.

Both the GSS and the GMSS make a part of the Stratigraphic Code of Russia (2006) and are mandatory for any geologic surveying and geologic prospecting throughout our country. Regretfully, since publication in 2000 (Khramov and Shkatova 2000) the GMSS has not been actualized, except for the Quaternary GMSS, updated to comply with the latest amendments in the MSS and the GSS (Guzhikov and Shkatova 2016). By comparison, the GSSP has been actualized four times after 2000: in 2004 (Gradstein et al. 2004), 2008 (Ogg et al. 2008), 2012 (Gradstein et al. 2012) and 2016 (Ogg et al. 2016).

The GSS structure has been substantially modified during that period. Among other things, tripartition of the Permian has been accepted (Stratigraphic … 2006), views on positions of many stage boundaries relative to geomagnetic reversal sequence have changed owing to adopting new global stratotype sections and points (GSSP, "golden spikes"). Thus, upon adopting the Tercis section from France as the GSSP, the lower boundary of the Maastrichtian stage was shifted relative to the paleomagnetic scale from the basis of the C32n2 polarity chron to the upper part of the C32n1 chron (Odin and Lamaurelle 2001) (Fig. 24.1a), and the lower boundary of the Thanetian, upon selection of the Zumaia section from Spain as the GSSP stage, was shifted from the C25r chron bottom to the basis of the C26n chron (Fig. 24.1b) (Schmitz et al. 2011).

At present, substantial differences are recorded in paleomagnetic structures of some coeval intervals of the GMSS and the GPTS. In certain instances, the differences are fundamental. For example, according to the GMSS, the Wenlock series and the Pragian stage are peculiar for alternated polarity (Fig. 24.2a, b), while in the

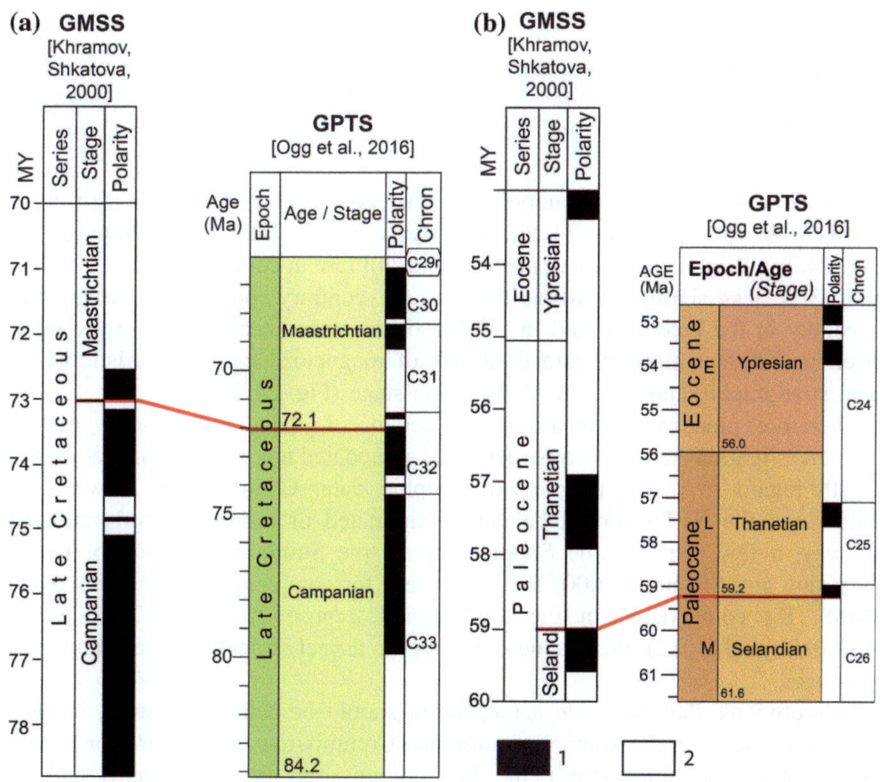

Fig. 24.1 Changes in the relations between the lower boundaries of Maastrichtian (**a**) and Thanetian (**b**) with geomagnetic reversals after the approval of the GSSP of stages. (1)—normal polarity, (2)—reverse polarity

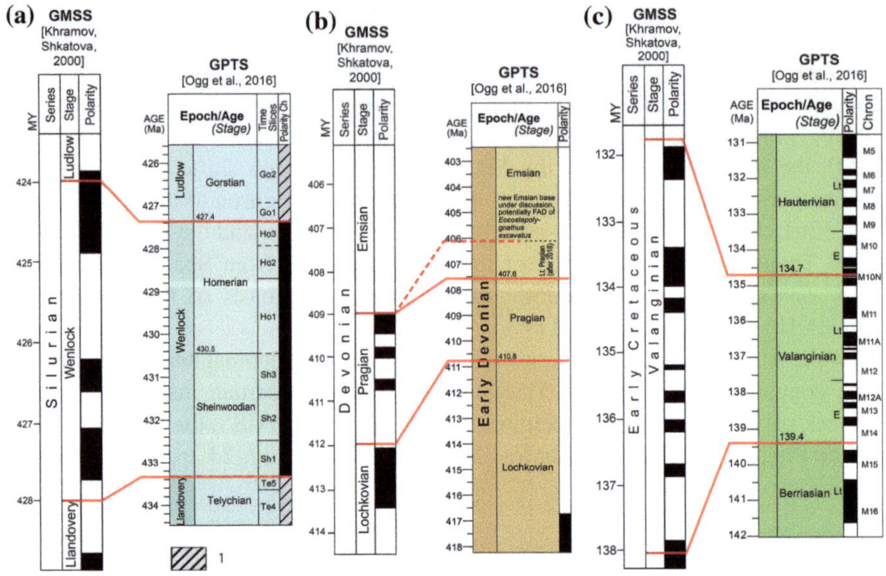

Fig. 24.2 Contradictions between GMSS and GPTS in the paleomagnetic structure of the Wenlock series (**a**), Pragian stage (**b**) and Valanginian stage (**c**). (1)—uncertain polarity. For other explanations see Fig. 24.1

GPTS they correspond to monopolar magnetozones: normal (N) polarity corresponds to the Wenlock (Fig. 24.2a), and reverse polarity (R)—to the Pragian stage (Fig. 24.2b). While similar contradictions are still rare in comparing the GMSS and the GPTS, less significant disparities, on the contrary, occur quite frequently, especially in the Paleozoic and in the Mesozoic scale intervals. Those are manifested in different numbers and durations of magnetozones characteristic of the same stratigraphic unit, e.g. the Valanginian stage (Fig. 24.2c).

Therefore, practical use of a domestic version of paleomagnetic scale for geologic investigations requires construction of an updated and verified GMSS version—with regard to recent magnetostratigraphic data. Contradiction between the GMSS and the GPTS should be either eliminated or thoroughly substantiated. Anyway, a few pages in the Supplements to the Stratigraphic Code of Russia (Khramov and Shkatova 2000) are insufficient for complete characteristics of the GMSS. By comparison, in the fundamental treatise "Geologic Time Scale" (Gradstein et al. 2012), the magnetostatigraphic aspect is dealt with on more than 100 pages.

It is obvious, that the scale actualization should be achieved jointly, in close cooperation between the Magnetostratigraphy Commission at the Interdepartmental Stratigraphic Committee (ISC), the ISC commissions on the systems and the Regional Interdepartmental Stratigraphic Commissions (RISC).

Simultaneously with the process of data verification, procedural problems should be solved that may arise in the course of further development of the General

Magnetostratigraphic Scale. Approaches to the GMSS construction and structuring should not indiscreetly simulate the experience of the GPTS authors.

The GMSS hierarchy is among its advantages. The ranks of super- and hyperzones reflect the principal stages in geomagnetic field evolution. It is necessary to continue improving the GMSS hierarchic structure on the quantitative basis (Molostovskii et al. 2007). Meanwhile, at D. M. Pechersky's suggestion (2003), quantitative reliability (quality) estimates for magnetostratigraphic data from various sites should be developed and introduced into the scale.

At present, the GMSS, except for the Quaternary, is incomparable with the GPTS in terms of detailed referencing of paleomagnetic and stratigraphic units. While the GPTS chrons are tied to the zonal (macro- and microfauna) scales of various paleobiochores, the GMSS magnetozones are tied only to the stage sequence. The Geologic Time Scale (Gradstein et al. 2012; Ogg et al. 2008, 2016), however, mechanically unites absolute dating and geomagnetic reversals with zonal standards, sequences, event levels (oceanic anoxic events, isotopic anomalies, etc.), superimposing stratigraphic boundaries of different nature and with known different spatial and temporal stabilities. This is methodically wrong, because the boundaries of all the biostratigraphic units are diachronic to a varying degree. Theoretically, this has been clear for a long time, but the documentary evidence and quantitative estimates of the time shift in some stratigraphic units (of both, macro- and microfauna grounding) from various regions have emerged only recently, owing to the use of geomagnetic reversal scale. Due to short durations of the reversals ($<10^4$ years), they may be regarded as practically isochronous references for thorough correlations of the deposits and for calibrating zonal scales of various paleobiochores. Detailed complex bio- and magnetostratigraphic correlations have revealed the ages of the Lower-Upper Barremian boundary in the Tethyan and Boreal super-region to differ by no less than one million years, judging from magnetic chron durations (Guzhikov and Baraboshkin 2006). A diachroneity value of similar order is recorded at the bottom of the NKT nanoplankton zone in the Jurassic-Cretaceous boundary interval from the Northern Mediterranean; moreover, this is observed in the sections spaced apart by less than 150 km (Channell et al. 2010). The boundaries of the Lower Aptian ammonite zones in the North Caucasia, Russian Plate and West Europe are also prone to time shifting, though to a lesser degree (Guzhikov and Baraboshkin 2006).

Since the time shift at comparing the maximally remote paleobasins or at calibrating scales based on various fauna groups amounts up to 10^6 years, integration of a paleomagnetic scale with detailed biostratigraphic units should be accompanied, whenever possible, with indicating boundary diachroneity in the provincial zones relative to geomagnetic reversals (Fig. 24.3). Horizontal lines as the boundaries of detailed stratigraphic units are appropriate, when the boundary isochroneity is proved (solid lines) or cannot be disproved at the present stage of investigation by means of independent methods (dotted lines).

Boundaries of the global stratigraphic units—stages, though they also serve as the zone boundaries, will remain in the GMSS as horizontal lines, because in the

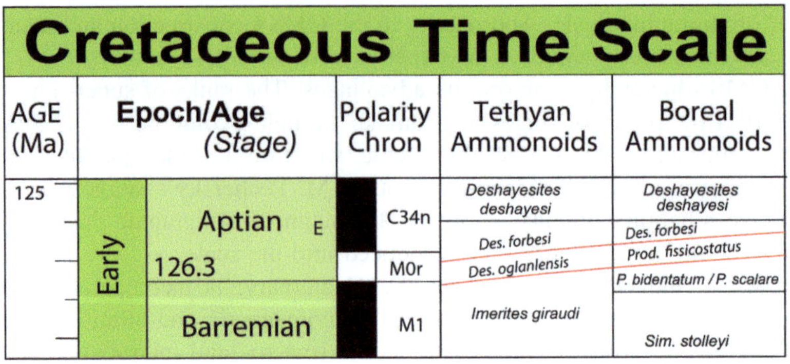

Fig. 24.3 The possibility of showing (red lines) of the ammonite zone boundaries diachronism while integrating a paleomagnetic scale with ammonite zone scales [on the example of the Lower Aptian interval of GTS-2012 (Gradstein et al. 2012)] in connection with the data on the diachroneity of the Lower Aptian ammonite zones boundaries relative to the M0 (Guzhikov and Baraboshkin 2006)

GSSP (to be provided for all the stages in the foreseeable future) the position of the stage bottom is fixed relative to the sequence of geomagnetic reversals.

As for synchronization of the stage boundaries in spatially separated areas, involvement of paleomagnetic criteria as the primary features for their substantiation seems to be quite attractive. This statement does not encroach upon the priority of the biostratigraphic method in distinguishing the GSS stratons and substantiating their boundaries. Selection of a paleomagnetic marker for the stage bottom is inevitably preceded by paleontologic grounding of the stratigraphic boundary; it is only afterwards, that a geomagnetic reversal may be selected—the nearest to the level of fauna assemblage change in a GSSP.

Irrespective of the lack of any strict theoretical grounding, the idea of using the paleomagnetic criteria for identifying the boundaries of common stratigraphic units has been put into practice. For example, the Matuyama-Bruhnes reversal is used to identify the GSS Eo- and Neopleistocene boundary (Calabrian-Ionian (Lower—Middle Pleistocene) in the International Stratigraphic Scale—ISS (Cohen et al. 2013; Guzhikov and Shkatova 2016). The magnetostratigraphic boundary of the Kiama-Illawarra hyperzones lies close to the lower boundary of the Tatarian series (Kotlyar et al. 2013). It seems quite assignable, that magnetic polarity features were primarily wanted to substantiate the Quaternary and the Upper Permian units, i.e. by the specialists who are frequently involved in correlations of heterofacies (marine and continental) deposits. Similar situation arises with correlations of marine deposits formed in maximally separated paleobiogeographic provinces. For example, the problem of global tracing of the Jurassic-Cretaceous boundary cannot be solved on the basis of exclusively paleontological data. Such situation may result from both diachroneity of the biostratigraphic boundaries [inclusive of the *C. alpina* (Lakova et al. 2017) level, which is insistently recommended as the primary marker of the Jurassic-Cretaceous boundary (Schnabl et al. 2015; Wimbledon 2016,

2017)], and the paleontological "muteness" of the deposits and impossibility of using the same faunal groups for substantiating the stratigraphic boundary in different regions. At the same time, the paleomagnetic chrons, irrespective of the relatively complicated alternated paleomagnetic zonality of the Jurassic-Cretaceous boundary interval, are basically identifiable even with the minimum paleontological description of the rock. The case of the Feodosiya section (Crimea) may be cited as the prime example. Scarce ammonite finds and meagre calpionella assemblages prevent one from determining the boundary level, but allow reliable identifications of magnetic chrons in the examined deposits (Guzhikov et al. 2012). The choice of geomagnetic reversal as the primary feature for substantiating the Jurassic-Cretaceous boundary [which was actually proposed in GTS-2012 (Gradstein et al. 2012)] would make it possible to solve this problem (Guzhikov 2013; Baraboshkin et al. 2013; Arkadiev et al. 2014).

Similar tendencies are characteristic of other GSS intervals, as well, since using a paleomagnetic feature as the leading criterion for boundary selection allows to abate the disputable matters arising from mismatch of stratigraphic boundaries recognized from different flora and fauna assemblages in real sections. For example, the bottom of the M0 chron was recommended by the international working group on the Aptian stage as one of the principal markers of the Barremian-Aptian boundary (Erba et al. 1996). Outlining the Santonian-Campanian boundary along the bottom of the C33r chron has been repeatedly proposed by the authors of the Geological Time scale (Ogg et al. 2016); expediency of matching the Campanian-Maastrichtian boundary with the top of the C32n2 chron has been stated at the X Cretaceous System Symposium in Vienna (Baraboshkin et al. 2017). The C27n bottom has already been officially chosen as the primary marker of the Thanetian stage base (Schmitz et al. 2011).

The importance of paleomagnetic features for the GSS structuring constitutes a separate aspect in discussing their stratigraphic role. The relations of deep geodynamic events, responsible for transformation of the polarity regime in a geomagnetic field, with the geologic events, inclusive of biotic changes that serve as the criteria for ranging stratons into series, stages and substages, predetermine the theoretical possibility of involving the paleomagnetic criteria into solving the problems of the GSS structuring. It has been repeatedly noted, that clear division of the Cretaceous interval of the magnetostratigraphic scale into three parts: that of alternating sign (Berriasian—lowermost Barremian), of dominant direct polarity (uppermost Barremian—lowermost Santonian) and another part of alternating sign, though with fewer reversals (uppermost Santonian—Maastrichtian) may be regarded as an argument in favor of tripartition of the Cretaceous system (Molostovskii and Fomin 2004). With all its disputability, the idea of using the features that determine the paleomagnetic scale structure for the GSS ranging deserves serious consideration and meticulous elaboration. It is not improbable, that the results of such approach will contribute to solving the urgent problems associated with choosing between dividing certain GSS systems and stages into two or three components.

References

Arkadiev V.V., Baraboshkin E.Yu., Guzhikov A.Yu. (2014). About Jurassic-Cretaceous boundary // In: Cretaceous system of Russia and the near abroad: problems of stratigraphy and paleogeography // Baraboshkin E.Yu., Markevich V.S., Bugdaeva E.V., Afonin M.A., Cherepanova M.V. (Ed.) (2014): *Proceedings of the Seventh Russian Scientific Conference with International Participation. 10–15 September 2014, Vladivostok.* – Vladivostok, Dalnauka. P. 27–30 (in Russian).

Baraboshkin E.Y., Arkadiev V.V., Benyamovski V.N., Guzhikov A., Kopaevich L., Jagt-Yazykova E.A. (2013). The stratigraphic scale of the Cretaceous of Russia: current state, main problems, ways of perfection // *Gladenkov Yu.B., Zakharov V.A., Ippolitov A.P. (Ed.) (2013) General Stratigraphie Scale of Russia: current state and ways of perfection. All-Russian meeting. May 23–25, 2013, Geological Institute of RAS, Moscow. – Collector of articles.* – Moscow: GIN RAS. P. 289–297 (in Russian).

Baraboshkin E.Yu., Benyamovski V.N., Guzhikov A., Aleksandrova S., Pervushov E., Selzer V. B., Ovechkina M.N., Kaljakin E.A., Kopaevich L., Vishnevskaya V.S., Guzhikova A., Pokrovsky B., Baraboshkin E.E., Iakovishina E.V. (2017) Integrated study of Campanian/ Maastrichtian boundary interval at Volga region (Russia) and Aktolagay Plateau (West Kazakhstan) of the Russian Platform // *Sames, B. (Ed.) (2017): 10th International Symposium on the Cretaceous – Abstracts, 21–26 August 2017, Vienna.* – Berichte der Geologischen Bundesanstalt, 120, Vienna. P.22.

Channell J.E.T., Casellato C.E., Muttoni G., Erba E. (2010) Magnetostratigraphy, nannofossil stratigraphy and apparent polar wander for Adria-Africa in the Jurassic–Cretaceous boundary interval // *Palaeogeography, Palaeoclimatology, Palaeoecology*, 293. P. 51–75.

Cohen, K.M., Finney, S.C., Gibbard, P.L. & Fan, J.-X. (2013; updated) The ICS International Chronostratigraphic Chart // *Episodes*, 36. P. 199–204.

Erba E., Aguado R., Avram E., Baraboschkin E.J., Bergen J.A., Bralower T.J., Cecca F., Channell J.E.T., Coccioni R., Company M., Delanoy G., Erbacher J., Herbert T.D., Hoedemaeker P.J., Kakabadze M., Leereveld H., Lini A., Mikhailova I.A., Mutterlose J. (1996) The Aptian Stage // *Bulletin de l'Institut Royal des Sciences Naturelles de Belgique, Sciences de la terre*, 66 (suppl.). P. 31–43.

Gradstein F.M., Ogg J.G., Smith A. (2004) A Geologic Time Scale 2004. Cambridge University press. 2004. 589 p.

Gradstein F., Ogg J.G., Schmitz M.D., Ogg G.M. (2012) The Geologic Time Scale 2012. Amsterdam, Elsevier. 1144 p.

Guzhikov A.Yu (2013) Solving unsolvable problems in stratigraphy (Comments to the paper "New data on the magnetostratigraphy of the Jurassic–Cretaceous boundary interval, Nordvik Peninsula (northern East Siberia)" by V.Yu. Bragin, O.S. Dzyuba, A.Yu. Kazansky, and B.N. Shurygin) // *Russian Geology and Geophysics*, 54. P. 349–354.

Guzhikov A.Y., Arkadiev V.V., Baraboshkin E.Y., Bagaeva M.I., Piskunov V.K., Rud'ko S.V., Perminov V.A. & Manikin A.G. (2012) New sedimentological, bio-, and magnetostratigraphic data on the Jurassic–Cretaceous boundary interval of Eastern Crimea (Feodosiya) // *Stratigr. Geol. Correl.* 20. P. 261–294.

Guzhikov A. Yu., Baraboshkin E.J. (2006) Assessment of Diachronism of Biostratigraphic Boundaries by Magnetochronological Calibration of Zonal Scales for the Lower Cretaceous of the Tethyan and Boreal Belts // *Doklady Earth Sciences*, 409A, 6, P. 843–846.

Guzhikov A.Yu., Shkatova V.K. (2016) About the implied changes in General magnetostratigraphic polarity scale of quaternary period // *Decisions of the Interdepartmental Stratigraphic Committee of Russia and its regular commissions. Release №44.* – St. Petersburg: VSEGEI publishing. 68 p (in Russian).

Khramov A.N., Shkatova V.K. (2000) General Magnetostratigraphic Scale of polarity of Phanerozoic // *Supplements to the Stratigraphic Code of Russia / A.I. Zhamoida (ch. Ed.)* – St. Petersburg: Izd. VSEGEI. P. 24–45 (in Russian).

Kotlyar G.V., Golubev V.K., Silantiev V.V. (2013) General stratigraphic scale of the Permian system: current state of affairs // *General stratigraphic scale of Russia: current state and ways of perfection. All-Russian meeting. May 23–25, 2013, Geological institute of RAS, Moscow. Collector of articles. / M.A. Fedonkin (resp. ed.), Yu. B. Gladenkov, V.A. Zakharov, A. P. Ippolitov (eds.).* – Moscow: GIN RAS. 408 p (in Russian).

Lakova I., Grabowski J., Stoykova K., Petrova S., Reháková D., Sobień K., Schnabl P. (2017) Direct correlation of Tithonian/Berriasian boundary calpionellid and calcareous nannofossil events in the frame of magnetostratigraphy: new results from the West Balkan Mts, Bulgaria, and review of existing data // *Geologica Balcanica,* 46, 2. P. 47–56.

Molostovskii E.A., Fomin V.A. (2004) Main features of the magnetic polarity structure of Cretaceous // *Second all-Russia meeting: Cretaceous system of Russia: Promblems of stratigraphy and paleogeography. School "Principles and methods of stratigraphic studies" (St Petersburg, April 12–15, 2004): Thesis of reports / V.V. Arkadiev (ch. Ed.).* – St. Petersburg. P. 55 (in Russian).

Molostovskii E.A., Frolov I.Yu., Pechersky D.M. (2007) Magnetostratigraphic timescale of the Phanerozoic and its description using a cumulative distribution function // *Izvestiya. Physics of the Solid Earth,* 43(10). P. 811–818.

Odin G.S., Lamaurelle M.A. (2001) The global Campanian-Maastrichtian Stage boundary // *Episodes,* 24, 4. P. 229–238.

Ogg J.G., Ogg G., Gradstein F.M. (2008). The concise geologic time scale. Cambridge, U.K., Cambridge University Press, 177 p.

Ogg J.G., Ogg G.M., Gradstein F.M. (2016). A Concise Geologic Time Scale: 2016. Amsterdam, Elsevier. 230 pp.

Pechersky D.M. (2003) About a Geomagnetic Plarity Scale of Phanerozoic // *Institute of Physics of the Earth RAS. Laboratory of the Main Geomagnetic Field and Petromagnetism. Inter-seminar.* URL: http://paleomag.ifz.ru/inter-seminar/pecherskiy-o-shkale.html (in Russian).

Schmitz B., Pujalte V., Molina E., Monechi S., Orue-Etxebarria X., Speijer R.P., Alegret L., Apellaniz E., Arenillas I., Aubry M.-P.6, Baceta J.-I., Berggren W.A., Bernaola G., Caballero F., Clemmensen A., Dinarès-Turell J., Dupuis C., Heilmann-Clausen C., Orús A. H., Knox R., Martín-Rubio M., Ortiz S., Payros A., Petrizzo M.R., von Salis K., Sprong J., Steurbaut E., Thomsen E. (2011) The Global Stratotype Sections and Points for the bases of the Selandian (Middle Paleocene) and Thanetian (Upper Paleocene) stages at Zumaia, Spain // *Episodes,* 34, 4. P. 220–243.

Schnabl P., Pruner P., Wimbledon W.A.P. (2015) A review of magnetostratigraphic results in the Tithonian-Berriasian of Nordvik. (Siberia) and possible biostratigraphic constraints // *Geologica Carpathica,* 66. P. 489–498.

Stratigraphic Code of Russia. Third edition (2006). SPb.: VSEGEI Press. 96 p (in Russian).

Wimbledon W.A.P. (2016) Resolving the positioning of the Tithonian/Berriasian stage boundary and the base of the Cretaceous System // *Michalik, J., Fekete K. (Ed) (2016): XII Jurassica Conference. Workshop of the ICS Berriasian Group and IGCP 632. Field Trip Guide and Abstracts Book, April 19–23, 2016, Smolenice, Slovakia* – Earth Science Institute, Slovak Academy of Sciences. Bratislava. 2016. P. 128–130.

Wimbledon W.A.P. (2017) The Tithonian/Berriasian stage boundary and the base of the Cretaceous System // *Sames, B. (Ed.) (2017): 10th International Symposium on the Cretaceous – Abstracts, 21–26 August 2017, Vienna.* – Berichte der Geologischen Bundesanstalt, 120, Vienna. P.290.

Chapter 25
Existence of the Reversal Polarity Zones in Turonian-Coniacian from the Lower Volga (Russia): New Data

A. A. Guzhikova, A. Yu. Guzhikov, E. M. Pervushov, I. P. Ryabov and A. M. Surinskiy

Abstract The detailed magnetostratigraphic study of the Turonian-Coniacian sections from the Volga right-bank region was conducted. The results on magnetic polarity from the Nizhnyaya Bannovka section (Saratov Region) (about 18 m thick) agree with the traditional data on exclusively normal regime of geomagnetic field polarity in the Cenomanian-Santonian ages. In the Kamennyi Brod section (Volgograd Region) between two zones of normal polarity an interval (about 16 m thick) occurs with no reliable paleomagnetic data due to extremely low magnetization of rocks. In the composite section Ozerki-Lipovka (Saratov Region) a long (thickness not less than 13 m) zone of reverse polarity is recorded. This contradicts the paleomagnetic structure as established in the Geomagnetic Polarity Time Scale (Ogg et al. 2016) but is partly consistent with the General Magnetostratigraphic Scale (Supplements … 2000) which records existence of a prolonged reversed-polarity epoch (R) in the Coniacian age (the Klyuyevskaya R-zone).

Keywords Magnetostratigraphy · Petromagnetism · Turonian · Coniacian Santonian · Geomagnetic polarity

The detailed magnetostratigraphic study of the Turonian-Coniacian sections from the Volga right-bank region was conducted. The results on magnetic polarity from the Nizhnyaya Bannovka section (Saratov Region) (about 18 m thick) agree with the traditional data on exclusively normal regime of geomagnetic field polarity in the Cenomanian-Santonian ages. In the Kamennyi Brod section (Volgograd Region) between two zones of normal polarity an interval (about 16 m thick) occurs with no reliable paleomagnetic data due to extremely low magnetization of rocks. In the composite section Ozerki-Lipovka (Saratov Region) a long (thickness not less than 13 m) zone of reverse polarity is recorded. This contradicts the paleomagnetic structure as established in the Geomagnetic Polarity Time Scale (Ogg et al. 2016) but is partly consistent with the General Magnetostratigraphic Scale (Supplements …

A. A. Guzhikova (✉) · A. Yu. Guzhikov · E. M. Pervushov · I. P. Ryabov · A. M. Surinskiy
Saratov State University, Saratov, Russia
e-mail: blackhole3@yandex.ru

D. Nurgaliev et al. (eds.), *Recent Advances in Rock Magnetism, Environmental Magnetism and Paleomagnetism*, Springer Geophysics,
https://doi.org/10.1007/978-3-319-90437-5_25

2000) which records existence of a prolonged reversed-polarity epoch (R) in the Coniacian age (the Klyuyevskaya R-zone).

Introduction

Currently, there is no universally accepted idea of the geomagnetic field behavior in Late Cretaceous: in the Geomagnetic Polarity Time Scale (GPTS) (Ogg et al. 2016), the Cenomanian-Santonian corresponds exclusively to normal (N) polarity (the upper part of the C34 magnetic chron), while the General Magnetostratigraphic Scale (GMSS) (Supplements … 2000) records a prolonged reversed-polarity epoch (R) in the Coniacian age (the Klyuyevskaya R-zone). There is an abundant evidence on the presence of reversed magnetozones within the Cenomanian-Coniacian from various regions (Eremin et al. 1995; Fomin 2003; Fomin and Eremin 1993; Fomin and Molostovskiy 2001; Krumsiek 1982; Makarova and Tsapenko 1971; Montgomery et al. 1998; Nairn et al. 1981; Pechersky 1970). Regretfully, a substantial part of the data does not comply with the current requirements to the magnetic data quality: some of the earliest publications provide only magnetic polarity columns without any information on paleomagnetic directions, magnetic cleanings, etc. In some cases R-magnetozones are either substantiated with insufficient number of samples or their recognition is based on paleomagnetic data of poor quality generally associated with extremely weak natural remanent magnetization values, at the threshold sensitivity level of available equipment. Sometimes, the stratigraphic ages assigned to the respective sedimentary formations seem dubious. Therefore, acquisition of reliable magnetostratigraphic data on the reference sections from various geostructural and paleobiogeographic areas is necessary to prove the reality of prolonged reverse-polarity epochs in the Cenomanian-Coniacian.

Integrated bio- and magnetostratigraphic explorations of the Turonian-Coniacian sections from the Volga right-bank region have been aimed at refining the magnetic structure of the Cretaceous Normal superchron by testing the hypothesis of the presence of reversed polarity epochs in the Turonian-Coniacian. On the whole, 3 sections were examined. Two of those (Ozerki-Lipovka and Nizhnyaya Bannovka) are situated in the Saratov Region and one (Kamennyi Brod) in the Volgograd Region (Fig. 25.1). According to the stratigraphic chart of the Upper Cretaceous in the East European Platform (Stratigraphic scheme … 2004), the Turonian-Coniacian in the Volga right-bank region in the vicinity of Saratov and Volgograd comprises a carbonate formation represented by marls and chalk, from 5 to 60 m thick, enclosed between the Cenomanian terrigenous rocks and the condensed (the so-called "sponge") horizon [base of Santonian, according to (Stratigraphic … 2004)]. In the Ozerki-Lipovka composite section, beside the carbonate formation, the siliceous-carbonate beds above the "sponge" horizon have been sampled (Figs. 25.2 and 25.3). Those beds have been traditionally referred to as Santonian (Stratigraphic … 2004), but the microfauna that is characteristic for Coniacian stage has been recently found there (Guzhikov et al. 2017b; Pervushov et al. 2017a, b). Until

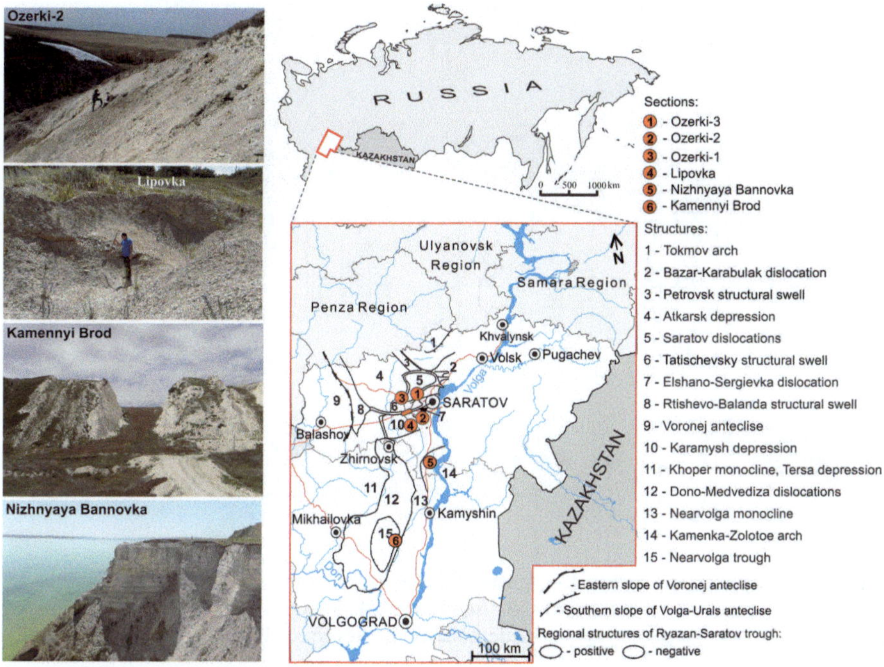

Fig. 25.1 The photos of the studied sections and their locations in the tectonic scheme of the Ryazan-Saratov trough (Shebaldin 2008; State Geological … 2009)

recently, there were practically no magnetostratigraphic descriptions of the Turonian-Coniacian either from the Volga right bank region or from the whole of the Russian Plate. This is primarily accounted for by weak magnetic properties of the rocks that were unmeasurable by means of the available equipment.

In terms of tectonic regionalization, the studied sections are situated in Ryazan-Saratov through (Fig. 25.1), the most part of plicative structures (flexures, brachyanticlines) within it are inherited and develop since Paleozoic to date. In all studied outcrops it was determined, the subhorizontal bedding, except for Lipovka, where layers dip to the north at an angle of 23 degree.

Research Technique and Investigation Results

The field study consisted of sample collection for various types of analyses according to the "sample-to-sample" system with concurrent geologic description and layer-by-layer macrofauna collecting. In all, 257 oriented hand samples from 256 stratigraphic levels have been collected. In the Ozerki-Lipovka section, 29 levels have been sampled from the 5.5 m-thick carbonate formation, and 26 levels from the sequence above the "sponge horizon" (Fig. 25.2). In Kamennyi Brod and

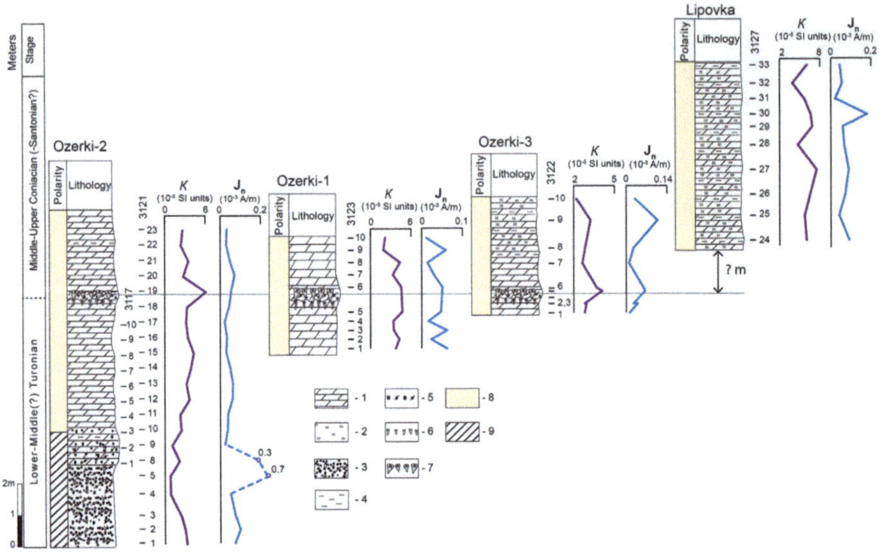

Fig. 25.2 Paleomagnetic and rock-magnetic characteristics of the sections Ozerki-1, -2, -3 and Lipovka. 1—marls, 2—siliceous rocks, 3—sandstones, 4—clays, 5—phosphorites, 6—bioturbation zones, 7—sponge fossils; geomagnetic polarity: 8—reversed (R), 9—anomalous directions

Nizhnyaya Bannovka sections, 109 and 87 levels have been sampled, respectively (Figs. 25.4 and 25.5). Each hand sample was sawn into 3–4 cubic specimens with 2 cm edge; these specimens were subsequently subjected to paleomagnetic and petromagnetic examinations.

The petromagnetic study consisted of examining magnetic susceptibility (K) and its anisotropy (AMS), measuring natural remanent magnetization ($\mathbf{J_n}$), experiments on magnetic saturation with determinations of remanent saturation magnetization ($\mathbf{J_{rs}}$) and remanent coercivity ($\mathbf{H_{cr}}$). K measurements were made with a MFK1-FB kappabridge, those of $\mathbf{J_n}$—with a JR-6 spinner magnetometer and a 2G-Enterprises cryogenic magnetometer (at the Schmidt Institute of Physics of the Earth of the Russian Academy of Sciences (IPE RAS), Moscow). The studies of the magnetic susceptibility dependences on temperature were conducted on the MFK1-FA kappabridge with a CS3 furnace (Trofimuk Institute of Petroleum Geology and Geophysics of Siberian Branch of Russian Academy of Sciences, Novosibirsk). The analysis of AMS data was carried out by means of the Anisoft 4.2 software.

All the examined samples were weakly magnetic: $K = 0\text{–}6 \times 10^{-5}$ SI units, $\mathbf{J_n} < 0.2 \times 10^{-3}$ A/m, with rare exceptions (Figs. 25.2, 25.4 and 25.5). From the magnetic saturation results, only a magnetically soft phase is found ($\mathbf{J_{rs}}$ is nearly saturated in the fields about 100–200 mT, while remanent coercivity is typically 20–40 mT) (Fig. 25.6a). This is characteristic of finely dispersed magnetite. In a number of samples a small but consistent increase in the magnetization up to 700

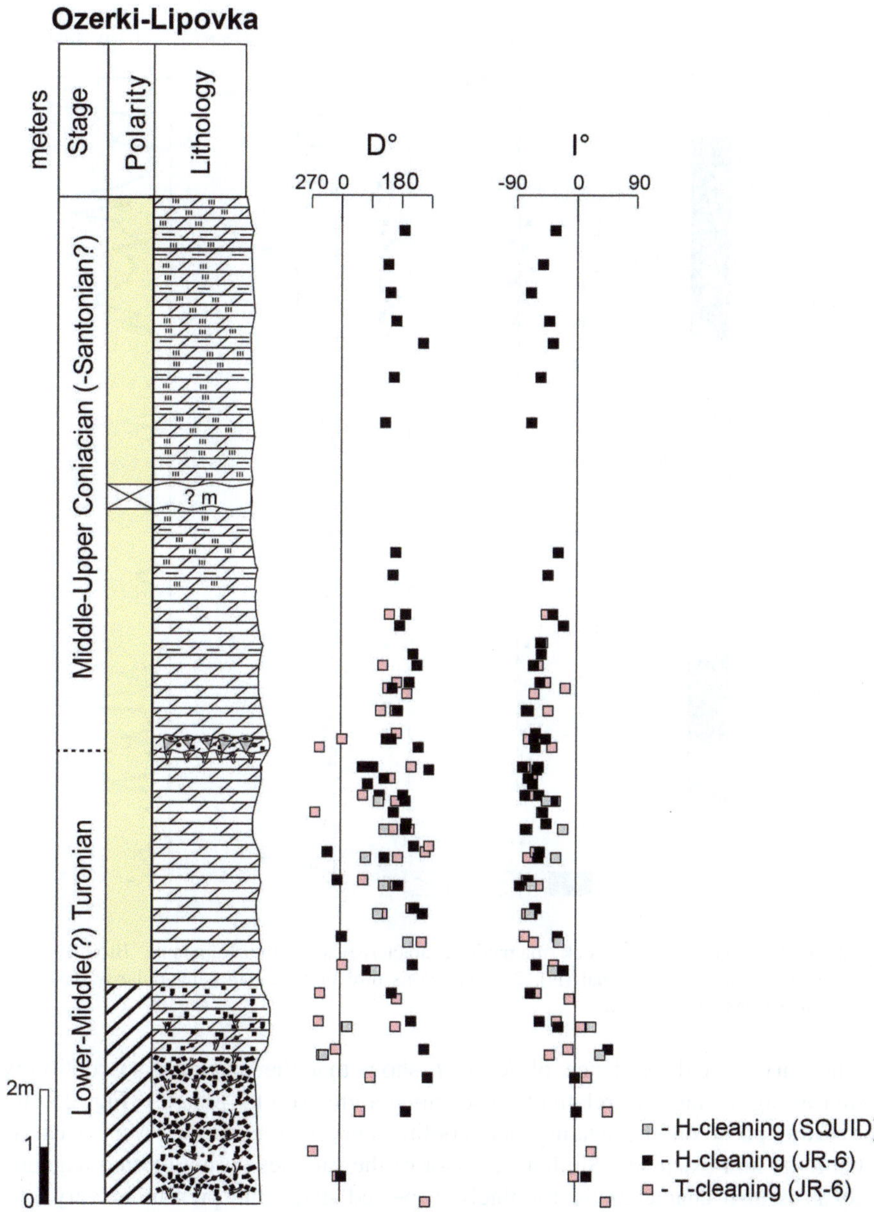

Fig. 25.3 Composite paleomagnetic section of Turonian-Coniacian (-Santonian?) from the Ozerki-1, -2, -3 and Lipovka. For legend see Fig. 25.2

mT was observed (Fig. 25.6a, sample 3124/3). A small increase, accounting for a few percent of the magnetization, reached at 100–200 mT is probably related to the presence of some amount of magnetically hard hydrous ferric oxides.

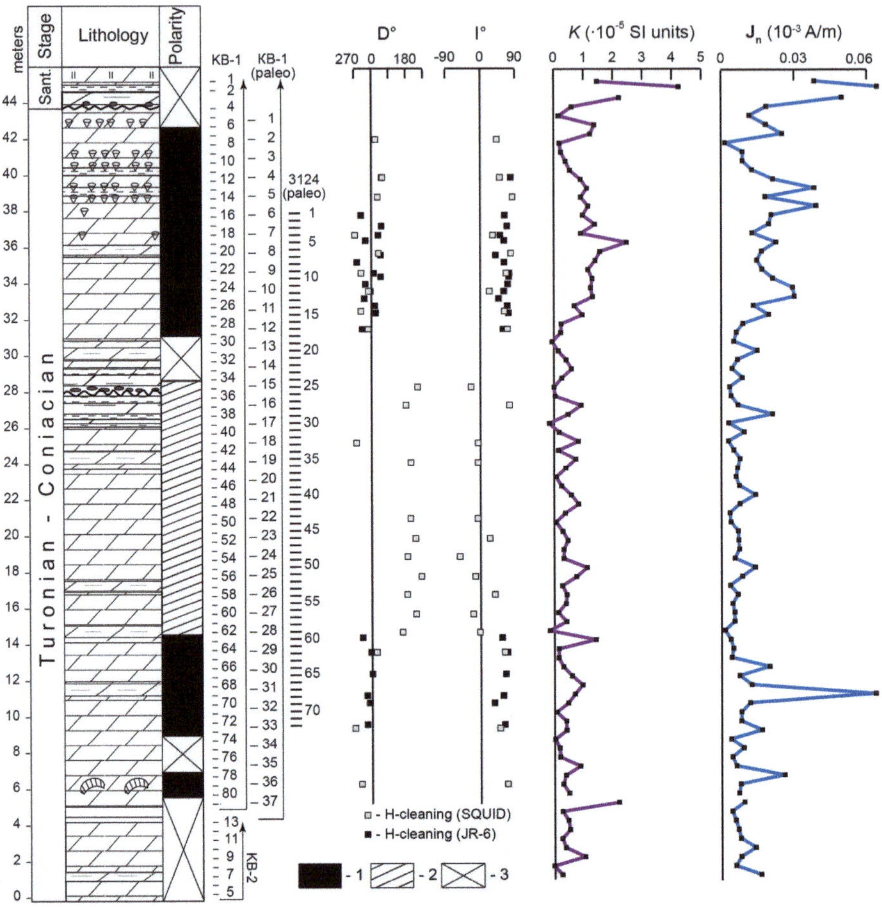

Fig. 25.4 Paleomagnetic and rock magnetic characteristics of the Kamennyi Brod section. Geomagnetic polarity: 1—normal (N), 2—anomalous directions, 3—missing polarity data. For other explanations see Fig. 25.2

The curves of dependence of *K* on *T* show that the magnetic susceptibility decreases significantly at relatively low temperatures (up to 400 °C) (Fig. 25.6b). The Curie point on the heating and cooling curves is expressed very weakly, down to its non-existence. Similar behavior of the samples during thermomagnetic analysis is also characteristic for finely-dispersed magnetite present in very low concentration (Burov and Yasonov 1979).

At a first glance, anisotropy of magnetic susceptibility in the deposits appears random (Fig. 25.7a). In the strongest magnetic varieties, however, it is close to a classic sedimentary magnetic texture, with the short axes of magnetic ellipsoids (K3) positioned vertically, and directions of the long (K1) and medium (K2) axes distributed over the stereogram margin (Fig. 25.7b). This would give some support towards the reliability of obtained paleomagnetic results. In the rest of samples,

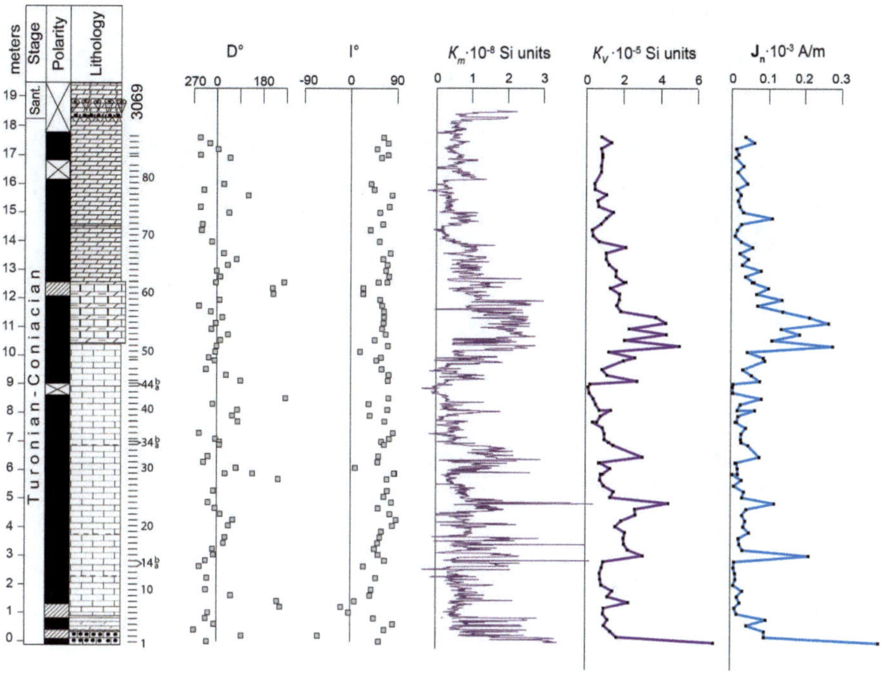

Fig. 25.5 Paleo- and petromagnetic characteristics of the Nizhnyaya Bannovka section. For legend see Figs. 25.2 and 25.4

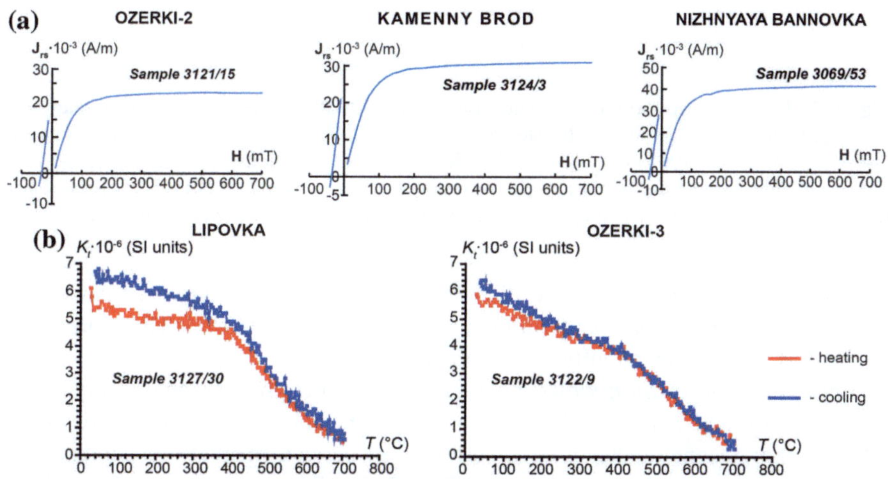

Fig. 25.6 Results of magneto-mineralogical analysis. **a** Curves of magnetic saturation and destruction, **b** curves of dependence of K on T

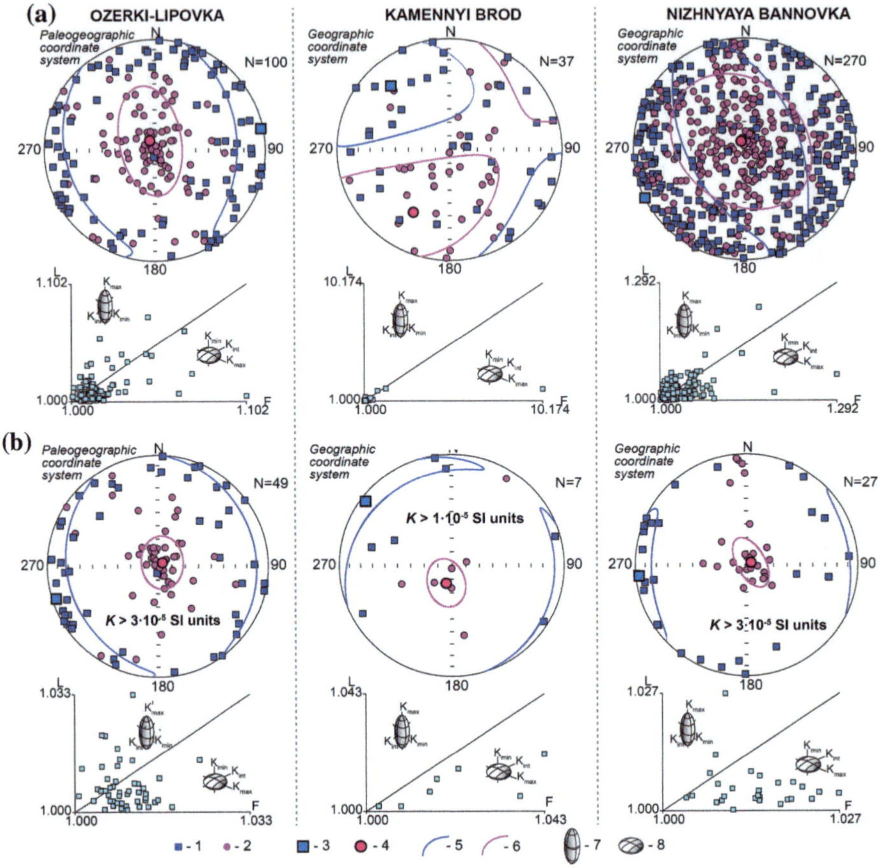

Fig. 25.7 Data on anisotropy of magnetic susceptibility for all samples (**a**) and for samples with the largest magnetic susceptibility (**b**): stereograms of long (K1) and short (K3) axes directions and L–F diagrams (Flinn 1965) (F = K2/K3, L = K1/K2). Axes directions: 1—long (K1), 2—short (K3); 3, 4—mean directions of K1 and K3, respectively; 5, 6—confidence ellipses for K1 and K3, respectively; 7, 8—schematic images of elongate and flat magnetic particles, respectively

however, a large variance of the directions of AMS axes is observed caused by small values of *K* comparable with the instrument sensitivity level. Therefore, most of the AMS data is apparently not informative.

Paleomagnetic examinations were carried out through the standard procedure (Molostovsky and Khramov 1997): the sample $\mathbf{J_n}$ values were measured with a JR-6 spinner magnetometer after a series of successive magnetic cleanings with alternating field (mostly up to 30–60 mT, in 2–5 mT increments) in a LDA-3 AF device and a 2G-Enterprises cryogenic magnetometer (SQUID) in IPE RAS. In the Ozerki-1, 2, 3 sections, temperature cleanings (from 100° to 200–350 °C in 25 °C increments) were made in a home build furnace constructed by Aparin. Thermal cleaning above 350 °C was impossible since the $\mathbf{J_n}$ values became lower than the

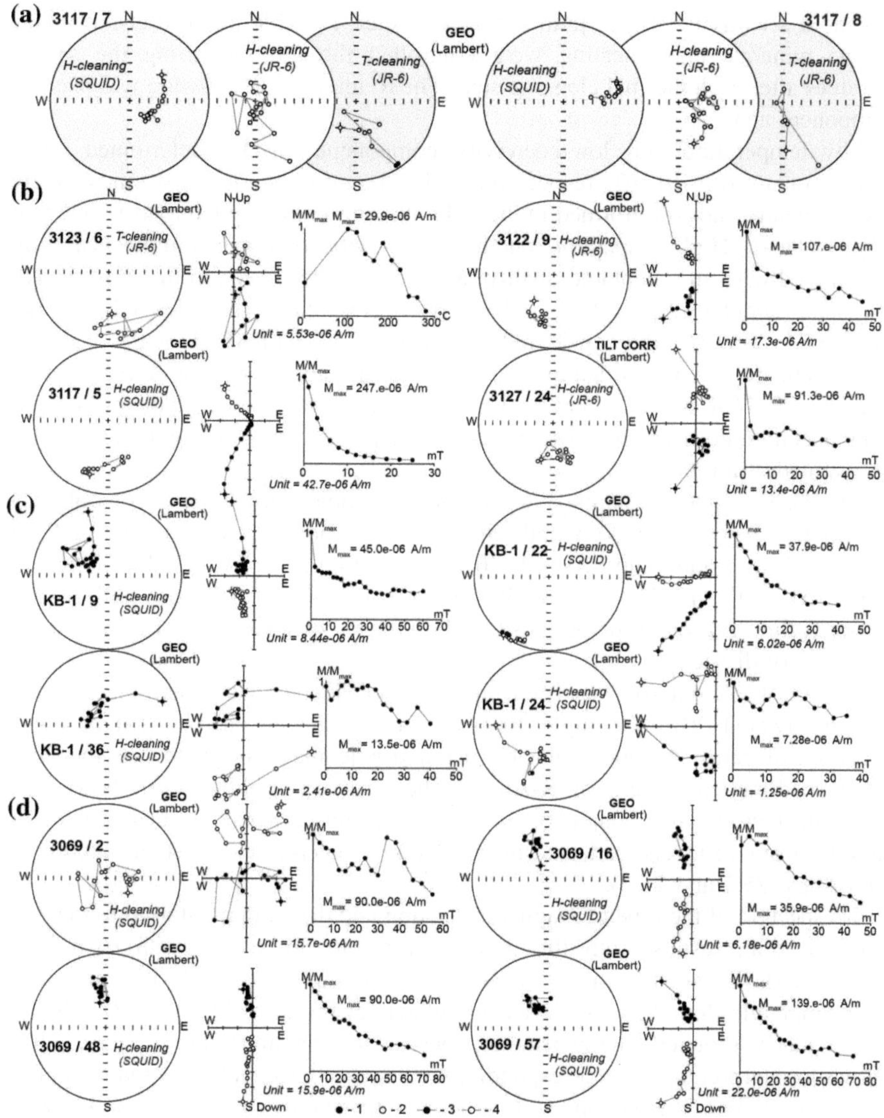

Fig. 25.8 Results of paleomagnetic studies. **a** comparison of H- and T-cleanings on a spinner magnetometer JR-6 and on a cryogenic magnetometer 2GEnterprises. The results of component analysis (left to right: stereographic presentation of $\mathbf{J_n}$ changes in the process of magnetic cleaning, Zijderveld diagrams, sample demagnetization plots) of samples from Turonian-Coniacian for sections Ozerki-Lipovka (**b**), Kamenniy brod (**c**), Nizhnyaya Bannovka (**d**). 1,2 - Projection of $\mathbf{J_n}$ on the lower hemisphere and upper hemisphere, respectively; 3,4 – Projection of $\mathbf{J_n}$ on the horizontal and vertical plane, respectively

threshold sensitivity of the spinner magnetometer. Possible phase transformations of the minerals upon heating were controlled through measuring the sample *K* values after each thermal cleaning step. The Remasoft 3.0. software was used for component analysis.

Low-temperature and low-coercivity components of $\mathbf{J_n}$ (determined up to 100 °C or up to 5–10 mT, respectively) of different directions have, most likely, viscous nature and were formed in the field, that present in the laboratory lodgment (Fig. 25.8a–d). High-temperature and high-coercivity components in most cases are determined with an acceptable quality (MAD < 10°–15°) even in the most weakly magnetic samples ($\mathbf{J_n} < 30\text{–}40 \times 10^{-3}$ A\m) (Fig. 25.8c, samples KB1/22 and Fig. 25.8d, 3069/16) and considered as characteristic (**ChRM**). In some cases full demagnetization in an alternating magnetic field did not occur (Fig. 25.8c, samples KB-1/9, KB-1/24, KB-1/24 and etc.). It is probably connected with the presence of magnetically hard hydrous ferric oxides, whose presence is consistent with the data on magnetic saturation.

The Ozerki-Lipovka composite section was studied in a series of 4 outcrops: Ozerki-1, Ozerki-2, Ozerki-3 and Lipovka, with roughly equal spaces of 2 km between them (coordinates: 51° 31′ 36.4″N, 45° 14′ 0.6″E, 51° 33′ 2.6″N, 45° 17′ 7.9″E, 51° 34′ 34.5″N, 45° 19′ 39.4″E and 51° 30′ 17.6″N, 45° 09′ 35.8″E, respectively). The outcrops lie close to the villages of Ozerki and Lipovka in the Lysye Gory district.

The Ozerki-2 outcrop has been already studied paleo- and petromagnetically (Pervushov et al. 2017a). Samples from 10 of its levels were used to produce a wide spectrum of petromagnetic characteristics and paleomagnetic directions that have established the existence of an R-zone on the basis of sampling from 7 levels. In a subsequent study, this magnetozone has been substantiated by examination of samples from 45 levels, collected in the Ozerki-2 outcrop and in other outcrops as well (Figs. 25.2 and 25.3).

Irrespective of the type of magnetic cleaning and the equipment used to measure magnetization, results from the four outcrops are in accord with each other (Fig. 25.8a). In most samples, characteristic components with the maximum deviation angles less than 15° have been reliably distinguished (Fig. 25.3). **ChRMs** are regularly grouped in the southern bearing of the upper hemisphere, which allows to interpret them as corresponding to the reverse (R) polarity of geomagnetic field (Fig. 25.9a). Terrigenous and carbonate-terrigenous rocks from the Ozerki-2 section (samples 3117/1–3117/3 and 3121/1–3121/9) are peculiar for low paleomagnetic quality, and the paleomagnetic vectors obtained there have anomalous directions preventing reliable determination of the polarity sign (Fig. 25.9b).

The Kamennyi Brod section lies close to the village of the same name in the Olkhovka district of the Volgograd Region (49° 44′ 16.6″N, 44° 23′ 11.3″E). The section is composed mostly of carbonate rocks (marl); its thickness is about 45 m (Fig. 25.4).

In most cases, neither a characteristic nor any other magnetization components could be distinguished from the results of sample measuring with a JR-6 spinner magnetometer. According to the results of measuring with a cryogenic

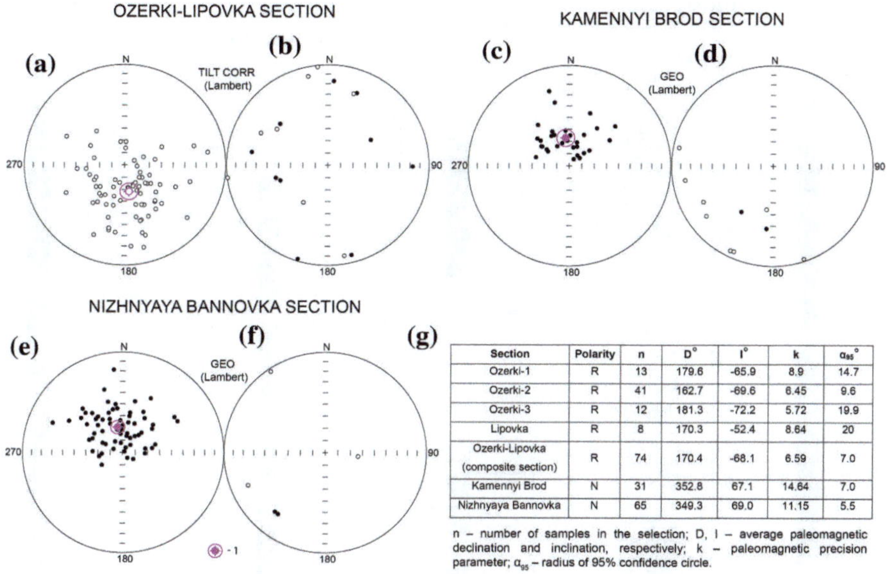

Section	Polarity	n	D°	I°	k	α_{95}°
Ozerki-1	R	13	179.6	-65.9	8.9	14.7
Ozerki-2	R	41	162.7	-69.6	6.45	9.6
Ozerki-3	R	12	181.3	-72.2	5.72	19.9
Lipovka	R	8	170.3	-52.4	8.64	20
Ozerki-Lipovka (composite section)	R	74	170.4	-68.1	6.59	7.0
Kamennyi Brod	N	31	352.8	67.1	14.64	7.0
Nizhnyaya Bannovka	N	65	349.3	69.0	11.15	5.5

n – number of samples in the selection; D, I – average paleomagnetic declination and inclination, respectively; k – paleomagnetic precision parameter; α_{95} – radius of 95% confidence circle.

Fig. 25.9 Results of paleomagnetic studies. Stereographic projections of $\mathbf{J_n}$ components of the studied sections: Ozerki-Lipovka (**a** magnetozone of reverse polarity in carbonate and siliceous-carbonate rocks, **b** intervals with anomalous directions of magnetization in sandy rocks in the bottom of the section), Kamennyi brod section (**c** magnetozone of normal polarity, **d** intervals with anomalous directions of magnetization), Nizhnyaya Bannovka (**e** magnetozone of normal polarity, **f** intervals with anomalous directions of magnetization). **g** Statistic paleomagnetic characteristics of the Turonian-Coniacian from the studied Sections. 1—average paleomagnetic direction and confidence circle; For other explanations see Fig. 25.8

magnetometer, **ChRM** directions corresponding to normal polarity were isolated in many samples (Figs. 25.4, 25.8c and 25.9c). N-zones have been recognized in the uppermost and lowermost parts of the section (Fig. 25.4). Between these two zones, an interval occurs where $\mathbf{J_n}$ components either cannot be resolved or have anomalous directions, or, in rare instances, **ChRM** corresponding to reversed polarity is recorded (Figs. 25.4, 25.8c and 25.9d). One may suppose that this section interval was formed during the reversed polarity epoch, but the substantiation of this hypothesis with reliable paleomagnetic data is still beyond the limits of current equipment.

The Nizhnyaya Bannovka section lies on the Volga right bank, 1–2.5 km southwards from the village of the same name in the Krasnoarmeisk district, Saratov Region. The outcrop coordinates: 50° 43′ 22.6″N, 45° 39′ 17.5″E. The thickness of the interval of Turonian-Coniacian age is 18.2 m (Fig. 25.5).

All the samples from the Nizhnyaya Bannovka section have been measured with a cryogenic magnetometer (Fig. 25.8d). In the vast majority of samples, the **ChRM**, characterized with the maximum deviation angle less than 15°, have been reliably isolated (Figs. 25.5 and 25.9e, f). Paleomagnetic directions are grouped in

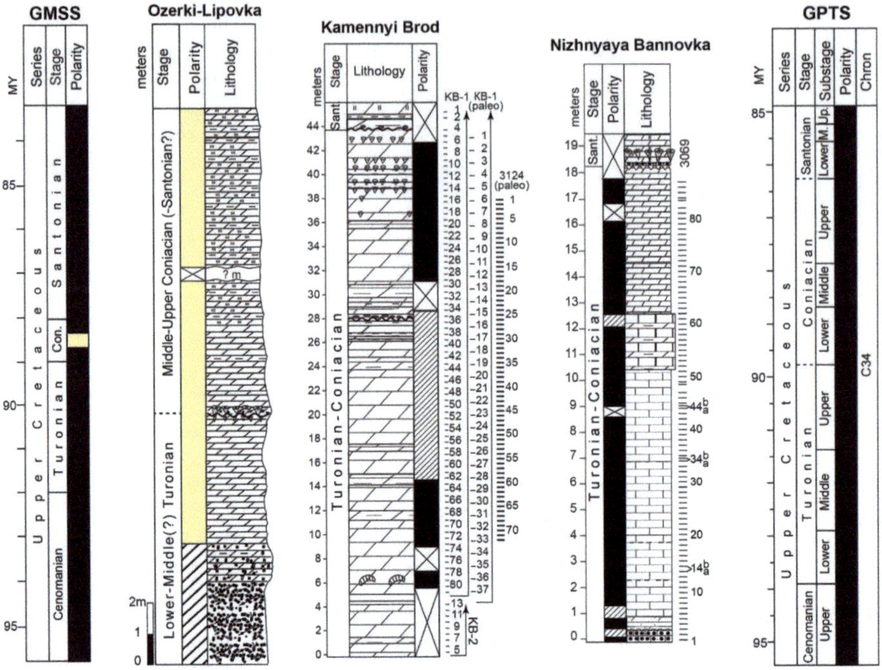

Fig. 25.10 Differences of the Turonian-Coniacian paleomagnetic interval in Geomagnetic Polarity Time Scale (GPTS) (Ogg et al. 2016) and in General Magnetostratigraphic Scale (Supplements … 2000), and the studied sections. For legend see Figs. 25.2 and 25.4

the northern bearings of the lower hemisphere (Fig. 25.9e), which makes it possible to interpret them as corresponding to the normal polarity (N) of geomagnetic field. Thus, the Nizhnyaya Bannovka Turonian-Coniacian section, unlike other coeval section sampled in the Volga region, is peculiar for dominating normal polarity.

Discussion

The results on the magnetic polarity from Nizhnyaya Bannovka and Kamennyi Brod sections (except for the interval with no reliable paleomagnetic data) agree with the existing data which indicate the exclusively normal polarity of geomagnetic field in Cenomanian to Santonian as recorded in the Geomagnetic Polarity Time Scale (Ogg et al. 2016) (Fig. 25.10). Presence of a reversed polarity interval in the Ozerki-Lipovka section however contradicts to this implying that reversed **ChRM** in these rocks has been acquired in some later geologic epochs.

The following circumstances, however, do not support remagnetization:

1. A reversed-polarity nagnetozone is traced in various rock types (marls, siliceous marls) in 4 sections, spaced by as much as 6 km (Fig. 25.2). Such lateral persistence makes a strong argument in favor of primary magnetization (Opdyke and Channell 1996; Supplements … 2000).
2. The reversal test performed for the samples from Ozerki-Lipovka and Nizhnyaya Bannovka sections (Fig. 25.9a, e, respectively), is positive: angle between average N and R vectors (γ) = 179.0°, critical angle (γ_c) = 9.0°, classification after (McFadden and McElhinny 1990)—B. The positive reversal test is in accord with primary magnetization, but theoretically, this may be accounted for by rock remagnetization, e.g. by remagnetization of the Ozerki section in the Matuyama epoch and that of the Nizhnyaya Bannovka section—in the Brunhes epoch. Practically, however, such selective remagnetization seems unlikely.

Mean directions of **ChRM**, corresponding to normal polarity of the field (Fig. 25.9g) are statistically undistinguishable from the direction of modern field in the study area (I = 67.7°–68.3°) (during the statistical testing of the vectors difference as a function of α_{95} (Debiche and Watson 1995) an average amplitude of secular geomagnetic variation—10°—was adopted as a α_{95} value for the modern field). But this circumstance does not contradict the hypothesis of primary nature of magnetization because the directions of late Cretaceous and modern field for the territory of lower Volga region are close. The obtained **ChRM** statistically coincide with the Campanian-Maastrihtian directions, obtained from the sediments in Saratov region (Guzhikov et al. 2017a) and also with directions for the study area, calculated from the coordinates of Turonian-Coniacian (90–80 Myr) paleomagnetic poles for the stable Europe (Besse and Courtillot 2002).

The sediments of the lower parts of Ozerki and Nizhnyaya Bannovka sections, for which the anomalous directions of magnetization are characteristic, represent the condensed horizons, complicated by coarsely sands and nodular phosphorites. In the condensed sediments $\mathbf{J_n}$ vectors are usually significantly different from the directions of geomagnetic field, synchronous to the formation of rocks because of mechanical displacement of rudaceous particles during the constant rewashing, numerous sedimentary gaps, and other factors (Baraboshkin et al. 2016; Guzhikov et al. 2017a). Therefore the anomalous character of magnetization in such sediments is more consistent with the hypothesis of old age of magnetization, rather than with the remagnetization of rocks. In the Kamenniy Brod section the anomalous character of paleomagnetic directions is, most likely, connected with exclusively small values of $\mathbf{J_n}$ (few 10^{-6} A/m), that are comparable with instrument sensitivity level.

3. The acquired data partially agree with a view on the existence of a reversed polarity interval within the Coniacian stage. This is reflected in the General Magnetostratigraphic Scale (Supplements … 2000).

In the Ozerki-Lipovka section the absence of upper Turonian and lower Coniacian is reliably established (Pervushov et al. 2017a, b). A reversed polarity zone covering middle-upper Coniacian may be acknowledged as a probable analogue of Kluevskaya zone. To accommodate all obtained bio- and magnetostratigraphic information on the studied sections it is also necessary to allow the presence of a reverse polarity epoch in the first half of Turonian age. Similar data, based on the of results obtained in different regions do exist (Eremin et al. 1995; Fomin 2003; Fomin and Eremin 1993; Guzhikov et al. 2007; Molostovsky and Khramov 1984; Nazarov et al. 1987; Montgomery et al. 1998; Pechersky 1970). From this point of view in the Nizhnyaya Bannovka section the lower part of Turonian and middle-upper Coniacian are absent, that does not contradict existing information on the biostratigraphy of the section.

The duration of Kluevskaya R-zone, on the basis of its correspondence to middle-upper Coniacian cannot be more than 2.3 Myr, based on the modern views on the duration of Coniacian age (Ogg et al. 2016) and equal time volume of Coniacian substages. Maximum duration of a hypothetic reverse polarity epoch in the first half of Turonian, calculated in a similar way, has the same range ($\sim$2 Myr). The minimal duration of Kluevskaya zone may be estimated as few hundreds of thousands of years, based on that several microfaunistic zones correspond to it. Turonian interval of reverse polarity is situated within the benthic foraminifers subzone (LC3b) (Pervushov et al. 2017a), and so it is entirely possible that it may be quite short-term (less than 10^5 years).

The data about the existence of the reverse polarity epochs definitely contradict generally accepted views about the monopolar regime of normal polarity in Turonian-Coniacian (Ogg et al. 2016), that are based on the magnetostratigraphic data (obtained, in the most part, from the carbonate formations of the Northern Mediterranean) as well as on the analysis of linear magnetic anomalies (LMA). However, it is important to note that the Northern Mediterranean data are not without faults as a source of information about the polarity of geomagnetic field in Turonian-Coniacian neither are the data on LMA (Guzhikov et al. 2003, 2007; Guzhikov 2011). For example, in the Mediterranean sections there is a high chance of regional remagnetization and/or missing of R-zones because of gaps in geological records. The LMA succession may be distorted by the domination of chemical magnetization over the previous thermoremanent magnetization in basalts (Gorodnitskii et al. 1996) and/or the reorganization of spreading patterns, that took place in the middle Cretaceous and the other factors (Larson and Olson 1991). Short epochs of reverse polarity (less than 10^5 years) may not be present in the LMA record bearing in mind insufficient resolution of marine magnetic surveys.

Conclusions

The main result of current study is the detection of a relatively long magnetozone of reverse polarity in the Turonian-Coniacian of Lower Volga region. Paleomagnetic data that form the basis for its determination in the sections Ozerki-1, -2, -3 and Lipovka satisfy the most part of confidence criteria, accepted for the estimation of reliability and quality of magnetostratigraphic materials: 7 out of 8 possible according to (Supplements … 2000) and 8 out of 10 according to (Opdyke and Channell 1996). This is why they cannot remain unnoticed despite the contradiction with the views about normal regime of polarity in Turonian-Coniacian generally accepted at the present time (Ogg et al. 2016).

In the Turonian-Coniacian section Nizhnyaya Bannovka a dominating normal polarity was obtained. In the Kamennyi Brod section on the background of predominantly normal polarity an interval with anomalous directions of magnetization that do not show the definite sign of polarity was detected.

Available biostratigraphic materials and new magnetostratigraphic data may be adjusted under the hypothesis about primary nature of magnetization assuming that the detected R-zone is in fact a superposition of two magnetozones: the lower, associated with the lower part of Turonian, and the upper corresponding to middle-upper Coniacian. The lower magnetozone, probably corresponds to a short (less than 10^5 years) reversed interval, while the upper magnetozone has a much longer duration, about 10^6 years, and represents an analogue of so called "Klyuyevskaya" R-zone, as present in the General Magnetostratigraphic scale (Supplements … 2000).

Substantiation of the idea of a prolonged reversed polarity epoch(s) in the Turonian-Coniacian would be of primary importance for both geophysics and stratigraphy. Acquiring refined information on the regime of the Late Cretaceous field is urgent for the theory of geomagnetic dynamo. Stratigraphically, new reversed polarity zone(s) would make most valuable reference levels of isochronous nature for inter-regional and global correlations.

Further magnetostratigraphic research is necessary for the ultimate solution of the problem relevant to the nature of reverse magnetization revealed in the Turonian-Coniacian sections from the Lower Volga Region.

Acknowledgements The research was financially supported by the RFBR (projects №16-35-00219-mol_a and №16-35-00339-mol_a.).

References

Baraboshkin E.Yu., Rogov M.A., Guzhikov A.Yu., Dzyuba O.S., Urman O.S., Shurygin B.N., Pestchevitskaya E.B., Manikin A.G. (2016) Kashpir section (Volga River, Russia), the proposed auxiliary section for the J/K interval in Subboreal Realm // XII Jurassica Conference. Workshop of the ICS Berriasian Group and IGCP 632. Field Trip Guide and Abstracts Book /

Ed. J. Michalik and K. Fekete. Earth Science Institute, Slovak Academy of Sciences. Bratislava. P. 109–112

Besse J., Courtillot V. (2002). Apparent and true polar wander and the geometry of the geomagnetic field over the last 200 Myr // J. Geophys. Res. 107 (B11), 2300. EPM6-1-31. https://doi.org/10.1029/2000jb000050

Burov B.V, Yasonov P.G. (1979). Introduction to differential thermomagnetic analysis of rocks. – Publishing of Kazan University. 160 p. [in Russian]

Debiche M.G., Watson G.S. (1995). Confidence limits and bias correction for estimating angles between directions with applications to paleomagnetism // J. Geophys. Res. V. 100(B12). P. 24405–24430

Eremin V.N., Nazarov H., Ramazanov S.A., Fomin V.A. (1995). Magnetostratigraphy of a key section of Upper Cretaceous in Western Kopet Dag (Kanavchay) // Proc. of Turkmenistan Academy of Science. 4. P. 163–169. [in Russian]

Flinn D. (1965). On the symmetry principle and the deformation ellipsoid // Geol. Mag. Vol. 102. No. 1. P. 36–45

Fomin V.A. (2003). Magnetostratigraphy of Upper Cretaceous sediments of Eastern Caucasus, Western Kopet Dag and Tuarkyr: abstract of geological science candidate dissertation. Saratov. 22 p. [in Russian]

Fomin V.A., Eremin V.N. (1993). Magnetostratigraphy of upper cretaceous sediments of south regions of USSR // Questions of Paleozoic Mesozoic and Cenozoic stratigraphy / Kuleva G.V. and Ocheva V.G. (Editors) Saratov: Saratov university publishing. P. 134–142. [in Russian]

Fomin V.A., Molostovskiy E.A. (2001). Magnetostratigraphy of Senomanian of Western Turkmenistan // Bulletin of Moscow Society of Naturalists, geological series. 76, 4. P. 62–70. [in Russian]

Gorodnitskii A.M., Shcherbakov V.P., Lamash B.E., Shcherbakova V.V. (1996) The Origin of Magnetic Anomalies and the Structure of the Oceanic Crust. – Moscow: VNIIRO. 283 p. [in Russian]

Guzhikov A. Yu. (2011). About relationship between geomagnetic, geodynamic and geological events in the Cretaceous superchron of normal polarity (Barremian – Campanian) // The current state of Earth Sciences. Materials of the International Conference dedicated to the memory of V.E. Khain (Moscow, February 1–4, 2011). Moscow, Publisher: Geological Faculty, Moscow State University. Pp. 527-532. [in Russian]

Guzhikov A.Yu., Molostovskii E.A., Nazarov Kh., Fomin V.A., Baraboshkin E. Yu., Kopaevich L. F. (2003). Magnetostratigraphic Data on the Upper Cretaceous of Tuarkyr (Turkmenistan) and Their Implications for the General Paleomagnetic Time Scale // Izvestiya, Physics of the Solid Earth, Vol. 39, No. 9. P. 728–740.

Guzhikov A.Yu., Fomin V.A., Baraboshkin E.Yu. (2007). Cretaceous Polarity Scale: Modern Status, Problems of Construction and Outlook of Development // Cretaceous system of Russia and CIS countries: problems of stratigraphy and paleogeography. / Ed. E.M. Pervushov. – Saratov: Saratov university publishing. P. 69–86. [in Russian]

Guzhikov A.Yu., Baraboshkin E.Yu., Beniamovsky V.N., Vishnevskaya V.S., Kopaevich L.F., Pervushov E.M., Guzhikova A.A. (2017a) New Bio- and Magnetostratigraphic Data on Campanian–Maastrichtian Deposits of the Classical Nizhnyaya Bannovka Section (Volga River Right Bank, Southern Saratov Region) // Stratigraphy and Geological Correlation, Vol. 25, No. 1, pp. 39–75

Guzhikov A., Guzhikova A., Pervushov E., Ryabov I., Surinskiy A., Seltser V., Beniamovski V. (2017b). The Reverse polarity zone in the Turonian-Coniacian interval of the Lower Volga region // Sames, B. (Ed.): 10th International Symposium on the Cretaceous – Abstracts, 21–26 August 2017, Vienna. – Berichte der Geologischen Bundesanstalt, 120, Vienna. P. 104

Krumsiek K. (1982) Cretaceous Magnetic Stratigraphy of Southwest Morocco. In: von Rad U., Hinz K., Sarnthein M., Seibold E. (eds.) Geology of the Northwest African Continental Margin. Springer, Berlin, Heidelberg. P. 475–497

Larson R.L., Olson P. (1991) Mantle Plumes Control Magnetic Reversal Frequency // Earth Planet. Sci. Lett. Vol. 107. P. 437–447

Makarova S.D., Tsapenko M.N. (1971). About rhythmostratigraphic and paleomagnetic correlation of Cretaceous formations in Northern and Eastern Fergana region // Report of Uzbek SSR Academy of Science. 8. P. 44–46. [in Russian]
McFadden P.L., McElhinny M.W. (1990). Classification of the reversal test in palaeomagnetism // Geophys. J. Int. 103. P. 725–729
Molostovsky E.A., Khramov A.N., (1984). Paleomagnetic scale of Phanerozoic and problems of magnetostratigraphy // Transactions of XXVII Internatioanl Geological Congress, Moscow, 1: 16–24. [in Russian]
Molostovsky E.A., Khramov A.N. (1997). Magnetostratigraphy and its importance in geology / Saratov: Saratov university publishing. 180 p. [in Russian]
Montgomery P., Hailwood E.A., Gale A.S., Burnett J.A. (1998). The magnetostratigraphy of Coniacian-late Campanian chalk sequences in southern England // Earth Planet. Sci. Lett. Vol. 156. Pp. 209–224
Nairn A.E.M., Schmitt T.J., Smithwick M.E. (1981). A paleomagnetic study of the Upper Mesozoic succession in Northern Tunisia // Geophys. J. Res. astr. Soc. 65. P. 1-18
Nazarov H., Mamedov M., Ramazanov S., et al. (1987). Paleomagnetic-stratigraphic studies of Mesozoic sediments of Turkmenistan territory // Geologic structure of Turkmenistan. Ashgabat. P. 161–179
Ogg J.G., Ogg G.M., Gradstein F.M. (2016). – A Concise Geologic Time Scale. Elsevier. 242 p
Opdyke N.D., Channell J.E.T. (1996). Magnetic Stratigraphy. – N.Y.: Academic press. 344 p
Pechersky D.M. (1970). Paleomagnetism and paleomagnetic correlation of Mesozoic sediments in North-East of USSR // Paleomagnetic and biostratigraphic characteristic of some Mesozoic and Cenozoic key sections of Northern Far East. – Magadan: Northeast complex Research and development Institute. 37. P. 58–99. [in Russian]
Pervushov E.M., Seltser V.B., Kalyakin E.A., Fomin V.A., Ryabov I.P., Ilinskiy E.I., Guzhikova A.A., Biryukov A.V., Surinskiy A.M. (2017a) Integrated bio and Magnetostratigraphic Study of the Ozerki Sections (Upper Cretaceous, Right Bank of the Volga in the Saratov Region). Paper 1. Section Characteristics, Results of Petromagnetic and Magnetic-Mineralogical Examinations // Izv. Saratov Univ. (N. s.), Ser. Earth Sciences. Vol. 17, No. 2. P. 105–116. [in Russian]
Pervushov E.M., Seltser V.B., Kalyakin E.A., Fomin V.A., Ryabov I.P., Ilinskiy E.I., Guzhikova A.A., Biryukov A.V., Surinskiy A.M. (2017b) Integrated bio and Magnetostratigraphic Study of the Ozerki Sections (Upper Cretaceous, Right Bank of the Volga in the Saratov Region). Paper 2. Oryctocoenosis Characteristics and Biostratigraphy // Izv. Saratov Univ. (N. s.), Ser. Earth Sciences. Vol. 17, No. 3. P. 182–199. [in Russian]
Shebaldin V.P. (2008). Tectonics of the Saratov region. – Saratov: publishing house of "Saratovneftegeofisica". 40 p. [in Russian]
State Geological Map of the Russian Federation. Scale 1 : 000 000. M–38 (Volgograd). – Editor: Zastrozhnov S.I. St. Petersburg, "Rosnedra", 2009. [in Russian]
Stratigraphic scheme of upper Cretaceous sediments of East European Craton. – St. Petersburg.: Cartographic factory of VSEGEI, 2004, 6 schemes on 10 sheets. [in Russian]
Supplements to the Stratigraphic Code of Russia / A.I. Zhamoida (ch. Ed.) (2000). St. Petersburg: Izd. VSEGEI. 112 p (in Russian). [in Russian]

Chapter 26
Magnetostratigraphy of the Key Loess-Palaesol Sequence at Roxolany (Western Black Sea Region)

D. V. Hlavatskyi and V. G. Bakhmutov

Abstract Roxolany Section (46° 10′N; 30° 27′E) is one of the most representative outcrops within Western Black Sea Lowland. Its 55-m thick loess-palaesol sequence was stratified by palaeomagnetic, palaeopedological, thermoluminescence and other methods, but the results are different. Thus in palaeomagnetic studies of Roxolany the Matuyama–Brunhes (M/B) boundary was placed in the diverse stratigraphic horizons. The authors of (Tretyak et al. Int geol nauk 50, 1987) found a number of zones of normal, reversed and anomalous polarity, but attributed the whole 55-m loess-soil sequence to the Brunhes chron of normal polarity. Later, based on the results of the complex investigation of this section and taking into account the palaeomagnetic data from other outcrops in this region, the position of the M/B boundary was drawn in the loess located below the PK_6 soil at a depth of ~34 m (Tsatskin et al. in Palaeogeogr Palaeoclimatol Palaeoecol 143:111–133, 1998), which corresponds to the Tyasmyn horizon according the stratification cited in (Bogucki et al. in Loess-covering of the North Black Sea Region. KARTPOL, Lublin, pp. 47–58, 2013). Our new palaeomagnetic investigations were carried out in a few laboratories with detailed measurements of specimens from duplicated outcrops after full AF and thermal demagnetization. We identified that the M/B boundary in the Roxolany section is located at a depth of 46.6 m at the contact between the buried soils of the Lubny and Martonosha horizons, which generally agrees with the present-day notions about the Quaternary stratigraphy of the south of Ukraine (Bakhmutov and Hlavatskyi in Geofiz Zhurn 4(38):59–74, 2016).

Keywords Magnetostratigraphy · Matuyama-Brunhes boundary
Palaeomagnetic method · Pleistocene · Loess-paleosol sequence
Roxolany section

D. V. Hlavatskyi (✉) · V. G. Bakhmutov
Institute of Geophysics by S.I. Subbotin Name, National Academy of Sciences of Ukraine, Kiev, Ukraine
e-mail: hlavatskyi@gmail.com

V. G. Bakhmutov
e-mail: bakhmutovvg@gmail.com

D. Nurgaliev et al. (eds.), *Recent Advances in Rock Magnetism, Environmental Magnetism and Paleomagnetism*, Springer Geophysics,
https://doi.org/10.1007/978-3-319-90437-5_26

Introduction

Thick loess-palaeosol sequences spanning the time interval from at least the late Matuyama chron through the Holocene have long been known in the western Black Sea area (Veklitch et al. 1967), similar to other non-glaciated areas of the southern East European plain (Velichko 1990). The loess-palaeosol series occupies more than 70% of Ukraine's plane area and include the stratigraphically complete sequences of the Lower and Middle Pleistocene. Cyclic alternation of loess and palaeosol units provided the bases for elaboration of the Pleistocene stratigraphic framework of Ukraine (Veklitch et al. 1967, 1984), which is presently substantiated by results from multidisciplinary study of more than 200 Quaternary sites in different areas of Ukraine. The recent review of the main stratigraphic units with considering of palaeoenvironmental evolution of region is presented by (Gozhik and Gerasimenko 2011).

Despite the details, the current schemes of Pleistocene stratigraphy in Ukraine basically have a significant flaw. Due to the change of thickness of the main loess/palaeosol units and abrupt changes of their sedimentation conditions the correlation even within individual provinces are not always straightforward. In addition, most of the continental Pleistocene strata have not fauna which complicates their stratigraphic correlation, and the results of thermoluminescence dating are still controversial and questionable (Gozhik et al. 1995; Fedorowicz et al. 2013).

Therefore the determination of key benchmark of Quaternary magnetostratigraphy—Matuyama/Brunhes boundary—in the loess/palaeosol sequences is one of the most important issue in Quaternary geology. The age of this boundary calibrated by an astronomical scale is 0.78 My and for the oxygen isotope scale it falls to the Marine oxygen isotope stage 19. Determination of the Matuyama/Brunhes boundary provides the only absolute age assignment.

The position of Matuyama/Brunhes (M/B) boundary on the different stratigraphic scales has been discussed for a long time. The "floating" position of the M/B boundary due to nature of the loess/palaeosol characteristic remanent magnetization (ChRM) leads to the fact that in Central Europe it was defined both in the interglacial palaeosols (Hungary, Czech Republic, Poland) and in loesses (Austria, Bulgaria, Ukraine) as well (Bakhmutov et al. 2005; Bakhmutov and Hlavatskyi 2016).

In the palaeomagnetic research of 1970–1990 yy the stratigraphic position of the M/B boundary on the territory of Ukraine belonged to Sula loess, Martonosha palaeosol, Shyrokino palaeosol (Tretyak and Volok 1976; Tretyak 1983; Tretyak et al. 1987, 1989, Tretyak and Vigilyanskaya 1994).

In the 1990s the multidisciplinary study of loess/palaeosol sequences in the Western Black Sea region, emplaced the M/B boundary in other stratigraphy horizon which was refined the pedostratigraphic scheme (Tsatskin et al. 1998, 2001). This data was contradicting the current schemes of Pleistocene stratigraphy in Ukraine (Gozhik and Gerasimenko 2011). According to review (Gozhik 2013) the M/B boundary attaches to Martonosha soil horizon of Ukrainian Quaternary framework.

The contradictions in positioning this boundary can be illustrated by the history of the paleomagnetic studies of the Roksolany section in the Western Black Sea region. The authors of (Tretyak et al. 1987; Tretyak and Vigilyanskaya 1994) distinguished a number of the zones of normal, reversed and anomalous polarity in this section but attributed the whole 55-m loess-soil section to the Brunhes chron of normal polarity.

In the mid-1990s, based on the results of the combined investigation of this section (Tsatskin et al. 1998; Sartori 2000; Tsatskin et al. 2001; Gendler et al. 2006), the position of the M/B boundary was drawn in the loesses located below the PK_6 soil at a depth of ~34 m, which corresponds to the Tyasmyn horizon according to the stratification cited in (Bogucki et al. 2013). These data formed the basis for constructing the new pedostratigraphic scheme for Western Black Sea region with its subsequent correlation to the oxygen isotope (OI) time scale (Tsatskin et al. 2001). Remarkably, none of the reversed polarity events within the Brunhes chron are reported. Nevertheless, in the other detailed studies of the upper part of the Roxolany section, the authors of (Sharonova et al. 2004; Pilipenko et al. 2005) identified the anomalous zones within the top 20 m of the sequence, which were attributed to the Mono, Blake, and Jamaica excursions (~30, 100, and 200 ka ago, respectively).

These discrepancies can be caused by difficulties in removal of the viscous component which produces a wide scatter of the cleaned NRM directions (Bakhmutov et al. 2017).

We performed an independent study of the Roxolany section focusing on the definition of the M/B boundary.

Geological Setting

Roxolany Section (46° 10′N; 30° 27′E) is located on the coast of the Dniester estuary west of the Roxolany village of the Ovidiopol region of Odessa oblast, Ukraine. This section is one of the most representative outcrops of the loess series in Black Sea Lowland. The terrace complex is composed of the Dniester terrace alluvium VII which erosionally overlays the Pontic deposits. These units are covered by almost a 55-m thick loess and loam stratum with a complex of fossil soils which was stratified in detail in its time by the paleopedological, thermoluminescence, and other methods: this stratum characterizes the very Pleistocene stage of the Quaternary period. The detailed description of the section which we were guided by in rock sampling is presented in (Bogucki et al. 2013).

Results

Sampling Method and Laboratory Measurements

A representative collection of samples have been collected over two field seasons in 2012–2013 yy. 127 oriented rectangular blocks and 203 mini-cores were taken from 9 outcrops down the profile. We started with a pilot collection, to make the necessary measurements that constrained the positions of the geomagnetic polarity zones, followed by additional sampling to gain higher precision. We pay attention to the more problematic levels, taking extra samples, with a sampling density of about every 5–15 cm. The above described economic sampling strategy finally yielded more than 500 oriented samples, which were mostly collected as hand samples. For palaeomagnetic measurements were cut standard cylinders (2.2 cm in length) or cubes (2.0 cm side).

Particular attention was paid to the zone of the expected localization of the M/B boundary at a depth from 34 to 54.5 m. Therefore, the palaeomagnetic studies in this depth interval were carried out for 128 samples (out of a total of 186 samples), i.e. the average sampling interval in this depth segment was 0.16 m. Seventy five of these 128 samples were subjected to T-demagnetization and 54 samples to H-demagnetization. These data also include several samples from a depth of 34–42 m, which were repeated in two other exposures (in one of them the group of the authors (Tsatskin et al. 1998; Sartori 2000) was working).

Palaeomagnetic measurements were carried out in the laboratory of the Institute of Geophysics of the National Academy of Sciences of Ukraine in Kyiv.

The vectors of characteristic remanent magnetization (ChRM) were isolated by both stepwise thermal (TD) and alternating field (AF) demagnetization. The procedures of demagnetization of specimens (TD or AF) and all measurements were made inside of the magnetically shielded rooms, to minimize the acquisition of present-day viscous magnetization.

Specimens were stepwise thermally demagnetized using an MMTD80 oven up to 300–400 °C. After each heating step, the magnetic susceptibility at room temperature was measured with a MFK1 Kappabridge to monitor possible mineralogical changes. Duplicate specimens were subjected to AF demagnetization with steps 5–20 mT up to 100 mT using a LDA-3A demagnetizer. The NRM of specimens was measured by JR-6 spin magnetometer with a sensitivity of 2.4 μA/m. For verification of demagnetization results, a duplicate specimens were measured in the palaeomagnetic laboratory of the Institute of Geophysics of the Polish Academy of Science, Warsaw (by 2G SQUID DC magnetometer accompanied by an AF demagnetizer).

Demagnetization results were processed by multicomponent analysis of the demagnetization path (Kirschvink 1980), using Remasoft 3.0 software (Chadima and Hrouda 2006). Anisotropy of magnetic susceptibility (AMS) was measured with a MFK-1 Kappabridge.

Palaeomagnetic Results and Determination of the Matuyama/Brunhes Boundary

The initial results of our palaeomagnetic study of pilot 131 specimens from Roxolany section were published in (Bakhmutov and Hlavatskyi 2014). The standard palaeomagnetic experiments were performed, consisting of the measurement of natural remanence (NRM) of specimens in their original state and, after each demagnetization step, during AF and TD procedures.

For now the total of 237 samples was treated using AF demagnetization (97 specimens) and thermal demagnetization (143 specimens). Since most of the specimens were fragile, they cannot be heated to temperatures above 300–400 °C and we were limited to a few steps demagnetization at temperatures 180, 210, 240, 270 °C and higher if the specimen did not broken. After each step, a specimen's magnetic susceptibility was measured at room temperature, to monitor possible mineralogical changes.

Multicomponent analysis of demagnetization paths reveals that the NRM of the samples is composed of two components. The low stability component was erased in the temperature range 20 °C to 180–210 °C (sometimes to 240 °C) or AF field in the interval 10–20 mT (sometimes to 30 mT). The more stable (characteristic) component in many palaeosol samples was less than 5% of the initial NRM while in losses it is about 10–20%. Since its value is low even for measuring by SQUID-magnetometers, ChRM-component is not displayed in some specimens or couldn't be distinguished by multicomponent analysis method.

Figures 26.1 and 26.2 show an example of AF and TD demagnetized specimens beginning with Dnipro loess and down from lower loess or soil horizon. Stereographic projection of the directions and NRM intensity decay during demagnetization (M/M_{max}) are presented. In addition the changes of magnetic susceptibility during thermal treatment, orthogonal projections of demagnetization paths (Zijderveld diagrams) on horizontal and vertical planes are given.

Incomplete removal of the viscous component produces a scatter of the ChRM directions. Nevertheless, a fairly clear magnetic polarity stratigraphy was appeared. In the upper and middle part of the section the ChRM-component is characterized by normal polarities (from the top to the depth of 46.6 m, Fig. 26.1). Only 8 specimens from Dnipro (dn), Orel (or), Zavadivka (zv) and Lubny (lb) units with reverse polarity were manifested (Fig. 26.3), but often these results were not confirmed by the other samples from each unit. Beginning with Martonosha palaeosol (mr) all specimens are definitely showed the reverse polarity of ChRM-components both by AF and TD demagnetization (Fig. 26.2).

The presence of a reversed direction in loess and alluvium is characterized by a small increase of intensity at a field of 15–30 mT above which the reversed ChRM direction is clearly depicted (Fig. 26.2). The same increasing of intensity is observed at the temperature of 210–270 °C. In the palaeosol samples the overprint is carried by low-coercivity minerals that can be demagnetized using AF and thermal demagnetization (Fig. 26.2a, b). The thermal demagnetization is clearly

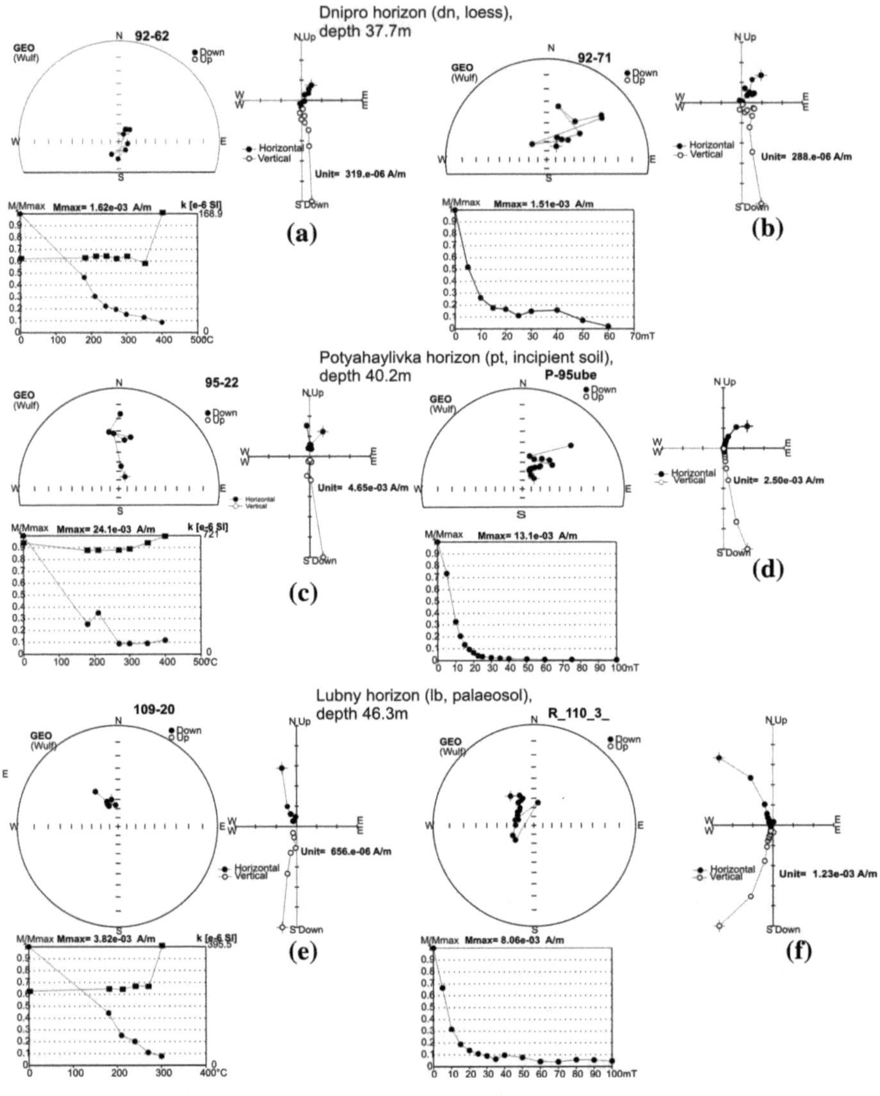

Fig. 26.1 Examples of stepwise thermal (**a**, **c**, **e**) and alternating field (**b**, **d**, **f**) demagnetization of typical loess (**a**, **b**) and palaeosol (**c**, **d**, **e**, **f**) specimens from middle part of the section

indicates the presence of a reversed characteristic direction which is overprinted by a normally oriented secondary component. The latter is completely demagnetized at 250 °C, revealing the ChRM and its reversed direction.

Since in some horizons the presence of molehills was recorded, in order to avoid errors due to mechanical disturbances of rocks, in addition to visual inspection, the anisotropy of the magnetic susceptibility parameters was analyzed. Anomalous

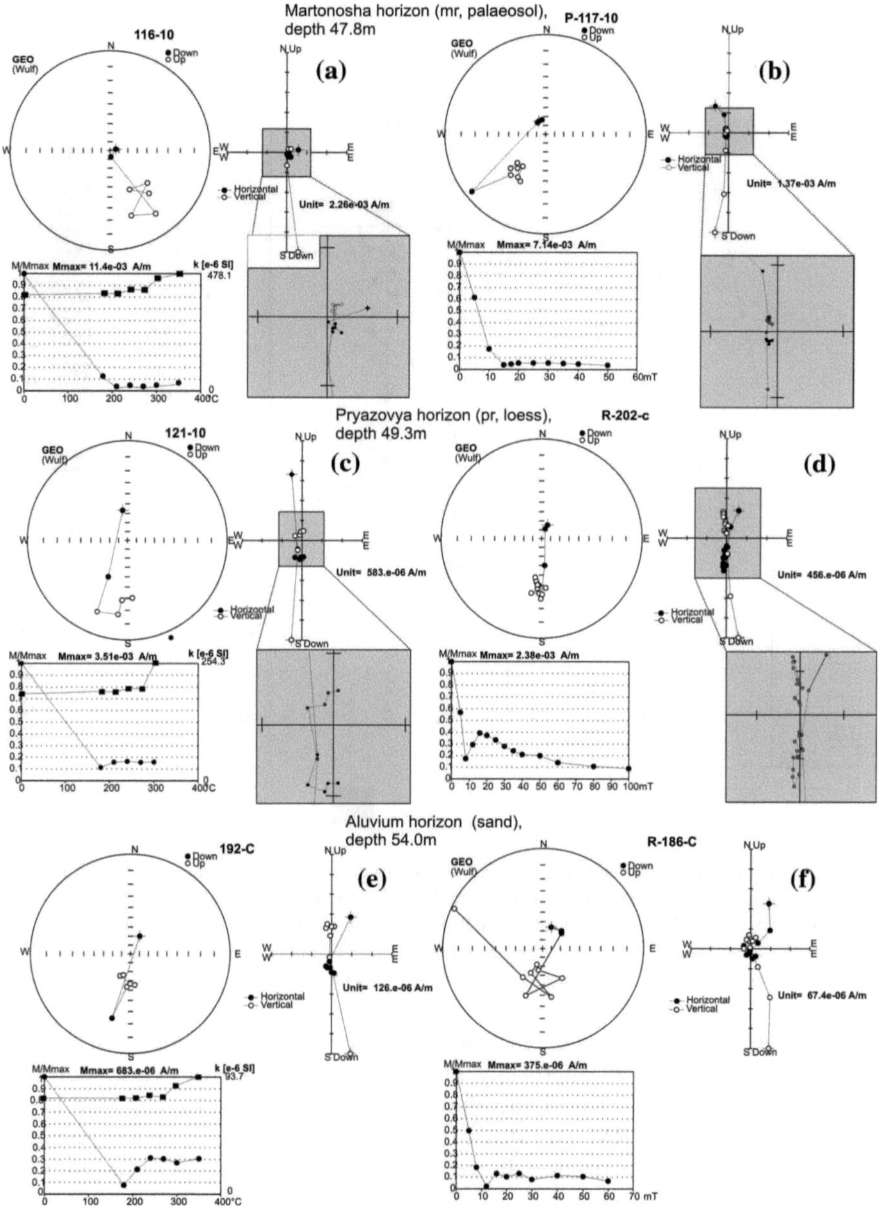

Fig. 26.2 Examples of stepwise thermal (**a**, **c**, **e**) and alternating field (**b**, **d**, **f**) demagnetization of typical palaeosol (**a**, **b**), loess (**c**, **d**) and sand (**e**, **f**) specimens from lower part of the section

deviations from a typical sedimentary structure were determined on some samples, and if palaeomagnetic results also showed abnormal directions, such samples were excluded in further interpretation.

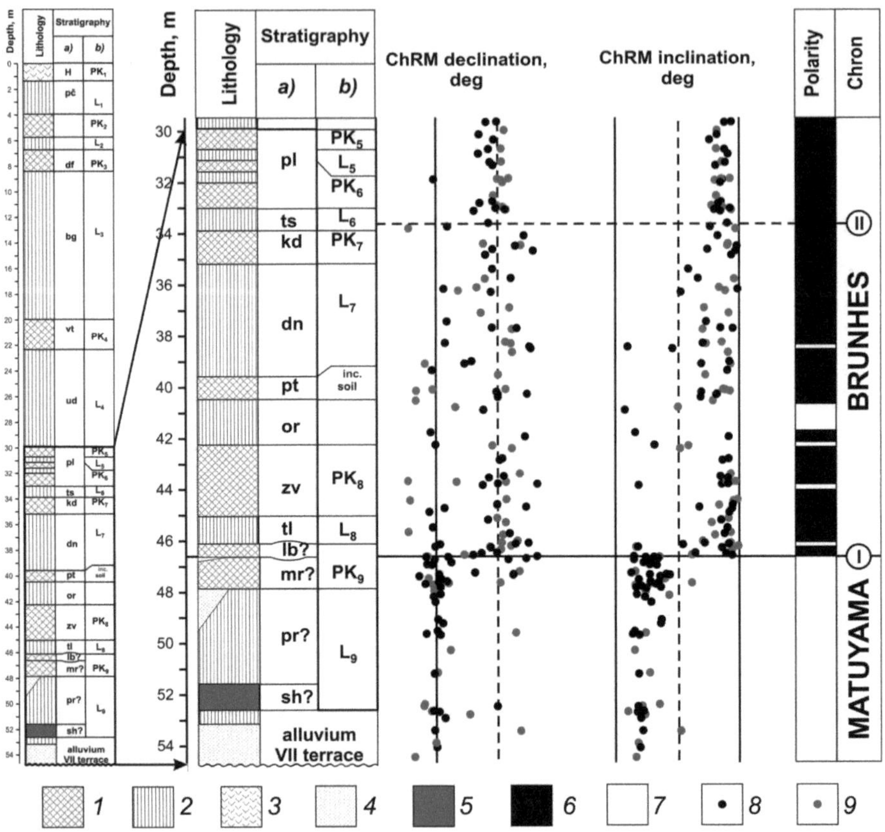

Fig. 26.3 Results of palaeomagnetic study of the lower part of Roxolany section. Position of the Matuyama/Brunhes boundary (right) according to: (I)—our data; (II)—data of (Tsatskin et al. 1998; Dodonov et al. 2006). Stratigraphy: **a** Bogucki et al. (2013); **b** Tsatskin et al. (1998). 1—palaesol, 2—loess and loess-like loam, 3—chernozem, 4—sand, 5—hydromorphic soil, 6—normal polarity, 7—reversed polarity, 8—thermal demagnetization, 9—A. F. demagnetization

The final results of the demagnetization of the samples by temperature at 112 levels (black dots) and an alternating magnetic field at 74 levels (gray dots) are shown on the Fig. 26.3. The figure shows the declination and inclination of the ChRM component of NRM, calculated from the results of the component analysis after TD and AF demagnetization. In the right-hand part of Fig. 26.3 is a composite palaeomagnetic section.

The M/B boundary is defined by us at the junction of the Lubny and Martonosha horizons at a depth of 46.6 m. The samples below demonstrate reversed polarity of the ChRM component. In the depth range of 38–46.6 m eight samples show reverse polarity (as a result of TD demagnetization). Reversed polarity zones corresponding to these samples are taken to the palaeomagnetic section, but aren't compared with the palaeomagnetic events, because there are not enough data for their unambiguous interpretation.

Discussion

If the results of the rock magnetic measurements obtained for the Roxolany section completely agree with the results of the previous studies (Bakhmutov et al. 2017), the results of our palaeomagnetic studies are different from previous data (Tsatskin et al. 1998, 2001; Sartori 2000; Dodonov et al. 2006). According to our results, the M/B boundary in Roxolany section is located at a depth of 46.6 m at the junction of two palaesols of Lubny and Martonosha horizons. This is consistent with modern concepts of the stratigraphy of the Quaternary strata of southern Ukraine (Gozhik 2013) and refutes data that the Matuyama-Brunhes boundary is at a depth of about 34 m in loess bellow the Pryluky horizon (PK_6) (Tsatskin et al. 1998, 2001; Dodonov et al. 2006).

The reasons for these discrepancies may be as follows. Laboratory work on previous palaeomagnetic studies of Roxolany section (Tsatskin et al. 1998, 2001; Gendler et al. 2006), are described in detail in the monograph (Sartori 2000). A total of 118 samples were demagnetized by an alternating magnetic field up to 100 mT and temperature only up to 150 °C (the samples were placed in plastic boxes). At this temperature the viscous component is not yet removed (Bakhmutov and Hlavatskyi 2014), the ChRM component is appeared at temperatures of 210–270 °C.

Therefore, temperature cleaning was carried out in another laboratory by J. Hus, and new data were included in (Sartori 2000). According to these data, the viscous component disappeared at 250 °C and characteristic component was isolated. Most of the samples below 34 m are shown the reversed polarity, the M/B boundary was placed in the Tyasmyn horizon at a depth of 34 m, bellow the overlying pedocomplex PK_6 [Pryluky horizon according to (Bogucki et al. 2013)].

Unfortunately, the Zijderveld diagrams showing reversed polarity are given only for two samples from depths of 46.3 and 46.6 m, and it is not possible to evaluate the direction of the characteristic component of the magnetization for samples in the depth range 34–46 m. Surprisingly, in the palaeomagnetic interpretation given in (Sartori 2000; Gendler et al. 2006) authors ignore the fact that in the depth range of 34–46 m about a third of the heated samples and half of the samples after AF demagnetization showed normal polarity. The authors do not specify on which statistical sample the location of the M/B boundary is based. Moreover, according to our results of the study of samples from outcrops which were sampled earlier by (Tsatskin et al. 1998; Sartori 2000) in the interval 34–42 m, nine samples from the ten indicate only the normal polarity direction of ChRM.

In other publication of palaeomagnetic studies of the Roxolany section the authors (Dodonov et al. 2006) draw the M/B boundary at the bottom of the PK_6 pedological complex (under Pryluky horizon) on depth of about 34 m, which coincides with (Tsatskin et al. 1998, 2001). According to (Dodonov et al. 2006), from 22 samples at a depth of 34–50 m (i.e., samples were taken at an interval of about 0.7 m) 20 samples shown the reversed polarity, while 2 samples with normal polarity were placed in the lower (Shyrokino?) soil and were correlated with the

Jaramillo episode. However, the authors (Dodonov et al. 2006) do not indicate what method of magnetic cleaning and what equipment was used and how the directions of the ChRM directions were calculated.

All of this allows us to consider the position of the M/B boundary in Roxolany section at a depth of 46.6 m at the junction of palaesols of the Lubny and Martonosha horizons as the most reliable.

Conclusions

1. The ChRM components of the magnetization were isolated both in soil and loess samples. This component has the features of the primary magnetization, its normal and reversed polarities were identifying both in loess and soil horizons due to detail measurements on the modern equipment.
2. The M/B boundary in Roxolany section is located at a depth of 46.6 m at the contact of two palaesols of Lubny and Martonosha horizons. This is consistent with modern concepts of the stratigraphy of the Quaternary strata of southern Ukraine (Gozhik 2013) and refutes data that the Matuyama-Brunhes boundary is at a depth of about 34 m in Tyasmyn loesses bellow the Pryluky horizon (PK_6) (Tsatskin et al. 1998, 2001; Dodonov et al. 2006).
3. Brunhes chron includes some subzones of reversed polarity that aren't assigned yet to any palaeomagnetic events, because there are not enough palaeomagnetic, geological or geochronological data for their unambiguous interpretation.
4. At sites where independent age control is lacking, the identification of reversed polarity remanence allows a minimum age assignment of 780 ka. Such minimum age data are extremely useful in correlation of subaerial deposits. Episodes of reversed polarity reported from the Brunhes chron are of short duration (up to a few thousand years), may not be global in extent and they are difficult to detect due to weak magnetic signal in basic sediments even using new precise equipment.

References

Bakhmutov V., Hlavatskyi D., 2014. New data about Matuyama-Brunhes boundary in Roxolany section. *Geol. Zhurn.* № 2 (347). – P. 73–84. (in Russian).

Bakhmutov V.G., Hlavatskyi D.V., 2016. Problems of magnetostratigraphy of Pleistocene loess-soil deposits of the South of Ukraine. *Geofiz. Zhurn.* 4 (38). – P. 59–74. (in Russian).

Bakhmutov V.G., Mokriak I.N., Skarboviychuk T.V., Yakukhno V.I., 2005. Results of palaeomagnetic studies of Danube terraces sections and problems of Pleistocene magnetostratigraphy of the west Black Sea region. *Geofiz. Zhurn.* T. 25, № 6. – P. 980–991. (in Russian).

Bakhmutov V.G, Kazansky A.Y., Matasova G.G., Glavatskii D.V., 2017. Petromagnetism and magnetostratigraphy of Ukrainian loess/palaeosol sequence (Roxolany, Boyanichi and Korshev sections). *Izvestiya, Physics of the Solid Earth (Fizika Zemli).* Vol 53, № 6. – P. 864–884.

Bogucki A., Lanczont M., Gozhik P., Komar M. Loess Roxolany section: location, history of studies, characteristic of deposits. *Loess-covering of the North Black Sea Region.* Lublin: KARTPOL s.c. Lublin, 2013. – P. 47–58. (in Ukrainian).

Chadima M., Hrouda F., 2006. Remasoft 3.0 a user-friendly paleomagnetic data browser and analyzer. *Travaux Geophysiques* XXVII. – P. 20–21.

Dodonov A.E., Zhou L.P., Markova A.K., Tchepalyga A.L., Trubikhin V.M., Aleksandrovski A.L., Simakova A.N., 2006. Middle-Upper Pleistocene bio-climatic and magnetic records of the northern Black Sea coastal area. *Quaternary International.* Vol. 149. – P. 44–54.

Fedorowicz S., Łanczont M., Bogucki A., Woźniak P.P., Wróblewski R., Adamiec G., Bluszcz A., Moska P., 2013. Datowania izotopowe w profilu Roksolany, 2011. *Loess-covering of the North Black Sea Region.* Lublin: KARTPOL s.c. Lublin. – P. 65–683. (in Polish).

Gendler T.S., Heller F., Tsatskin A., Spassov S., Du Pasquier J., Faustov S.S., 2006. Roxolany and Novaya Etuliya – key sections in the western Black Sea loess area: Magnetostratigraphy, rock magnetism, and paleopedology. *Quaternary International.* Vol. 152–153. – P. 78-93.

Gozhik P., 2013. Study questions of Roxolany section. *Loess-covering of the North Black Sea Region.* Lublin: KARTPOL s.c. Lublin. – P. 17–33. (in Ukrainian).

Gozhik P., Gerasimenko N., 2011. The Lower and Middle Pleistocene of Ukraine. *Quaternary studies in Ukraine.* Kyiv. – P. 9–26.

Gozhik, P., Shelkoplyas, V., Khristoforova, T., 1995. Development stages of loessial and facial formation in Ukraine (Stratigraphy of loess in Ukraine). Lublin. *Annales Universitae Mariae Curie-Sclodowska.* Sec. B. 50, 65–74.

Kirschvink J.L., 1980. The least squares line and plane and the analysis of palaeomagnetic data. *Geophys. J.* Roy. Astron. Soc. – №62. – P. 699–718.

Pilipenko O.V., Sharonova Z.V., Trubikhin V.M., Didenko A.N., 2005. Fine structure and evolution of the geomagnetic field recorded in the Roksolany (Ukraine) loess-soil section 75–10 ka. *Fizika Zemli,* № 1. – P. 66–73 (in Russian).

Sartori M., 2000. The Quaternary climate in loess sediments: Evidence from rock and mineral magnetic and geochemical analysis. *Doctor of Natural Sciences Thesis.* Zürich/ –231 p.

Sharonova Z.V., Pilipenko O.V., Trubikhin V.M., Didenko A.N., Feyn A.G., 2004. Reconstruction of the geomagnetic field over the last 75000 years from paleomagnetic records in the Roksolany loess-soil section (the Dniester river, Ukraine). *Fizika Zemli* № 1. – P. 4–13. (in Russian).

Tretyak, A.N., Volok, Z.E., 1976. Paleomagnitnaya stratigrafiya pliotsen-chetvertichnykh osadochnykh tolshch Ukrainy [Paleomagnetic stratigraphy of Pliocene and Quaternary sediments in the Ukraine). Naukova Dumka, Kiev. (in Russian).

Tretyak A.N., Shevchenko A.I., Dudkin V.P., Vigilyanskaya L.I., 1987. Palaeomagnetic stratigraphy of the late Cenozoic key sections in the south of Ukraine. Kiev: AN USSR, In-t geol. nauk. – 50 p. (in Russian).

Tretyak A.N., 1983. Remanent magnetization and the problem of paleomagnetic stratigraphy in sedimentary sequences., Kiev: Naukova Dumka. – 254 p. (in Russian).

Tretyak A.N., Vigilyanskaya L.I., Makarenko V.N., Dudkin V.P., 1989. Fine structure of the geomagnetic field in the late Cenozoic. Kiev: Naukova Dumk. – 156 p. (in Russian).

Tretyak A.N., Vigilyanskaya L.I., 1994. Pleistocene magnetostratigraphic scale of Ukraine. *Geofiz. Zhurn.* №2, t. 16. – P. 3–14. (in Russian).

Tsatskin A., Heller F., Gendler T.S., Virina E.I., Spassov S., Du Pasquier J., Hus J., Hailwood E. A., Bagin V.I., Faustov S.S., 2001. New scheme of terrestrial paleoclimate evolution during the last 1.5 Ma in the western Black Sea region: integration of soil studies and loess magmatism. *Phys. Chem. Earht (A).* Vol. 26., № 11–12. – P. 911–916.

Tsatskin A., Heller F., Hailwood E.A. Gendler T.S., Hus J., Montgomery P., Sartori M., Virina E. I., 1998. Pedosedimentary division, rock magnetism and chronology of the loess/palaeosol sequence at Roxolany (Ukraine). *Palaeogeography, Palaeoclimatoligy, Palaeoecology.* Vol. 143. – P. 111–133.

Veklitch, M. F., Artyushenko, A. T., Sirenko, N. A., Dubnyak, V. A., Mel'nichuk, I. V., and Parishkura, S. I., 1967. Key Sections of the Anthropogene of Ukraine [Opornye geologicheskie razrezy antropogena Ukrainy]. Naukova dumka, Kiev. (in Russian).
Veklich M.F., Sirenko N.A., Matviyishyna Z.M., 1984. Pleistocene palaeogeographic stages and detail stratigraphic disjunction of Ukraine. Kiev: Naukova Dumka. – 32 p. (in Russian).
Velichko, A.A., 1990. Loess-paleosol formation on the Russian Plain. *Quat. Int.* 7/8. – P. 103–114.

Chapter 27
Magnetic Polarity Stratigraphy of the Upper Cenozoic Deposits of Near-Sea Dagestan (Russia) and the Age of the Early Paleolithic Site Rubas-1

A. Yu. Kazansky, A. A. Anoikin, A. P. Derevianko, G. G. Matasova and V. Yu. Bragin

Abstract We studied magnetic polarity in four Upper Neogene (Akchagylian stage) sections of Near-Sea Dagestan. Paleomagnetic studies have shown that Shor-Dere section is the most complete section of Late Neogene—Late Quaternary in this territory. The section Rubas-1, which contains Paleolithic artifacts, corresponds to the upper part of the Shor-Dere section, while sections Ajinour and Rubas River are likely short fragments, so their correlation with the reference section is ambiguous. A composite magnetic polarity section of Near-Sea Dagestan was constructed on the base of correlation between Shor-Dere, Rubas-1 and, in part, Rubas River sections. Magnetic polarity pattern of the composite paleomagnetic section is well correlated with the polarity zonation of Akchagylian stage for the neighboring regions of the Caspian basin (Azerbaijan and Turkmenistan), the south of the European part of Russia and the Transcaucasia (the Northern Caspian and the

A. Yu. Kazansky (✉)
Geological Department, Lomonosov Moscow State University, Moscow, Russia
e-mail: kazansky_alex@mail.ru

A. Yu. Kazansky
Geological Institute of RAS, Moscow, Russia

A. A. Anoikin · A. P. Derevianko
Institute of Archaeology and Ethnography of the SB RAS, Novosibirsk, Russia

A. A. Anoikin · A. P. Derevianko
Altai State University, Barnaul, Russia

A. A. Anoikin · A. P. Derevianko
Novosibirsk State University, Novosibirsk, Russia

G. G. Matasova · V. Yu. Bragin
Laboratory of Geodynamics and Paleomagnetism, Trofimuk Institute of Petroleum Geology and Geophysics of the SB RAS, Novosibirsk, Russia

D. Nurgaliev et al. (eds.), *Recent Advances in Rock Magnetism, Environmental Magnetism and Paleomagnetism*, Springer Geophysics,
https://doi.org/10.1007/978-3-319-90437-5_27

Lower Volga region). Such correlation of the magnetic polarity with those from startotype regions of Akchagylian stage made it possible to identify magnetic polarity zones in the composite section with the magnetic polarity chrons of the Neogene and Quaternary Time Scale (2016). We suggest that the upper part of the composite section can be correlated with Matuyama Chron including Olduvai and Reunion subchrons, while the lower part attributes to Gauss Chron with Kaena subchron. The proposed correlation provides the basis for approximate age estimates for the studied strata and supposes the age of the stone industry from Rubas-1 site not later earlier than 2 million years ago.

Keywords Magnetostratigraphy · Near-Sea Dagestan · Akchagylian Paleolithic

Introduction

The discovery of the unique early Paleolithic sites in Near-Sea Dagestan caused a heated discussion about the age of these ancient cultures. The lack of reliable paleontological data and the accurate results of absolute dating do not allow determining the age of the archaeological locations with acceptable accuracy. However, this problem can be solved on the base of paleomagnetic data. Earlier we have obtained a preliminary paleomagnetic data form the section Rubas-1, which contains Paleolithic artifacts most likely belonging to Akchagylian regional stage. This data allowed to give the first age estimation for the culturally-containing layer as in the base of the section as ≈2.2–2.3 million years (Derevianko et al. 2015).

Nevertheless, the absence of paleomagnetic data on Akchagylian reference sections in Near-Sea Dagestan significantly complicates the confirmation of this age estimation due to ambiguous correlation of stratigraphic succession of Rubas-1 section with Geomagnetic Polarity Time Scale (GPTS). At present, there is no consensus on the stratigraphic volume and duration of the Akchagylian regional stage as well as its magnetic polarity pattern (Van Baak et al. 2013), therefore, the only way to confirm our age estimates is to compare the paleomagnetic section of Rubas-1 archaeological site with the most complete paleomagnetic section in Near-Sea Dagestan and then through correlations with available paleomagnetic sections from neighboring regions to compare it with GPTS. One of the most complete sections of Akchagylian regional stage in Near-Sea Dagestan is an outcrop in Shor-Dere canyon located 20 km from the section Rubas-1 and it is not paleomagnetically studied yet.

The purpose of this paper is to gain new paleomagnetic data from the reference section of Akchagylian stage in Shor-Dere canyon and in the valley of the Rudas river for confirmation and refinement of previously obtained age estimation of Early Paleolithic site. In addition, the question was considered of whether the interbeds of fossil volcanic ash are isochronous and can be used as stratigraphic markers in this region.

Geological Background and Sampling

Paleomagnetic study was carried out on 4 different sections: (i) key section in Shor-Dere canyon (Shor-Dere section); (ii) a fragment of the section in the Ajinour depression with a layer of fossil volcanic ash (Ajinour section); (iii) Arhaelogical site Rubas-1 (Rubas-1 section); (iv) a fragment of the section in Rubas river valley with layers of fossil volcanic ash (Rubas River section). Geographical position of the sections is shown in Fig. 27.1.

Shor-Dere section is located in the upper part of Shor-Dere canyon on its left bank (41° 45′ 32.9″N, 048°16′44.1″E). The section was investigated and described by S. V. Leshchinsky and V. N. Zenin in 2007 (Leschinsky 2012). According to the description the section is composed of slightly tilted (Dip direction 64°, dip 8°) thin-layered sandy-clayey deposits of the Akchagylian stage more than 80 m in thickness which are represented by layers of dark greenish-brown (to gray) silt and yellow-brown clay with thin interlayers of fine-grained sand. A horizon of white volcanic ash 0.25 m in thickness can be traced at a depth of ~45 m from the top of the deposits. An upward gradual color transition from the general gray tone to light brown and yellowish brown accompanied with the increase of sand content is observed from the depth of ~20.5 m (Fig. 27.2). Akchagylian deposits with deep erosion are covered with gravel-pebble deposits of the Bakunian (?) stage. The lower boundary of Achagylian stage is nor exposed here.

Paleomagnetic samples were collected from three outcrops complementing each other. A total of 98 samples (one from the ash horizon) were taken at an average spacing of 50 cm. The total thickness of the section studied from the base of the Bakunian deposits to the bottom of the canyon was 53 m (Fig. 27.2).

Fig. 27.1 Shematic map of Near-Sea Dagestan showing location of the sections

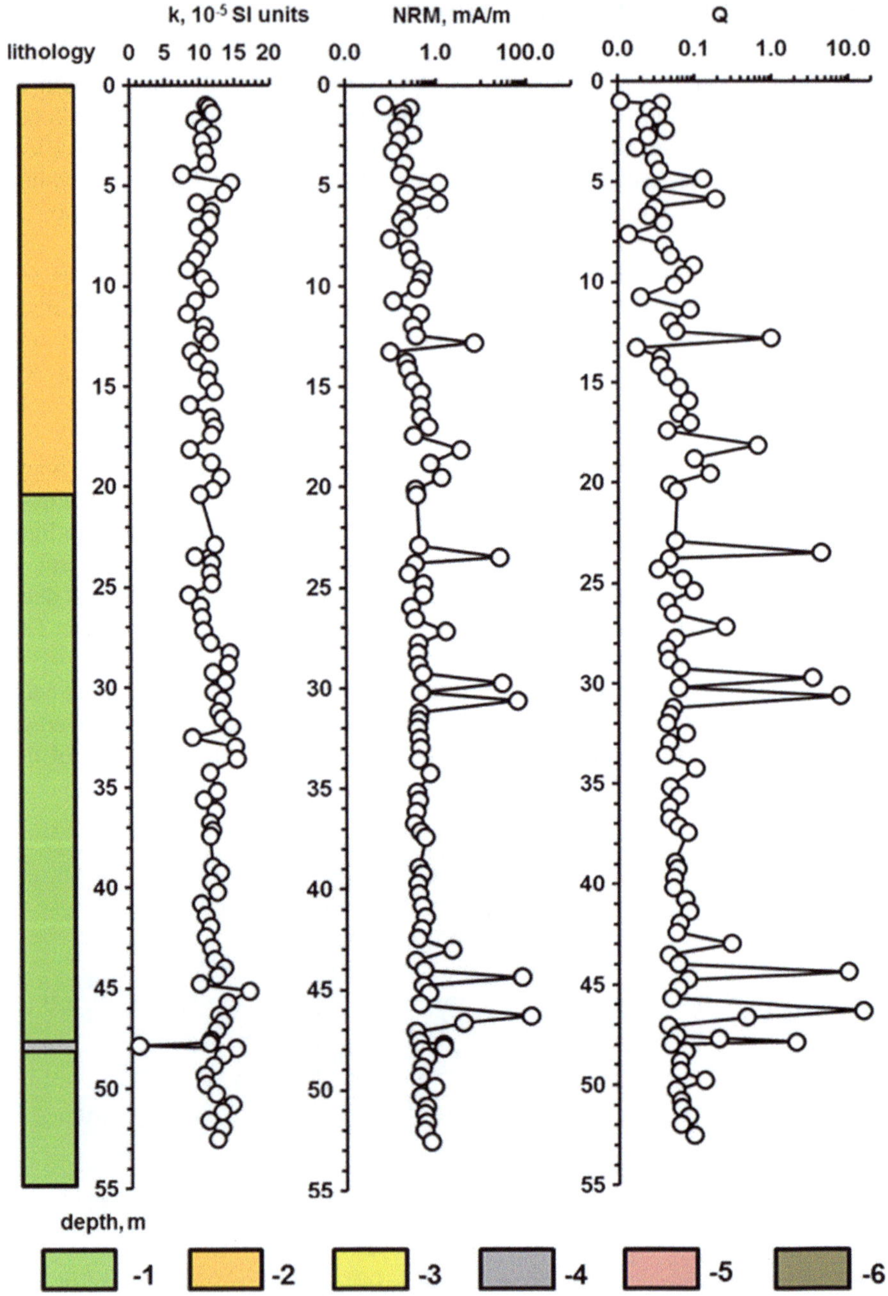

Fig. 27.2 Lithostratigraphy and rock-magnetic properties of Shor-Dere section. *Legend* 1—greenish-gray clayey silt; 2—yellowish-gray sandy silt; 3—sand; 4—volcanic ash; 5—redeposited volcanic ash; 6—silt, enriched with plant detritus

Ajinour section is located in the gully in the right side of the Shor-Dere canyon, in front of Shor-Dere Section (41° 44′ 58.1″N, 048° 17′ 09.3″E). The section was described by S. V. Leschinsky and V. N. Zenin in 2008. Greenish-brown silt the Akchagylian Stage is exposed in the gully slopes with apparent thickness up to 2 m. In the middle part of the outcrop there is a horizon of volcanic ash ~0.1 to 0.15 m in thickness (Fig. 27.3). It is assumed that it corresponds to the horizon of volcanic ash in the Shor-Dere section. Totally 5 paleomagnetic samples were collected from the section: 2 from overlapping silt 2 from ash horizon and 1 from underlying silt in order to determine the synchronicity of ash horizons.

Rubas-1 section. Archaeological monument (41° 53′ 21″N, 048° 07′ 35″E). Rubas-1 is located on the right bank of Rubas River ~3 km upstream from the Chulat village. The monument is located in the body of an ancient large landslide of the block type, now having the form of an extended terrace-like ledge with a flat slightly rising table-like surface and steep slopes. The resulting composite stratigraphic section of the site (depth—up to 20 m from the surface) is represented by four main packets of deposits. Packet 1 (layer 6, Tortonian age, N^2_{1tr}) is a marine mud. Packet 2 (layers 5 and 4, Late Akchagylian time, N^3_{2ak}) was formed in subaerial and subaqueous conditions of the sea coast (backshore, foreshore and shoreface zones). The section is composed of shingle beds with sand and silt. The culturally-containing layer itself (layer 5, Early Paleolithic) is represented by a gravel-pebble interlayer with a thickness of up to 0.4 m, with a silty-sandy matrix. Layer 4 is composed of fine-horizontally and wavy-layered fine and medium-grained carbonate sand and silt. In the lower part (up to 1 m) it is enriched with plant detritus (Leshchinsky 2012). The layers 5 and 4 are slightly tilted: dip direction 227°, dip 10°. Packet 3 (layers 3 and 2, late Pleistocene) is river alluvium. Middle Paleolithic archaeological material was found in those layers. Packet 4 (layer 1, Late Pleistocene—Holocene) was formed under subaerial conditions. Several stratigraphic levels of stone artifacts belonging to the late stages of the Middle—Upper Paleolithic have been revealed in those layers (Derevianko et al. 2012).

Archaeological characteristics of the site. The most ancient findings at the site are associated with layer 5 and represent the small-tool industry of the early stages of the Early Paleolithic (157 specimens), which is characterized by a large number of small-sized objects (~2 to 4 cm), frequent use of non-flaking bases (splinters, spals),

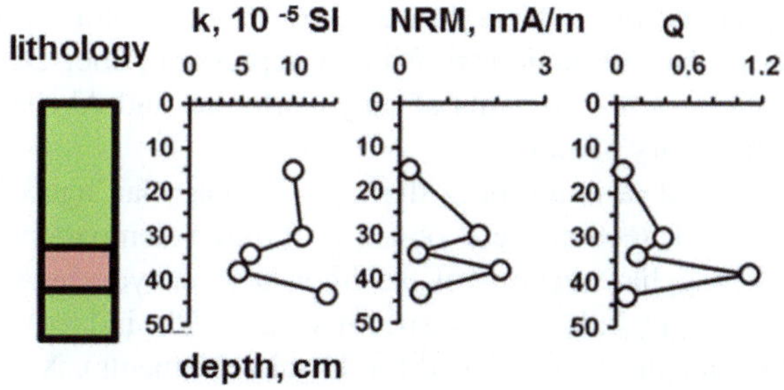

Fig. 27.3 Lithostratigraphy and rock-magnetic properties of Ajinour section. Here and after the legend is the same as in Fig. 27.2

and also a weak typological severity and instability of tool forms (Fig. 27.4). Among the latter, objects with various notches, spike-like protrusions and scraper-like products predominate.

Initially, stone assemblages, which can be classified as the small tools due to their size, appear in East Africa and are recorded at the Omo 57 and Omo 123 sites, whose age, according to geological data, is 2.3–2.4 million BP (Torre et al. 2003). Such complexes are widely spread from Central Europe to China and the territory of Eurasia, in the chronological interval from about 1 million to 300 thousand years ago.

Herewith, the Middle East (Bisat Ruhama and others), Central Europe and the Apennine Peninsula (Isernia-la-Pineta, Vėrtesszölös, etc.), Central Asia (Kuldara and others) and the Northern China (Dongutuo etc.) are the main areas, where Early Palaeolithic small-tool industries are found (Lower Palaeolithic … 2003; Burdukiewicz 2003; Derevianko 2009). Thus, the discovery of small-tool assemblages in the territory of Dagestan in sediments, presumably having a significant age, allow to propose the earlier appearance of similar industries in Eurasia.

59 paleomagnetic samples were taken from the layer 4 in three partially overlapping outcrops. The total thickness of layer 4 was determined as 5.2 m; the average space between samples makes up 10 cm (Fig. 27.4).

Rubas River Section is located on the right side of the valley of Rubas River, about 1 km upstream of Rubas-1 monument (41° 53′ 50.5″N, 048° 07′ 02.2″E). The section is represented by dark greenish-brown silt and clay with a layer of gray sand in the upper part of the section. A double layer of dark gray volcanic ash (thickness of the interlayers is 0.10 and 0.14 m) with signs of redeposition was found at the depth 0.12–0.38 m in the silts. Another layer of light gray ash with a thickness of 25 cm was observed at the depth 1.15–1.40 m. The total thickness of the section is 2.5 m (Fig. 27.5). 12 paleomagnetic samples were taken with average space from 15 (ash interlayers) to 60 cm (silt underlying the lower ash layer). Among them 5 samples are represented by ash tuffs, 1—by sand and 6—by silt (Fig. 27.5).

Geological evidences of the Akchagylian age of the studied sediments. It should be noted that all available geological data on Pliocene-Quaternary sediments of Near-sea Dagestan are based, mainly, on the data obtained in the first half of the last century. For example, the results for the basin of Rubas river are based of geological mapping by V. D. Golubyatnikov, conducted in the 30th years of the XX century, and then served as the basis for the geological map 1: 200,000 scale, approved in 1958 (Golubyatnikov 1933; Geological map … 1961). According to these data all sections in Ajinour depression, Shor-Dere canyon and in Rubas River valley contain deposits of only Tortonian and Akchagylian age overlapped by late Quaternary sediments.

The data confirming the Late-Akchagylian age of the sediments of layer 5 in Rubas-1 section were obtained through comparative analysis of the Rubas and Akchagylian sections of the Shor-Dere canyon in Ajinour depression. Laboratory studies included mass spectroscopy with inductively coupled plasma ICP-MS (31 samples were studied for 45 microelements), X-ray fluorescence analysis XRD

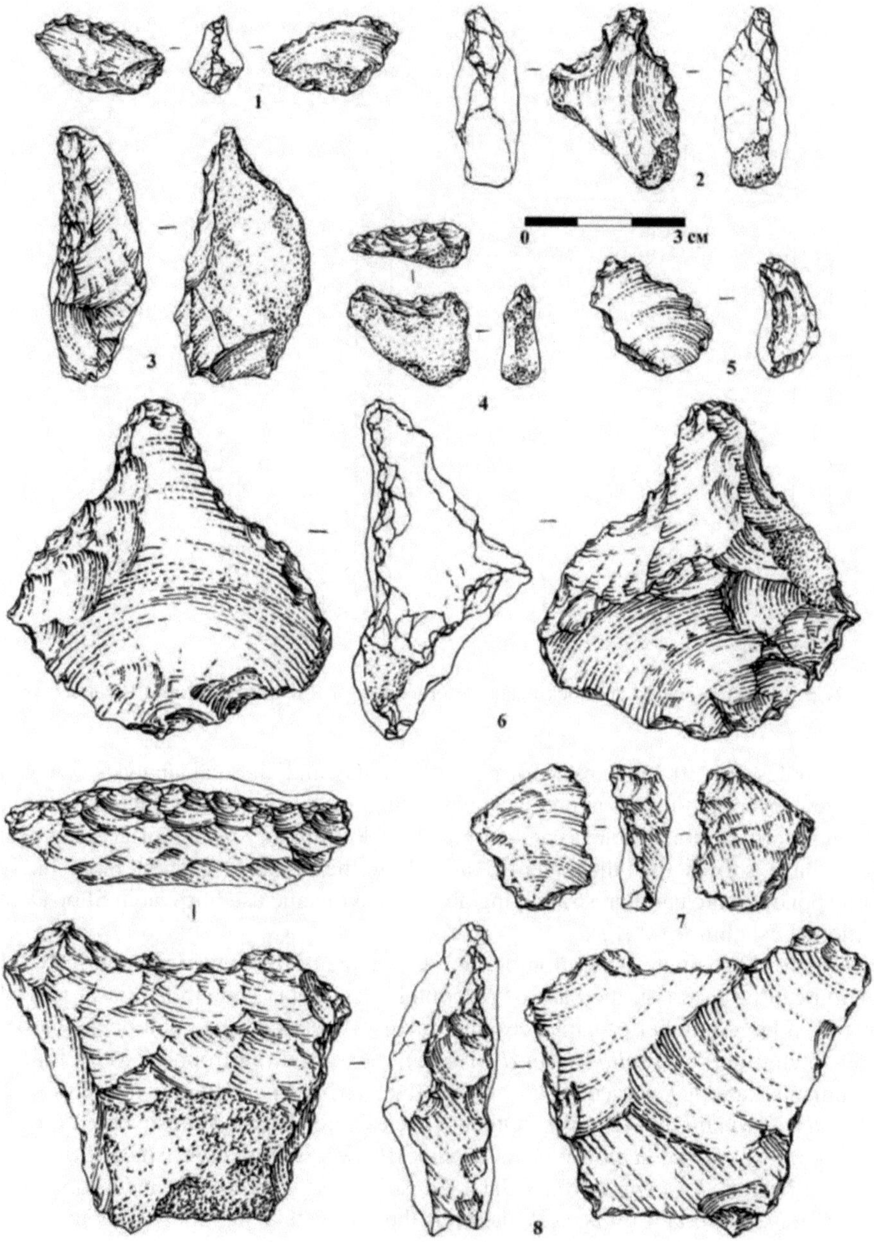

Fig. 27.4 Artifacts from layer 5 of Rubas-1 site. 1, 5, 7—flakes; 2, 6—spur-like implement, 3—side-scrapers; 4—end-scrapers atypical, 8—notched tool

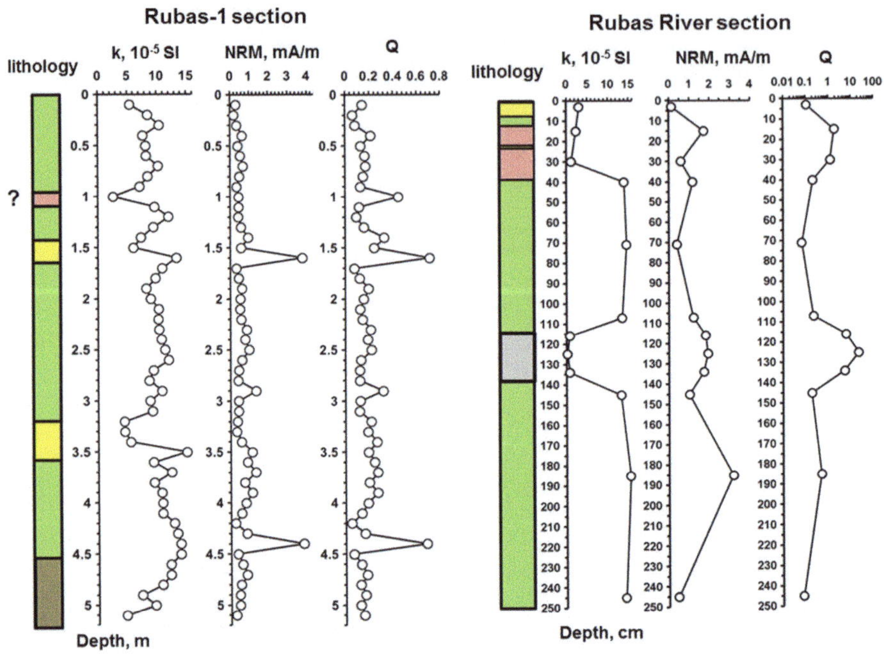

Fig. 27.5 Lithostratigraphy and rock-magnetic properties of Rubas-1 and Rubas River sections

(36 samples for 10 main rock-forming elements) and thermal analysis TA (30 samples for polymineral aggregates consisting of a mixture of layered aluminosilicates (hydromicas, clays) and carbonates). Preliminary results of geochemical study have shown that the layers 5 and 4 of the Rubas-1 section most likely correspond to Akchagylian strata lying above the volcanic ash horizon in Shor-Dere section (Leshchinsky 2012).

A more substantiate estimation of the relative age of the lower layers in Rubas-1 section can be made on the base of paleontological data. Thus in the basal part of layer 4, a large number of carapaces and valves of ostracods of good preservation, rare foraminifers (*Rotalia beccari* (Linne), etc.) and gastropods were found. Dominants among ostracods are euryhaline *Cyprideis littoralis* (Brady) and *Cyprideis punctillata* (Brady). Other species are brackish-freshwater forms: *Limnocythere* aff. *luculenta* Livental, *Limnocythere tenuireticulata* Suzin, *Candona abichi* Livental, *Ilyocypris bradyi* Sars, *Eucypris* aff. *puriformis* Mandelstam.

The most important for age estimation is the presence of significant amount of *L. tenuireticulata* Suzin and *L. aff. luculenta* Livental The first species appears in the Akchagylian time and is typical species for this stage. Distribution of the second one ranges from Akchagylian to Bakunian time, where it fades away. The species *C. punctillata* (Brady) is known sediments of vast time span from Miocene to Apsheronian. Carapaces of Akchagylian species *C. abichi Livental* and *E. aff. puriformis* Mandelstam have indicators of local transportation *C. littoralis* (Brady) и *I. bradyi* Sars are species of wide distribution. Thus, ostracod composition and lithology of sediments suggest that sedimentation of layer 4 occurred in desalinated coastal sea waters precisely during the Akchagylian time (Leshchinsky et al. 2009).

Methods

Depending on the type of deposits, a different sampling procedure was applied. In of Shor-Dere, Ajinour and Rubas River sections silty rocks were sampled as oriented blocks, which were cut in the laboratory into standard 8 cm^3 specimens. Loose sands and volcanic ash were sampled into standard 7 cm^3 plastic boxes. Sediments in Rubas-1 selection were sampled using the sampler designed by Minyuk (1982). The orientation of the samples was carried out using magnetic and solar compasses.

Laboratory studies were carried out on the basis of the Laboratory of the Main Geomagnetic Field and Petromagnetism of the IPE RAS (Moscow) and the laboratory of the Paleomagnetic Center of IPGG SB RAS (Novosibirsk) using standard methods (Tauxe 2010). Low field magnetic susceptibility (k) was measured using either a Bartington MS2B sensor or an AGICO MFK-1 Kappabridge. Measurements of Natural Remanent Magnetization (NRM) were made using a 2G Enterprises 755R cryogenic magnetometer. Koenigsberger ratio was calculated as Q = NRM/Hlab, where Hlab is Earth magnetic field in the laboratory. Specimens from each horizon were subjected to stepwise AF demagnetization, using a build in Automatic Degaussing System 600 in 3–10 mT steps up to 70 mT (for pilot specimens up to 130 mT).

The specimen demagnetization results were analyzed using orthogonal plots (Zijderveld 1967) and principle component analysis (Kirschvink 1980) implemented in a computer program package by Enkin (1994).

The magnetic mineralogy of representative lithologies was studied using thermal dependence of magnetic susceptibility. Susceptibility was measured with the help of an AGICO MFK-1 with CS4 High Temperature Furnace up to 700 °C in argon atmosphere.

Results

Magnetic properties (k, NRM, Q). Variations of magnetic characteristics versus depth for the studied sections studied are shown in Figs. 27.2, 27.3 and 27.5. All deposits are weakly magnetized. However, an obvious differentiation in magnetic susceptibility values can be observed: the smallest values of k is observed in the volcanic ash horizons (Shor-Dere and Rubas River sections, Figs. 27.2 and 27.5) and in layers of redeposited volcanic ash (Ajinour and Rubas River sections, Figs. 27.3 and 27.5). Sand interlayers in the Rubas-1 and Rubas River sections,

Fig. 27.5, have a somewhat higher susceptibility values. The highest susceptibility values are attributed to silt and clay layers. Therewith variations of the parameters in the clayey-silty sediments of the sections are of a smooth character (Figs. 27.2, 27.3 and 27.5).

In this case, the changes in the parameters in the clayey-silty sediments of the section are of a smooth character (Figs. 27.2, 27.3 and 27.5) except several samples from the Shor-Dere section with an increased NRM intensity (more than 10 mA/m), clearly falling out of the general distribution of magnetic properties (Figs. 27.2 and 27.6).

Q value of these samples significantly exceeds unity (4.5–15.5). Most likely, the magnetization of these samples has a secondary chemical nature. For all other samples Q values do not exceed one, and for 50% of the samples Q < 0.1. On the bilogarithmic plot of k-NRM (Fig. 27.6), the distribution of sample points areas from silty-clay deposits of the Shor-Dere, Ajinour, Rubas-1 and Rubas River sections practically coincide, thus indicates their close sedimentation environments. In general, the values of the scalar magnetic parameters of sedimentary deposits do not depend on the lithology of the rocks and the position in the section. The magnetization of upper (more sandy) part of Shor-Dere section is relatively lower, while in Rubs-1 and Rubas River sections there are samples with a lowered magnetic susceptibility. The latter, possibly, is associated with a large content of diamagnetic pyroclastic material due to redeposition of volcanic ash in the coastal zone.

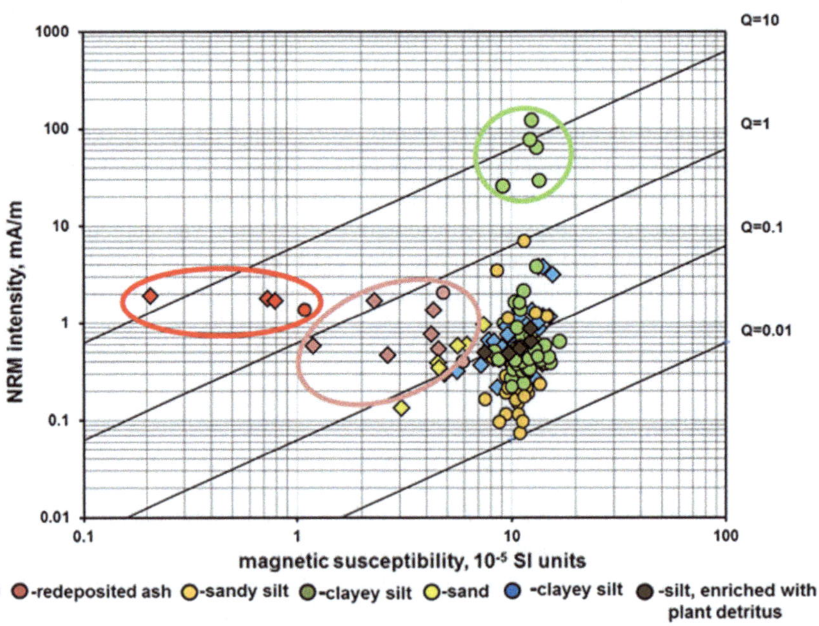

Fig. 27.6 Logarithmic biplot of magnetic susceptibility and NRM intensity. See text for details. *Legend* circles—samples from Shor-Dere and Ajinour sections; diamonds—samples from Rubas-1 and Rubas River sections

Nevertheless, low values of magnetic parameters (especially Q) and the absence of regularities in their variations versus depth may indicate possible changes in primary magnetic minerals (dissolution and leaching). High Q values in ash horizons (2–23.2) are associated with a low value of magnetic susceptibility (Fig. 27.6) due to the high content of diamagnetic volcanic glass. The distribution of the points for ash horizons on the k-NRM biplot differs significantly from those for clayey-silty sediments (Fig. 27.6).

Redeposited ash also contains a significant proportion of volcanic glass, so their susceptibility is lower, and NRM is somewhat higher than that of clayey-silty sediments (Fig. 27.6). Such a contrast of magnetic properties makes it possible to identify different lithotypes, and in some cases also to refine the structure of the sections. Thus, a layer of redeposited ash in Rubas-1 section was determined on the base of its magnetic properties, even though it was not observed during the field description (Fig. 27.5).

Magnetic minerals. In most of the samples magnetic susceptibility is carried by paramagnetic fraction so their temperature behavior (especially on cooling) significantly depends on transformation on paramagnetic minerals. The distinct magnetite Curie points were observed only in ash samples where paramagnetic content is negligible (Fig. 27.7, sample 15r008) and susceptibility increment does not exceed 1.5 times from initial value. Sandy silt samples show a similar behavior on heating but at cooling from 700 °C magnetic susceptibility increases dramatically due to a chemical transformation of paramagnetic (clay?) minerals to magnetite (Fig. 27.7, sample 15S010). In clayey silt samples with Q less than 1 the transformation begin during heating above 450 °C suggesting the presence of pyrite which is gradually transforming to magnetite with sharp increase of susceptibility (Fig. 27.7, sample 15S098). The same behavior was observed for redeposited tuff, however the total increase of magnetic susceptibility in redeposited tuff is lower (Fig. 27.7, sample 15R015). Pyrite was also found in clayey silt samples with Q more than 1. Along with pyrite a probable presence of maghemite and hematite were recognized (Fig. 27.7, sample 15S086).

Demagnetization results. The low NRM values caused a significant dispersion of magnetization directions during the demagnetization process ("noisy" orthogonal diagrams) and incomplete separation of magnetization components. Nevertheless, one can distinguish three main groups of samples: (1) samples of normal polarity with a steep positive inclination; (2) samples of reverse polarity with negative inclination; (3) anomalous samples with a low (less than 35°) inclination and/or reverse declination. Typical orthogonal diagrams are shown in Fig. 27.8.

The best preservation of magnetization is observed in the upper part of the silty-sandy stratum of the Shor-Dere section (Fig. 27.8a) and the lower part of the Rubas-1 section (Fig. 27.8b). Here, along with viscous component, which completely removes at 20–25 mT, only the characteristic magnetization component is present, which remains until the end of the demagnetization process (60–78 mT). The similar behavior of magnetization is observed in ash horizons a (Fig. 27.8c) and redeposited ash layers (Fig. 27.8d), where magnetization is practically one-component with the exception of the viscous one.

In other samples, the situation is more complicated. Due to the presence of a secondary component, which is close in stability to the characteristic component of magnetization the complete separation of those components is not possible (Fig. 27.8e) or significant oscillations of the direction are observed in high fields

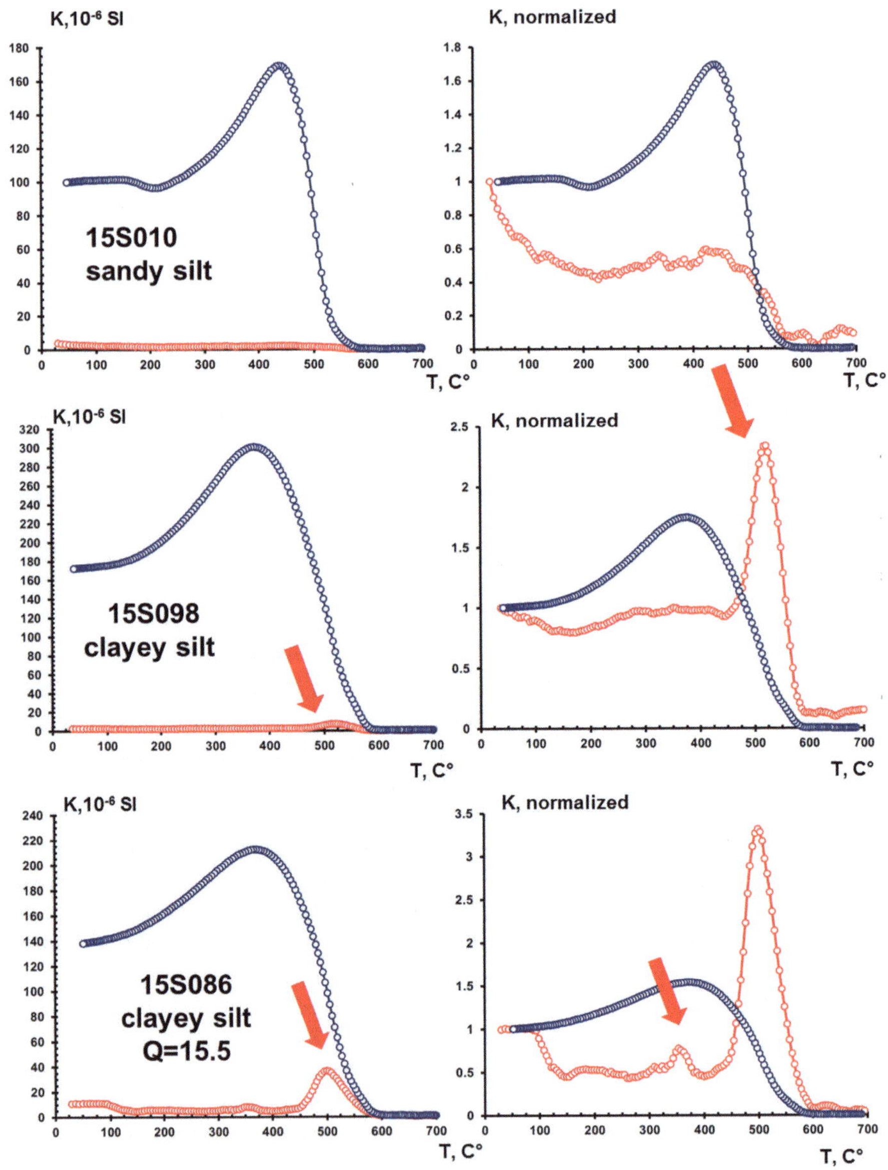

Fig. 27.7 Temperature dependence of magnetic susceptibility. See text for details. Red curve—heating; blue curve—cooling. Left panels are normalized curves

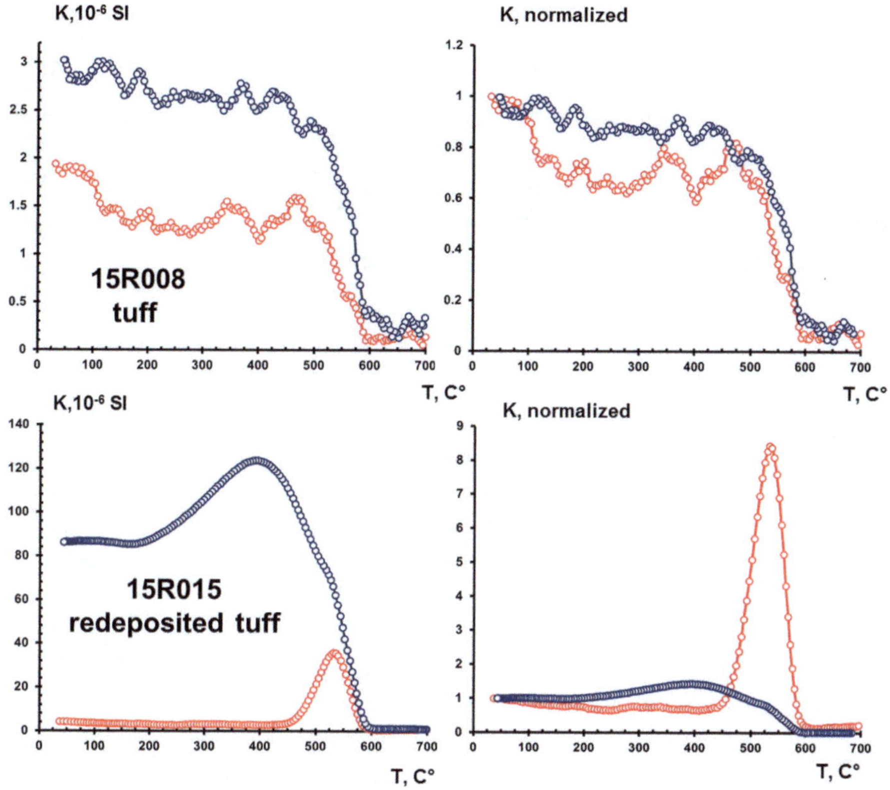

Fig. 27.7 (continued)

(Fig. 27.8f). As a rule, the contribution of the viscous magnetization in these samples is more sufficient. Nevertheless, even in such cases it is possible to determine the direction of the stable magnetization, at least to within a sign of polarity.

Samples with high values of Q (excluding ash and redeposited ash samples) have a completely different demagnetization character. Here one can observe two components with considerably overlapping coercive spectra which cannot be separated (Fig. 27.8g, h). The shape of NRM decay curves in those samples is more fiat than in the other samples. These samples can be both of normal and reverse polarity, but since the direction their magnetization differs from neighboring samples above and below in the section, we suggest a secondary chemical nature of their magnetization and data on these samples were excluded from further consideration.

Directions of the stable component of magnetization. Stereoplots of normal and reverse directions distribution in all the studied sections are given in Fig. 27.9. Stereoplots show that both normal and reverse directions can have different confidence intervals from 2–3 to 10°–15°, which indicates a different degree of preservation of magnetization in samples. In general, there is a similarity between the

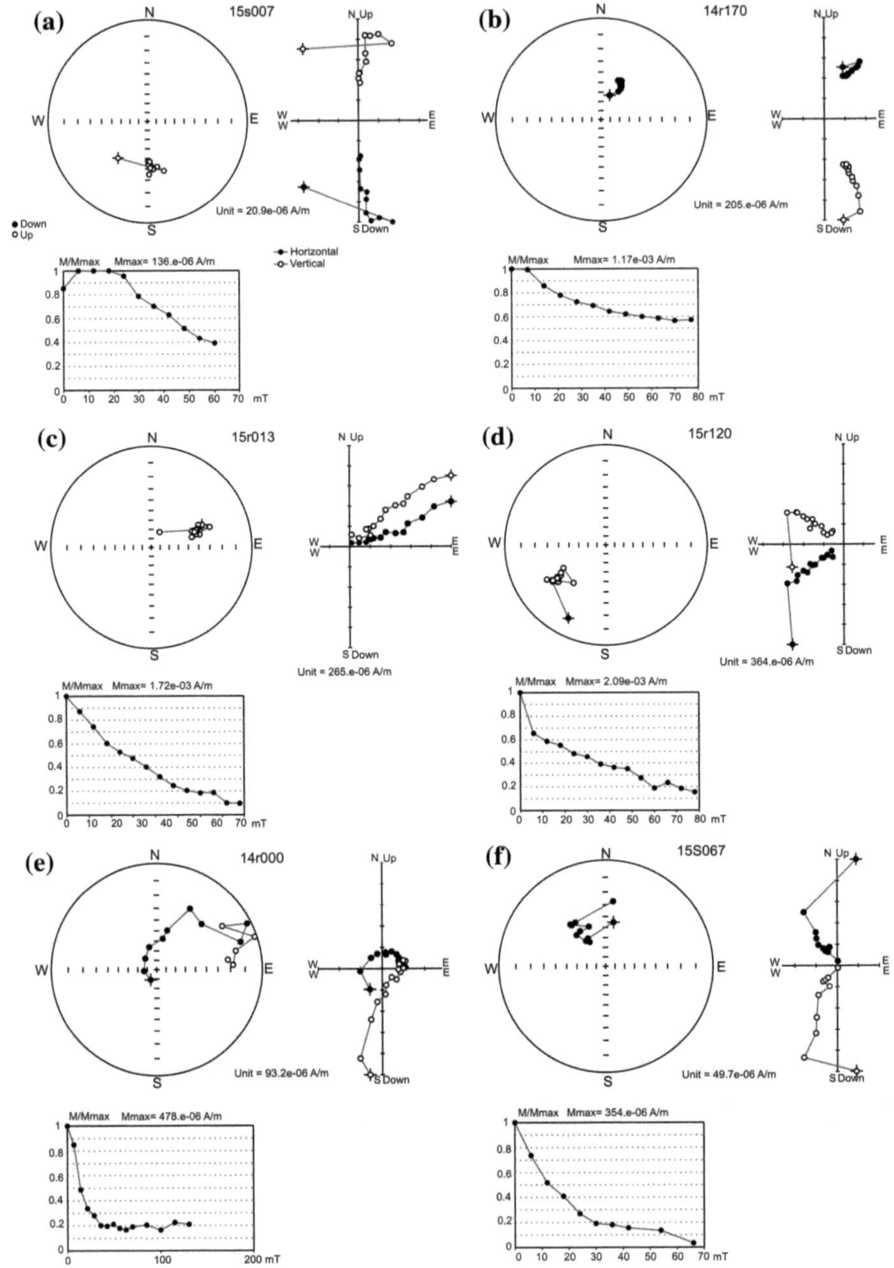

Fig. 27.8 Zijderveld, stereoplots and demagnetization plots of six specimens, showing typical behavior during AF demagnetization. **a** and **b**—sandy silt; **c**—tuff; **d**—redeposited tuff; **e** and **f**—clayey silt ($Q < 1$); **g** and **h**—clayey silt ($Q \gg 1$)

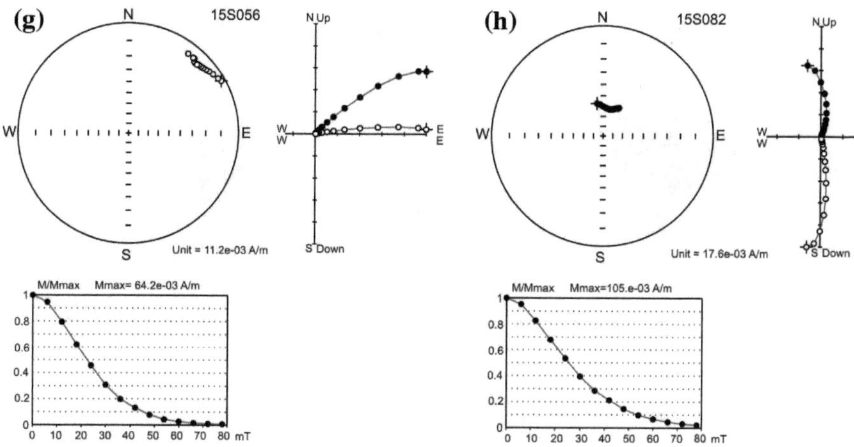

Fig. 27.8 (continued)

distributions of different polarities in the sections. Directions of normal polarity form fairly compact groups in the first and fourth quarters of the stereoplot (compare Fig. 27.9a, b). Reverse directions demonstrate significant scatter throughout the stereoplot area (Fig. 27.9c, d). In the most cases the "reversed" directions are travelling from lower to upper hemisphere along great circles as it is shown in Fig. 27.8e. Due to overlapping of corcitive spectra the reversed component can't be completely isolated so the directions form elongated areas on the stereoplot (Fig. 27.9c, d). However, for the upper part of the Shor-Dere section, where magnetization is more stable, reverse polarity directions concentrate in the area practically antipodal to normal polarity (Fig. 27.9c). Several samples with anomalous directions—with a low positive inclination <20° and reverse declination from 207 to 253° (Fig. 27.9c, d) are present in both groups of sections. For interpretation such samples were assumed as reverse, since the neighboring samples are always reversely magnetized.

Due to the large scatter of reverse polarity directions (Fig. 27.9c, d; Table 27.1), the reversal test was not possible to perform, therefore, in order to estimate the age of the selected components of the magnetization with respect to the folding time, a fold test (McFadden 1990) was performed for samples from Shor-Dere and Rubas-1 section. The test is positive for normal polarity directions (DC slope 0.754 ± 0.247). The stepwise unfolding shows a maximum of precision parameter at 88.2% of unfolding, that is, the pre-folding close to the primary component, predominates. For the reverse polarity directions fold test is inconclusive (DC slope 0.367 ± 0.031). The stepwise unfolding shows maximum of precision parameter at 36.7% of unfolding, that is, the magnetization is synfolding. Thus, for samples with reverse polarity, the primary magnetization was not completely isolated; it is significantly complicated by the secondary post-folding component of normal polarity most likely of a chemical nature.

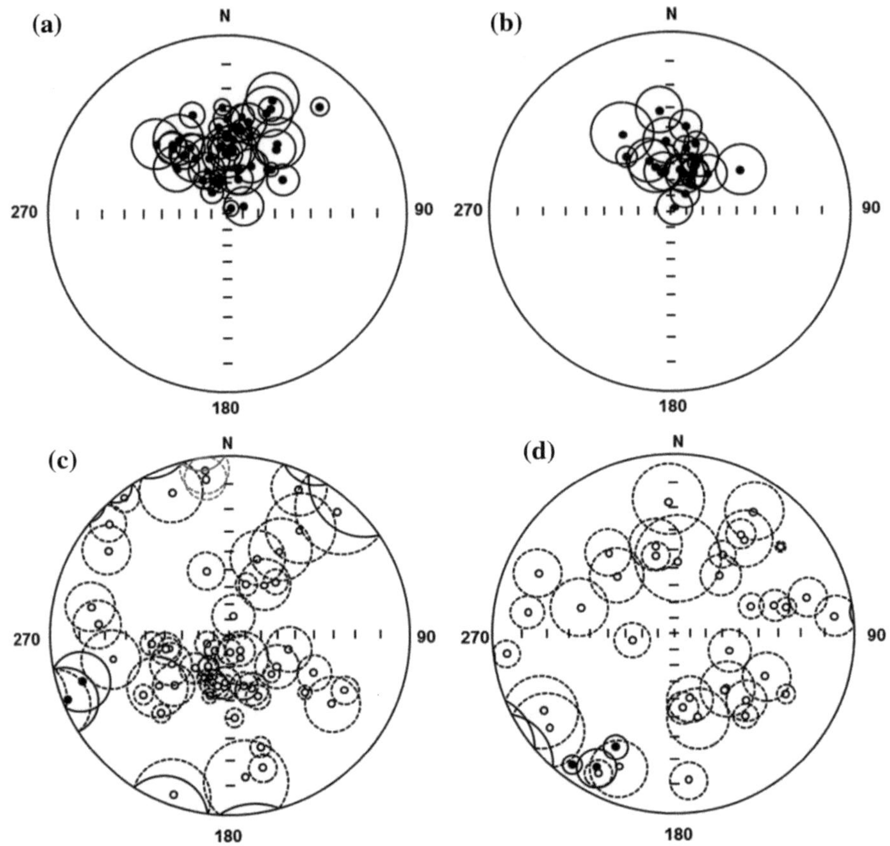

Fig. 27.9 Stereoplots of normal and reverse directions in the studied sections. **a** and **c**—Shor-Dere and Ajinour sections; **b** and **d**—Rubas-1 and Rubas River sections

Table 27.1 Normal and reverse polarity directions from Shor-Dere and Rubas-1 sections

Section	Polarity	N	Dg	Ig	Ds	Is	Ks	a95s
Shor-Dere	N	41	350.5	53.0	0.0	50.1	11.9	6.7
	R	56	196.1	−84.6	224.3	−77.7	2.38	16.1
Rubas-1	N	19	15.1	49.7	7.1	57.8	20.8	7.5
	R	32	189.4	−73.9	147.8	−79.1	1.7	29.6

Note Polarity: N—normal; R—reverse. D—declination (in degrees); I—inclination (in degrees), lower index corresponds to the coordinate system (g—geographic, s—stratigraphic, k—precision parameter of the Fisher distribution (Fisher 1953), a95—confidence interval (in degrees)

Magnetic polarity stratigraphy and composite paleomagnetic section. Variations of the angular parameters of the stable magnetization (declination, inclination, and confidence interval) for all sections are shown in Figs. 27.9, 27.10 and 27.11. It is obvious that the intervals of normal and reverse polarity in the

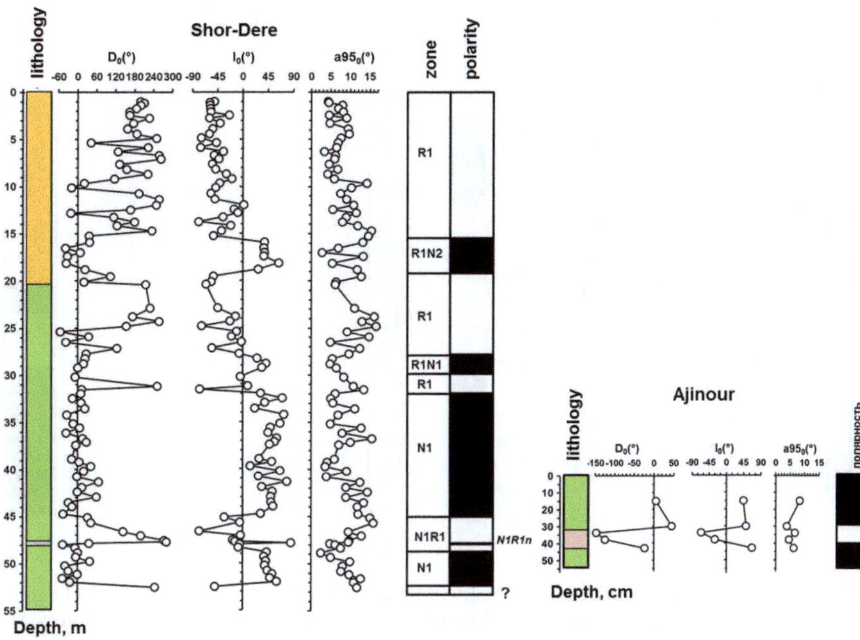

Fig. 27.10 Paleomagnetic reversal stratigraphy of Shor-Dere and Ajinour sections. Inclination, declination and precision parameter of characteristic remanent magnetization (ChRM) are plotted as a function of depth. The letters and numbers at the right of the polarity column indicate the nomenclature of magnetic polarity intervals. Here and after the legend is the same as in Fig. 27.2

studied sections occupy a completely definite position (Figs. 27.10 and 27.11). The most complete magnetic polarity pattern was recorded in Shor-Dere section (Fig. 27.10). Here, in accordance with the generally accepted nomenclature (Stratigraphy … 1982), two large intervals of magnetic polarity (N1 and R1) are distinguished, which most likely correspond to orthozones, which are complicated by intervals of opposite polarity, corresponding to subzones (N1R1, R1N1 and R1N2). The lowest part of the section has reverse polarity. In the middle part of the subzone N1R1 a micro-interval of normal polarity N1R1n corresponding to the horizon of fossil ash was determined. Since these intervals are represented only by single samples, their rank cannot be established, and their reliability is not high.

Ajinour section (Fig. 27.10) is represented by a normal polarity interval with a reverse polarity subzone inside, which corresponds to the horizon of redeposited ash. Taking into account the small thickness of the studied part of section the rank of these intervals remains unclear, and their correlation with Shor-Dere section is ambiguous.

A large interval of reverse polarity (R1) is distinguished in Rubas-1 section (Fig. 27.11); its rank is comparable orthozone. From below and from above this interval is framed by intervals of normal polarity of smaller rank (R1N1 and R1N2). Most likely, these polarity zones are analogues of the R1, R1N1 and R1N2 polarity

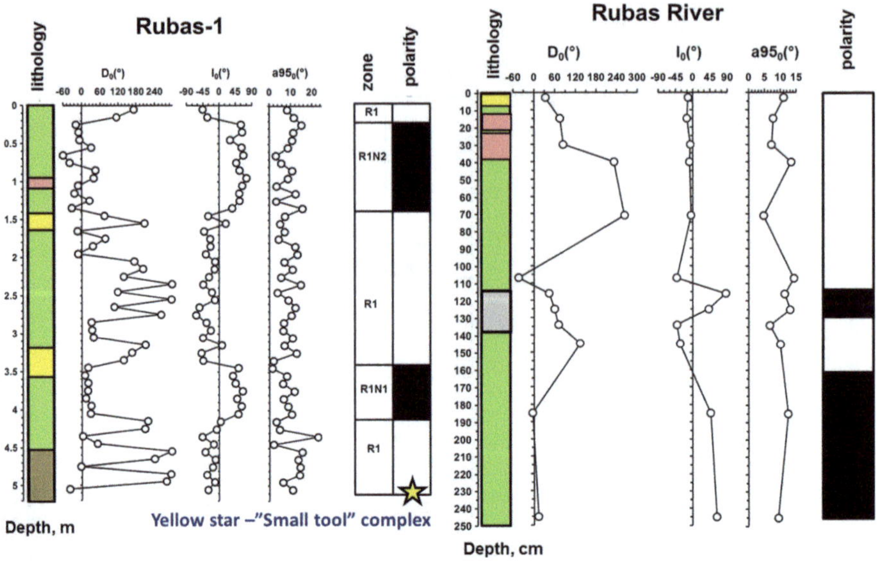

Fig. 27.11 Paleomagnetic reversal stratigraphy of Rubas-1 and Rubas River sections. Inclination, declination and precision parameter of characteristic remnant magnetization (Charm) are plotted as a function of depth. The letters and numbers at the right of the polarity column or Rubas-1 section indicate the nomenclature of magnetic polarity intervals

zones in Shor-Dere section. The horizon of redeposited volcanic ash in the upper part of Rubas-1 section is characterized by reverse polarity (Fig. 27.11).

The lower part of Rubas River section (Fig. 27.11) is characterized by normal polarity while the upper part of the section is characterized by reverse polarity which is complicated by an interval of normal polarity of a lower rank. This interval corresponds to the upper part of the fossil ash horizon, while its lowest part is reversely magnetized. The horizons of redeposited ash in the upper part of Rubas River section are also characterized by reverse polarity.

Based on the proposed correlation of magnetic polarity intervals, a composite paleomagnetic section of Near-Sea Dagestan was compiled using paleomagnetic patterns from Shor-Dere, Rubas-1 and the Rubas River sections (Fig. 27.12). The data from Ajinour section were not taken into account, since the studied interval here has a small thickness and the rank of the allocated intervals cannot be established, that makes ambiguous its correlation with the other sections. As for the Rubas River section, its correlation with Shor-Dere section is based on the assumption of a synchronous deposition of fossil ash horizons in these sections. In both sections, this horizon is predominantly characterized by normal polarity y and is located in the middle part of reverse polarity interval, which is underlain by normal polarity interval (compare Figs. 27.10 and 27.11). This version of correlation confirms the existence of a micro interval of normal polarity N1R1n inside the N1R1 polarity zone of Shor-Dere section.

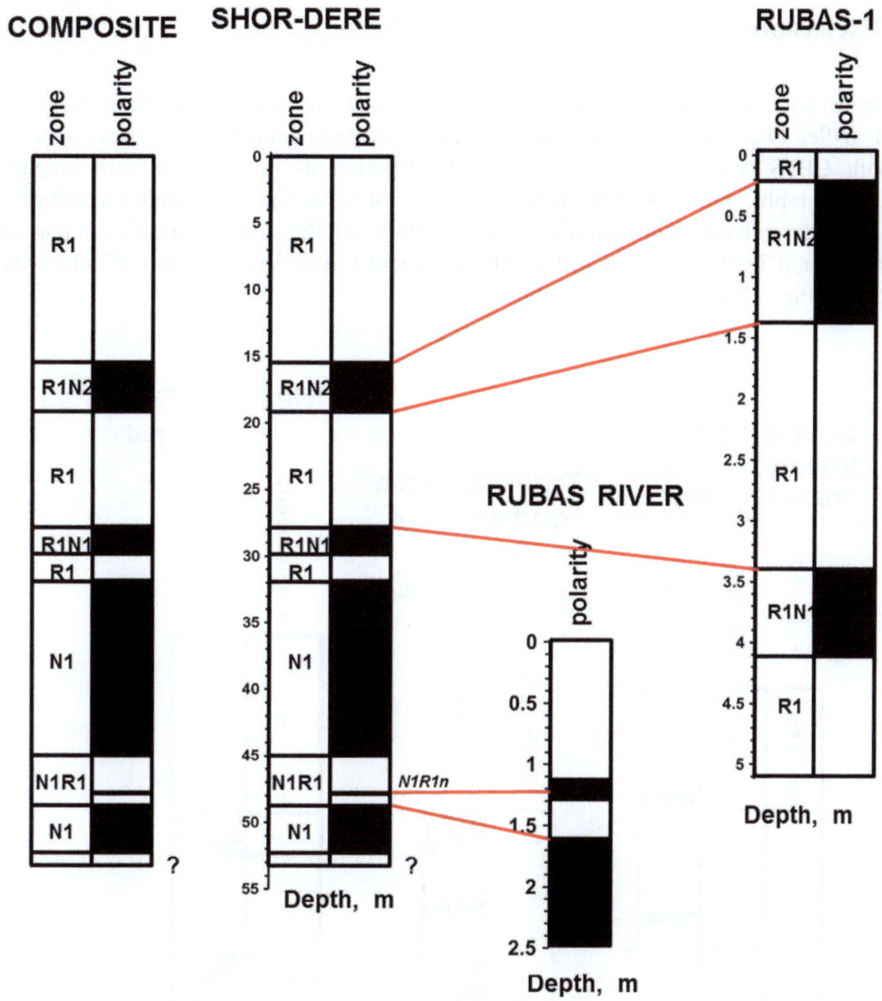

Fig. 27.12 Preferred correlation between Shor-Dere, Rubas River and Rubas-1 sections and the composite magnetic polarity section of Akchagylian stage of Near-Sea Dagestan

The magnetic polarity pattern of the Shor-Dere section was used as the basis for the composite paleomagnetic section of the Akchagylian Stage of Near-Sea Dagestan, because it contains all normal and reverse polarity intervals recorded in Rubas-1 and Rubas River sections are present. The composite section is represented in a conventional scale, with the ratio between the thicknesses of magnetic polarity zones being preserved (Fig. 27.12). The reverse polarity interval at the very base of Shor-Dere section was not taken into account, since it was supported by only one sample.

Discussion

Up to now, there has been no paleomagnetic data on the Akchagylian Stage of Near-Sea Dagestan, therefore, the correlation of composite paleomagnetic section with GPTS was possible only indirectly, through its comparison with magnetostratigraphic sections of the neighboring regions: the Caspian region (Azerbaijan and Turkmenistan) (Gurary and Trubikhin 1980) and the south the European part of Russia and Transcaucasia (Northern Caspian and Lower Volga region) (Grebenyuk 2004) (Fig. 27.13).

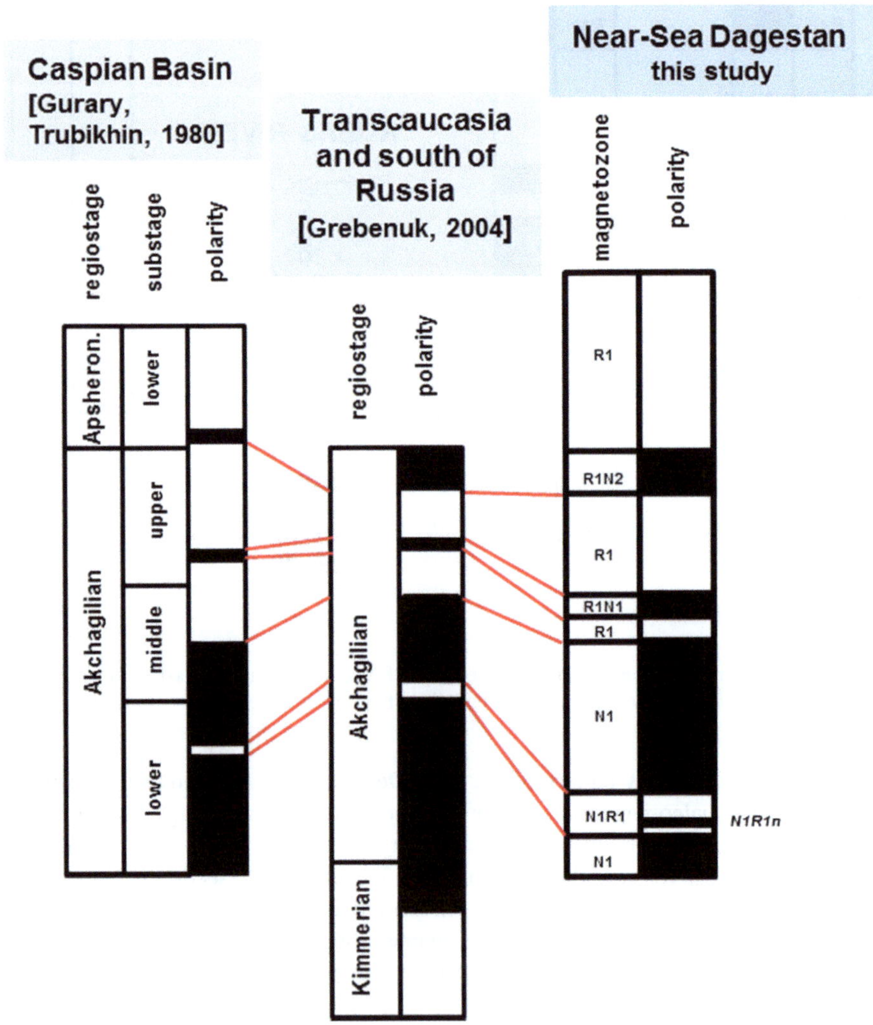

Fig. 27.13 Comparison of the composite magnetic polarity section of Akchagylian stage of Near-Sea Dagestan with Akchagylian polarity patterns from adjacent regions

Obviously, the Akchagylian Stage in both magnetostratigraphic sections has a similar magnetic polarity zonation, the differences are observed only in the relative duration of magnetic polarity intervals (Fig. 27.13), due to difference in sedimentation rates.

In general case Akchagylian interval is represented by the two major polarity intervals complicated by a series of intervals of lower rank. The first normal polarity orthozone encompasses the whole lower Akchagylian and the lower part of the middle Akchagylian (in case of its ternary subdivision); the lower boundary of this orthozone is traced at the top of the Cimmerian stage. It is generally agreed (Gurary and Trubikhin 1980; Grebenyuk 2004), that this normal polarity interval corresponds to Gauss Chron of GPTS (Pillans and Gibbard 2012; Hilgen et al. 2012). A sub-zone of reverse polarity in the middle part of Lower Akchagylian most likely corresponds to Kaena subchron.

The second orthozone is of reverse polarity and covers the upper part of Middle Akchagylian and the entire Upper Akchagylian and Apsheronian. In the Upper Akchagylian, it includes two subzones of normal polarity, attributed to the basal part of Upper Akchagylian and the border between of Akchagylian and Apsheronian. The most probable analogues of these subzones are Reunion and Olduvai subchrons.

Such correlation of the magnetic polarity zonation of the composite section of Near-Sea Dagestan with magnetostartigraphic schemes of neighboring regions (Fig. 27.13) allows to perform a direct correlation of magnetic polarity zones identified in the section with chrons of GPTS (Pillans and Gibbard 2012; Hilgen et al. 2012) (Fig. 27.14). In this case, the inverse polarity orthozone R1 is an analog of Matuyama Chron, and the subzones R1N2 and R1N1, respectively, are analogues of the Oludvai and Reunion subchrons. Normal polarity orthozone N1 is an analog of the Gauss Chron, and the subzone N1R1 is an analog of the Kaena subchron. The N1R1n microzone within the N1R1 subzone does not have its analog in GPTS and, apparently, is cryptochron within Kaena subchron.

The proposed version of identification of magnetic polarity intervals of the composite paleomagnetic section of Near-Sea Dagestan with the chrons of GPTS makes it possible to give some age estimates for the boundaries of magnetic polarity zones in Shor-Dere and Rubas-1 sections and to calculate sedimentation rates. Figure 27.15 shows age-thickness biplots for the Shor-Dere and Rubas-1 sections (where more than two paleomagnetic boundaries are present which are reliably correlated with the boundaries of GPTS chrons).

Obviously, the sedimentation rates in the both sections vary significantly. The average sedimentation rate in the Shor-Dere section varies from 0.8 to 5.4 cm/thousand years, averaging ~2.3 cm/thousand years (Fig. 27.15a), while in Rubas-1 section from 0.6 to 1.2 cm/thousand years, averaging ~0.9 cm/thousand years (Fig. 27.15b).

Such variations in the rates are apparently resulted from the differences in sedimentary environments. Sediments in Shor-Dere section were deposited in the deeper part of the basin, as evidenced by their homogeneous lithological composition Sediments in Rubas-1 section are characterized by a more variegated

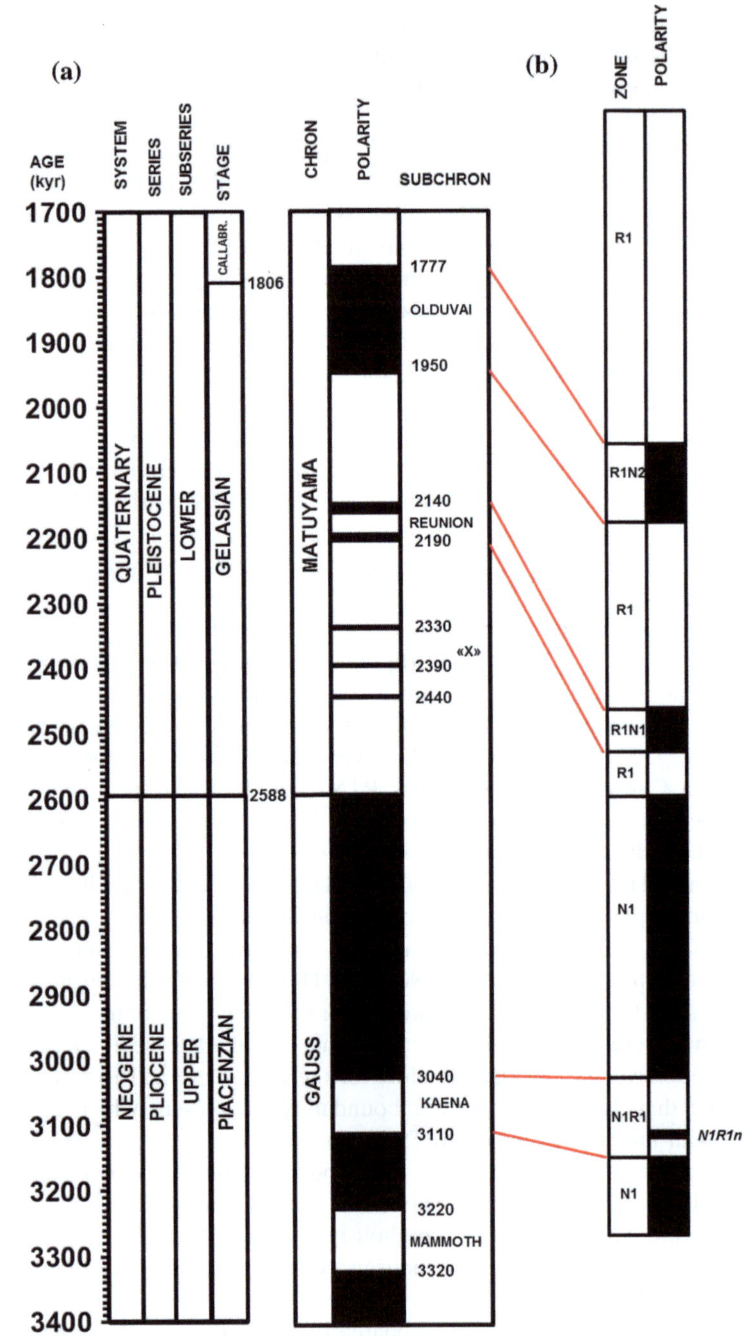

Fig. 27.14 Preferred correlation of the composite section of Near-Sea Dagestan (b) with (a) geomagnetic-polarity time scale (GPTS) modified from (Pillans and Gibbard 2012; Hilgen et al. 2012)

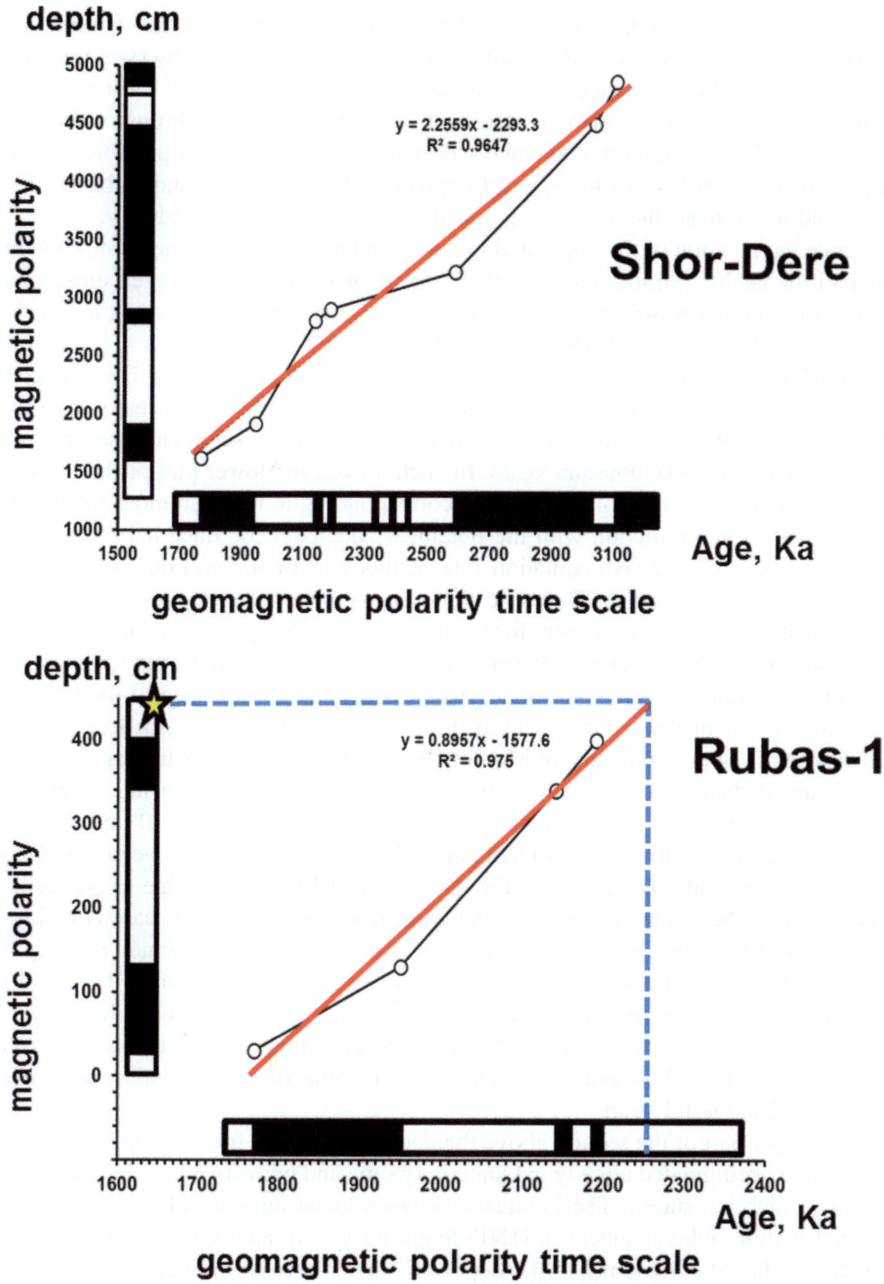

Fig. 27.15 Sedimentation rates for Shor-Dere and Rubas-1 sections determined using depth-age biplots. The points on curves correspond to paleomagnetic boundaries. Red lines are linear approximations of sedimentation rates. Yellow star shows the position of "Small tolls" complex. GPTS is modified from (Pillans and Gibbard 2012; Hilgen et al. 2012)

lithological composition and were deposited in shallow water conditions. Nevertheless, a general trend in sedimentation rates variations is observed for both sections: during Olduvai subchron, sedimentation rates were somewhat lower than those in the interval between Olduvai and Reunion subchrons (compare Fig. 27.15a, b). Calculation of average sedimentation rates makes it possible to give approximate estimates of time span of deposition for Shor-Dere and Rubas-1 sections and to estimate the age of geological events and the stone industry.

Since the position of the upper and lower boundaries of the Akchagylian stage in GPTS is debatable (Stratigraphy … 1982; Van Baak et al. 2013), all age estimations in the present paper will be given according to Neogene-Quaternary time scale (Pillans and Gibbard 2012; Hilgen et al. 2012).

Shor-Dere section. The sedimentation in the studied part of Shor-Dere section has begun during the Gauss Chron before the beginning of Kaena subchron. Approximate age of section base estimated by average sedimentation rate as comprises about 3260 thousand years. In such a case, the lower part of the section should contain a reverse polarity interval corresponding to the Mammoth subchron. This comes in contradiction with the obtained paleomagnetic data, and this is not surprising because the sedimentation rate in this part of the section substantially increases (see Fig. 27.15a). Therefore, for a correct estimate, we took the average sedimentation rate calculated only for the lower part of the section, i.e. including the upper and lower boundaries of Kaena subchron, and Gauss/Matuyama boundary. The obtained rate of 3.1 cm/thousand years gives an age estimation of the base of the studied part of the section ~3220 thousand years, that is, it practically coincides with the upper boundary of Mammoth subchron. This estimation agrees with the obtained data, since the lowest horizon of the studied section has a reverse polarity, although it is represented by only one sample (Figs. 27.10 and 27.12). Thus, the sedimentation in the studied part of Shor-Dere section has begun ~3220 thousand years and corresponds to the upper half of Piacenzian stage of Neogene. The boundary between the Neogene and Quaternary systems in this case coincides with Gauss/Matuyama boundary (2588 thousand years) which is located at a depth of 32.23 m from the upper edge of the Shor-Dere section. The age of the f fossil ash horizon was estimated from the sedimentation during Kaena subchron, where it was determined as 5.4 cm/years. Such rate supposes an estimate age of the ash horizon within 3090–3095 thousand years ago, and the time of its deposition does not exceed 1–2 thousand years

The upper part of the section above the depth of 32.23 m from the upper edge of the section accumulated already in Quaternary time and correspond to Gelasian and, in part, Calabrian stages. The boundary between Gelasian and Calabrian stages is located within Olduvai subchron (1806 thousand years), and according to our calculations, should correspond to the depth of ~16.5 m from the upper edge of the section. It should be noted that the change in sedimentation environment from clayey silt to sandy silt at had occurred approximately 2040 thousand years ago.

Sedimentation in the Shor-Dere section had completed approximately 1160 thousand years ago, somewhat earlier than the lower boundary of the Jaramillo subchron. Subchrons Gilsa and Cobb Mountain were not recorded in Shor-Dere section.

Section Rubas-1. Sedimentation in Rubas-1 section had been going on only in Quaternary time. It has begun ~2260 thousand years ago, which roughly corresponds to the middle of the Gelasian stage. (It should be noted that the absence in the section of the upper part of the "X" subchron with an age of 2330 thousand years (Pillans and Gibbard 2012) is in good agreement with this estimation). The deposition was terminated ~1750 thousand years ago, which corresponds to the lower part of Calabrian stage. The boundary between Gelasian and Calabrian stages (1806 thousand years) is located inside the magnet zone R1N2 (Olduvai subchron) at the depth of ~60 cm below the upper boundary of the studied section. The upper boundary of the deposits in Rubas-1 section is estimated as ~1750 thousand years, somewhat earlier than the lower boundary of the Gilsa subchron.

The proposed age estimation suggest that the archaeological material from the layer 5 the Rubas-1 section, corresponding to the beginning of deposition in Rubas-1 section, can be estimated approximately more than 2260 million years ago, which allows us to consider the early Paleolithic complex of Rubas-1 as one of the oldest archaeological industries in the Caucasus and Eurasia. This statement is in agreement with the recent finding that the waves of migration of early hominids had reached the edge of southeastern Europe by about 2 Ma and a route of migration was located along the western shore of the Caspian Sea (Amirkhanov et al. 2014).

Magnetic polarity intervals isolated in Ajinour and Rubas River sections are not uniquely identified with GPTS chrons and it is impossible to obtain reliable age estimates for those sections.

Conclusions

- Shor-Dere section is the most complete late Neogene-Quaternary section in Near-Sea Dagestan. It was deposited from ~3220 to ~1160 thousand years ago.
- The boundary between Neogene and Quaternary (2588 thousand years ago) was recorded in Shor-Dere section at the depth of 32.23 m from the upper edge of the section.
- The boundary between Gelasian and Calabrian stages in the Shor-Dere section is assumed at the depth of 16.5 m from the upper edge of the section.
- The change in sedimentation environment in the Shor-Dere section had occurred ~2040 thousand years ago.
- The age of the fossil ash horizon in Shor-Dere section is estimated as ~3090 to 3095 thousand years; According to paleomagnetic data, this layer is practically

an isochronous reference level in the studied area, while the redeposited ash horizons correspond to different stratigraphic levels.

- On the base of paleomagnetic data Rubas-1 section can be uniquely correlated with Shor-Dere section. Sedimentation in Rubas-1 section had been going on only in Quaternary time from ~2260 to ~1750 thousand years ago.
- The position of Gelasian and Calabrian stages in Rubas-1 section is assumed at the depth of 60 cm from the upper edge of the section.
- The age of archaeological material found in the layer 5 of tubas-1 section and corresponding to the beginning sedimentation in the section, can be estimated approximately more than 2260 million years ago.

Acknowledgements We are very grateful to A. Yu. Guzhikov for the useful comments to our manuscript. The work was supported by RFBR grants № 10-06-00085-a and № 11-06-12000-ofi-m-2011 and by RSF grant №. № 14-50-00036. This study has also been carried out following the plans of the scientific research of the Geological Institute of RAS (for AYK, project no 0135-2016-0009).

References

Amirkhanov H. A., Ozherel'ev D. V., Gribchenko Yu. N., Sablin M. V., Semenov V. V., Trubikhin V. Early Humans at the eastern gate of Europe: The discovery and investigation of Oldowan sites in northern Caucasus // C. R. Palevol. -2014. V. 13. pp. 717–725

Burdukiewicz J.M. Technokompleks mikrolityczny w paleolicie dolnym środkowej Europy. Wroclaw: WERK, 2003. 374 p. (in Polish)

Derevianko A.P., Anoykin A.A., Kazansky A.Yu., Matasova G. G. New data on the justification of the age of the Early Paleolithic complex of artifacts of the location of Rubas-1 (Primorsky Dagestan) // News of Altai State University 2015, Vol. 2. № 3 (87). P. 78–83. (in Russian)

Derevianko A.P. The earliest human migrations in Eurasia in the Early Paleolithic. Novosibirsk: IAET SB RAS Press, 2009. 232 p. (in Russian)

Enkin R. J. A computer program package for analysis and presentation of paleomagnetic data. Pasific Geoscience Centre // Geol. Surv., Canada, 1994, 16 p

Fisher R.A. Dispersion on a sphere // Proc. R. Soc., 1953, L. A217, pp. 295–305

Geological map of the USSR scale 1: 200000. Series Caucasian [sheet K-39-XIX, XX] / Explanatory note. - M.: Gosgeoltekhizdat, 1961. 76 p. (in Russian)

Golubyatnikov V.D. Geological structure of the Tertiary deposits of Southern Dagestan between the rivers Rubas-Chay and Samur. - L.-M.: Georazvedizdat, 1933. - Proceedings of the All-Union Geological Exploration Association of the NKTP of the USSR. - Ser. 278. 43 p. (in Russian)

Grebenyuk L.V. Magnetostratigraphy of Neogene sediments in the south of European Russia and Transcaucasia. Abstract of diss.... cand. geol.- min. sciences. Saratov. 2004. 22 pp. (in Russian)

Gurary G.Z., Trubihin V.M. Cyclical development of the West-Turkmen part of the Paleo-Caspian in the Late Cenozoic and the paleomagnetic scale // Neogene and Quaternary Border. Moscow: Nauka, 1980. pp. 3–7. (in Russian)

Hilgen F.J., Lourens L.J., Van Dam J.A. The Neogene Period. In: The Geologic Time Scale. Eds. F.M. Gradstein, J.G. Ogg, M. Schmitz, G. Ogg. Elsivier, 2012. pp. 923–978

Kirschvink J.L. The least-square line and plane and the analysis of paleomagnetic data. Geophys. J. R. Astron. Soc., 1980, v. 62, pp. 699–718

Leschinsky S.V, Konovalova V.A., Burkanova E.M., Babenko S.N. Justification of the relative age of the Early Paleolithic sites Darvagchay-1 and Rubas-1 (Southern Dagestan) // The oldest human migration in Eurasia: Materials of the International Symposium. Novosibirsk: Izdatelstvo IAET SB RAS, 2009. pp. 140–149. (in Russian)
Leschinsky S.V. Justification of the relative age of the lower cultural horizon of the location of the Rubas-1 // Derevianko AP, Amirkhanov Kh.A., Zenin VN, Anoykin AA, Rybalko AG Problems of the Paleolithic of Dagestan. Novosibirsk: Izdatelstvo IAET SB RAS, 2012. pp. 277–289. (in Russian)
Lower Palaeolithic Small Tools in Europe and the Levant. J. M. Burdukiewicz, A. Ronen (ed.): BAR International Series. № 1115. Oxford, 2003. 239 p
McFadden P.L. A new fold test for palaeomagnetic studies // Geophys. J. Int. 1990, V. 103. pp. 163–169
Minyuk P.S. A.S. 900163 USSR. Sampler for weakly durable soil No 292381422-26; claimed. 08.05.80; publ. 23. 01. 82, 1982, Bul. No. 3. 3 p. (in Russian)
Pillans B., Gibbard P. The Quaternary Period. In: The Geologic Time Scale. Eds. F.M. Gradstein, J.G. Ogg, M. Schmitz, G. Ogg. Elsivier, 2012. pp. 979–1010
Stratigraphy of the USSR. Quaternary system (semivolume1). M., 1982. 443 p. (in Russian)
Tauxe L. Essentials of Paleomagnetism. University of California Press, Berkeley., Ca. 2010, 512 p
Torre de la I., Mora R., Dominguez Rodrigo M., Luque de L., Alcala L. The Oldowai industry of Peninj and its bearing on the reconstruction of the technological skills of lower Pleistocene hominids // Journal of Human Evolution. 2003. Vol. 44(2). pp. 203–224
Van Baak C.G.C., Vasiliev I., Stoica V., K.F. Kuiper K.F., A.M. Forte A.M., Aliyeva E., Krijgsman W.A magnetostratigraphic time frame for Plio-Pleistocene transgressions in the South Caspian Basin, Azerbaijan //Global and Planetary Change. 2013 V. 103. pp. 119–134
Zijderveld J.D.A. A.C. demagnetization of rocks: analysis of results. In: Methods in paleomagnetism. Eds. Collinson D.W., Creer K.M. and Runkorn S., Elsevier, Amsterdam, 1967, pp. 254–286

Chapter 28
Magnetic Stratigraphy of the Bazhenov Suite of Western Siberia and the Surrounding Deposits

A. G. Manikin, V. A. Grishchenko, A. Yu. Guzhikov and V. V. Kolpakov

Abstract A paleomagnetic and petromagnetic study of the Jurassic-Cretaceous boundary in three wells, located on the territory of Imilorskoye and Severo-Yeguryahskoe deposits of Western Siberia was conducted. The sediments of Bazhenov Formation and underlying, as well as overlaying, deposits are detected in each section. Paleomagnetic and petromagnetic studies were carried out on partially (up-down) oriented core samples. According to the results of alternating field magnetic cleaning characteristic components of magnetization were determined and a paleomagnetic column was compiled, showing the dominant normal polarity in the interval from Bajocian-Bathonian to Berriasian-Valanginian. This contradicts the existing data on the Middle-Jurassic to Early-Cretaceous geomagnetic field regime, but still does not prove the secondary magnetization of rocks. For the definitive solution of the problem of magnetization genesis in oil and gas bearing formations of Western Siberia it is necessary to continue their magnetostratigraphic studies. Anisotropy of magnetic susceptibility (AMS) of 659 samples from the sections of Bazhenov Formation and overlying deposits was studied on the territory of Imilorskoye and Severo-Yeguryahskoe deposits of Western Siberia in 6 wells. The half of the wells represent the classical section of Bazhenov Formation (organic-rich siliceous shales), and another wells represent the anomalous section of

A. G. Manikin (✉) · V. A. Grishchenko · A. Yu. Guzhikov
Geological Faculty, Saratov State University, Saratov, Russia
e-mail: agmanikin@mail.ru

V. A. Grishchenko
e-mail: grishenko-vladimir@bk.ru

A. Yu. Guzhikov
e-mail: aguzhikov@yandex.ru

V. V. Kolpakov
LUKOIL-Engineering, KogalymNIPIneft, Kogalym, Russia

D. Nurgaliev et al. (eds.), *Recent Advances in Rock Magnetism, Environmental Magnetism and Paleomagnetism*, Springer Geophysics,
https://doi.org/10.1007/978-3-319-90437-5_28

Bazhenov Formation (alternation of black shales and sand-silty rocks). The results of AMS study support an idea of close age of bituminous argillites and terrigenous rocks from anomalous section of Bazhenov Formation.

Keywords Bazhenov suite · Western Siberia · Magnetic susceptibility
Anisotropy of magnetic susceptibility · Oil and gas bearing formations
Paleomagnetism · Petromagnetism

Introduction

Owing to its hydrocarbon prospects, the Bazhenov Formation (BF) is currently among the most intensively studied subsurface objects in Western Siberia. Nevertheless, the BF exact stratigraphic age is still open to discussion, which is accounted for by the formation confinement to the boundary interval of the Jurassic and Cretaceous systems. Detailed Boreal-Tethyan correlation of this interval remains one of the most urgent problems in modern stratigraphy. Due to the separation between Tethyan and Boreal paleobasins at the Jurassic-Cretaceous boundary, unequivocal comparison of the zonal scales from different regions cannot be based only on paleontological data, without involvement of independent methods, specifically, the paleomagnetic one. Moreover, the latest and the only wide-scale magnetostratigraphic study of the Bazhenov Formation was carried out over 40 years ago (Paleomagnetism … 1976). Recently, only a single reference section covering the Jurassic-Cretaceous boundary in the Boreal domain (Nordvik Cape) has been paleomagnetically studied at the present-day level (Bragin et al. 2013; Hosha et al. 2007).

Another debatable problem concerns the genesis and the age of the sandy-silty interlayers in the so-called Bazhenov anomalous section (BAS). BAS is a section of BF, in which a characteristic bundle of dark-colored bituminous argillites is either delaminated by sandstones, siltstones, or completely replaced by them (Nezhdanov et al. 1985). Irrespective of being difficult-to-develop at the present state of petroleum geology in Western Siberia, these interlayers are the most promising objects in terms of oil and gas contents. The current hypotheses of the BAS generation may be divided into two principal types: the "even-aged" models, assuming that the sandy-silty interlayers were shaped in the Volgian time (Zubkov and Portmeister 2005), and the "uneven-aged" models, assuming the Neocomian age of the terrigenous varieties, with their subsequent intrusion into bituminous clays (Grishkevich et al. 2017). Paleontological examinations are still unable to substantiate or disprove any of the standpoints unequivocally; therefore, it is urgent to involve other independent methods to clarify the question. A major part in solving this problem may be assigned to the data on the deposit anisotropy of magnetic susceptibility (AMS), determined by the textural features (shapes, relative positions) of magnetic particles. Magnetic textures of the rocks, formed under different conditions, differ significantly, providing a rationale for using AMS data to determine the genesis of the anomalous section of Bazhenov formation. In particular, the primary magnetic texture of bazhenites would be inevitably changed upon aleurolite and sandstone intrusion.

Examination Procedure

The total of 311 partially oriented (top-bottom) and 1730 non-oriented core samples from 6 well Sections (4 well from the Imilorskoye and 2 well from the Severo-Eguryakhskoye fields) were selected for paleomagnetic and petromagnetic measurements. In the wells 280, 405 and 412 from the Imilorskoye field, the Bazhenov anomalous section is represented, in other wells the BF is composed only of classic clayey-bituminous deposits. Samples for paleomagnetic measurements were collected every 0.5–0.75 m, those for petromagnetic studies—every 0.3 m. Selection gaps of several meters occasionally occurred because of the lack of core material.

Petromagnetic measurements were aimed at estimating the sample usability for paleomagnetic determinations and for acquiring an additional information on the deposit material composition and generation environments. Those included kappametry—measuring of initial magnetic susceptibility (*K*), thermokappametry—measuring the initial magnetic susceptibility upon sample incremental heating up to 500 °C for one hour in a muffle furnace ($dK = Kt - K$, with *Kt*—magnetic susceptibility measured after heating), measuring the anisotropy of magnetic susceptibility and natural remanent magnetization ($\mathbf{J_n}$), differential thermomagnetic analysis (DTMA) of selected samples.

Oriented cores were sawn into two or three 2-cm cubic samples. Later on, those samples were subjected to kappametry, thermokappametry, paleomagnetic measurements and AMS measurements. Non-oriented samples of random shape were subjected only to kappametric and thermokappametric examinations.

Paleomagnetic examination consisted of the $\mathbf{J_n}$ measurements after magnetic cleanings with alternating field (h-cleanings), from 5 mT to 50–60 mT, in 5 mT increments, and temperature (t-cleanings) from 100 to 300 °C, in 50 °C increments. No cleanings with higher temperatures were made because of the sample laboratory magnetic biasing resulting from the phase transformations of iron sulfides above 300 °C. The acquired data were subjected to component analysis (Kirschvink 1980) with the purpose of distinguishing the characteristic remanent magnetization (**ChRM**) and providing interpretations in terms of its magnetic polarity.

Magnetic susceptibility was measured by means of a MFK1-FB kappabridge, remanent magnetization—with a JR-6 spin-magnetometer. Thermal magnetic cleanings were made in a home-made furnace designed by V. P. Aparin. Magnetic cleanings with alternating field were made with a LDA-3AF unit. DTMA was made with a TAF-2 thermoanalyzer of magnetic fractions (« magnetic balance »). The AMS data was analyzed by means of the Anisoft 4.2 software, component analyses of paleomagnetic materials were made by means of the Remasoft 3.0 software.

Research Results

<u>**Petromagnetism**</u>. Magnetite presence was recorded in the samples from the DTMA data. Magnetite was diagnosed from magnetization loss close to the Curie temperature of 578 °C (Fig. 28.1). At the first heating, after 600 °C a new ferromagnetic phase forms in the samples. This follows from an increase of magnetization after 600 °C and from the Curie point above 700 °C (probably, the iron formed by reduction of magnetite in the presence of organic matter). Iron sulfides of the pyrrhotine and/or pyrite type, clearly recorded in the second-heating thermomagnetic curves from magnetization growth during the phase transformations of those minerals to magnetite close to the temperatures of 300–350 °C and/or 400–450 °C (Fig. 28.1), may represent the products of interaction of the newly formed ferromagnetic phase with the hydrogen sulfide released upon organic matter burning.

Sections of the three most complete well logs are petromagnetically well-differentiated. The petromagnetic characteristics acquired from them allow more elaborate well-section divisions and correlations (Manikin et al. 2017; Samarin et al. 2016) (Fig. 28.2).

Anisotropy of magnetic susceptibility was measured in 659 paleomagnetic samples.

Short axes of magnetic ellipsoids (K3) in the *Sortymskaya and the Megionskaya formations*, overlying the BF, tend towards vertical (projection of the K3 average direction statistically coincides with the stereogram center), while the long (K1) and the medium (K2) axes are distributed over the stereogram margins (Fig. 28.3a).

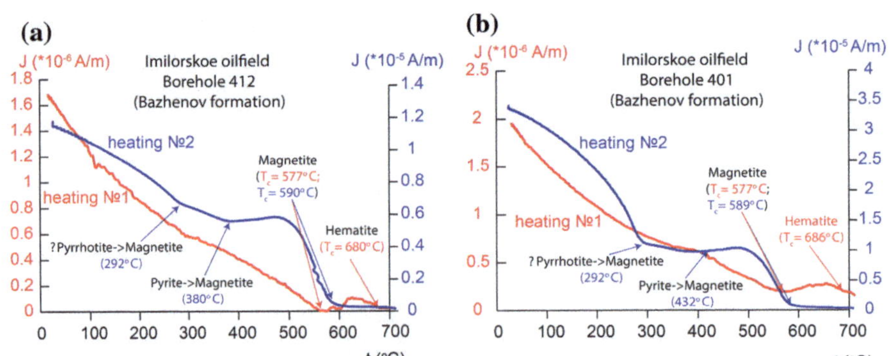

Fig. 28.1 Results of differential thermomagnetic analysis (DTMA)

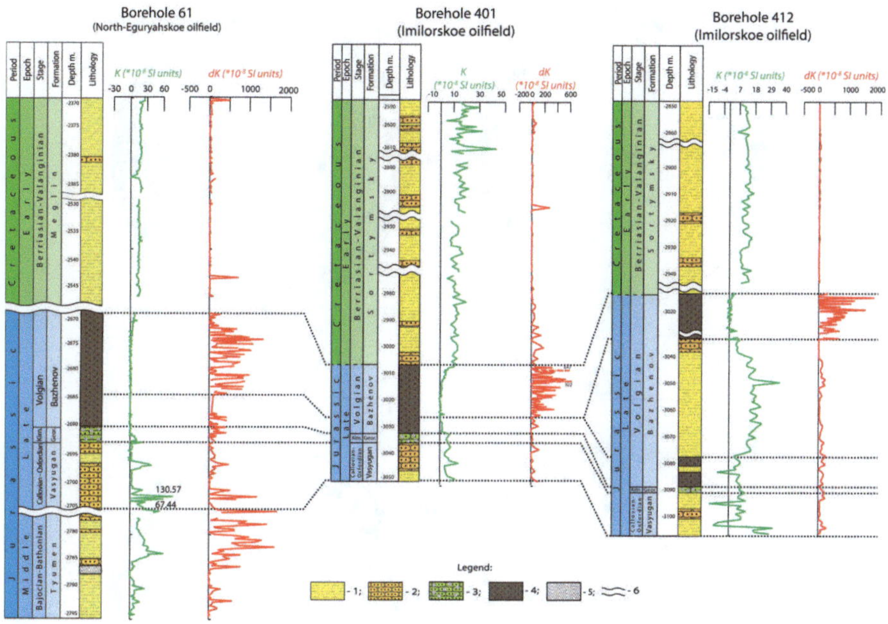

Fig. 28.2 Correlation of studied wells using petromagnetic parameters. Legend: 1—siltstone; 2—sandstones; 3—glauconitic sandstone; 4—bituminous mudstones; 5—coals; 6—stratigraphic breaks

Similar magnetic texture is characteristic of sediments formed in quiet hydrodynamic conditions (Tarling and Hrouda 1993). Positive values of the T-shaped parameter are indicative of flattened ferromagnetic particles. The effect of flatness is likely to be associated with magnetic mineral aggregation on the scale of clayey minerals. The fraction of elongated (cigar-shaped) ferromagnetic particles does not exceed 7% (Fig. 28.3a).

Clayey-bituminous deposits from the Bazhenov Formation. Distribution of the AMS-axes projection in the stereograms, corresponding to bituminous argillites (in the zones of both classical and anomalous BF sections), is also indicative of their generation in quiet hydrodynamic conditions: the K3 axes tend towards vertical, while the K1 and K2 axes are distributed over the stereogram margins. The unchanged classic sedimentary texture of the rocks is incompatible with the hypothesis of the BAS formation proposed by A. A. Nezhdanov and co-authors (1985), which assumes that terrigenous interlayers result from the activity of the Valanginian landslides, highly pressurized and wedged into the underlying deposits, producing contortions and ruptures.

Meanwhile, there are levels with abnormally large K3 axes shifts from the vertical, from 15° to 89° (Fig. 28.3b, c), in the clayey-bituminous beds, both in classic BF sections and in the BAS. The "anomalous" samples (their share may be as high as 11%) are not confined to the boundaries between bituminous argillites

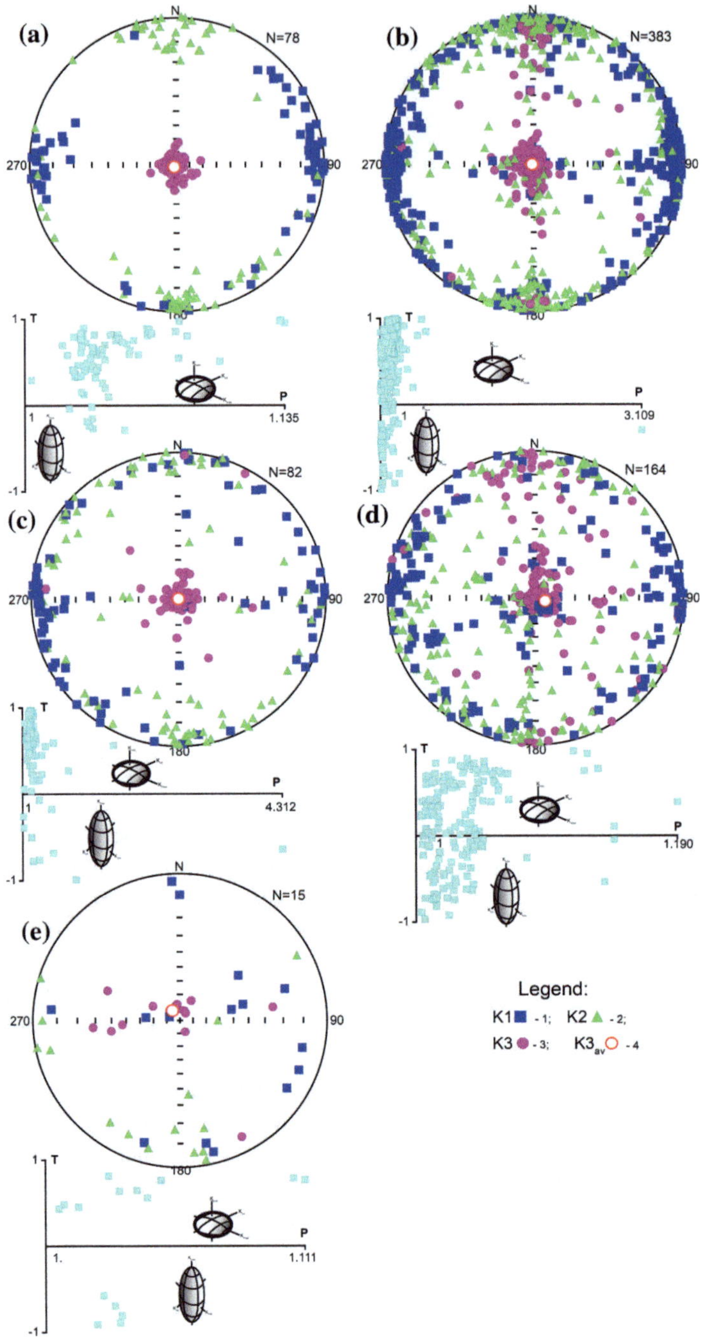
(a)
N=78
(b)
N=383
(c)
N=82
(d)
N=164
(e)
N=15
N
270
90
180
T
P
1.135
3.109
4.312
1.190
1.111
Legend:
K1 - 1; K2 - 2;
K3 - 3; K3av - 4

◂**Fig. 28.3** AMS stereograms and P-T AMS diagrams as an indicator of magnetic ellipsoid shape of overlaying formations (**a**), clayed-bituminous sediments of classical (**b**) and anomalous (**c**) sections of Bazhenov formation, of terrigenous layers in anomalous section of Bazhenov formation wells 405 and 412 (**d**) and well 280 (**e**). Legend: 1,2,3—directions of long (K1), medium (K2), and short (K3) axis of magnetic ellipsoids, respectively; 4—mean direction of short axis ($K3_{av}$) of magnetic ellipsoids

and terrigenous interlayers, but occur sporadically within bituminous argillites. Their magnetic texture may be determined by siderite or single-domain magnetite in the ichnofossil infillings. The latter statement is supported by the highly elongated shapes of ferromagnetic particles, recorded from the *T* negative values and the *P* (T = (2lnK2−lnK1−lnK3)/(lnK1−lnK3)—shape parameter; P = K1/K3—degree of anisotropy) increase up to 4.31 (Fig. 28.3b, c).

In the BAS sand-and siltstones, the magnetic texture differs drastically from both the BF bituminous argillites and terrigenous beds of the Sortymskaya and the Megionskaya formations. In the AMS stereograms from the sections of wells 405 and 412, in more than a half of the samples, strong shifts of the K3 axes from the vertical to the first and second quadrants of the stereogram are recorded, and the projection of the average K3 axis is shifted by 5.3° eastwards from the vertical (Fig. 28.3d). The *T* shape parameter indicates the presence of both flat and cigar-shaped magnetic particles, the latter ones comprising up to 36% from the total sample number. Such distribution of the short axes directions may be associated with an effect of strong storm waves on the non-lithified sediment. Their activity is expected to roil the bottom sediment and disturb its primary magnetic texture (Lanza and Meloni 2006).

Anisotropy of magnetic susceptibility of the BAS terrigenous interlayers from the well 280 reveals the K3 sublatitudinal distribution (Fig. 28.3e). Regular "stretching" of the AMS short axes directions along the E-W line, alongside with the shift of their average position from the vertical is observed during the sediment transfer as high-density turbidite flows (Lanza and Meloni 2006; Popov and Zhuravlev 2012). Diagrams of the dependence of the *T* shape parameter from the anisotropy degree *P* indicate the possible presence of both disc-shaped and elongated magnetic particles (Fig. 28.3e).

The AMS data does not comply with the sedimentological model by O. M. Mkrtchyan and co-authors (1987). According to this model, formation of the band-shaped sand- and siltstone bodies is associated with the avandelta and bar channels. Neither the regularities in the K1 orientation, associated with the near-bottom currents in the avandelta channels, nor the specific magnetic textures of the cross-bedded deposits from the sea bars are observed in the ABS terrigenous rocks.

Paleomagnetism. 317 samples from the sections of the wells 401, 412 (the Imilorskoye field) and 61p (the Severo-Eguryakhskoye field) were examined paleomagnetically. The h-cleaning results are peculiar for good quality: in most samples, the characteristic remanent magnetizations (**ChRM**) are isolated in the fields from 5 to 40 mT, their maximum deviation angle is <15° (Fig. 28.4a).

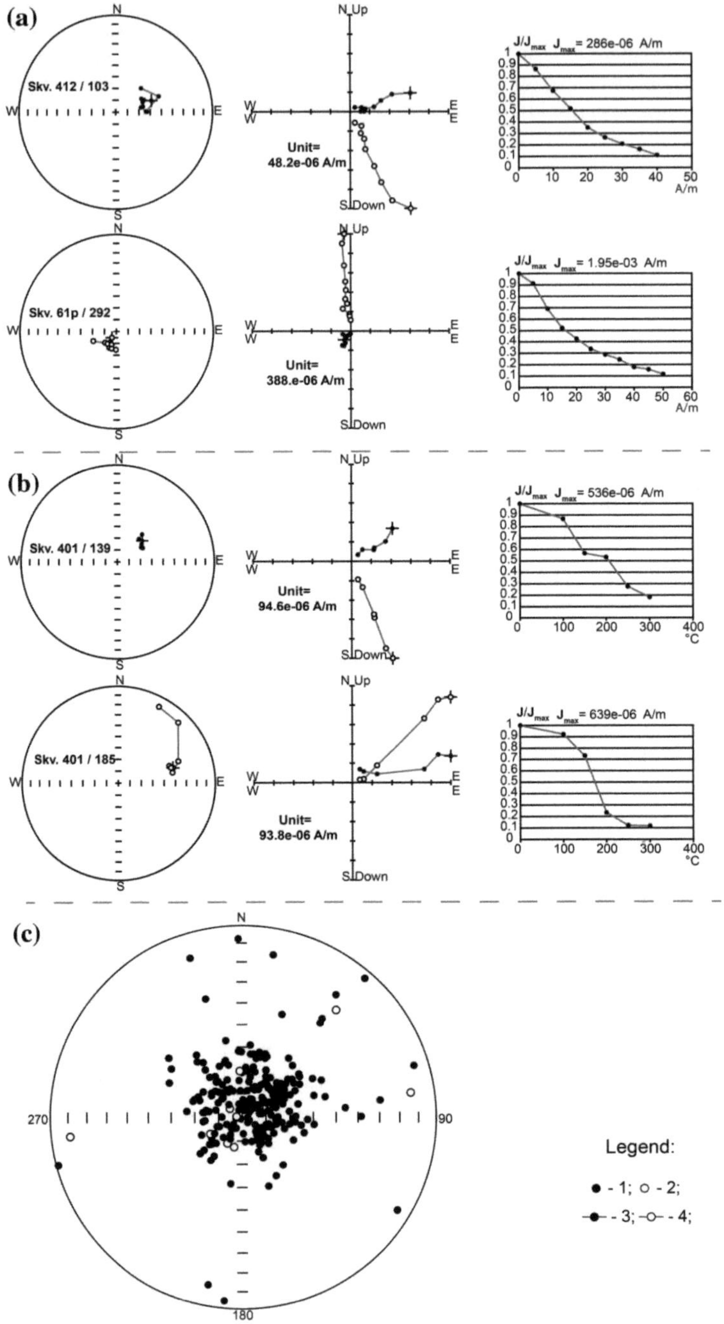

Fig. 28.4 Results of magnetic cleaning with alternating field (**a**), temperature (**b**) and summary stereograms of **ChRM** (**c**) on the studied wells. Legend: 1,2—stereographic projections of $\mathbf{J_n}$ directions on the lower semisphere and upper semisphere respectively; 3,4—projection on the horizontal and vertical plane respectively

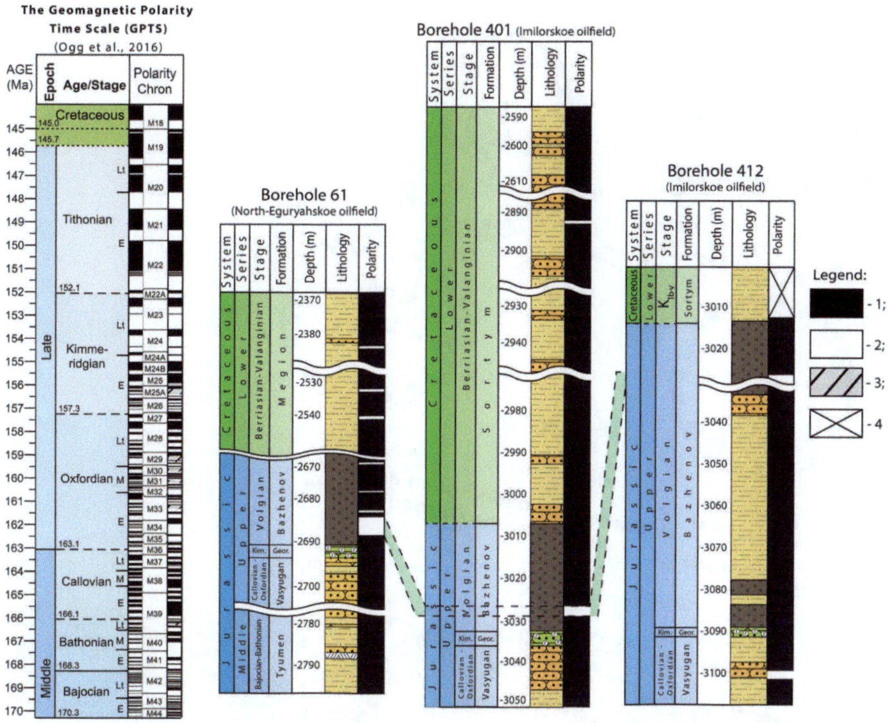

Fig. 28.5 Comparison of magnetostratigraphic data of well material with the Geomagnetic Polarity Time Scale (GPTS). Legend: 1,2—normal and reverse polarity respectively; 3—uncertain polarity; 4—missing of paleomagnetic data. Other legend symbols: see Fig. 28.2

The results of T-cleanings up to 300 °C agree well with the demagnetization by alternating field (Fig. 28.4b).

The **ChRM,** isolated in the examined samples, correspond to both normal (N) and reverse (R) polarities, with inclinations (I) up to +88° and up to (−84°), respectively (Fig. 28.4c). But negative inclinations are recorded only in 13 stratigraphic levels (in $\sim$4% of samples), and the paleomagnetic columns in all three wells are peculiarly dominated by the normal polarity (Fig. 28.5). Infrequent reversed polarity intervals are substantiated with samples from 1 or 2 levels, and therefore they cannot be formally recognized as independent magnetozones (according to Khramov and Sholpo (1967), a magnetozone must be substantiated with samples from at least three levels).

Dominant normal polarity, recorded in the studied sections, does not comply with the current notion of complicated alternating-sign paleomagnetic structure of the Bajocian-Berriasian (Ogg et al. 2016) (Fig. 28.5). Therefore, irrespective of the good quality of the paleomagnetic data, secondary remagnetization seems to be the most likely. Some circumstances, however, prevent from its absolute acceptance:

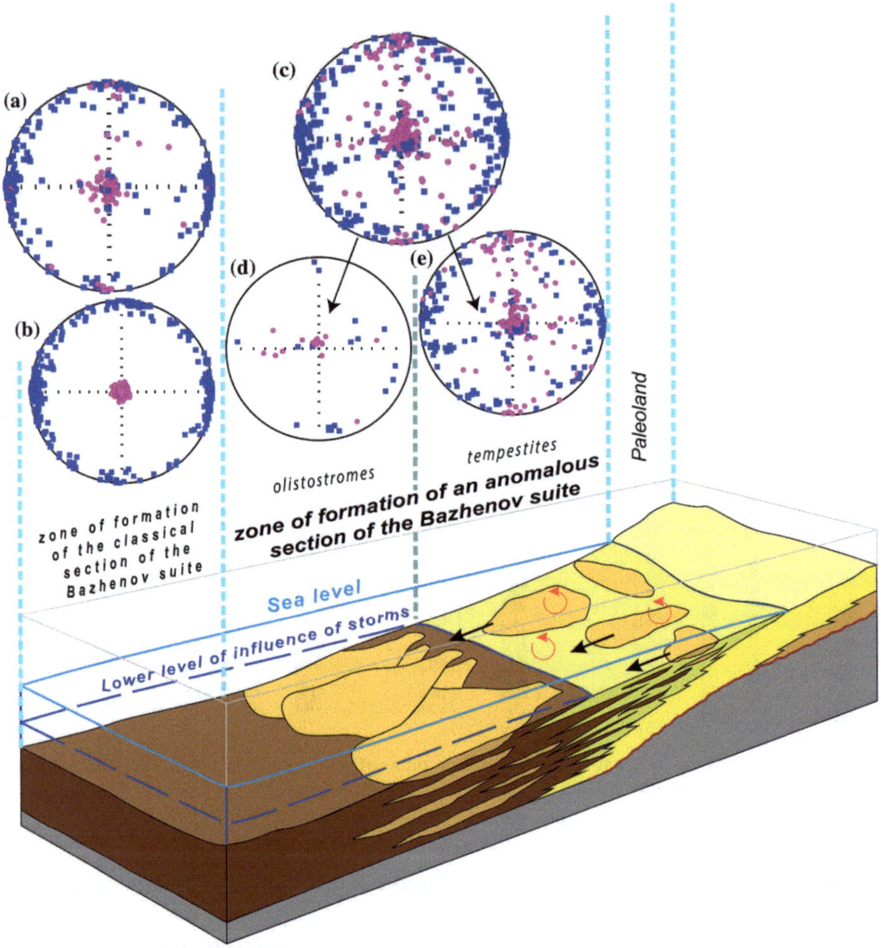

Fig. 28.6 The proposed model of Bazhenov formation (BF) development on the base of data obtained from AMS. **a**, **b** classical section of BF in well 401: all samples (**a**), after the exclusion of supposedly bioturbated samples (**b**); AMS of anomalous section of BF: all samples (**c**), samples from well 280 (**d**), samples from wells 405 and 412 (**e**)

1. The reversed polarity interval in the lower part of classical Bazhenov formation is observed in the sections from Northern Yeguriahskoe (well 61p) and Imilorskoye (well 401) oilfields that are 180 km away from each other. It is not improbable, that the only reverse interval in the Bazhenov anomalous section (well 412) is coeval to them (Fig. 28.5). Similar lateral persistence is indicative of the **ChRM** ancient nature, and the small thicknesses of the reversed intervals in bituminous argillites may be explained by the sediment high water contents and low lithification rates in the Bazhenov Formation (Bordyug et al. 2010).

The water, squeezed by compaction into the sediment upper layers, could have contributed to reorientation of magnetic particles after a geomagnetic reversal. The normal polarity regime, dominant in the Tithonian, could have caused domination of normal magnetization in the Bazhenov Formation.

2. The sections of oil-and-gas bearing formations of Siberia are condensed (Baraboshkin 2008), and the attribution of sediments to a particular formation does not necessarily yield the exact age of sediments (The regional … 2004). Furthermore, a continental or subcontinental genesis of the Tyumen suite implies extremely uneven sedimentation rates (Kontorovich et al. 2010). Owing to these reasons, a paleomagnetic structure of studied sections may differ cardinally from the real reversal sequence.
3. The acquired data are in good accord with G. A. Pospelova's (Paleomagnetism … 1976) materials on the Middle Ob Region which register only the normal polarity in the Volgian stage and dominantly normal polarity in the Bajocian-Kimmeridgian. Meanwhile, numerous thick reversed magnetozones are recorded in the Lower Cretaceous (Ogg et al. 2016), which makes the remagnetization rather unlikely.

Conclusions

Detailed correlations of 3 wells separated by the distance from 20 up to 180 km were made from the kappametry and thermokappametry data.

Results of examining the anisotropy of magnetic susceptibility are at variance with all the versions of "uneven-aged" models of genesis of the Bazhenov anomalous section, but agree with the idea of close ages of terrigenous rocks and bituminous argillites. The acquired AMS data may be used to propose a model whose principal factors are the bottom relief of the Bazhenov basin and the sea-level fluctuations (Fig. 28.6). The essence of the model is that the tempestite sediments, formed during regressions in the shallow-water shelf zone close to the local elevation zones (paleoislands), were partially flushed into the deep-water zone as high-density turbidite flows, and were overlaid at the transgressive stages with clayey silts enriched in organic matter.

Regretfully, the acquired paleomagnetic data does not allow to prove either ancient or recent age of magnetization. Therefore, further magnetostratigraphic studies are required to obtain the magnetic polarity characteristics of the Bazhenov and other oil-and-gas formations in Western Siberia.

Acknowledgements The research was financially supported by the RFBR (projects №17-05-00716-a).

References

Baraboshkin E. Yu., (2008) Condensed sections: terminology, types, conditions of generation // *V. N. Staroverov (ed.) Essays on regional geology. Collection of scientific articles. Saratov.* P. 20-33. (in Russian).

Bordyug M.A., Slavkin V.S., Gavrilov S.S., Potryasov A.A. (2010) Peculiarities of structure and formation of anomalous section of Bazhenov suite with reference to Nort-Konitlorskoe field // *Oil and gas geology.* 1. P. 32–40. (in Russian).

Bragin V., Dzyuba O.S., Kazansky A., Shurygin B.N. (2013) New data on the magnetostratigraphy of the Jurassic-Cretaceous boundary interval, Nordvik Peninsula (northern East Siberia) // *Russian Geology and Geophysics.* 54, 3. P. 335–348.

Grishkevich V.F., Lagutina S.V., Panina E.V., Dolmatova S.S., Laptey A.G., Toropov E.S., Starikov V.S., Khoroshev N.G., Blinkova A.V. (2017) Geomechanical model of abnormal sequences formation of Bazhenov suite: physical simulation and practical application // *Geology, Geophysics and Development of Oil and Gas Fields.* 3. P. 33–47. (in Russian).

Hosha V., Pruner P., Zaharov V.A., Kostak M., Shadima M., Rogov M.A., Shlehta S., Mazuh M. (2007) Boreal-Tethyan correlation of the Jurassic-Cretaceous boundary interval by magneto-and biostratigraphy // *Stratigraphy and geological correlation.* 15. 3. P. 3–75.

Khramov A.N., Sholpo L.E. (1967) Palaeomagnetism // *Leningrad, Nedra.* 224 p. (in Russian).

Kirschvink J.L. (1980) The least-squares line and plane and the analysis of palaeomagnetic data // *Geophysical Journal International.* 62. P. 699–718.

Kontorovich A.E., Vakulenko L.G., Kazanenkov V.A., Skvortsov M.B., Yan P.A., Bykov V.V., Popov A.Yu. and Saenko L.S. (2010) Sedimentogenesis and resource potential of middle–upper bathonian reservoirs in the Middle Ob' Region // *Russian Geology and Geophysics.* 51. 2. P. 147–158. (in Russian).

Lanza R., Meloni A. (2006) The Earth's Magnetism: An Introduction for Geologist // *Springer.* 278 p.

Manikin A.G., Grishchenko V.A., Guzhikov A.Yu. (2017) Paleomagnetic data on the Jurassic-Cretaceous boundary interval of Western Siberia. // *V.A. Zakharov, M.A. Rogov, E. V. Shchepetova (eds.). Jurassic System of Russia: Problems of stratigraphy and paleogeography. Seventh all-Russian Conference. September 18–22.* P 132 - 136. (in Russian).

Mkrtchyan O.M., Trusov L.L., Berklin N.M., Degtev V. A. (1987) Seismogeological analysis of oil and gas content in the sediments of Western Siberia // *Nauka.* 126 p. (in Russian).

Nezhdanov A.A., TumanovN. N., Kornev V.A. (1985) Anomal sections of Bazhenov formation and their seismo-geological characteristic // *Seismic survey for lithology and stratigraphy: Writings of West Siberian Geological Research and Development Oil Institute. Tyuman.* P. 64–71. (in Russian).

Ogg J.G., Ogg G.M., Gradstein F.M. A (2016) Concise Geologic Time Scale // *Elsevier.* 242 p.

Paleomagnetism of Mesozoic and Cenozoicos Siberia and Far East (1976) / *Fotiadi E.E. (ed.). Novosibirsk. Siberian Branch of the Russian Academy of Sciences.* 142 p. (in Russian).

Popov V.V., Zhuravlev A.V. (2012) Anisotropy of various magnetic parameters use for determining directions of material transfer in the study of turbidite currents // *Petroleum Geology - Theoretical and Applied Studies.* 7. 1. 21 p.

Samarin S.V, Isaev P.V, Manikin A.G., Pustoshkin R.V. (2016) Results of petromagnetic studies of the core material of the Imilor oil field // *Vagarin VA (ed.) New technologies in the gas industry: articles of the correspondence scientific conference of young scientists and specialists of the gas industry and educational institutions of the Saratov region (Saratov, December 21, 2016).* P. 35–39. (in Russian).

Tarling D.H., Hrouda F. (1993) The magnetic anisotropy of rock // *Chapman & Hall.* 217 p.

The regional stratigraphic correlation scheme of Callovian and upper Jurassic sediments of Western Siberia (2004) // *F.G. Gurari et all. (eds.). 6-th Interdepartmental stratigraphic meeting awards on the consideration and acceptation of specified stratigraphic schemes of Mesozoic sediments of Western Siberia. Novosibirsk.* P. 77–113. (in Russian).

Zubkov M.Yu., Portmeister Ya. A. (2005) Clinoform construction of Neocomian and “anomalous” sections of Bazhenov formation within Kalchinskoe oilfield (based on a data of seismic survey and tectonic modeling) // V.I. Karasev, E. A. Ahpatelov, V.A. Volkova *(eds.) ways to implement oil-and-gas potential KhMAD. Khanty-Mansiysk.* P. 305–318. (in Russian).

Part IV
Extraterrestrial Magnetism

Chapter 29
Metallic Iron in the Planets Interior: Generalization of Thermomagnetic and Microprobe Results

D. M. Kuzina and D. M. Pechersky

Abstract The thermomagnetic and microprobe data for metallic iron in terrestrial rocks (xenoliths from upper mantle hyperbasites, Siberian trapps, and oceanic basalts) compared to the data for metallic iron from sediments, lunar basalts and meteorites are shown. It was found that the metallic iron particles contained in all groups terrestrial and extraterrestrial rocks are similar in composition, shape, and grain size. This similarity suggests the similar sources of these particles origin. This means that the terrestrial conditions were close to the conditions that existed at the source planets of the meteorites, e.g., the bodies from the asteroid belt were subsequently disintegrating, and crushing into cosmic interplanetary dust, which fell into the terrestrial sediments. This similarity originates from the homogeneity of the gas-dust cloud at the early stage of the Solar System and obviously other star-planetary systems and subsequent gravitational differentiation in the process of all planet formation. The predominance of extraterrestrial metallic iron in sediments is accounted for by the fact that the interplanetary dust is mainly contributed by the upper mantle material of the source planets of cosmic dust.

Keywords Metallic iron · Terrestrial rocks · Extraterrestrial rocks
Thermomagnetic analysis · Microprobe analysis

D. M. Kuzina (✉) · D. M. Pechersky
Institute of Geology and Petroleum Technologies, Kazan Federal University, Kazan, Russia
e-mail: di.kuzina@gmail.com

D. M. Pechersky
e-mail: diamarmp@gmail.com

D. M. Pechersky
Schmidt Institute of Physics of the Earth, Russian Academy of Sciences, Moscow, Russia

D. Nurgaliev et al. (eds.), *Recent Advances in Rock Magnetism, Environmental Magnetism and Paleomagnetism*, Springer Geophysics,
https://doi.org/10.1007/978-3-319-90437-5_29

Introduction

The aim of this paper is the synthesis of thermomagnetic and microprobe data on metallic iron and iron-nickel alloys for terrestrial and extraterrestrial rocks.

In recent years we've published a large amount of data on metallic iron particles in the epicontinental sediments of different regions and ages in Eurasia (Grachev et al. 2009; Pechersky et al. 2006, 2009a, b, 2010, 2011), in the oceanic sediments of the North-West Atlantic (Pechersky et al. 2013; Murdmaa et al. 2015), in lacustrine sediments of Lake Baikal (Pechersky et al. 2013) and Mongolia (Pechersky et al. 2013), in meteorites (Pechersky et al. 2012); and other researchers have studied lunar basalts (Doell et al. 1970; Nagata et al. 1970, 1974; Reid et al. 1970; Runcorn et al. 1971; Strangway et al. 1970). In addition to the listed thermomagnetic, electron microscopic and microprobe studies of samples terrestrial rocks were studied: (1) upper mantle hyperbasites that carried at the Earth's surface by basalt lavas in the form of xenoliths from Antarctica, Mongolia, Primorye (Russian Far East), Syria, Spitsbergen and the Vitim Plateau (Russia), (2) trapps of Angarsk, Maimecha-Kotui and Norilsk provinces of the Siberian platform, and (3) oceanic basalts that form the floor of the Atlantic, Pacific and Indian Ocean, the Red Sea (Pechersky et al. 2015; Pechersky 2015). These results are compared with previously obtained for sediments, meteorites and lunar basalts.

Methods

The main (mass) method of research is thermomagnetic analysis (TMA). TMA measurements were conducted at the paleomagnetic laboratory of Kazan University using Curie express balance constructed in this laboratory (Burov et al. 1986), and at the laboratory of the geomagnetic field and petromagnetism of Schmidt Institute of Physics of the Earth, Russian Academy of Science (IPE RAS) using thermal vibromagnetometer constructed by N. M. Anosov and Yu. K. Vinogradov, TMA measurements in IPE RAS was performed by G. P. Markov. The TMA included the measurements of magnetic moment of samples in the magnetic field of 400–500 mT and the temperature dependences of magnetic moment in the temperature interval from 20 to 800 °C (Fig. 29.1). The temperature limit was specified at 800 °C in order to overlap the Curie points of metallic iron, which may reach 770 °C (pure iron). As for now, TMA is the only method suitable for the fast and mass estimation of the magnetic mineral content without extracting these minerals from the rock. The Curie points were estimated as accurately as within 5°–10°. The heating rate was $\sim$100°/min. Heating-cooling cycle was repeated twice. The sensitivity of the device for the specific magnetization was $\sim$0.0001 Am^2/kg. The specific saturation magnetization of iron at 20 °C is 217.5 Am^2/kg (Bozorth 1951), hence the lower limit of measuring iron concentration (in wt%) is $0.0001 \times 100 / 217.5 = 4.6 \times 10^{-5}$ %. TMA was conducted in the air. Thermomagnetic curves M (T) (Fig. 29.1) allow us to determine

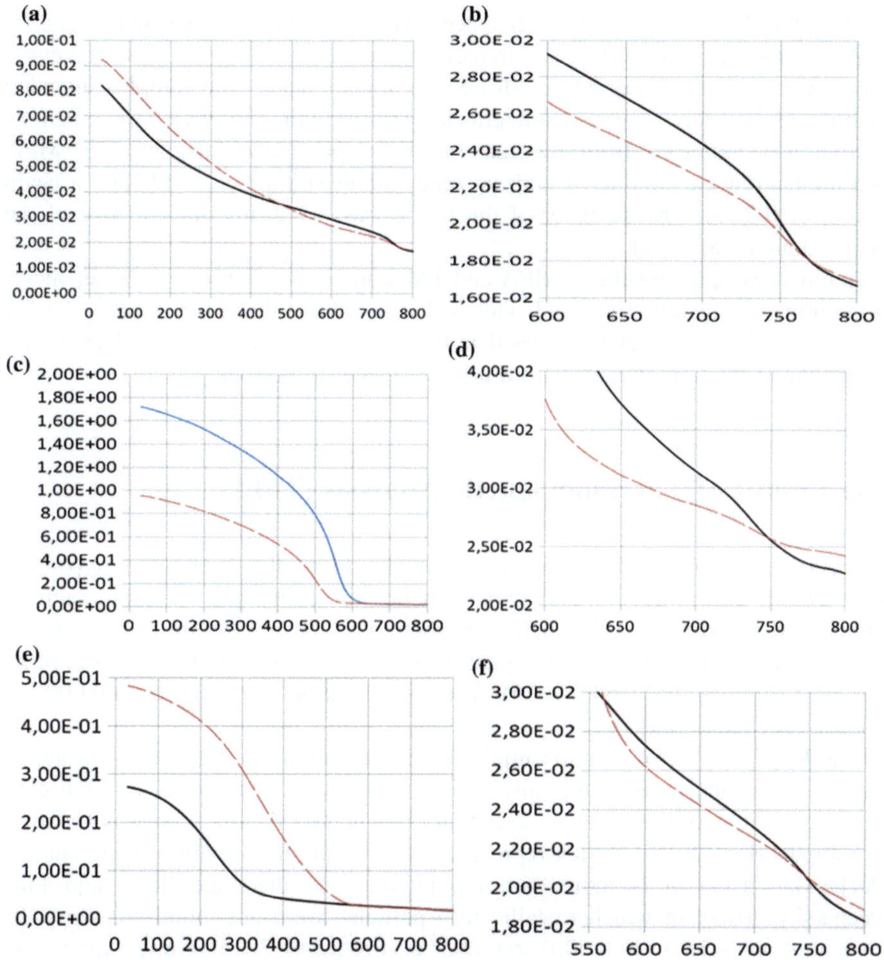

Fig. 29.1 Examples of TMA results. The x-axis is the temperature in degrees of Centigrade; the vertical axis is the magnetic moment, Am^2. The solid line indicates the first heating of the sample, dashed line indicates the second heating. **a, b** sample CB-462, lherzolite, Primor'e. Typical curve M (T) for hyperbasites has the form of a hyperbola. **a** M (T) in the temperature range 20–800 °C, **b** a high-temperature interval of the curves M (T) 600–800 °C, metallic iron is recorded with T_c = 770 °C. **c, d** sample A-10, lherzolite, Syria. **c** The curve M (T) of Q-type. **d** a high-temperature interval of the curves M (T) 600–800 °C, metallic iron is recorded with T_c = 765 °C. **e, f** sample 2793-3, basalt, Reykjanes Ridge, a typical curve M (T) for oceanic basalt lava; **e** M (T) in the temperature range 20–800 °C, curve M (T) fixed titanomagnetite with T_c = 320 °C; **f** a high-temperature interval of the curves M (T) 500–800 °C, metallic iron is recorded with T_c = 760 °C

the Curie temperature in the sample, firstly in the vicinity of the Curie temperature 700–780 °C. This is a metallic iron with various impurities.

The microprobe analysis (MPA) and other microscopic studies were carried out at the Borok Geophysical Observatory of the Schmidt Institute of Physics of the

Earth of the Russian Academy of Sciences. MPA measurement was performed by V. A. Tselmovich. The MPA measurements were conducted on the Tescan Vega 2 microprobe with an X-ray energy-dispersive spectrometer. The samples intended for microprobe analysis were disintegrated by ultrasound, after which the magnetic fraction was extracted from the material by a permanent magnet. The MPA measurements were conducted with an accelerating voltage of 20 kV and a beam current of 0.2 nA. The diameter of the probe was ~ 0.2 μm, and the size of the analyzed area was 2 μm.

In most cases, the results of MPA and TMA agree, but significant discrepancies of the iron content were observed. This is due to the fact that a tiny specimen were used for TMA whose weight is less than 0.1 g, whereas MPA was conducted with the magnetic fraction extracted from a larger piece of the rock.

The Results of Thermomagnetic Analysis (TMA)

The curve M (T) for ultramafic rocks usually takes the form of a hyperbola (Fig. 29.1a), indicating that the paramagnetic iron-containing minerals such as olivine and pyroxene are the main contributors to the magnetization. The Curie temperature of iron is detected in 14 samples of hyperbasites (Fig. 29.2a) out of 50 measured samples (Pechersky 2015). Iron concentration is ranging from 0.0001 to 0.003% (Pechersky 2015).

The M (T) curves for all trap samples have parabolic form (Q-type, for example, Fig. 29.1c), which is usual for fully crystallized basalts and due to the high concentration of titanomagnetite and secondary magnetite. Metallic iron was found in 16 samples (Fig. 29.2b), which represented half of the trap samples being studied (Pechersky 2015), it is more common in sills and there are flows and tuffs (Pechersky 2015), in which metallic iron particles were not found by TMA. Its concentration varies from 0.0001 to 0.0055% (Pechersky 2015). Curie temperature distribution is similar to hyperbasites (Fig. 29.2a, b).

The M (T) curves of oceanic basalts reflect the conditions for underwater lava effusion; they are characterized by a hyperbolic form because appreciable amounts of titanomagnetite grains had not yet been formed at the time of lava effusion. Data of microscopic observations show that among titanomagnetite grains studied in oceanic basalts the skeletal forms prevail. At the same time, ocean basalts do not differ from other terrestrial magmatic formations in terms of the appearance of metallic iron (Fig. 29.2). Consequently, the iron particles they contained have time to form before lava flows and the conditions of their appearance are similar to trapps and hyperbasites. Metallic iron was found in 27 samples (Fig. 29.2c), which represented half of the samples of oceanic basalts being studied (Pechersky 2015), and the concentration of iron in them varies from 0.0002 to 0.018% (Pechersky 2015). Distribution of the Curie temperature is similar to trapps and hyperbasites (Fig. 29.2).

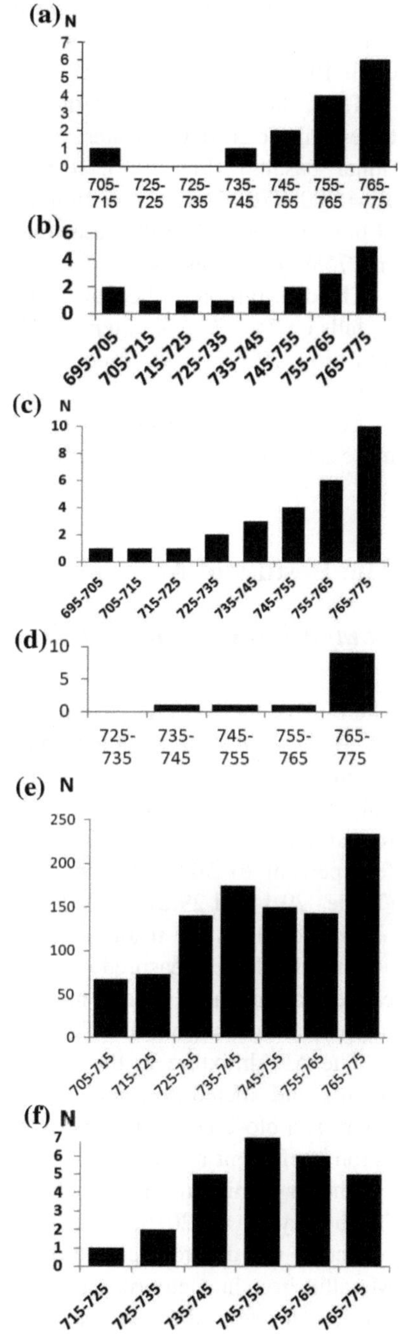

Fig. 29.2 Histograms of metallic iron Curie temperature (°C) distribution in hyperbasites **a** in trapps **b**, in oceanic basalts **c**, in lunar basalts **d** (Nagata et al. 1974), in sediments **e** (Pechersky 2015), and in meteorites **f** (Pechersky et al. 2012). N is number of samples with a given Curie temperature

Close to pure metallic iron prevails in lunar basalts even more sharply than in the terrestrial rocks (Fig. 29.2d), its concentration is $\leq 0.1\%$, mode is 0.05% (Nagata et al. 1974).

Thus, the distribution of the Curie temperature of iron is very similar, as in terrestrial rocks (hyperbasites, trapps and oceanic basalts) and extraterrestrial rocks (lunar basalts) (Fig. 29.2). It is an increase in the number of samples from T_c = 700–770 °C with a maximum in the range of pure iron (Fig. 29.2). Percentage of pure iron is 31% of the cases (oceanic basalts), 36% (trapps), 43% (hyperbasites) and 75% (lunar basalts).

The distribution of the Curie temperature in meteorites is different: the maximum T_c falls on 745–755 °C interval (27% of samples), and 19% of samples fall on pure iron (Fig. 29.2f). It is important to emphasize that the pure iron is present in meteorites though in relatively smaller amounts.

The particles of metallic iron with T_c = 730–760 °C are found in the sediments more than in hyperbasites, trapps and basalts (Fig. 29.2e) (Pechersky 2012, 2015).

The Results of Microprobe Analysis (MPA)

Composition of Metallic Iron Particles

The bulk of the particles belongs to the pure iron and iron containing impurities of Si, Al, Cr, et al., but not containing Ni (Table 29.1): there are 70% of such particles in sediments, 77%—in hyperbasites, 85%—in oceanic basalts, 86%—in trapps, and only 8% of nickel-free metallic iron particles can be found in meteorites. The remaining 92% of the particles in meteorites are the Fe–Ni alloys containing from a few percent to 50% or more of nickel, schreibersite and other intermetalloids (Tables 29.1 and 29.2).

The Si admixture in the metallic iron was detected in all samples studied, and about half of the measured particles are nickel-free metallic iron (Table 29.1). Its concentration varies widely, reaching 6% in sediments and 3–4%—in hyperbazites, trapps and oceanic basalts (Pechersky 2015).

The Al admixture in the metallic iron reaches 3–4% in hyperbazites, oceanic basalts and meteorites, and 9%—in sediments (Table 29.1) (Pechersky 2015). There is a close correlation between the Si and Al concentration, the linear correlation coefficient r = 0.87.

The Cr admixture in the metallic iron of sediments and basalts form two groups (Pechersky et al. 2015; Pechersky 2012, 2015): 0.2–6% and 8–20%, probably reflecting the different conditions of formation of these two groups of particles. Metallic iron in meteorites and hyperbasites refers to the first group (Pechersky 2015). The Mn admixture is noticeable in some particles of metallic iron from oceanic basalts, its concentration reaches 8–10% and Mn is usually presented together with Cr (Pechersky 2015).

Table 29.1 The number of the metallic iron particles of different composition (MPA data)

	N	Σ	Fe	Si	Al	Ti	Cr	Mg	Other	FeNi
Sediments										
Total	10	55	38	27	10	2	10	53	39	99
%			70	49	20	5	18	10	7	18
Hyperbasites										
Total	16	16	12	80	25	2	21	53	10	23
%			77	48	15	1	13	32	6	14
Vitim	3	28	27	21	4	5	8	7	0	1
Mongolia	2	22	14	5	5	0	0	4	0	2
Primor'e	7	72	57	36	10	1	7	24	9	10
Syria	3	33	23	11	6	5	6	16	1	5
Antarctica	1	11	6	7	0	0	1	2	0	5
Trapps										
Total	21	72	62	40	12	5	12	14	21	2
%			86	55	17	7	17	19	29	2.8
Angara	6	23	22	15	7	3	3	3	6	0
Maymecha	11	33	26	14	3	2	8	7	12	1
Norilsk	2	7	5	6	0	0	3	2	2	0
Tuff layer	1	9	9	5	0	0	0	0	1	0
Oc. basalts										
Total	14	79	67	29	28	1	19	8	33	8
%			85	38	35	1	24	10	42	10
Meteorites										
Total	44	89	68	50	34	0	2	18	18	823
%			8	6	4	0	0.2	2	2	92

Note N is the number of samples in a given group of the rocks; *Σ* is the total number of particles of native iron and Fe–Ni alloys (Fe + FeNi) in a given group of the rocks; *Fe* is the number of Ni-free native iron particles; *Si, Al, Ti, Cr, Mg* are the number of iron particles containing one of the elements; *other* means the other admixture elements in iron particles (Mn, P, S, etc.); *FeNi* is the number of kamacite particles and other Fe–Ni alloys and the alloys with more complex composition; % is the percentage of iron particles containing the admixtures of different elements with relative to the Σ

Some iron particles in basalts and hyperbasites contain titanium and magnesium impurities (Table 29.1). This can be explained by "contamination" of the material with surrounding minerals, mainly silicates and titanomagnetite. The admixture of titanium is not detected in the metallic iron of the meteorites (Table 29.1) (Pechersky et al. 2012).

The content of nickel in metallic iron particles varies considerably regardless of terrestrial or extraterrestrial origin of iron (Table 29.1), (Pechersky et al. 2015; Pechersky 2015).

For example, the content of nickel impurities in iron particles of stone meteorites forms three groups (Fig. 29.3): (1) pure iron without admixture of nickel, these

Table 29.2 Particles of metallic iron and Fe–Ni Alloys on Planets (Strangway et al. 1970; Bozorth and 1951; Pechersky 2012)

N	Σ	Fe	FeNi	Fe/FeNi	SH	CFe
Trapps (crust)						
21	72	62	2	31	0	
%		86	2.8		0	≤0.01
Oceanic basalts (crust and uppermost mantle)						
14	79	67	8	8.4	0	
%		85	10		0	≤0.01
Lunar basalts (crust and upper mantle)				~10	0	0.05
Hyperbazites (upper mantle)						
16	166	128	23	5.6	0	
%		77	14		0	≤0.01
Sediments (upper mantle)						
105	552	338	99	3.9	2	
%		70	18		0.4	≤0.01
Stone meteorites (bottom mantle)						
16	267	34	232	0.147	6	
%		12.8	87.3		2.2	>0.1–9
Iron meteorites (core)						
28	535	68	466	0.146	90	
%		12.7	87.1		16.8	>90

Note *N* is the number of samples in given group of rocks; Σ is the total number of metallic iron particles and Fe–Ni alloys (Fe + FeNi) in given group of rocks; *Fe* is the number of Ni-free metallic iron particles; FeNi is the number of particles of Fe–Ni alloys, Fe/FeNi—their ratio; *SH* is the number of schreibersite particles (Fe, Ni)$_3$P; % is the percentage of Ni-free iron particles, Fe–Ni alloys, and schreibersite relatively to the total number Σ.; *CFe* is the metallic iron concentration (%), TMA data

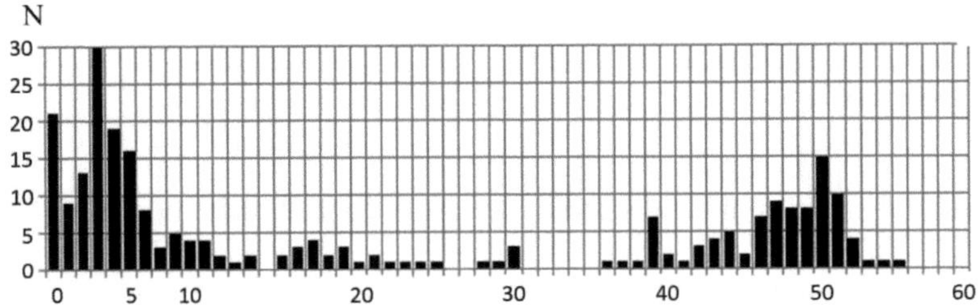

Fig. 29.3 Nickel content (%) in the metal part of stone meteorites. N—number of measurement. MPA data (Pechersky et al. 2012)

particles form a separate group, not a continuous series of solid solution with the second group, (2) group with mode of 4% Ni and smoother decline to 10% Ni, 3) group of Fe–Ni alloy having a nickel content of 40–53% with mode of 50% Ni. This dividing of the iron particles into three groups obviously indicates a various conditions of formation of pure iron, nickel containing iron and Fe–Ni alloy.

Form and Size of Metallic Iron Particles

The great bulk of the metallic iron particles in the hyperbasites, trapps, ocean basalts, meteorites and sediments (Fig. 29.4) have the same configuration: angular shape, chips, flakes, spirals, dendrites, etc., regardless of terrestrial or extraterrestrial origin.

A simple explanation of such a "strange" complicated forms of metallic iron particles suggests: it is the result of "pollution" of samples during their processing (crushing, etc.).

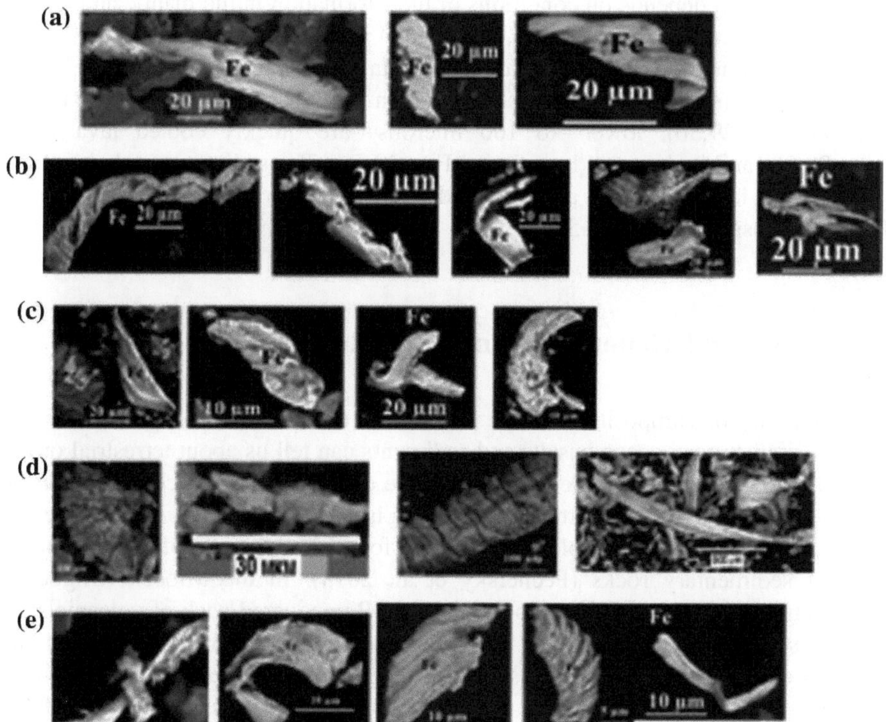

Fig. 29.4 Examples of metallic iron grain forms. **a** hyperbasites; **b** trapps; **c** the ocean basalts; **d** meteorites; **e** sediments

However, firstly, a Fe–Ni alloys are often present, it is very unlikely that they are also the result of samples processing; secondly, the treatment might have an affect only on samples for MPA, as for TMA just pieces of rocks were taken, and finally, the particle shape of native gold, silver, etc. is substantially similar to metallic iron, but it cannot be assumed that such particles of gold, silver had changed during processing. Such a variety of forms and at the same time their repetition in all studied groups of rocks reflect the obviously uniform primary magmatic conditions of their formation, more simply - a common origin and they all got together in the sediments.

The similarity of metallic iron particles in the sediments, in the mantle hyperbasites, in the trapps and basalts is expressed in the identity of their size distribution: it is a wide range of variation of particle size from submicron up to 300–600 microns with the same mode in the range of 10–20 microns (Fig. 29.5). It relates to particles of metallic iron containing and not containing nickel impurities. Particles less than 100 microns prevail. These dimensions correspond to the extraterrestrial particles, most of which are preserved when passing through the Earth's atmosphere and the particles larger than 100 microns are almost completely burned (Florenskii 1965). But this does not explain the same size distribution in the case of iron particles of known terrestrial mantle and crust origin. Obviously, the particle size range of the iron depends on conditions of their formation in the mantle and crustal igneous rocks such as hyperbasites, continental trapps and oceanic basalts.

The durations of metallic iron particles crystallization are reflected in the "tails" of the histogram: the largest particles found in hyperbasites with a slow crystallization (Fig. 29.5b), lower—to 100 microns—are quickly cooled lava flows (Fig. 29.5d) and underwater lava flows (Fig. 29.5a). The sediments include all the particles, but larger particles, as mentioned above, burn, arriving from space into the Earth's atmosphere (Fig. 29.5a).

Discussion and Generalization of the Results

The similarity of composition, shape and size of the metallic iron particles in the hyperbasites, trapps, ocean basalts and sediments can tell us about terrestrial origin of such particles in the sediments. Look at one of the possible mechanisms for the global spread of terrestrial particles of metallic iron on the Earth; it is the scattering of volcanic ash. Layers of volcanic ash were found in only two of the 30 studied sites of sedimentary rocks (Pechersky et al. 2013). In the Miocene-Oligocene sediments of the Atlantic the wellbore 386 DSDP uncovered pack of volcaniclastic turbidites of ~ 160 m thickness. 22 particles of metallic iron are found in the samples of this pack, which is less than in neighboring clastic sediments (33 iron particles) and in the neighboring hole 391A in clastic-carbonate deposits of the Miocene, not containing volcanic material (50 particles of metallic iron) (Pechersky et al. 2013). Metallic iron particles have not been found in interlayers of volcanic ash of Permo-Triassic sediments Meishan (China) (Korchagin et al. 2010), even

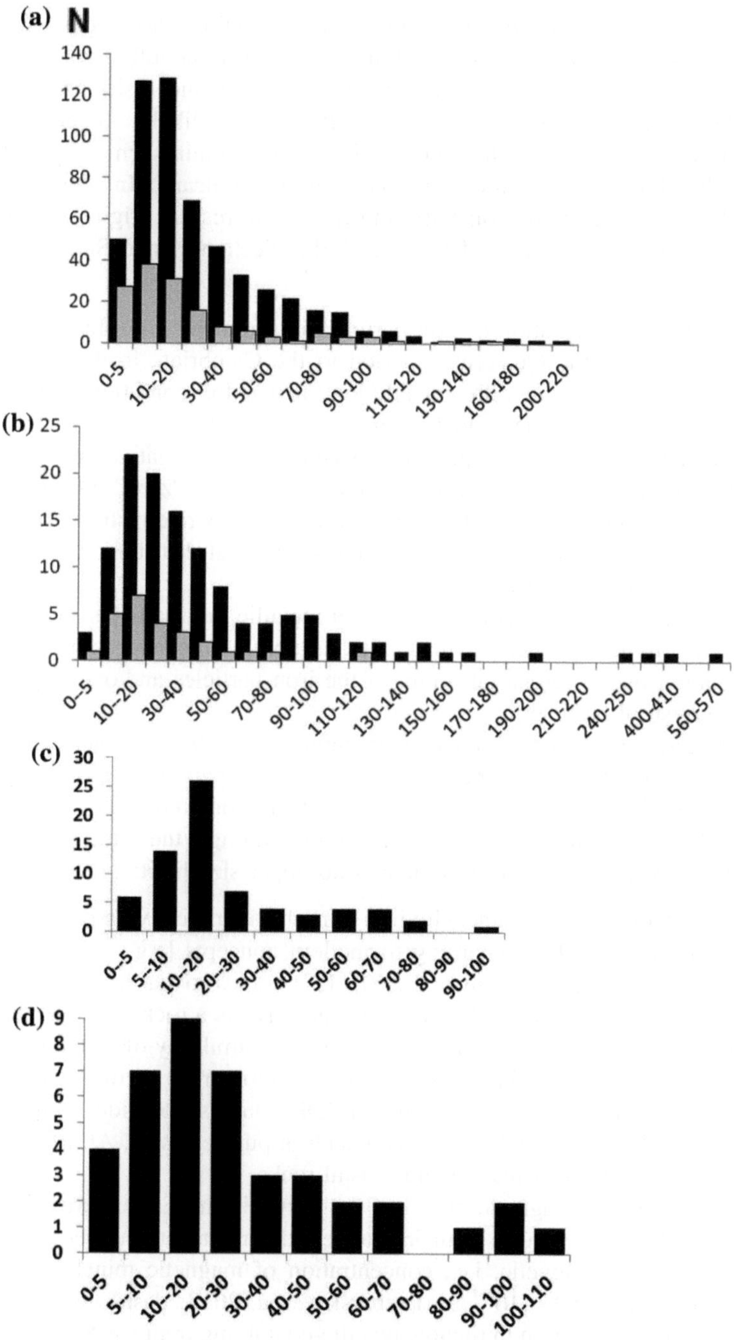

Fig. 29.5 The size distribution (microns) of metal iron particles, containing (gray) and not containing (black) nickel admixture: **a** sediments, **b** hyperbasites, **c** trapps, **d** the ocean basalts

though they are found in the surrounding clastic sedimentary rocks (Korchagin et al. 2010). Metallic iron has not been found by TMA in the tuff interlayer between trap flows (Pechersky 2015). Consequently, volcanogenic (ash) version of the global distribution of the metallic iron particles is unlikely. Other options of metallic iron particles global distribution of terrestrial origin such as hydrothermal, metamorphic, biogenic et al. are even less likely than volcanic. In addition, a set of statistical features clearly demonstrates mostly extraterrestrial origin of the metallic iron particles in the sediments (Pechersky 2012; Pechersky and Sharonova 2012, 2013):

(1) wide (global) distribution of iron particles in sediments in different regions of the world, from the Quaternary period to the Cambrian, in the sediments of lakes, seas and oceans, different lithology and redox conditions. The concentration of particles varies widely—from none to 0.05%;
(2) a bimodal distribution of particle concentrations of native iron obligatory "zero" group, in which the iron particles are absent; "zero" group represents sites of sediment, where metallic iron particles of extraterrestrial origin do not fall, because their quantity is limited (such bimodal distribution is not typical for iron particles of terrestrial origin);
(3) correlation between the concentrations of metallic iron particles and the earth's magnetic minerals, such as magnetite, is absent; the appearance of such a correlation means terrestrial origin of the iron particles and/or redeposition of extraterrestrial particles;
(4) the inverse dependence of the concentration of metallic iron particles from the sedimentation rate; this dependence for terrestrial particle is absent;
(5) the predominant particle size (less than 100 microns) corresponds to extraterrestrial particles that are preserved passing through the earth's atmosphere, whereas for the earth particles there is no upper size limit;

Global distribution of the metallic iron particles is logically associated with the interplanetary dust (IDP). According to modern concepts [e.g., (McFadden et al. 2007) and (Encyclopedia Britanica 2012)], IDP comes to Earth from the asteroid belt, which is located between Jupiter and Mars. IDP is a rock dust, formed by the collision of asteroids, comets and other bodies. The similarity of metallic iron in all the investigated objects as in terrestrial and extraterrestrial, allows us to conclude that the same conditions of formation of IDP sources coincide with terrestrial conditions. Nickel-free metallic iron with such impurities as Si, Al, Mg, and Cr is related to terrestrial upper mantle and crustal rocks.

According to petromagnetic model of the Earth, in the Earth's crust distributed titanomagnetites and products of their changes (magnetite); the upper mantle rocks are practically non-magnetic, i.e., concentration of magnetic minerals in there is below devices sensitivity ($10^{-4}\%$) (Pechersky et al. 2012; Pashkevich 1994). The average content of titanium in titanomagnetite-containing trapps is 8.7%, in oceanic basalts 10.5%, and in basalts containing ultramafic xenoliths 12.5% (Pechersky 2015). Such titanium content corresponds to the depth of the last equilibrium state of basaltic magma, i.e. its chamber, 30–50 km (Pechersky et al. 1975), which is within the Earth's crust and uppermost mantle.

According to our data (Table 29.2; Pechersky 2015), the ratio of a number of particles of Ni-free metallic iron (Fe) to Fe–Ni alloy (Fe/FeNi) in trapps is 31, in oceanic basalts—8.4. The situation on the Moon is similar: the metallic iron particles from volcanic glass and basalt are composed according to the MPA and TMA of practically pure iron, nickel content is from zero to less than 0.5 wt%, and much less to 20% Ni. The Curie temperature is often close to 770 °C and much less drops to 750 °C, the ratio of Fe/FeNi $\geq$ 10 (Doell et al. 1970; Nagata et al. 1970, 1974; Reid et al. 1970; Runcorn et al. 1971; Strangway et al. 1970).

The hyperbasites have ratio Fe/FeNi = 5.6; in stone and iron meteorites this ratio is, respectively, 0.147 and 0.146 (Table 29.2). The sediments have this ratio at 3.9 (Table 29.2). It is close to the hyperbasites, i.e. to the rocks of the upper mantle. Accordingly, it can be assumed that the sediments accumulate particles of metallic iron reflecting the ratio of Fe/FeNi in the upper mantle of the planet (planets) which is the main source of IPD.

We emphasize that the concept of "upper mantle" has no seismological significance. In our case it is called the mantle part in which the content of particles Fe + FeNi corresponds to sediments and hyperbasites that at least two orders of magnitude lower than in stony meteorites of the lower mantle (Table 29.2).

Coincidence of ratios Fe/FeNi in stone and iron meteorites (0.147 and 0.146) and the presence of schreibersite indicate that a sharp boundary between the lower mantle and core is absent and smooth transition between them exists.

The concentration of Ni-free iron + Fe–Ni alloys Fe + FeNi in the studied samples of stony meteorites according to TMA is 0.3–9%, average 2.4%, while the concentration of Fe + FeNi in basalts and hyperbasites very rarely reaches 0.01% (Pechersky et al. 2012; Pechersky 2015). Therefore, our samples of stony meteorites belong most probably to the low (bottom) of mantle. The foregoing highlights the schreibersite distribution: it is typical for the iron meteorites, where the number of schreibersite particles is 17% (Table 29.2), and practically absent in hyperbasites and sediments, i.e., in the rocks of the upper mantle (Pechersky et al. 2012; Pechersky 2015).

Results obtained suggest that the material of the upper mantle of the planets is easier to destroy than the material of the lower mantle and core. This means, that the material of the upper mantle was originally contained in the interplanetary dust. At the same time within the upper mantle there is a visible trend of decreasing the Fe/FeNi from uppermost mantle (hyperbasites Fe/FeNi = 5.6) to the whole of the upper mantle (sediments Fe/FeNi = 3.9). The material of the lower mantle and core falls on Earth's surface mainly in the form of relatively large fragments of meteorites (Fe/FeNi = 0.146–0.147).

Described trends can be traced in the total content of nickel in rocks: Ni concentration in iron meteorites is about 10% and it drops sharply to 1–0.05% in chondrites; the average concentration of nickel is 0.01% in carbonaceous chondrites (McFadden et al. 2007; Encyclopedia Britanica 2012; Marakushev et al. 1992).

Thus the tendency of the distribution of particles of Fe–Ni alloys is clearly visible in the planetary interior: high, more than 90%, in the core, the prevalence in the lower mantle, the relatively low proportion (14%) in the upper mantle and the practical absence in the basic igneous rocks, intrusive and volcanic crystallized in the Earth's and the Moon's crust and on the surface. At the same time the total percentage of Ni-free metallic iron particles and particles of Fe–Ni alloys in the core (iron meteorites, Table 29.2) and in the lower mantle (stone meteorites, Table 29.2) is the same, but the percentage of particles schreibersite varies considerably: from 17% in the core (iron meteorites) to less than 3% in the lower mantle (stony meteorites) (Table 29.2) and to the sporadic grains in the upper mantle of the terrestrial rocks.

Conclusions

The results of the thermomagnetic and microprobe studies of metallic iron in the terrestrial rocks (sediments, mantle hyperbasites, oceanic basalts, Siberian trapps), and extraterrestrial rocks (meteorites and lunar basalts) are generalized. This analysis revealed the similarity in the compositions, grain shapes and sizes of the particles of Ni-free metallic iron, Fe–Ni alloys in all of these groups of rocks. The particles of metallic iron in sediments (Pechersky 2010, 2012, 2015; Pechersky and Sharonova 2012, 2013) in most cases has extraterrestrial origin and their ubiquity is associated with interplanetary dust. The marked similarity of iron particles suggests the similar conditions of formation of such particles in the Earth and other planets, as a result of formation of homogeneous gas-dust clouds at an early stage of existence of Solar system and other star-planetary systems, and at the stage of subsequent gravitational differentiation on the planets (McFadden et al. 2007; Encyclopedia Britanica 2012; Marakushev et al. 1992; Zharkov 2013; Hubbard 1984).

According to existing modern data, the gas-dust clouds of close (if not the same) composition are widespread in the Universe, and the star-planetary systems, including our Solar system, were formed from gas-dust clouds. Consequently, it is possible to create a general model of distribution of native iron on all the planets. Main positions are as follows.

(1) Titanomagnetite is the main magnetic mineral of planets crust. Its composition varies within wide limits, particularly depending on the redox conditions. For example, in terrestrial rift basalts on the continents and oceans, where redox conditions are homogeneous, titanomagnetite contains 12–14% of titanium, while redox conditions of arc volcanics vary widely, and titanomagnetite composition varies from similar to rift basalts to magnetite. The situation close to terrestrial rift is typical for the Martian basalts; in lunar basalts an ulvospinel (titanomagnetite with Fe^{2+} only) and other ferrospinels containing only Fe^{2+} crystallized because of highly reducing conditions.

Small volume of crustal material compared with the rest of the planets explains the absence of crustal rocks among the studied meteorites; the concentration of metallic iron as containing or not containing nickel is $\leq 0.01\%$, in lunar basalts $\leq 0.1\%$; the ratio of cases of Ni-free metallic iron (Fe) to Fe–Ni alloys (FeNi) in terrestrial and lunar basalts is Fe/FeNi $\geq$ 10.

(2) Titanomagnetite is absent in uppermost mantle, but visibly increases the proportion of particles of Fe–Ni alloys, so in hyperbasites of uppermost mantle Fe/FeNi = 5.5; concentration of metallic iron particles is $\leq 0.01\%$, as in crustal rocks.

(3) The total characteristic of the entire upper mantle is expressed in the value of Fe/FeNi = 3.9 in sediments, that is close to hyperbasites, i.e. to the rocks of the Earth's upper mantle. We assume that this material mainly falls into the interplanetary dust, in contrast to the material of the lower mantle, which falls on the Earth mainly in the form of relatively large fragments of meteorites. The rocks of the upper mantle are characterized by absence of the schreibersite.

(4) The bottom of the mantle is reflected in the studied samples of stony meteorites: their concentration in Fe + FeNi alloys is 0.2–9%, the ratio of Fe/FeNi = 0.147; schreibersite ($\sim 2\%$) appears in metal part of lower mantle.

(5) In the core of planets-sources of meteorites and probably in the Earth's core the ratio of Fe/FeNi = 0.146, the concentration of Ni-free iron (Fe) + Fe-Ni alloys is $\sim 90\%$. Schreibersite presents appreciable fraction (17%) among the Fe–Ni alloys. There is no sharp boundary between the lower mantle and the core of the planet, there is a smooth growth in concentration of metallic iron, Fe–Ni alloys, in particular, that is schreibersite, between stony meteorites, stony-iron meteorites, and iron meteorites.

(6) The generality of the distribution of metallic iron in the planets in the form of Ni-free and Fe–Ni alloys is obviously determined by the action of general physical law, such as the law of gravity, that works both at the stage of the existence of gas-dust cloud and subsequent gravitational differentiation in the process of all planet formation.

Acknowledgements Our gratitude to colleagues for helping us with our study: G. N. Bazhenova and A. Ya. Saltykovskii for providing the samples of the mantle xenoliths from the basalts of Primor'e, Mongolia, Syria, and other localities; I. Ashchepkov for the mantle xenolith samples from the Vitim basalts; A. Latyshev for the samples of the Siberian trapps; and V. Matveenkov and S. Silant'ev for the samples of the oceanic basalts. Many thanks to G. P. Markov for help with the thermomagnetic research and to V. A. Tselmovich for the fulfillment of microprobe analyze. The work was supported by Russian Government's Program of Competitive Growth of Kazan Federal University.

References

Grachev, A.F., Borisovsky, S.E., Kollmann, H.A., Korchagin, O.A, Nourgaliev, D.K., Pechersky, D.M., Sharonova, Z.V. and Tse'lmovich V.A. (2009) « The K/T boundary of Gams (Eastern Alps. Austria) and the nature of terminal Cretaceous mass extinction » . Ed. A.F.Grachev. Abhandlungen der geologischen bundesanstalt. band 63, 89–134.

Pechersky, D.M., Nurgaliev, D.K., and Sharonova, Z.V. (2006) Magnetolithologic and magnetomineralogical characteristics of sediments at the Mesozoic/Cenozoic boundary: the Koshak section (Mangyshlak peninsula). *Izvestiya, Physics of the Solid Earth*, 42, 957–970.

Pechersky, D. M., Asanidze, B.Z., Nourgaliev, D. K. and Sharonova, Z. V. (2009a) Petromagnetic and Paleomagnetic Characterization Deposits at Mesozoic/Cenozoic Boundary: The Tetritskaro Section (Georgia). *Izvestiya, Physics of the Solid Earth, 45, 134–149.*

Pechersky, D. M., Nurgaliev, D. K. and Sharonova, Z. V. (2009b) Magnetic Properties of the Boundary Layer at the Cretaceous/Tertiary Boundary in the Gams Section, Eastern Alps, Austria. *Izvestiya, Physics of the Solid Earth, 45, 482–494.*

Pechersky, D.M. (2010) Metallic iron and nickel in Cretaceous and Cenozoic sediments: the results of thermomagnetic analysis. *Journal Environmental Protection*, 1, 143–154.

Pechersky, D. M., Nurgaliev, D. K., Fomin, V. A., Sharonova, Z. V. and. Gil'manova, D.M (2011) Extraterrestrial Iron in the Cretaceous–Danian Sediments. *Izvestiya, Physics of the Solid Earth*, 47, 379–401.

*Pechersky, D. M., Gil'manova, D. M., Markov, G. P., Murdmaa, I. O., Nurgaliev, D. K., Tsel'movich, V. A. and Sharonova Z. V. (*2013) Native Iron and Other Magnetic Minerals in the Sediments of the Northwestern Atlantic: Thermomagnetic and Microprobe Evidence. *Izvestiya, Physics of the Solid Earth,* 49, 426–447.

Murdmaa, O.I., Pechersky, D. M., Nourgaliev D. K., Gil'manova, D. M. and Sloistov, S.M. (2015) Extraterrestrial Native Iron in Deep-Water Sediments of the NW Atlantic: Evidence from Thermomagnetic Analyses. *Lithology and Mineral Resources, 50, 117–133.*

Pechersky, D.M., Gil'manova, D.M., Ivanov, E.V., Kuz'min, M.I., Markov, G.P., Nurgaliev, D.K., Tsel'movic, V.A. (2013) Native iron in the sediments of Lake Baikal (*borehole BDP-98*): results of thermomagnetic analysis. *Russian Geology and Geophysics*, 54, 1044–1054.

Pechersky, D.M., Gil'manova, D.M., Kazansky, A.Yu., Krivonogov, S.K., Nurgaliev, D.K., Tsel'movich, V.A. (2013) Native iron in Quaternary deposits of the Darhad Basin (*northern Mongolia) // Russian Geology and Geophysics*, 54, 1503–1518.

Pechersky, D.M., Markov, G.P., Tsel'movich, V A. and Sharonova Z.V. (2012) Extraterrestrial Magnetic Minerals. *Izvestiya, Physics of the Solid Earth, 48, 653–669.*

Doell, K.B., Gromme, C.S., Thorne, A.N., Sentfle, F.E. (1970) Magnetic studies of Lunar samples. *Science*, 167, 695–697.

Nagata, T., Ishikawa, Y., Kinoshita, H., Kono, M., Syono, Y. and Fisher, R.M.·(1970) Magnetic properties of Lunar samples // *Science*, 167, 703–704.

Nagata, N., Sugiura, N., Fisher, R.M., Schwerer, F.C., Fuller, M.D. and Dunn, J.R. (1974) Magnetic properties of Apollo 11–17 lunar materials with special reference to effects of meteorite impact. *Proceedings of the Fifth Lunar conference*, 3, 2827–2839.

Reid, A.M., Ir, C.M., Harmon, R.S. and Bret, R. (1970) Metal grains in Apollo 12 igneous rocks. *Earth Planetary Science Letters*, 9, 1–5.

Runcorn, S.K., Collinson, D.W., O'Reilly, W., Stephenson, A., Greenwood, N.N., Battey, M.H. (1971) Magnetic properties of Lunar samples // *Proc. R. Soc. London*, A325, 157–174.

Strangway, D.W., Larson, E.E., Pearce, G.H. (1970) Magnetic properties of Lunar samples. *Science*, 167, 691–693.

Pechersky, D. M., Kuzina, D. M., Nurgaliev, D. K. and Tsel'movich V. A. (2015) The Common Nature of Native Iron in Terrestrial Rocks and Meteorites: Microprobe and Thermomagnetic Data. *Izvestiya, Physics of the Solid Earth*, 51, 748–763.

Pechersky, D.M. (2015) The distribution of particles of native iron and Fe-Ni alloys on the planet. Saarbrücken: Palmarium Academic Publishing. 56 p. (in Russian).

Burov, B.V., Nurgaliev, D. K., Yasonov, P.G. (1986) Paleomagnetic analysis. Kazan: Kazan state University. 186 pp. (in Russian).
Bozorth, R.M. Ferromagnetism. (1951) D.Van Nostrand Comp. Inc.. Toronto-New York-London. 968 pp.
Pechersky, D.M. (2012) Native Iron and Other Magnetic Minerals in Sediments: Thermomagnetic Signs of Extraterrestrial Origin), Saarbrucken: Palmarium Academic Publishing. (in Russian).
Florenskii, K.P. On the initial stage of differentiation of the Earth's matter. (1965) *Geokhimiya*, 8, 1–18. (in Russian).
Korchagin, O.A., Tsel'movich, V.A., Pospelov, I.I., and Bian Qiantao (2010) Cosmic magnetite microspherules and metallic particles near the Permian–Triassic boundary in a global stratotype section (stratum 27, Meishan, China). *Doklady RAS, Earth Sciences*, 432, 631–637.
Pechersky, D.M. and Sharonova, Z.V. (2012) Thermomagnetic evidence of native iron in sediments. *Izvestiya, Physics of the Solid Earth*, 48, 320–325.
Pechersky, D.M. and Sharonova, Z. V. (2013) A Relationship between the concentration of Native Iron Particles in Sediments and the Rate of Their Accumulation: A Synthesis of Thermomagnetic Data. *Izvestiya, Physics of the Solid Earth, 49, 718–724.*
McFadden, L., Weissman, P.R. and Johnson, T.V. (2007) Encyclopedia of the Solar System. 992 pp.
Encyclopedia Britanica (2012). www.britanika.com.
Petromagnetic Model of the Lithosphere. Ed. Pashkevich, I.K. (1994) Kiev: Naukova Dumka. (in Russian).
Pechersky, D.M., Bagin, V.I., Brodskaya, S.Yu. and Sharonova, Z.V. (1975) Magnetism and Formation Conditions of Igneous Rocks. Moscow: Nauka (in Russian).
Marakushev, A.A., Granovskii, L.B., and Zinov'eva, N.G. (1992) Cosmic Petrology, Moscow: Moscow State University, 325 pp. (in Russian).
Zharkov, V.N. (2013) The internal structure of the Earth and planets. M.: Nauka, 415 pp. (in Russian).
Hubbard, W. (1984) Planetary Interiors, New York: Van Nostrand, 328 pp.

Chapter 30
Magnetic Properties of Tektite-like Impact Glasses from Zhamanshin Astrobleme, Kazakhstan

Vyacheslav A. Starunov, Andrei Kosterov, Elena S. Sergienko, Svetlana Yu. Yanson, Gennady P. Markov, Petr V. Kharitonskii, Aleksandr S. Sakhatskii, Irina E. Lezova and Evgenii V. Shevchenko

Abstract We measured the low-field magnetic susceptibility and natural remanent magnetization (NRM) for an extensive collection of irghizites, tektite-like impact glasses from the Zhamanshin meteorite crater, Kazakhstan. For a small subset of samples, more detailed magnetic measurements, scanning electron microscopy and microprobe analysis have been carried out. In agreement with a previous study (Rochette et al. 2015), ~95% of samples showed relatively low magnetic susceptibility ranging from 45 to 300×10^{-9} m^3/kg. Combining susceptibility and NRM data, we are able to classify irghizites into "normal" (susceptibility $<300 \times 10^{-9}$ m^3/kg, NRM $< 4 \times 10^{-6}$ A m^2/kg) and "anomalous" groups. Detailed rock magnetic experiments and microscopic observation confirm this dichotomy. Normal irghizites contain very little ferrimagnetic material which, in addition, is predominantly in superparamagnetic state, implying an extremely small, on the order of a few nm, grain size. In contrast, anomalous samples exhibit magnetic properties consistent with much larger, μm to 10s of μm, ferrimagnetic grains, which were observed microscopically as well. We propose that this difference between two groups of samples is related to the initial temperature of the impact melts from which they were formed. Normal irghizites likely originate from the spray of a strongly overheated melt with initial temperatures up to 2500–3000 °C, cooling extremely fast, while anomalous ones could have been formed from less hot and slower cooling melt.

V. A. Starunov
Institute of Terrestrial Magnetism, Ionosphere, and Radiowave Propagation, St. Petersburg Branch, St. Petersburg, Russia
e-mail: slava.starunov@gmail.com

A. Kosterov (✉) · E. S. Sergienko · S. Yu. Yanson · P. V. Kharitonskii
A. S. Sakhatskii · I. E. Lezova · E. V. Shevchenko
St. Petersburg State University, St. Petersburg, Russia
e-mail: a.kosterov@spbu.ru

G. P. Markov
Institute of Physics of the Earth, Moscow, Russia

D. Nurgaliev et al. (eds.), *Recent Advances in Rock Magnetism, Environmental Magnetism and Paleomagnetism*, Springer Geophysics,
https://doi.org/10.1007/978-3-319-90437-5_30

Keywords Zhamanshin · Impact structure · Tektites · Magnetic susceptibility
Natural remanent magnetization · Low-temperature magnetometry
Scanning electron microscopy · Microanalysis

Introduction

Tektites and other tektite like impact glasses [named tektoids by Rochette et al. (2015)] are quenched products of melting the target rocks which occur during high-energy meteorite impacts (Dressler and Reimold 2001). Geochemically, they are acid amorphous formations with narrow compositional range, containing 68–82 wt% SiO_2, 10–15 wt% Al_2O_3 and extremely little water, about 0.02 wt% (Koeberl 1986, 1990; Glass 1990). Tektites and impact glass in general, differ from magmatic rocks having excessive SiO_2 and Al_2O_3 and low alkalis that evaporate at high temperatures produced by an impact. At these temperatures, metals are believed to be in a strongly reduced state; in particular, Fe^{2+} predominates over Fe^{3+}. Tektites have been preserved during quenching due to their high thermal conductivity and low thermal expansion, which are due to the high content of silica and low alkalis.

Magnetic properties of tektites were investigated in numerous studies (Senftle and Thorpe 1959; Senftle et al. 1969, 2000; Werner and Borradaile 1998; Rochette et al. 2015). However, they were mostly focused on the magnetic susceptibility. Little is reported on their natural remanent magnetization (NRM) (de Gasparis 1973; de Gasparis et al. 1975; Donofrio 1977; Gus'kova 1980) and on magnetic hysteresis properties (Werner and Borradaile 1998; Rochette et al. 2015). Most tektites show very low magnetic susceptibility and NRM; however, in rare cases both can be higher than mean values by one-two orders of magnitude.

The object of our study was tektite-like impact glasses collected on the surface of Zhamanshin meteorite crater, Kazakhstan (48° 24′N, 60° 48′E). The ~13 km diameter crater was formed about 1 My ago (Florensky and Dabizha 1980). A remarkable feature of the Zhamanshin crater is the simultaneous presence of tektite-like material (locally called irghizites), impact glasses (zhamanshinites) and microtektite like material referred to as microirghizite (Glass et al. 1983). Such a complex suite related to a single impact structure is unique allowing to study in detail the behavior of chemical elements in the impact process (Koeberl et al. 1985). Irghizites typically occur as sprays, droplets, dumbbells, strips, twigs, fragments up to 3–5 cm long and weighing from about 50 mg up to 3–4 g. Geochemically, irghizites can be classified as acidic glass, containing >70–75% SiO_2 and resembling in this aspect classical tektites. It is believed that irghizites were formed from the spray of high-temperature impact melt, with initial temperatures above the liquidus. Some irghizites could have been formed during the condensation of a gaseous-plasma cloud arising at the instant of shock-explosive event (Florensky and Dabizha 1980). Recent observation of condensate glass in some irghizites (Gornostaeva et al. 2016) lends further support to this hypothesis. Melt glass is characterized by SiO_2 content >80 wt% and presence of large amorphous inclusions, while condensate glass contains 80–100% silica as well as crystal nuclei 5–20 nm in size. Noteworthy is the presence of 15–20-nm crystals of native iron, as well as of amorphous inclusions close in composition to titanomagnetite (Gornostaeva et al. 2016). This suggests that irghizites containing condensate glass formed at the highest possible temperatures, higher than the initial temperature at which the main volume of melt glass has been formed.

Samples and Experimental Methods

The samples used in this study were collected in mid-1970s during a geological-geophysical expedition to Zhamanshin crater organized by the USSR Academy of Sciences. Samples were taken on the surface at different localities in the south-eastern part of the crater. The collection consists of samples of various shape, with different, variably chipped surface, totaling 840 samples with masses from 50 mg to 4 g and size up to 3–5 cm. Most samples (70–80%) have a shiny surface, the rest are the samples with weathered, matte surface, and fragments. It might be expected that impact glass would inherit to a certain degree the chemical composition of parent rocks, which are classified into four groups according to the silica content: low-, moderate-, high-, and extra high-silica rocks with an average SiO_2 content of about 55, 65, 73, and 93 wt%, respectively (Feldman and Sazonova 1993).

For over 700 samples, the initial magnetic susceptibility has been measured at 976 Hz frequency, 200 A/m field amplitude, using a MFK-1FA susceptibility bridge (AGICO, Czech Republic). Out of these, natural remanent magnetization (NRM) intensity was measured for 245 samples using a SRM 755 SQUID magnetometer (2G Enterprises, US, noise level $\sim 5 \times 10^{-12}$ A m^2). For 30 selected samples, NRM was demagnetized by alternating field up to 100 mT, followed by similar demagnetization of the anhysteretic remanent magnetization (ARM) acquired in a 100 mT af, 50 μT DC field.

For a further selection of 10 samples, more detailed magnetic experiments and microscopic observations have been performed. Blocking temperatures, considered as a proxy for Curie temperatures, have been estimated using thermal demagnetization of the composite IRM acquired in 0.1, 0.3, and 1 T fields, applied along three orthogonal axes respectively (Lowrie 1990). Thermal demagnetization was carried out using a TD-48S furnace. Maximum temperature reached was 775 °C, in an attempt to detect a possible contribution of native iron to the magnetization values.

Magnetic hysteresis loops at room temperature have been traced in a 1.5 T maximum field, using a PMC VSM 3900 vibrating sample magnetometer. For two samples showing a non-negligible remanent magnetization, First Order Reversal Curves (FORC) have been measured with the same instrument. FORC distributions were evaluated using VARIFORC v. 2.0 software (Egli 2013). For these two samples, as well as for five others, hysteresis loops were measured at 295 K in a 7 T maximum field, and magnetic properties were investigated at cryogenic temperatures, between 2 and 300 K.

Low-temperature experiments included the following. The saturation isothermal remanent magnetization (SIRM) acquired in a 5 T field at 2 K after zero field cooling (ZFC) and cooling in a strong (5 T) field (FC), respectively, was traced during the subsequent warming to 300 K in a zero field. SIRM acquired at 300 K was measured during the cooling-warming cycle between 300 and 2 K, also in a zero field. Both experiments were carried out using a Quantum Design (US) MPMS 3 instrument in the VSM mode, with temperature sweeping at 2 K/min. A (near-) zero field to acquire the remanent magnetization has been produced by setting the 2.8 mT nominal field in the MPMS 'No overshot' mode. Measurements of Pd standard sample at 298 K showed that this procedure generally yields a residual field below 10 μT. The temperature and frequency dependences of the complex magnetic susceptibility $\chi = \chi' - i\chi''$, where χ' and χ'' are the real and imaginary parts of the susceptibility, respectively, were measured between 2 and 300 K at 0.5 K increment between 2 and 10 K, 2 K increment between 10 and 100 K and 5 K increment above 100 K, using a Quantum Design PPMS instrument. Measurements were performed in a driving field of 250 μT at seven frequencies between 11 Hz and 7 kHz selected so that the step in log frequency would be approximately uniform.

To determine the chemical composition, morphological characteristics, and the degree of alteration in magnetic minerals, the same 10 samples were investigated with a scanning electron microscopy (SEM) using a system with focused electronic and ion probes QUANTA 200 3D (FIA, the Netherlands) with a Pegasus 4000 analytical complex (EDAX, USA). For SEM, samples were mounted in polystyrene pucks using a Struers CitoVac vacuum pouring system, and polished thin sections were prepared with a Struers TegraPol grinding and polishing machine (Struers, Austria).

Results

Initial Susceptibility and NRM Intensity

Irghizites show generally very low but positive initial magnetic susceptibility (χ) ranging from 45 to 1000 $\times 10^{-9}$ m^3/kg. In Fig. 30.1a we plot susceptibility values of the whole collection as a histogram, which shows that over 95% of samples have susceptibilities below 300 $\times 10^{-9}$ m^3/kg, in agreement with previous observations on a smaller collection of irghizites (Rochette et al. 2015). Figure 30.1b shows a similar histogram for NRM intensity, which ranges from 2.7×10^{-8} to 2.3×10^{-4} A m^2/kg. Again, for over 90% samples NRM values are less than 4×10^{-6} A m^2/kg. Furthermore, both low and high values of susceptibility and NRM intensity, respectively, correlate rather well (Fig. 30.2a), yielding a natural criterion to classify all samples into a "normal" ($\chi < 300 \times 10^{-9}$ m^3/kg, NRM $< 4 \times 10^{-6}$ A m^2/kg) and an "anomalous" groups. In terms of Koenigsberger factor (Q), the two groups also appear different (Fig. 30.2b). The normal samples are characterized by low Q values, generally below 0.5 while in the anomalous ones Q tends to be somewhat higher, reaching several units in some cases. This implies that

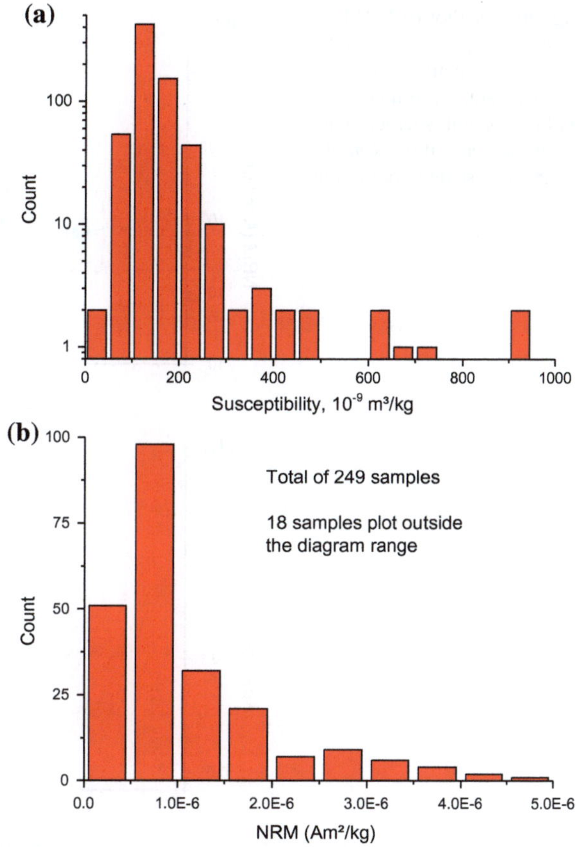

Fig. 30.1 Histograms of the initial magnetic susceptibility (**a**) and NRM (**b**) values of normal irghizites

the normal samples are dominated by paramagnetism, while in the anomalous ones a considerable amount of ferrimagnetic material may be present.

NRM and ARM Demagnetization

In Fig. 30.3 we compare the af demagnetization curves for NRM and ARM, respectively. According to the behavior during demagnetization, most samples appear either magnetically soft (NRM median destructive field about 5–6 mT), or magnetically hard (NRM mdf >30 mT). Somewhat unexpectedly, both soft and hard samples are encountered in normal and anomalous groups. ARM demagnetization curves are in some cases (e.g. samples 166a and 216b) similar to those of NRM, but more often this similarity holds only for a certain part of coercivity spectra. We therefore estimated NRM/ARM ratios using linear parts of the respective NRM versus ARM plots (shown as insets in Fig. 30.3). NRM/ARM

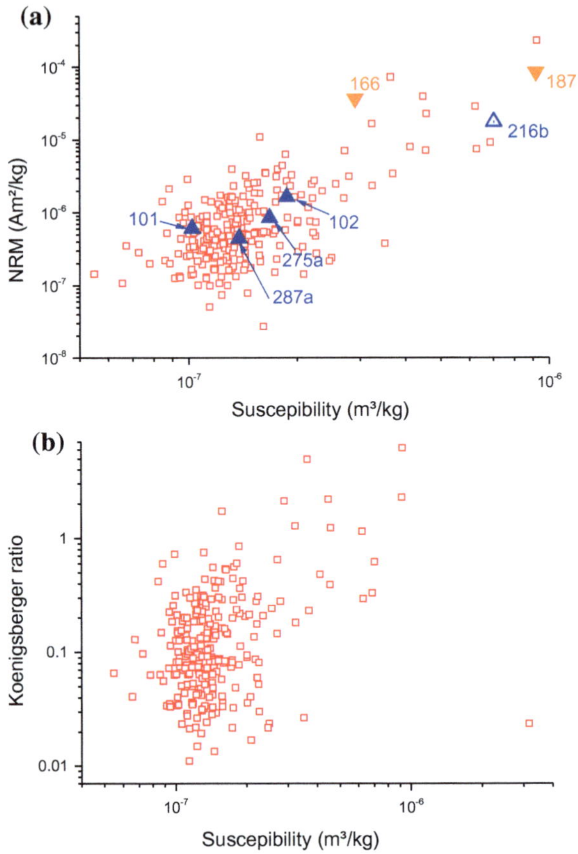

Fig. 30.2 **a** Plot of NRM versus susceptibility. Samples used in the detailed rock magnetic study are marked. **b** Plot of Koenigsberger factor versus susceptibility. Note the logarithmic scale in both plots

ratios in most samples proved to be very low, generally less than 0.5. Even bearing in mind that the geomagnetic field intensity at the time of crater formation could have been lower by a factor of 2 than the 50 μT value used to acquire ARM, it is still <1. Only in 3 samples out of 30 NRM/ARM ratio is somewhat larger, about 1, translating to a value about 2 if the difference between ancient and laboratory field intensity is considered. On the other hand, for thermoremanent magnetization TRM/ARM ratio is about 2.5–3 for (titano)magnetite-bearing volcanic rocks (Sholpo 1977; Borisova and Sholpo 1985) and even higher, of the order of 10, for synthetic submicron magnetite grains (Sugiura 1979; Dunlop and Argyle 1997). We suggest therefore that NRM in irghizites is unlikely to be TRM, and alternative mechanisms of its acquisition have to be invoked.

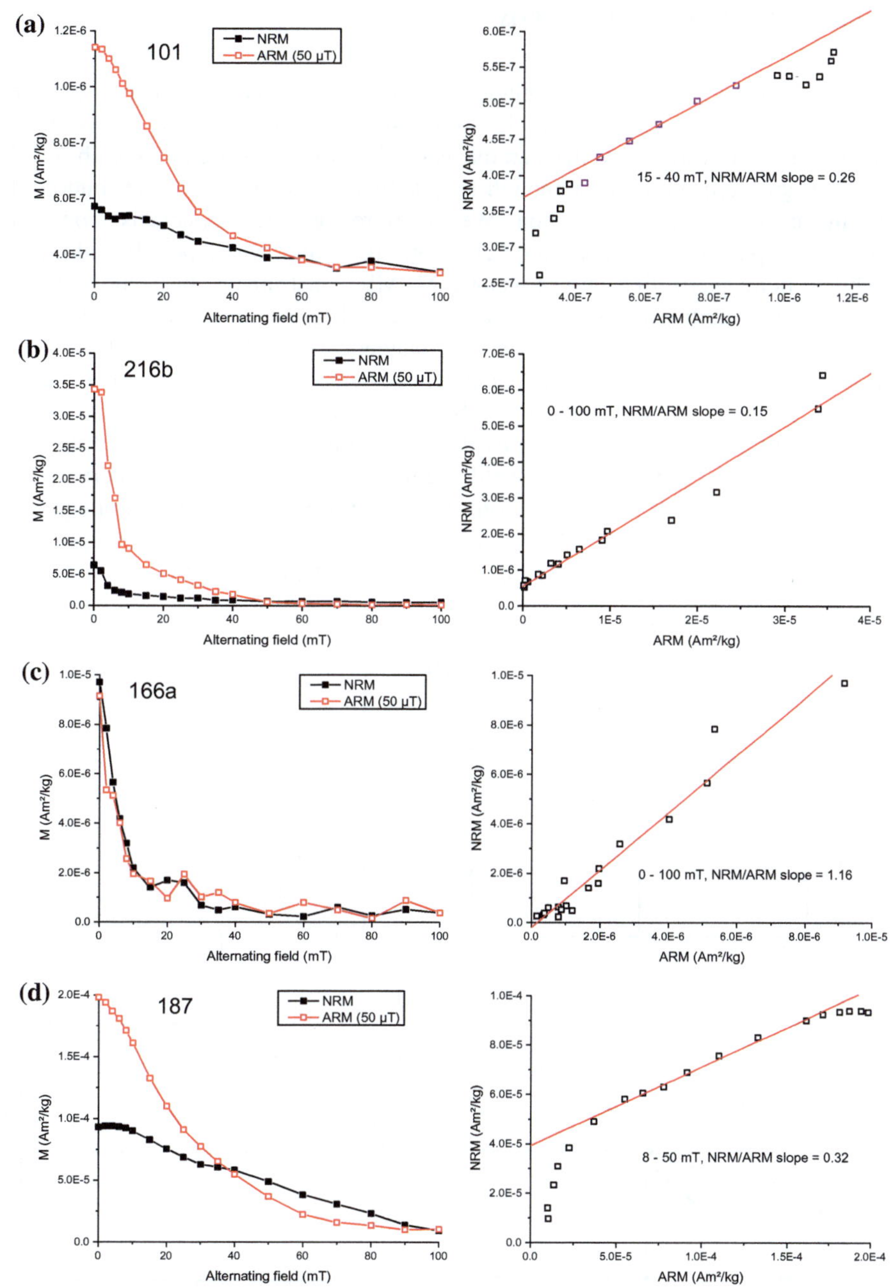

Fig. 30.3 Comparison of af demagnetization curves of NRM (black) and ARM (red). Right panels show the same data replotted in ARM versus NRM coordinates. NRM/ARM ratios were calculated as slopes of linear segments in NRM-ARM plots

Room Temperature Hysteresis

Magnetization versus magnetic field measurements at room temperature have shown that in normal irghizite samples hysteresis is absent, and nearly 99% of total magnetization is carried by the paramagnetic fraction. Measurements with VSM however appeared not precise enough to quantitatively extract the contribution of ferromagnetic (sensu lato) fraction. We therefore measured hysteresis loops for selected samples in a 7 T maximum field (Fig. 30.4a) and, where possible, also M_{rs} backfield demagnetization curves using a MPMS 3 instrument. Inspection of differential susceptibility (dM/dH) field dependences (Fig. 30.4b) shows that the latter does not become constant even in a 7 T magnetic field. We interpret this as an evidence for a predominantly superparamagnetic behavior of ferromagnetic material in these samples. Very fast decay of SIRM given at 1.8 K (see Fig. 30.7a, b) confirms this hypothesis. This means that evaluating paramagnetic susceptibility (χ_{HF}) by formally fitting the high-field segment of the magnetization curve by a straight line will result in a slight overestimate. However, since the paramagnetic contribution overwhelms the ferromagnetic one, even this small amount would be sufficient to noticeably distort the high-field part of magnetization curves obtained

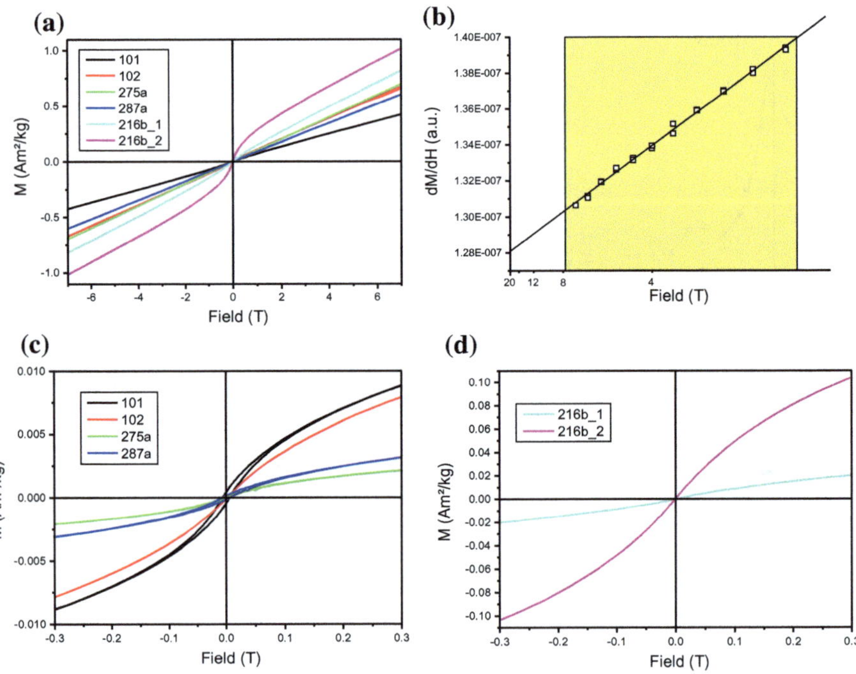

Fig. 30.4 Hysteresis loops for representative samples from the normal group. **a** Measured full-range (±7 T) loops, **c**, **d** middle part of the loops corrected for high-field susceptibility as illustrated in (**b**) and explained in the text

Table 30.1 Hysteresis parameters measured at 295 K with a MPMS 3 instrument

Sample ID	Mass (mg)	M_s (mA m^2/kg)	M_{rs} (mA m^2/kg)	M_{rs}/M_s	H_c (mT)	H_{cr} (mT)	H_{cr}/H_c	K_{hf} (m^3/kg 10^{-9})
Normal samples								
101	24.97	31.08	0.439	0.0141	5.72	17.21	3.01	70.5
102	17.85	26.95	0.092	0.0034	1.83	–	–	115.7
275a	27.38	5.990	0.084	0.0141	3.40	9.63	2.83	124.3
287a	9.22	10.25	0.156	0.0152	6.47	17.84	2.76	106.3
Sample with high initial susceptibility having predominantly SP magnetic								
216a_F1	12.15	81.89	0.372	0.0045	2.66	14.00	5.26	132.5
216a_F2	30.23	278.2	0.390	0.0014	0.64	24.67	38.55	132.5
Samples with high initial susceptibility having predominantly non-SP magnetic								
166	44.43	40.96	3.693	0.090	4.82	10.08	2.09	123.9
187	65.32	118.2	22.98	0.194	19.31	49.40	2.56	143.6

K_{hf} is the high-field susceptibility determined by extrapolating (dM/dH) versus (1/H) dependences to a 20 T field as explained in the text

by subtracting the $\chi_{HF}\cdot H$ term. As an alternative, we attempted to evaluate χ_{HF} by linearly extrapolating experimental dependences (dM/dH)(1/H) to a 20 T field (Fig. 30.4b). This field was selected as a compromise, where saturation of superparamagnetic contribution is nearly achieved while true paramagnetic susceptibility is reduced by less than 0.5% compared to a zero-field value. Values of χ_{HF} determined in this way are just about 2% lower than the values derived from the magnetization curves slope in >5 T. Nevertheless, even this small difference in χ_{HF} produces a noticeably different correction, resulting loops not showing saturation artefacts (Fig. 30.4c, d). We therefore evaluate saturation magnetization of the ferromagnetic component from these loops. As expected, corrected hysteresis loops yield very low coercivity and M_{rs}/M_s ratios (Table 30.1).

Hysteresis measurements for the sample 216b, showing initial susceptibility and NRM well outside the range of normal samples (Fig. 30.1a) nevertheless reveal the predominantly superparamagnetic nature of its ferromagnetic fraction, similar to that in normal samples. However, the observed M_s values for two chips from this sample are much larger than for other studied samples indicating the correspondingly higher concentration of magnetic material in them, which is in addition rather variable at the mm scale. This accounts for elevated NRM and initial susceptibility of this sample. Note that its high-field susceptibility, which is a measure of concentration of paramagnetic material in a sample, is squarely within the range characteristic of normal irghizites.

Samples 166a and 187 belonging to the anomalous group expectedly show totally different hysteresis properties (Fig. 30.5). In both samples hysteresis loops are well developed, the respective M_{rs}/M_s and H_{cr}/H_c ratios being characteristic of pseudo single-domain (PSD) behavior, even tending towards multidomain (MD) in the case of 166a. FORC diagrams (Fig. 30.5c, d) confirm this assessment,

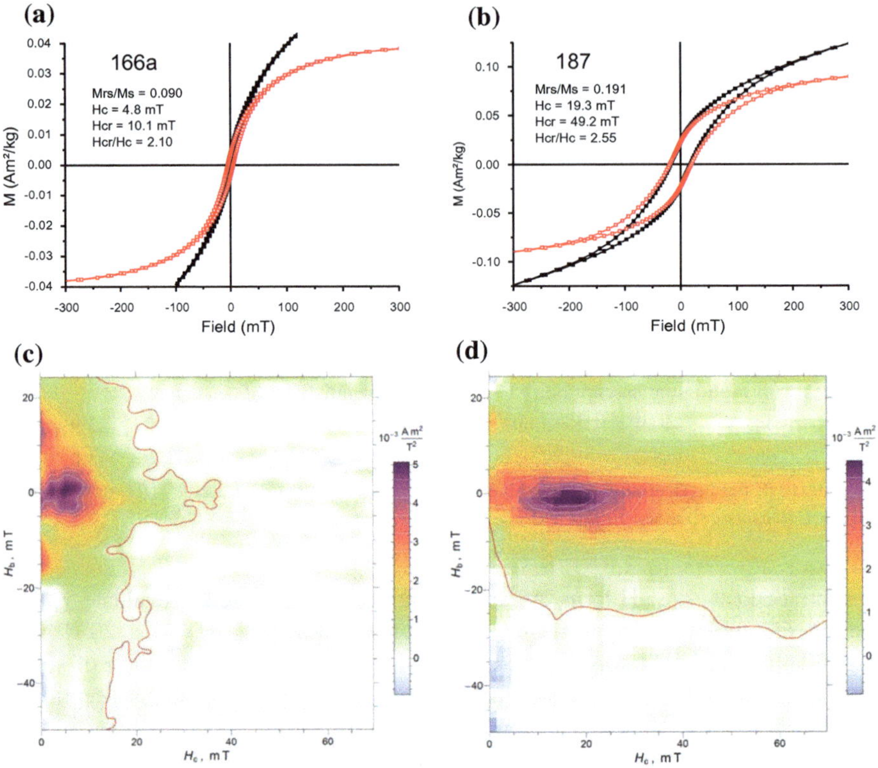

Fig. 30.5 Hysteresis loops (**a**, **b**) and FORC diagrams (**c**, **d**) for two representative samples from the anomalous group. Original loops are shown in black, loops corrected for high-field susceptibility—in red. FORC distributions were calculated using smoothing factor of 4 and stretching parameter (Egli 2013) of 0.07. Contours corresponding to the signal-to-noise ratio of 3 are shown in red

indicating relatively weakly interacting, small pseudo single-domain grains as main magnetization carrier in sample 187, and larger PSD-MD grains in 166a.

Lowrie 3-Axis IRM Test

Examples of the Lowrie test results are plotted in Fig. 30.6. The low-field IRM (100 mT) is the dominant component in most samples and unblocks between 350–450 °C. In some cases an intermediate component IRM (300 mT) unblocks between 500–575 °C, and in three samples a measurable hard component IRM (1 T) occurs, unblocking only at 675 °C. Therefore, the carrier of the low-coercivity IRM component may be titanomagnetite and/or magnetite, while the high-coercivity component is due to a hematite-like phase. In some irghisites a hematite-like phase

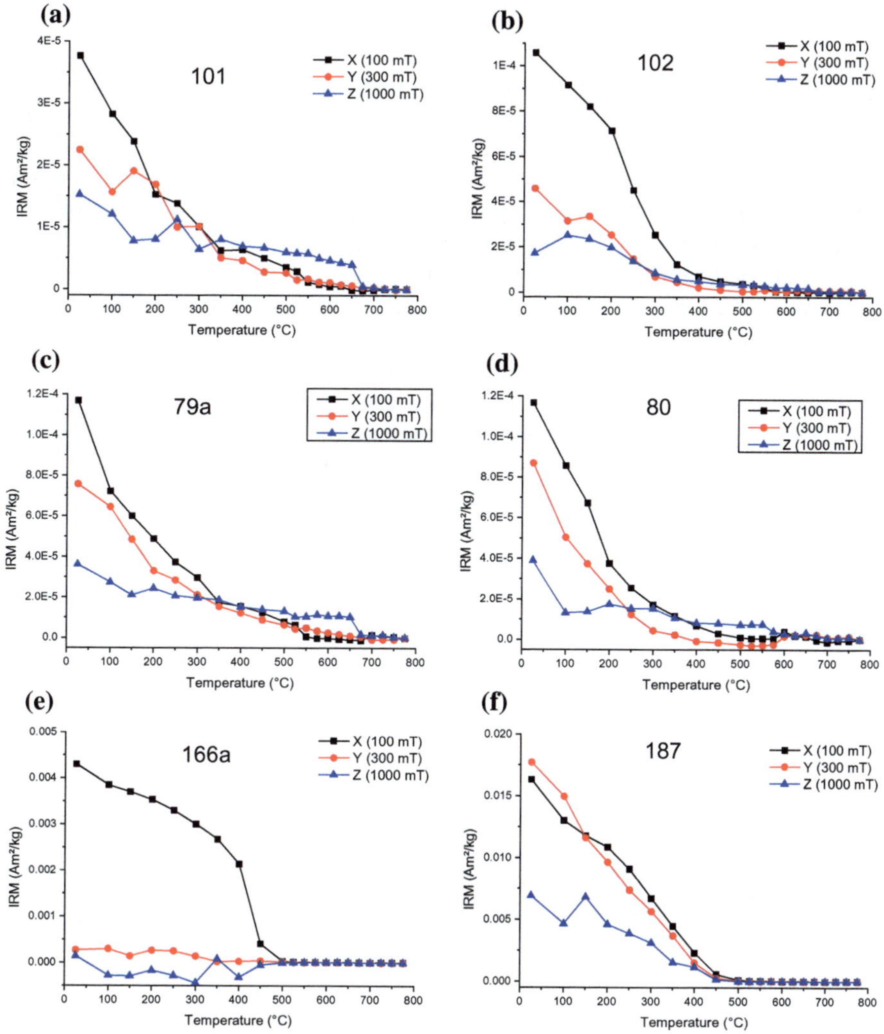

Fig. 30.6 Thermal demagnetization of the 3-axis IRM (Lowrie 1990) for selected samples

has been reported on the basis of microscopic observations (Badjukov et al. 1996) and magnetic measurements (Rochette et al. 2015). IRM unblocking above 675 °C was never observed, implying that the native iron, if present in the samples, is not detectable by magnetic methods.

Low-Temperature Magnetic Properties

Plots of remanence versus temperature for representative samples are shown in Fig. 30.7. The shape of low-temperature SIRM demagnetization curves is quite different for normal and anomalous samples, respectively. Normal samples acquire a considerable, of the order of several hundredths A m^2/kg, SIRM at 2 K, but this remanence is extremely thermally unstable decreasing by a factor of 100 between 2 and 25–50 K, dependent on the sample. This decrease shows further differences between studied samples in the amount of remanence lost at the lowest temperatures (below ca. 5 K), which ranges from <50% (sample 101, Fig. 30.7a) to >80% (sample 287a, Fig. 30.7b) of the initial value. This is consistent with their predominantly superparamagnetic behavior at room temperature. It is worth noting however that in sample 101 on this small SIRM remainder a clearly pronounced Verwey transition with the transition temperature of 97 K could still be observed (see inset in Fig. 30.7a). Shape of the 300-2-300 K zero-field cycle of SIRM acquired at 300 K, where cooling and warming branches diverge at about 85 K also suggest presence of a small amount of magnetite-like phase in this sample.

Anomalous samples 166a and 187 show a completely different low-temperature behavior (Fig. 30.7c, d). For 187, the respective LT-SIRM and RT-SIRM curves resemble some of those reported for Popocatepetl andesites where the main

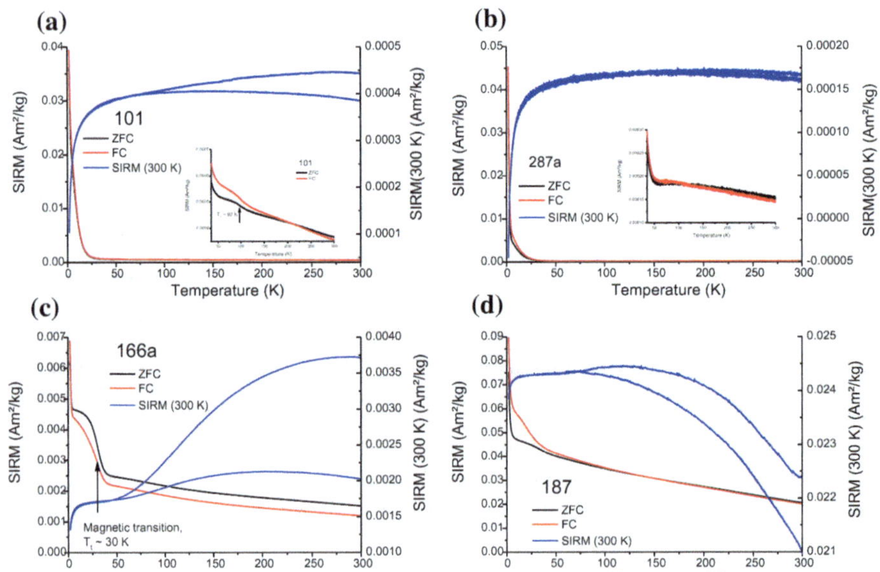

Fig. 30.7 Demagnetization curves of SIRM acquired at 2 K (ZFC—black, FC—red) and 300-2-300 K cycles of SIRM acquired at 300 K (blue). Insets in **a**, **b** show the behavior above 35 K for samples 101 and 287a. Note the Verwey transition at 97 K in sample 101

magnetic mineral is titanomagnetite of composition like TM10-15 (Kosterov et al. 2009). The results of the 3-axis Lowrie test for this sample yielding maximum unblocking temperature of 450 °C (Fig. 30.6e) agree with this interpretation. A relatively minor remanence loss in the RT-SIRM 300-2-300 K cycle is compatible with rather fine grain size implied by room temperature hysteresis and FORC data. In the case of 166a, several lines of evidence suggest that its low-temperature behavior is also controlled by a titanomagnetite-like phase but with considerably larger, than for 187, grain size. First, LT-SIRM ZFC and FC curves show a magnetic transition at ~30 K, similar to a transition-like feature reported for intermediate titanomagnetites (Moskowitz et al. 1998; Church et al. 2011) though occurring at a somewhat lower temperature. We note in passing that monoclinic pyrrhotite, which also has a magnetic transition in this temperature range (Besnus transition, Dekkers 1989; Dekkers et al. 1989; Rochette et al. 1990), is never observed in our samples either microscopically or by means of thermomagnetic analysis. Furthermore, ZFC curve runs somewhat higher than FC, and loss of remanence in the RT-SIRM 300-2-300 K cycle exceeds 45% of the initial value, pointing to a larger grain size than in 187.

In Fig. 30.8 we plot the temperature dependences of complex magnetic susceptibility in the 2-300 K range for the same four samples whose remanence is

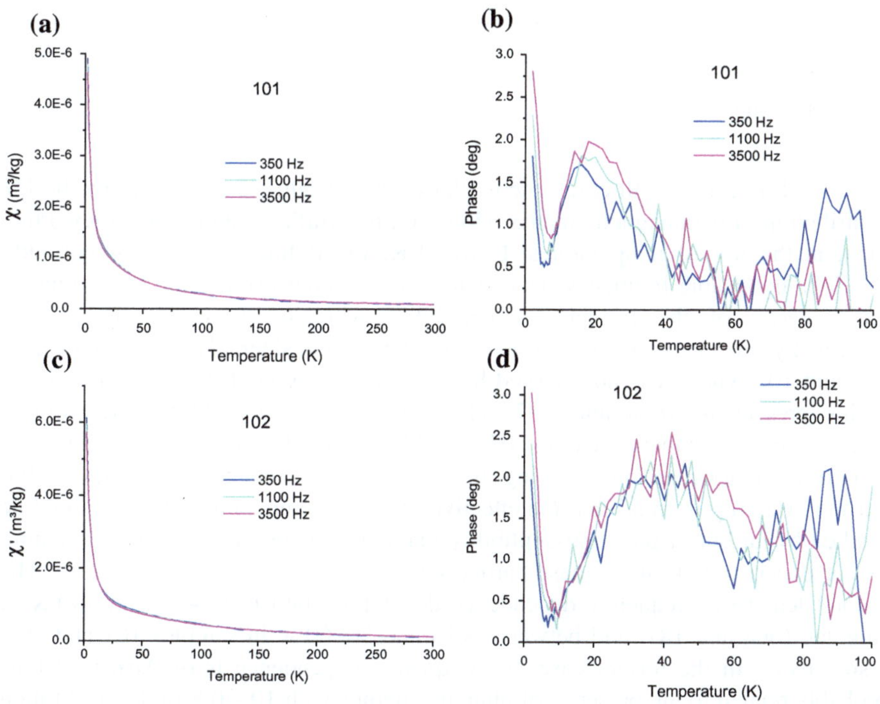

Fig. 30.8 Temperature dependences of the initial magnetic susceptibility in the 2–300 K range. For clarity, only curves measured at three frequencies indicated in the plots are shown. Insets in **e**, **g** illustrate frequency-dependent behavior above 50 K for samples 166a and 187

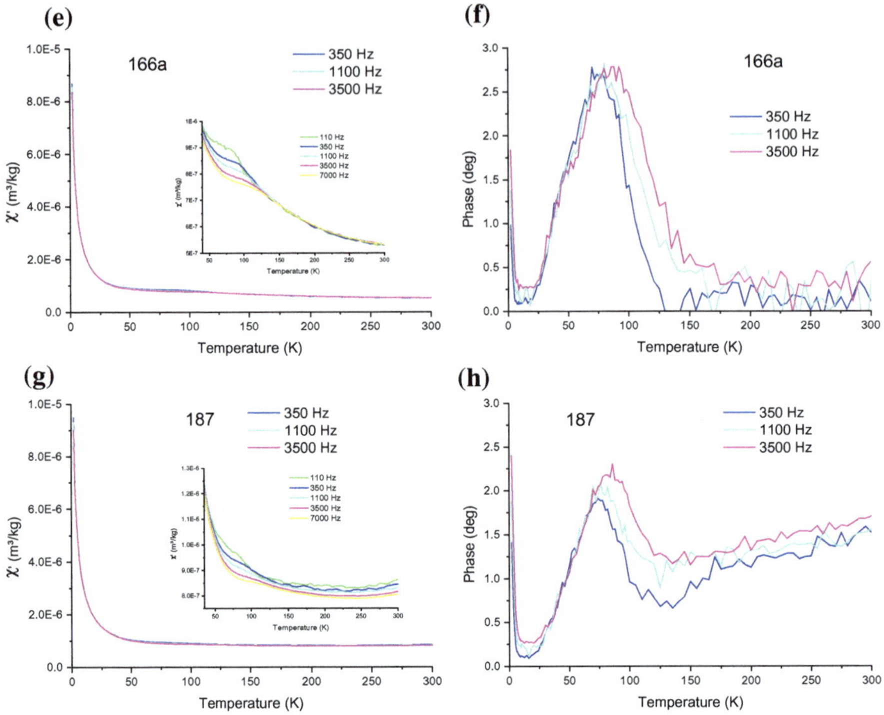

Fig. 30.8 (continued)

shown in Fig. 30.7. In all cases, the in-phase susceptibility is dominated by the 1/T contribution of paramagnetic matrix. However, two further features are noteworthy. First, at the lowest temperatures, below 4 K, in all four samples susceptibility develops noticeable frequency dependence, manifesting itself in an approximately linear decay of the in-phase susceptibility with log frequency (Fig. 30.9). Expectedly, in relative measure this decay appears considerably larger in samples 101 and 102 where it is not masked by a non-decaying contribution from coarser grains present in 166a and 187. The out-of-phase susceptibility accordingly increases in this temperature range (Fig. 30.8b, d, f, h). This behavior is most likely caused by magnetization blocking in very fine ferromagnetic particles and/or clusters. Bearing in mind that the effective blocking temperature, which should be marked by an out-of-phase susceptibility maximum, is below 2 K, the lower limit of our experimental range, these particles/clusters must be ~1–3 nm in size. The second feature is frequency dependence developing between ~50 and 100 K in samples 166a and 187, and between ~15 and 50–60 K in samples 101 and 102, respectively. In the former case, the frequency dependence is of Arrhenius type probably related to the presence of titanomagnetites with 10–20% of Ti substitution as observed previously for both synthetic samples (Radhakrishnamurty and

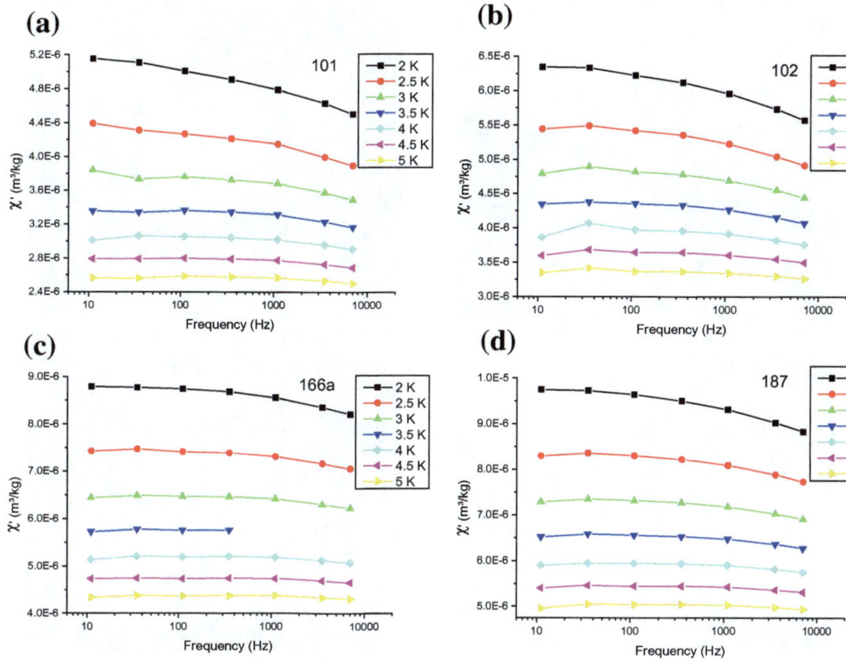

Fig. 30.9 Frequency dependences of susceptibility in the 2–5 K range. Note that frequency dependences are nearly flattened by 5 K

Likhite 1993; Church et al. 2011) and titanomagnetite bearing rocks (Kosterov et al. 2009, 2018). For the latter, peculiar susceptibility frequency dependence may simply be due to magnetization unblocking in somewhat larger yet very fine particles, seen also in SIRM (T) curves.

SEM Observations

Inspection of SEM images of polished sections (Fig. 30.10) confirms the distinction between the anomalous samples 166a and 187, on the one hand, and the rest, normal samples, on the other. The former exhibit abundant ore mineral inclusions, while in the latter inclusions are much rarer, and occur mostly as spherulites. The microprobe analysis shows that inclusions consistently have markedly lowered oxygen content.

In samples 166a and 187, large (up to 50 μm) grains of an unusual habit are found (Fig. 30.10a). Their "block" structure is not typical for any of the variants of magnetite genesis described in the literature. At the same time, structural chemical formulas of minerals calculated according to the oxygen method (Bulakh et al. 2014) identify these grains as Mg–Al–Ti substituted magnetite

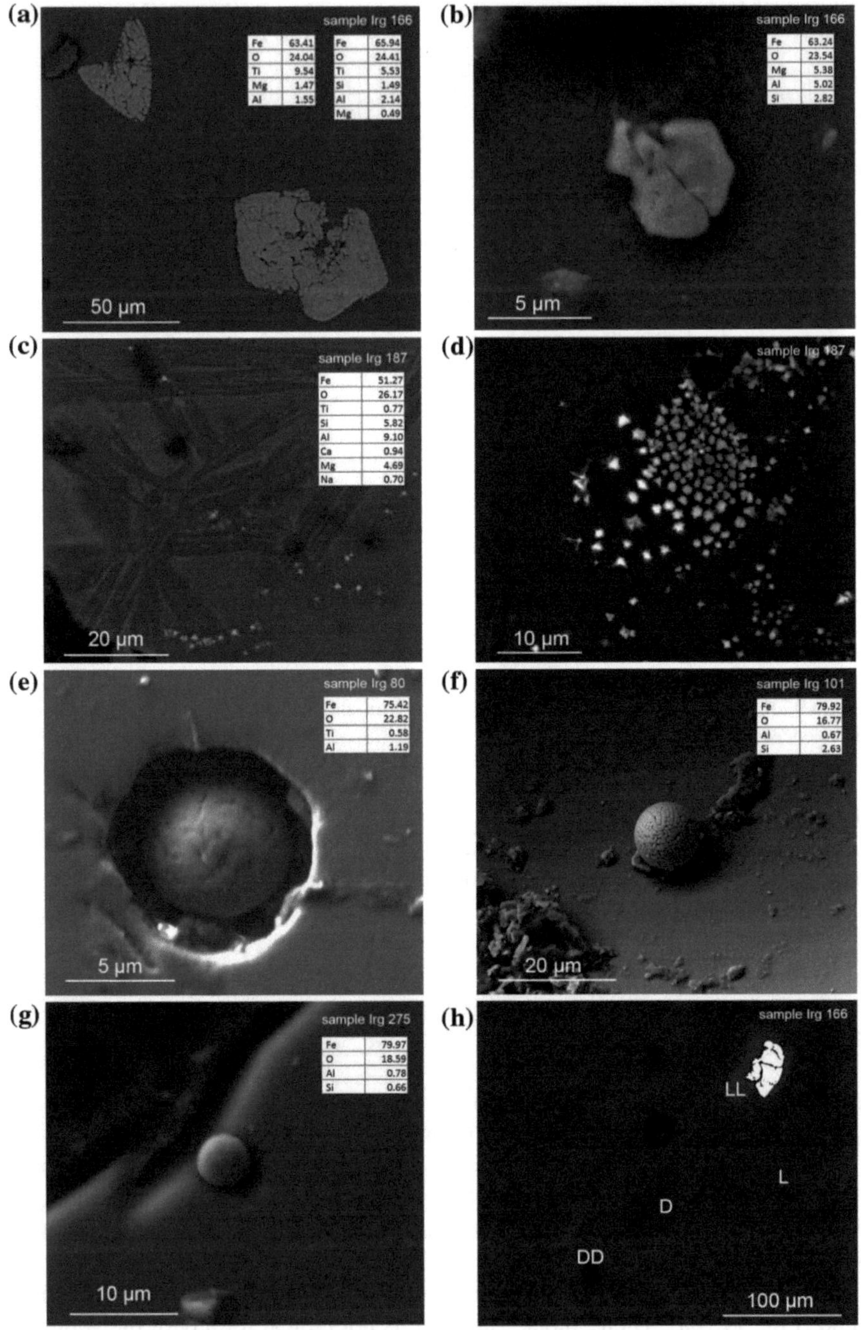

Fig. 30.10 Electron microscope images of the surface of polished thin sections of the samples. Tables in insets show the compositions of inclusions determined by microprobe

$\{(Mg_{0.25}, Fe^{2+}{}_{0.75})(Al_{0.26}, Fe^{3+}{}_{1.59}, Ti_{0.15})_2O_4)\}$. It is worth noting that often these grains have a drop-like (elongated) shape, and grains long axes are subparallel. In addition, individual small, shapeless grains were found, 1–10 μm in size, often with traces of resorption, and with impurities of magnesium, aluminum, silicon, and even copper (Fig. 30.10b–d).

Spherulites measured about 5–10 μm or less (Fig. 30.10e) were found in rounded cavities (bubbles) opened during polishing. Microspheres were also found freely laying on the polished surface (Fig. 30.10f, g). Most likely, they have been knocked out of pores by an electron beam during scanning. The surface of microspheres has a specific dendritic microstructure. Occasionally, we observe also flake-shaped inclusions of intermetallides (FeCr, FeCrNiCu), and of native metals (Bi, W).

Investigation of the nonmagnetic matrix shows the presence of four types of glass. In SEM images, these are manifested as areas of different color saturation, from the lightest (LL) to the darkest (DD), the main matrix (L) also containing spots of intermediate hue (D). The LL areas are localized around the inclusions of ore minerals, while DD areas—around the accumulations of pores and bubbles. The difference in chemical composition between the glass types is expressed as a monotonous decrease in the silica content from the lightest to the darkest varieties with a simultaneous increase of the impurities content. For each glass type, microanalysis shows stable results with fairly small scatter.

For the normal samples (Nos. 79, 80, 101, 102, 216, 275, 287) more than 90% of the non-magnetic matrix is composed of L-type glass containing 75.0 ± 3.0 wt% SiO_2. Other oxides include Al_2O_3 (6.3–11.1%), FeO (7.0–9.0%), MgO (3.1–3.9%), K_2O (2.0–2.5%), CaO (2.0–2.5%), Na_2O (0.4–1.2%), and TiO_2 (0.1–1.0%). There is also a small amount of D-type glass with SiO_2 content of 96.0 ± 4.0 wt%. For samples 166a and 187 (Fig. 30.10h), the matrix is differentiated into four types. L-type areas with SiO_2 content of 60.0 ± 4.0 wt% predominate, but D- (SiO_2 71.0 ± 2.0 wt%), DD- (SiO_2 100.0 wt%), and LL- (SiO_2 54.0 ± 2.0 wt%) areas are also present.

Discussion

The characteristic feature of irghizites is that their magnetic susceptibility forms rather narrow distribution around a median value of 120×10^{-9} m^3/kg (Fig. 30.1a) consistent with the 136×10^{-9} m^3/kg value in a previous study (Rochette et al. 2015). The susceptibility range (45 to 1000×10^{-9} m^3/kg) is somewhat wider than 103 to 791×10^{-9} m^3/kg range reported by Rochette et al., but the latter results were obtained on a much smaller collection. However, irghizite samples having susceptibility >300×10^{-9} m^3/kg are clearly anomalous according to both magnetic properties and microscopic structure and constitute <5% of the whole collection. We therefore feel justified to use the value of 120×10^{-9} m^3/kg as characteristic of normal irghizites. The narrow susceptibility distribution implies, in turn, a high degree of homogeneity of the impact glass material in terms of magnetic minerals concentration and composition. For a homogeneous material susceptibility is expected to be proportional to the concentration of the magnetism carriers as, e.g. in impactites where the content of iron is anticorrelated with that of silica (Sazonova et al. 1992). Therefore, the extremely low susceptibility values correspond to high-silica glasses, which are genetically related to high-silica target rocks such as loess, some Cretaceous sandstones and Paleogene clays (Feldman and Sazonova 1993).

Hysteresis and low-temperature magnetic properties strongly suggest that the ferrimagnetic fraction in normal irghizites is dominated by superparamagnetic grains with relaxation times less than few seconds. Magnetically stable component is very small being carried by <1% of total ferrimagnetic fraction, which is itself present in trace amounts, as follows from M_s values of the order of several tens of mAm^2/kg. Similar observations have been made recently on urengoite glass of possible tektite origin (Bezaeva et al. 2017). However, thermomagnetic studies of the stable ferrimagnetic component revealed, somewhat unexpectedly, the presence in some samples of small amounts of magnetite and hematite with unblocking temperatures 550–570 and 675 °C, respectively. Presence of magnetite is also confirmed by observation of the Verwey transition in sample 101. Hematite might possibly be a relic remaining after the shock metamorphism of target rocks, e.g. Paleogene clays, where it is common (Florensky and Dabizha 1980), as suggested previously (Badjukov et al. 1996). The dominant ferrimagnetic mineral with the unblocking temperatures 350–450 °C could be titanomagnetite, possibly enriched also in Al and Mg. The chemical composition of the main amount of ferrimagnetic fraction, which is in a superparamagnetic state, remains uncertain. However, in some irghizites presence of very fine amorphous particles with a composition close to titanomagnetite was recently demonstrated by transmission electron microscopy (Gornostaeva et al. 2016).

Origin of anomalous irghizite samples ($K > 300 \times 10^{-9}$ m^3/kg, $NRM > 4 \times 10^{-6}$ $A\ m^2/kg$) deserves special consideration. These samples are generally indistinguishable from the normal ones by their outlook, but differ from them both magnetically and microscopically. From the high- and low-temperature thermomagnetic behavior, main magnetic mineral in these samples is titanomagnetite possibly containing also Al and Mg. Hysteresis loops and FORCs measured at room temperature indicate that the ferrimagnetic fraction in pseudo single-domain or even in multidomain state. Relatively large titanomagnetite grains of an uncommon habit are observed by SEM (Fig. 30.10a, c). A possible explanation of this dichotomy might be a difference in the initial temperature of the impact melts, from which irghizites were formed. The normal samples could have originated from the high-temperature splashes of melt, and the anomalous ones from the low-temperature ones. A similar behavior has been observed for zhamanshinites (impact melts cooled on the Earth's surface) from the same crater (Starunov et al. 2016). Irghizites bear a considerable similarity to the latter, apart from having a much higher content of superparamagnetic grains.

Consistently low (<1) NRM/ARM ratios appear incompatible with the thermoremanent origin of irghizites' NRM. For small pseudo single-domain or single-domain magnetite grains TRM/ARM ratio may be as high as 7 (Sugiura 1979) or even exceed 10 (Dunlop and Argyle 1997), while for (titano)magnetite-bearing volcanic rocks it is somewhat lower, about 2.8 (Borisova and Sholpo 1985). One possible option to resolve this paradox is to hypothesize that a transformation of paramagnetic minerals into ferrimagnetic ones takes place in irghizites well below Curie temperature of the latter, and therefore irghizites' NRM is crystallization remanent magnetization (CRM) rather than TRM. Appearance of such source paramagnetic minerals, for example, ülvospinel, is quite likely as a result of a shock-explosive event creating a strongly reducing environment. Since the subsequent cooling is very fast, almost instantaneous, growth of the ferrimagnetic grains could proceed down to temperatures well below the Curie point. A related mechanism would invoke conversion of magnetite or iron-bearing silicates to wüstite during the shock metamorphism caused by impact. Wüstite is stable at high temperatures but decomposes into magnetite and native iron below the Curie temperature of magnetite (Wasilewski 1973). This mechanism could be operating in anomalous irghizites which apparently cooled down at considerably slower rate than normal ones.

Conclusion

The magnetic fraction of normal irghizites ($K < 300 \times 10^{-9}$ m^3/kg, NRM $< 4 \times 10^{-6}$ A m^2/kg) is dominated by superparamagnetic grains with very short relaxation time. These irghizites were most likely formed at the time of a shock-explosive event from the fine spray of a strongly overheated impact melt with an initial temperature of 2500–3000 °C, cooling extremely fast, at rates up to ~200 °C/s. Presence of strongly oxidized, up to Fe(III), iron forms in them is somewhat unexpected, bearing in mind that the latter should have been formed in a high-temperature, strongly reducing environment. Resolving this inconsistency would require further detailed studies; however, one may speculate that the amount of ferromagnetic minerals in irghizites is very low, and thus, only a tiny fraction of total iron needs to be in the Fe(III) form to explain the observed behavior.

The anomalous irghizites accounting for ~5% of their total number in our study contain significantly larger, pseudo single-domain to multidomain grains of titanomagnetite with the admixture of Al and Mg, characterized by unblocking temperatures in the 350–450 °C range. This suggests a different formation mechanism invoking considerably slower cooling rates.

The extremely low (<1) NRM/ARM ratios suggest that irghizites' NRM cannot be of thermoremanent origin, and is likely a CRM acquired by the grains grown into their final blocking volumes well below the Curie temperatures of the respective magnetic minerals.

Acknowledgements Samples used in this study have been collected during a geological-geophysical expedition to Zhamanshin crater organized by the Academy of Sciences of the USSR in the mid-1970s. Experimental research has been carried out using the facilities of the St. Petersburg University Scientific Park. We thank an anonymous reviewer for a detailed and helpful review. This work has been partially supported by RFBR, project 18-05-00626.

References

Badjukov, D. D., Brandstaetter, F., Petrova, T. L., Kurat, G., 1996. Iron oxides in irghizites. *Lunar Planetary Science Conference*, 27, 51–52.

Bezaeva, N. S., Rochette, P., Masaitis, V. L., Badyukov, D. D., and A. Kosterov (2017), Magnetic properties and petrography of urengoites and South-Ural glass. *80th Annual Meeting of the Meteoritical Society*, abs. # 6072.

Borisova, G. P., and L. E. Sholpo (1985), The statistical approach to the paleointensity evaluation, *Izv. Acad. Sci. USSR Fizika Zemli* (7), 71–79. (in Russian).

Bulakh, A. G., Zolotarev, A. A., and V. G. Krivovichev, *Structure, isomorphism, formulae, classification of minerals*. St. Petersburg, St. Petersburg University Publishers, 2014. (in Russian).

Church, N., J. M. Feinberg, and R. Harrison (2011), Low-temperature domain wall pinning in titanomagnetite: Quantitative modeling of multidomain first-order reversal curve diagrams and AC susceptibility, *Geochem. Geophys. Geosyst.*, 12(7), Q07Z27, https://doi.org/10.1029/2011gc003538.

de Gasparis, A. A. (1973), *Magnetic properties of tektites and impact glasses*, x, 173 pp., PhD thesis, University of Pittsburgh.

de Gasparis, A. A., M. Fuller, and W. A. Cassidy (1975), Natural remanent magnetism of tektites of the Muong-Nong type and its bearing on models of their origin, *Geology*, 3, 605–607.

Dekkers, M. J. (1989), Magnetic properties of natural pyrrhotite II: High- and low-temperature behaviour of Jrs and TRM as a function of grain size, *Phys. Earth Planet. Inter.*, 57, 266–283.

Dekkers, M. J., J. L. Mattei, G. Fillion, and P. Rochette (1989), Grain-size dependence of the magnetic behavior of pyrrhotite during its low-temperature transition at 34 K, *Geophys. Res. Lett.*, 16, 855–858.

Donofrio, R. R. (1977), *The magnetic environment of tektites*, xvii, 183 pp., PhD thesis, The University of Oklahoma.

Dressler, B. O., and W. U. Reimold (2001), Terrestrial impact melt rocks and glasses, *Earth Sci. Rev.*, 56, 205–284.

Dunlop, D. J., and K. S. Argyle (1997), Thermoremanence, anhysteretic remanence and susceptibility of submicron magnetites: Nonlinear field dependence and variation with grain size, *J. Geophys. Res.*, 102, 20199–20210.

Egli, R. (2013), VARIFORC: An optimized protocol for calculating non-regular first-order reversal curve (FORC) diagrams, *Global and Planetary Change*, 110, Part C, 302–320.

Feldman V. I., and L. V. Sazonova (1993), Formation and freezing conditions of impact melts in the Zhamanshin astrobleme, *Petrologiya*, 1(6), 596–614.

Florensky P. V., and A. I. Dabizha, *The meteorite crater Zhamanshin*, Moscow, Nauka, 1980. (in Russian).

Glass, B. P., Fredriksson, K., and P. V. Florensky (1983), Microirghizites recovered from a sediment sample from the Zhamanshin impact structure. *J. Geophys. Res.*, 88, Suppl. 1, B319-B330.

Glass, B. P. (1990), Tektites and microtektites: key facts and inferences, *Tectonophysics*, 171, 393–404.

Gornostaeva, T. A., Mokhov, A. V., Kartashov, P. M., and O. A. Bogatikov (2016), Condensate glasses from the Zhamanshin crater. I. Irghizites. *Petrology*, 24(1), 1–20.

Gus'kova, E. G. (1980), Magnetic properties of some tektites, *Meteoritika*, 39, 90–94. (in Russian).
Koeberl, C., Kluger, F., and W. Kiesl (1985), Zhamanshin and Aouelloul impact glasses: major element chemistry, correlation analyses, and parent material, *Chem. Erde*, 44, 47–65.
Koeberl, C. (1986), Geochemistry of tektites and impact glasses, *Annu. Rev. Earth Planet. Sci.*, 14, 323–350.
Koeberl, C. (1990), The geochemistry of tektites: an overview, *Tectonophysics*, 171, 405–422.
Kosterov, A., G. Conte, A. Goguitchaichvili, and J. Urrutia-Fucugauchi (2009), Low-temperature magnetic properties of andesitic rocks from Popocatepetl stratovolcano, Mexico, *Earth Planets Space*, 61, 133–142.
Kosterov, A., E. S. Sergienko, P. V. Kharitonskii, and S. Yu. Yanson (2018), Low-temperature magnetic properties of basalts containing titanomagnetite with composition near ~ TM30, *Isvestiya, Phys. Solid Earth*, 54, 134–149.
Lowrie, W. (1990), Identification of ferromagnetic minerals in a rock by coercivity and unblocking temperature properties, *Geophys. Res. Lett.*, 17, 159–162.
Moskowitz, B. M., M. Jackson, and C. Kissel (1998), Low-temperature magnetic behavior of titanomagnetites, *Earth Planet. Sci. Lett.*, 157, 141–149.
Radhakrishnamurty, C., and S. D. Likhite (1993), Frequency dependence of low-temperature susceptibility peak in some titanomagnetites, *Phys. Earth Planet. Inter.*, 76, 131–135.
Rochette, P., G. Fillion, J. L. Mattéi, and M. J. Dekkers (1990), Magnetic transition at 30–34 Kelvin in pyrrhotite: insight into a widespread occurrence of this mineral in rocks, *Earth Planet. Sci. Lett.*, 98, 319-328.
Rochette, P., J. Gattacceca, B. Devouard, F. Moustard, N. S. Bezaeva, C. Cournède, and B. Scaillet (2015), Magnetic properties of tektites and other related impact glasses, *Earth Planet. Sci. Lett.*, 432, 381–390.
Sazonova, L. V., N. N. Korotaeva, and A. G. Simakin (1992), Conditions of melted impactites formation, *Geokhimiya* (6), 871–880. (in Russian).
Senftle, F. E., and A. Thorpe (1959), Magnetic susceptibility of tektites and some other glasses, *Geochimica et Cosmochimica Acta*, 17(3–4), 234–240.
Senftle, F. E., A. N. Thorpe, and S. Sullivan (1969), Magnetic properties of microtektites, *J. Geophys. Res.*, 74, 6825–6833.
Senftle, F. E., A. N. Thorpe, J. R. Grant, A. Hildebrand, H. Moholy-Nagy, B. J. Evans, and L. May (2000), Magnetic measurements of glass from Tikal, Guatemala: Possible tektites, *J. Geophys. Res.*, 105, 18921–18925.
Sholpo, L. E., *Applications of rock magnetism for the solution of geological problems.*, Leningrad, Nedra, 1977. (in Russian).
Starunov, V. A., Kosterov, A. A., Sergienko, E. S., and Kharitonskii, P. V. (2016), Magnetic properties of impact melts from Zhamanshin astrobleme, in: *Problems of Geocosmos, Proceedings of the 11th International School and Conference*, pp. 81–90. (in Russian with English abstract).
Sugiura, N. (1979), ARM, TRM and magnetic interactions: concentration dependence, *Earth Planet. Sci. Lett.*, 42, 451–455.
Wasilewski, P. J. (1973), Shock remagnetization associated with meteorite impact at planetary surfaces, *The Moon*, 6, 264–291.
Werner, T., and G. J. Borradaile (1998), Homogeneous magnetic susceptibilities of tektites: Implications for extreme homogenization of source material, *Phys. Earth Planet. Inter.*, 108, 235–243.

Chapter 31
Background Magnetic Component of Holocene Cosmic Dust Allocated from Peat

V. A. Tselmovich and A. Yu. Kurazhkovskii

Abstract Cores taken from peat deposits were used to study the variations cosmogenic dust on the Earth's surface. In accordance with the amount of cosmogenic dust in various layers of the peat deposits, the dust flux intensity was conventionally divided into low (background) and high (burst) by means of saturation remanence intensity. Further, the mineralogical composition of the background component of cosmic dust has been studied by energy dispersive scanning electron microscopy on magnetic separates. It was found that its composition is mainly limited to the presence of magnetite spherules and iron. In contrast, when bursts occur, cosmic dust consists of a wide range of metallic particles as well as particles containing carbon and silicon. Probably, the sources of background and cosmic matter bursts are different.

Keywords Cosmic dust · Microprobe analysis · Peat · Native iron
Magnetite · Microsphere

Introduction

The Earth is exposed to a continuous flux of cosmic dust particles (e.g. Plane 2012) which accumulate on the surface. It is known that during certain geohistorical periods, the cosmic dust flux (CDF) increased sharply (Grachev et al. 2015). Such events are usually classified as catastrophic and are the subject of comprehensive research, such as for instance the Tunguska event of 1909 (Kolesnikov et al. 1999). There is also a certain minimum level of CDF intensity which can be called

V. A. Tselmovich (✉) · A. Yu. Kurazhkovskii
Borok Geophysical Observatory IPE RAS, bldg. 142, 152742 Borok,
Nekouzsky District, Yaroslavl, Russia
e-mail: tselm@mail.ru

D. Nurgaliev et al. (eds.), *Recent Advances in Rock Magnetism, Environmental Magnetism and Paleomagnetism*, Springer Geophysics,
https://doi.org/10.1007/978-3-319-90437-5_31

background. The study of the background flux may be of interest in terms of evolutionary changes of matter in the solar system and in our galaxy. The present work is an attempt to study the mineralogical composition of background flux of cosmic dust deposited on the Earth's surface.

Research Technique

In our opinion, peat deposits situated in high marsh zones are most convenient for studying the dynamics of the CDF on the Earth's surface (Tselmovich 2015). The amount of cosmogenic mineral particles in such peat deposits clearly exceeds the amount of terrigenous mineral particles (Tselmovich et al. 2016a, b). There is also no lateral transfer of mineral particles during peat layer formation. Therefore concentration changes of cosmic dust are related only to the CDF intensity variations.

Cosmic dust consists usually of particles that can acquire a remanent magnetization, such as for instance iron, iron alloys and magnetite spherules; thus magnetically weak materials in general. Relative changes of cosmic dust concentration in peat layers were identified through variations of the saturation remanent magnetization (Irs). Five cores were taken from different peat deposits with an "Instorf" sampling probe on the territory of Vologda (59° 56′N, 36° 3′E), Kemerovo (54° 47N, 88° 16′E) and Yaroslavl regions (57° 45′ 15″N, 38° 35′ 39″E and 57° 56′ 2″N 37° 29′ 12″E), Ayano-Maysky district of Khabarovsk Territory (58° 44′N, 130° 42″E). The thickness of the investigated peat deposits was between 2.5 and 5 m. The sampling locations are far from volcanoes and technogenic pollution emitting sources.

In order to measure the Irs variations with depth, cubic sub-samples of 2 cm edge length were taken from each core. They were placed in plastic containers and magnetized in a magnetic field of 1T. The Irs parameter was measured with a JR-4 magnetometer. After normalizing the measured magnetic moments by volume, the Irs values were plotted along the length of the cores (Fig. 31.1). The Irs signal was conditionally subdivided into two categories with low values (up to 0.012 A/m), or extremely large (more than 0.05 A/m) and associated with cosmogenic background and burst flux, respectively.

Further, magnetic extracts were taken from samples with low or extremely enhanced Irs values. This separation was done with the help of a powerful large hand-held neodymium magnet. The size of the magnet was 40 × 20 × 10 mm. The magnet was fixed on the outer surface of a plastic container, which was filled with a mixture of distilled water and sample. The container was then placed in an ultrasonic bath. After the magnetic particles accumulated on the wall of the container due to the magnetic force, the water was poured out of the container. The large magnet was then carefully removed and the sample was dried. Next a double-sided conductive carbon tape was glued on a small (transport) neodymium magnet with a diameter of 10 mm with a telescopic handle. With the help of this magnet the particles were fixed on the tape. Then the tape was pasted on a standard sample holder for scanning electron microscopy.

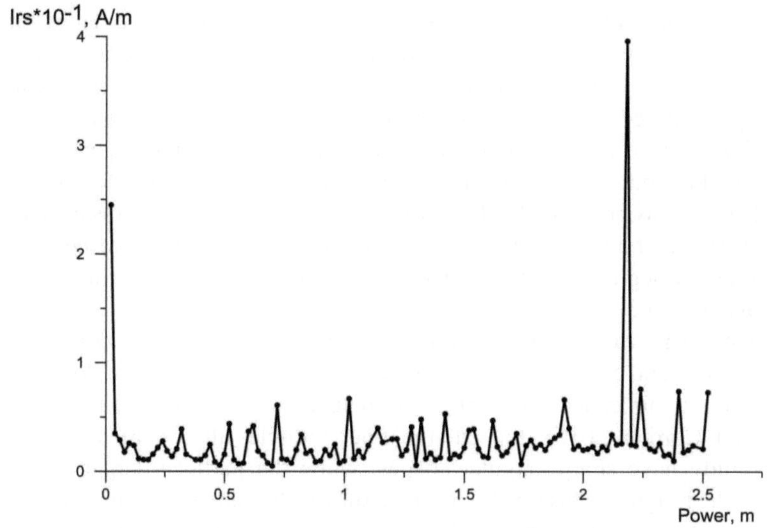

Fig. 31.1 Variation of saturation remanence versus peat bog thickness for the "Tundra" peat bog, from the Kemerovo region

These separates were later analyzed by scanning electron microscopy (SEM) "TESCAN VEGA2" with an attachment for energy-dispersive analysis. In the course of this study, the elemental composition of mineral particles was determined and their number was counted.

Within the framework of this preliminary study presented here, it was primarily of interest to investigate the mineralogical composition of samples having low Irs values, i.e. (up to 0.012 A/m).

Results

As it can be seen from Fig. 31.1, the Irs changes cyclically along the core length and covers a wide range of values up to 4×10^{-1} A/m. Although the sampling locations were carefully selected to be far from technogenic pollution, the 2.5×10^{-1} A/m Irs peak at the top of the core is most probably associated with long-distance pollution and therefore not taken into consideration for this study. Samples with low Irs values were observed at different levels and were used to study the composition of the background component of cosmic dust. Thus, samples from the layers at 50, 70, 100, 160, 228 cm depth were examined in the selected core taken in the Kemerovo region (Fig. 31.1).

Microprobe analyses of samples with a low Irs revealed that they do not contain any terrigenous magnetic material such as for instance titanomagnetites or ilmenites. Only magnetite microspheres (Mt), native Fe and a small number of other native metals and alloys (FeCr, FeW) were detected (Fig. 31.2). These findings were attributed to the class of cosmic dust (CD) and micrometeorites (MM). The bulk of the observed Mt microspheres has a well developed detrital surface and can be clearly identified as "cosmic" spherules, as described in hundreds of scientific publications, see e.g., (Zhang et al. 2014). These characteristic surfaces were formed at the time of ablation, i.e. during the flight of meteorites through the atmosphere. It was proposed by Tselmovich (2013), that such spheres develop from troilite in the melting meteorite crust. Particles consisting of native Fe could also have formed through crust melting. The mechanism of this process was described in (Tselmovich 2016c).

Particles of Fe can be conditionally divided into two groups according to their size: ≤ 50 μm and >50 μm. It is known that particles up to 50 μm do not heat up by friction with air molecules when micrometeorites travel through the atmosphere (Folco and Cordier 2015). Indeed, the finer particles that were detected in the present study have sharp, angular shapes, while the particles larger than 50 μm manifest clear signs of fusion. It can be assumed that large and small background cosmic dust has the same source. Perhaps one sees the product of crushing of the same solid cosmic body and the background cosmic dust is basically of interplanetary origin (e.g., Carrillo-Sánchez et al. 2016).

The Irs variations observed in the Kemerovo core (Fig. 31.1) show one extremely high Irs value at 2.18 m (Fig. 31.1). The SEM analysis of the magnetic separate extracted from this depth reveals a wide range of minerals. Native metals and their alloys were detected such as Sn, Zn, CuZn, FeCrNi, W, FeCr and on the microspherules also Kamacite FeNi. Figure 31.2a shows an overview micrograph of magnetic microparticles isolated from peat at level of 50 cm. Similar results were obtained at the levels 60, 70, 100 and 120 cm (Fig. 31.2b–d). A significant number of magnetite spherules, native Fe and FeCr intermetallics was found in all levels. Micro- and nanoscale particles of native W are visible on some Fe-particles, but their origin remains unclear (Fig. 31.2f).

The burst component of Irs is due to an increased total amount of magnetic materials, which is typical for an impact event. During such, material is ejected into the air, and there is evidence that also microstructures are formed during ejection. Films of metals deposited on cold objects can be good indicators of an impact process. So in Fig. 31.3a, b thin films of nickel and iron are shown. In Fig. 31.3a, the molten lyophilic Ni is covered with a layer of SiO_2 (probably of plant origin). In Fig. 31.3b a thin film of iron covers a detrital magnetite of terrestrial origin. Ni in the form of the films and in the form of a tube was found. A cylindrical shape could have been arisen when molten nickel had hit rod-shaped plant residues. These sticks contain carbon or silicon oxide. The nickel alloy covered completely the rods with a thin film due to its lyophilic properties.

Minerals found in samples with enhanced Irs indicate that the background (Fig. 31.2) and the burst (Fig. 31.3) components of cosmic dust are different. In this case, the composition of the background component of cosmic dust (samples with small values of Irs) did not depend on the depth of occurrence of the investigated layers.

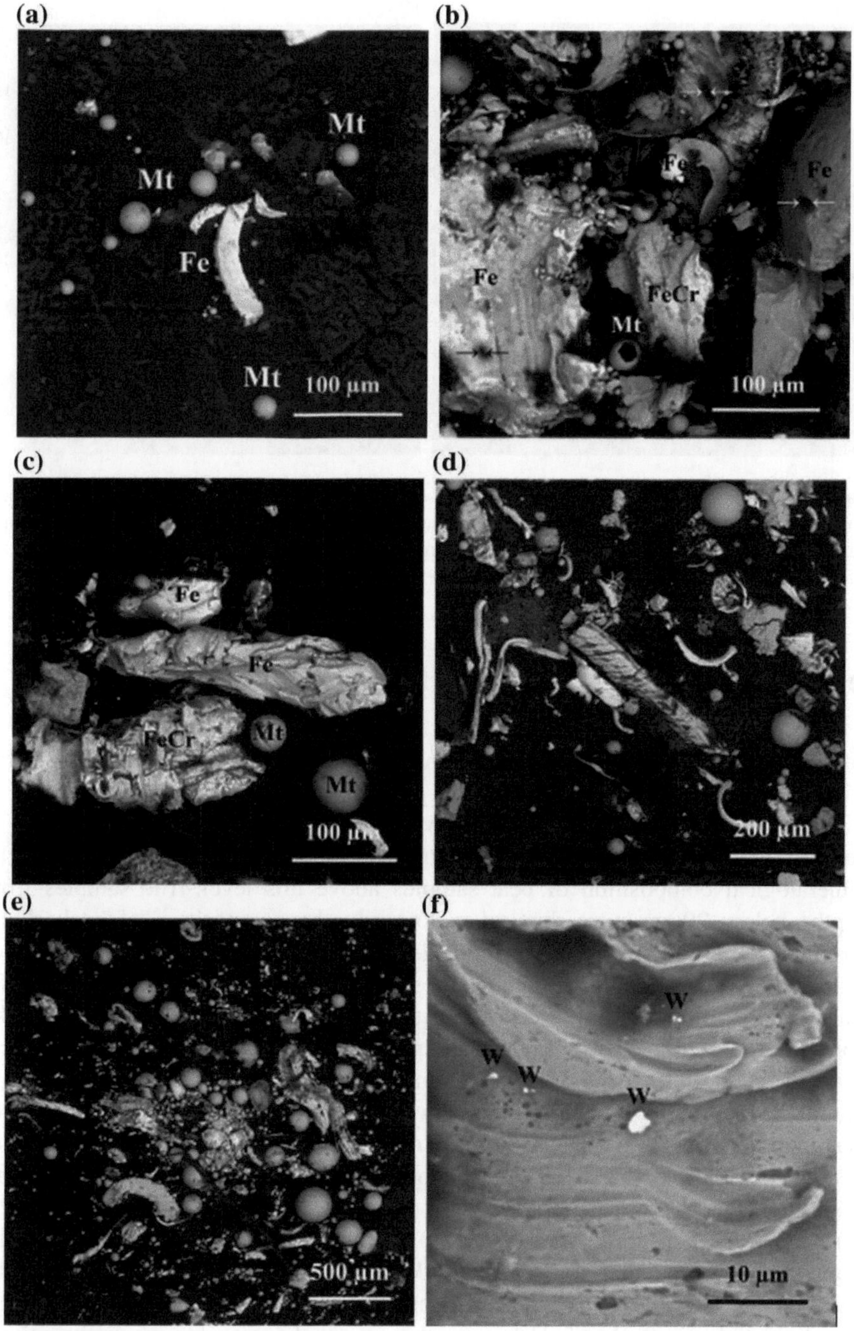

Fig. 31.2 Micrographs of background cosmic dust extracted from of magnetic separates of the "Tundra" peat bog from the Kemerovo region containing magnetite spherules and microspherules (Mt) and native iron (Fe). **a** level 50 cm: Mt, Fe; **b** level 60 cm: Mt, Fe, FeCr; **c** level 70 cm: Mt, Fe, FeCr; **d** level 100 cm: Mt, Fe, FeCr; **f** level 120 cm: Mt, Fe, FeCr; **e** level 120 cm: W particles on Fe

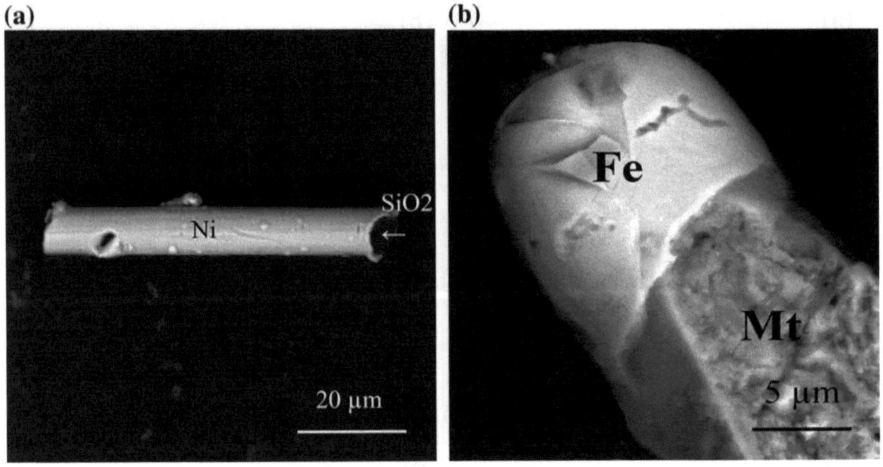

Fig. 31.3 Micrographs of the burst component extracted from of magnetic separates of the peat bog in the Ayano-Maysky district of the Khabarovsk Territory showing microstructures formed during impact. **a** a Ni film on SiO2; **b** a film of Fe on Mt (Ayano-Maysky district of the Khabarovsk Territory, expedition of Ivan Amelin in 2016 (Amelin et al. 2017), photo by V. Tselmovich)

Discussion

Samples from the uppermost layers show high Irs values which were associated with industrial pollution. The uppermost peat samples can therefore not be used to study cosmic dust. The samples with low values of Irs were observed at depths of below 20 cm. It is probable that industrial pollution can significantly affect the mineralogical composition of peat samples above this level. The samples from depths below 20 cm were thus taken to study the composition of background cosmic dust. The small variety of minerals found in the background component of cosmic dust, is in our opinion not associated with a random uneven distribution of a small number of particles between samples with low Irs. It should be noted that the volume of the samples was usually about 8 cm^3. In addition micromineralogical analyses of particles from a large volume of peat (1000 cm^3) were carried out, sampled at Kemerovo region in a pit between 45 and 50 cm depth. About 5 mg of magnetic material was obtained only, which also turned out to consist mainly of microspheres, native Fe, FeCr, and nanometric particles of native W. The composition of the detected mineral particles from peat levels with low Irs values did not change fundamentally with respect to the volume of the peat samples, i.e. the composition was similar in 8 cm^3 samples and the 1000 cm^3 sample. It can be assumed that Fe and FeCr in peat samples originate from material which the sampling probe is made of. However, this is not the case for magnetite microspheres Mt and native W. In addition, samples from the central part of the core were selected for the study, which had no apparent contact with the sampling probe.

Consequently, it is rather unlikely that the samples were contaminated during the sampling procedure. Detailed studies about the origin of native iron, presented in (Pechersky et al. 2017), allow concluding about the cosmic origin of the iron particles that were discovered in the present study.

Figure 31.1 (level 2.18 m) shows an Irs peak value. The analyzed samples from this depth manifest a mineralogical composition that is characterized by a wide range of cosmogenic minerals. This is not the case for other layers. In the context of this paper, the reasons for the appearance of these features were not investigated. It has to be said that such samples are rare and their composition is different from other samples.

In their study, Pechersky et al. (2017) discussed the cosmic origin of iron particles. In the present study, iron and magnetite particles with apertures (holes) between 5 and 10 μm were detected (Fig. 31.4a). It is argued here that these holes could have a origin in far space and result from collisions between small (i.e. sub-micrometer and micrometer size) high-speed particles and iron particles measuring some tens of micrometers. The holes found in some iron-target particles manifest specific shapes supporting the argument that the sub-micrometer particles entered the target at various angles.

Similar holes were observed by during the analysis of the meteorite 7428 (San Julio De Morayra, http://geo.web.ru/db/meteorites/card.html?id=10962) from the collection of the Geological Museum of Vernadsky Institute of Sciences (Fig. 31.4b), which was done for comparison. There is a high probability, that those microparticles which create the holes can have galactic origin rather than an interplanetary.

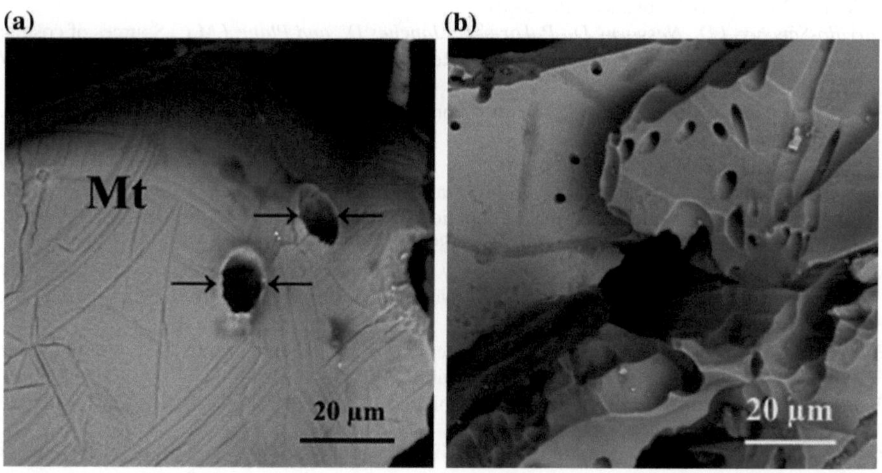

Fig. 31.4 **a** Holes in particle Mt, found in the peat bog from the Kemerovo region; **b** for comparison: Holes in the particle from the shreibersite meteorite San Julio De Moreira

Conclusions

The study of cores of peat deposits showed that the cosmic dust flux intensity at the Earth's surface can be can be conditionally subdivided into low (background) and high (burst) by means of saturation remanence. The mineralogical composition of background cosmic dust and bursts differs. Background cosmic dust consists mainly of magnetite spherules, iron and FeCr alloy (6–20% Cr). In contrast, burst is characterized by a wide spectrum of metallic particles, as well as compounds of carbon and silicon. Probably, the sources of background and burst are different. Material from the uppermost peat layer should not be used for cosmic dust study because it can be influenced by anthropogenic pollution or natural material of terrestrial origin.

Acknowledgements The work was carried out on the budget topic "Spatial-temporal structure of the ancient geomagnetic field and petrophysics of magnetic minerals as indicators of geological and geophysical processes State Registration No. AAAA-A17-117040610183-6", Microprobe studies were carried out with the support of the Russian Foundation for Basic Research project No. 16-05-00703a.

We thank Simo Spassov for assistance in evaluating this paper.

References

Amelin I.I., Gusiakov V.K., Tsel'movich V.A., Kurazhkovsky A.Yu. Expedition study of area of big meteorite falling in Uchur river basin / Zababakhin Scientific Talks International Conference. March 20–24, 2017 Abstracts. P. 15–16.

Carrillo-Sánchez J.D., Nesvorný D., Pokorný P., Janches D., and Plane J.M.C. Sources of cosmic dust in the Earth's atmosphere // Geophys. Res. Lett. 2016. V. 43. P. 11,979–11,986. https://doi.org/10.1002/2016gl071697.

Folco Luigi and Cordier Carole. Micrometeorites // EMU Notes in Mineralogy. 2015. V. 15 Chapter 9. P. 253–297.

Grachev, A.F., Korchagin O.A., Kollmann H. A., Pechersky D. M., and Tsel'movich V.A. A new look at the nature of the transitional layer at the K/T boundary near Gams, Eastern Alps, Austria, and the problem of the mass extinction of the biota // Russ. J. Earth Sci. 2015. V. 7, ES6001, https://doi.org/10.2205/2005es000189.

Kolesnikov E.W., Boettger T. and Kolesnikova N.V. Finding of probable Tunguska Cosmic Body material: isotopic anomalies of carbon and hydrogen in peat // Planet. Space Sci. 1999. V. 47. P. 905–916.

Pechersky D. M, Kuzina D.M., Markov G.P., Tselmovich V.A. Native iron in the Earth and space Sep 2017 // Izvestiya Physics of the Solid Earth. 2017. № 5. P. 44–62.

Plane J.M.C. Cosmic dust in the Earth's atmosphere // Chem. Soc. Rev. 2012. V. 41. P. 6507–6518.

Tselmovich V.A. Space balls on the surface of the Chelyabinsk meteorite / Materials and reports of the International Scientific and Practical Conference "Asteroids and comets. The Chelyabinsk event and the study of the fall of the meteorite into Lake Chebarkul" (Chebarkul, June 21–22, 2013). 2013. P. 140–147.

Tselmovich V.A. Possibilities of microscopic diagnostics of cosmic dust in peat. Materials of the international conference and school of young scientists "CHEBARKUL 2015". Publishing house "TETA", Chelyabinsk. 2015. P. 193–196.

Tselmovich V.A., Kurazhkovskii A.Yu., Kazansky A.Yu., Shchetnikov A.A. Research of the holocene events // EXPERIMENT In GeoSciences: Mineralogy.Petrology.Geochemistry. Crystallography. Geophysics. Isotopy. ISSN 0869-2904. 2016a. V. 22. № 1. P. 21–23.

Tselmovich V.A., Kurazhkovskii A.Yu., Kazansky A.Yu., Shchetnikov A.A., Blyakharchuk T.A., Amelin I.I. Catastrophic events in the Holocene and their registration in peat deposits // Proceedings of the 11th Intl School and Conference "Problems of Geocosmos" (Oct 03–07 2016, St.Petersburg, Russia). 2016b. P. 91–98.

Tselmovich V.A., Korzinova A.S. Features of the composition and microstructure of the thick crust of melting of the Chelyabinsk meteorite fragment. Materials of the IV International Conference of Young Scientists "Meteorites, asteroids, comets". Ekaterinburg. Fort Dialogue-Iset. 2016c. P. 141–145.

Zhang Hua, Shen Shu-zhong, Chang-qun Cao, Zheng Quan-feng. Origins of microspherules from the Permian–Triassic boundary event layers in South China // Lithos. 2014. V. 204. P. 246–257.

Part V
Geomagnetic Field and Magnetic Surveys

Chapter 32
Preliminary Results of the Geohistorical and Paleomagnetic Analysis of Marine Magnetic Anomalies in the Northwestern Indian Ocean

S. A. Ivanov and S. A. Merkuriev

Abstract We present preliminary results of the geohistorical and paleomagnetic analysis of marine magnetic anomalies in the northwestern Indian Ocean. The original data consisting of the series of magnetic profiles in the Arabian and Eastern Somalian basins are studied and the anomalies corresponding to the chrons C24–C26 are identified. These anomalies have arisen in Paleocene- Early Eocene Age. At this time we have the last stage of Gondvana continent fragmentation consisting of the separation of the India from the Seychelles continent and the drift of the India to the north. Interval half spreading rates estimated along the magnetic profiles in the Arabian and Somalian basins. The asymmetries of the crustal accretion has been revealed in the Arabian and Somalian basins. Also new estimates of paleomagnetic pole locations derived from the skewness of marine magnetic anomalies are obtained.

Keywords Marinemagnetic anomalies · Geochronological analysis
Paleomagnetic analysis · Spreading model · Magnetic modeling

Introduction

The collision of the Indian plate with Eurasian Plate is one of the most important events in the Earth history beginning with the Cretaceous period what permanently attracts attention of geologists and geophysicists. Recently several papers where published in the Earth sciences, which are initiated by new paleomagnetic, stratigraphic, and tectonic data obtained in the region of the collision, see, e.g., (Aitchison et al. 2007; Yang et al. 2015).

S. A. Ivanov (✉) · S. A. Merkuriev
Marine Geomagnetic Investigation Laboratory, SPbF IZMIRAN,
St. Petersburg, Russia
e-mail: Sergei.A.Ivanov@mail.ru

S. A. Merkuriev
Institute of Earth Sciences,
St. Petersburg State University, St. Petersburg, Russia

D. Nurgaliev et al. (eds.), *Recent Advances in Rock Magnetism, Environmental Magnetism and Paleomagnetism*, Springer Geophysics,
https://doi.org/10.1007/978-3-319-90437-5_32

To solve the connected geodynamic problems it is necessary to determine the age and location of the first contact and also the velocity of the Indian Plate before and after the collision. At the first stage of our investigations we estimate the parameters of the India Plate motion via the geohistorical analysis of the ancient marine magnetic anomalies formed at the beginning of the spreading and observed in the northwestern Indian Ocean. Because the fast spreading and high linearity of anomalies we have identified the whole sequence of anomalies including some short-wavelength anomalies with the low amplitudes (referred to as 'tiny wiggles').

At the second step we estimate the effective inclination of the ancient magnetic field what gives the paleolatitude of the ancient spreading center.

The apparent polar-wander path obtained from palaemagnetic analysis of conjugate magnetic anomalies in Arabian and Somalian basins is now interpreted to be the result of drifting of the Indian plate.

Geophysical Investigation of the Region

Geohistorical analysis of marine magnetic anomalies is a tool of investigation of plate motion with respect to the spreading centers. One of the first model of the evolution of the northwestern Indian Ocean was presented in (McKenzie and Sclater 1971). In this paper the first found key anomalies in the Arabian and Eastern Somalian basins were identified and the Euler poles of rotation for the Indian and Somalian Plates have been determined. These first investigations have initiated the creation two-stage scheme of the evolution of this region: the ancient fast spreading and the slow modern one. As a result two systems of magnetic anomalies were formed: the ancient sub latitudinal A23–A27 and the recent A1–A5 of the NW-SE direction along the Carlsberg ridge. This model has remained practically unchanged since this time. Usually the end of the fast spreading stage is connected with the first collision of India and Eurasia. The beginning of the spreading rate decreasing obtained in papers lies between 40 and 65 Ma. Along with this the different approaches give a fairly wide range both of spreading rates (White and Lister 2012).

Also attempts were made to estimate not only relative motions of the Indian and Somalian Plates but also the absolute location of the Indian Plate. The authors of the mentioned paper (McKenzie and Sclater 1971) have estimated the paleolatitude of the Carlsberg ridge, what is the divergent plate boundary. The estimate 10°S was obtained by the analysis of the shape of the older magnetic anomalies in the Arabian and Eastern Somalian basins. The paleolatitude estimates from other papers have a wide range. The authors of the paper (Besse and Courtillot 1988) have obtained the estimate 17°S from paleomagnetic data of India and from the paleogeographical reconstruction. Our estimate derived from paleomagnetic investigation of seamounts is 5°S, i.e. much farther north (Brusilovsky et al. 1991).

The systematic magnetic investigations of the northwestern Indian Ocean which were conducted by the Russian research vessels since the beginning of the 80s during more than a decade allow us to create the anomaly magnetic map of the region and to compile the $1' \times 1'$ magnetic grid.

The map of isolines of the anomaly magnetic field of the northwestern Indian Ocean with chrones, axes of the anomalies, plate boundary, and the magnetic anomaly profiles is shown in Fig. 32.1. The analysis of the map confirms the existence of two systems of linear anomalies. The recent magnetic anomalies of the NW-SE direction along the Carlsberg ridge have arose at the slow spreading stage. These anomalies have been investigated in (Merkouriev and DeMets 2006) what gives the detailed kinematic model of the Indian and Somalian plate rotation and reveals changes of the spreading rate for the last 20 million years unknown before.

In the older parts of the Arabian and Eastern Somalian basins there are linear anomalies corresponding to the first stage of the fast spreading. The study of this stage have been published in the papers (Miles et al. 1988; Miles and Roest 1993; Dyment 1998; Chaubey et al. 1998) where the authors supposed that the asymmetry of the spreading and the oblique faults are connected with the Deccan-Reunion plume.

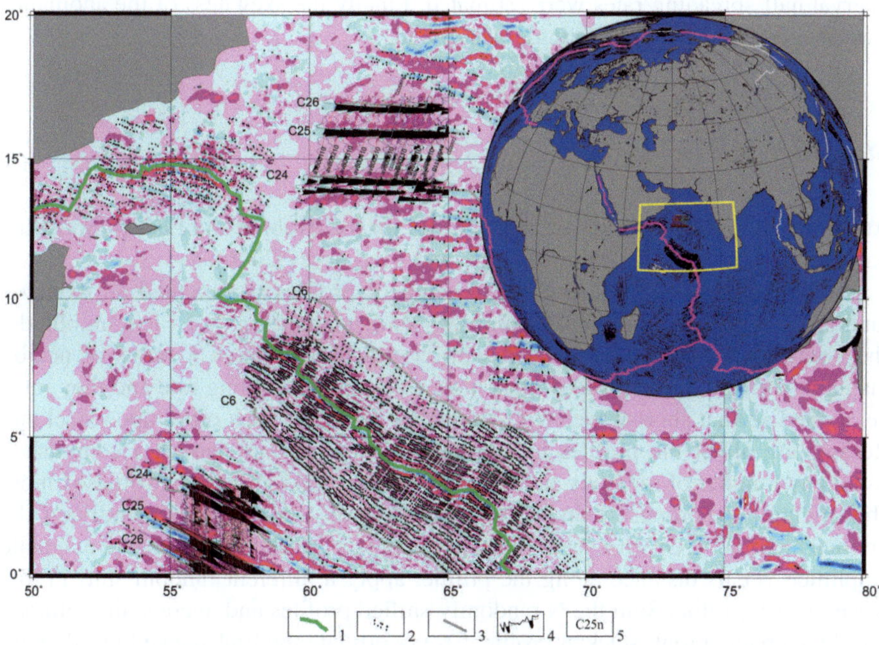

Fig. 32.1 Magnetic anomaly grid of the northwestern Indian Ocean. Map of linear magnetic anomalies according to (Seton et al. 2014) and magnetic anomaly profiles, crossing anomalies A24, A25, and A26. The plate boundary (1), crossings of MA (2), magnetic anomalies picks (axes of linear magnetic anomalies) (3), magnetic anomaly profile, negative anomalies are shaded (4), and polarity chron denotation (5).Yellow box in the inset map outlines the region shown by the larger map

The traces of the spreading axis propagation and its jumps were identified also at the current stage of spreading in (Merkouriev and Sotchevanova 2003a, b). Some of the pseudofaults can be recovered not only from magnetic data but also from bathymetry data especially in the Eastern Somalian basin where the thickness of the sedimentary cover is less than one in the Arabian basin. Using these data and the kinematic parameters of the spreading the authors recovered the trace of the pseudo-fault in the Eastern Somalian basin and its along-strike extension in the Arabian basin. The observed direction of the pseudo-fault in the Arabian basin coincides with the estimate of the direction what confirms the hypothesis about the ridge propagation at chrons C24–C21 (Merkouriev and Sotchevanova 2008).

Data, Methods, and Results

The first problem in the study of the ancient magnetic anomalies was to find the average spreading rate. After this the location of reversals were found and the interval half spreading rates were estimated. Finally the skewness of the anomalies was obtained and the paleolatitude was found.

Stacking

At the first step we found the average profiles for the Arabian and Eastern Somalian basin using a special algorithm (Ivanov and Merkuryev 2016).

We now discuss and test stacking of magnetic profiles as a method for reducing random noise in individual profiles and enhancing of low-amplitude, short-wavelength signals associated with cryptochrons. Given a series of nearby magnetic profiles, these profiles must have a common sampling rate and they must be aligned so that short-wavelength anomalies reinforce each other upon stacking. Resampling all of the profiles to a common sampling rate can be accomplished using any number of methods and is not treated here. Aligning the profiles is more challenging. We propose using cross-correlation as a method for identifying the optimal alignment. Given a set of N nearby magnetic profiles that measure magnetic amplitude versus distance along the profile, apply a different random shift in distance to each profile. Sum the N randomly-shifted profiles and average the summed profile to create a trial, stacked profile. Cross-correlate the trial stacked profile with each of the N randomly-shifted profiles to find a summed cross-correlation coefficient C for the N profiles. Repeat the process with numerous, other random profile shifts to identify the shifts and the stacked profile that maximize C, the cross-correlation coefficient.

Identification

At this step we choose the average spreading rate to find the best fitting with the simulated profile based on the scale of reversals (Cande and Kent 1995).

Our early estimates of the ages of the ancient anomalies coincide with the estimates of the other authors up to the using scale of reversals. Nevertheless we develop the algorithm to find the average spreading rate to obtain the best fitting. The idea of the algorithm is to find a part of the scale and the spreading rate which give the maximal correlation between the observed and the model profile. As the observed profile we take the stack obtained at the previous step. The model profile was calculated for the given parameters: the inclination, the declination, the direction of the anomalies, the depth to top of basalt layer (from the bathymetry). We set the thickness of the layer equal to 0.4 km. For the fixed spreading rate we generate the model profile includes the anomalies A1–A28. Then we find the correlation coefficient between this part of the model profile and the observed one. For the fixed spreading rate we find the part of the scale with the maximal correlation. Then we choose the spreading rate, corresponding to the maximum of the obtained coefficients of correlation. In the bottom of Fig. 32.2 we show the application of this algorithm. The observed (blue line) and calculated profiles

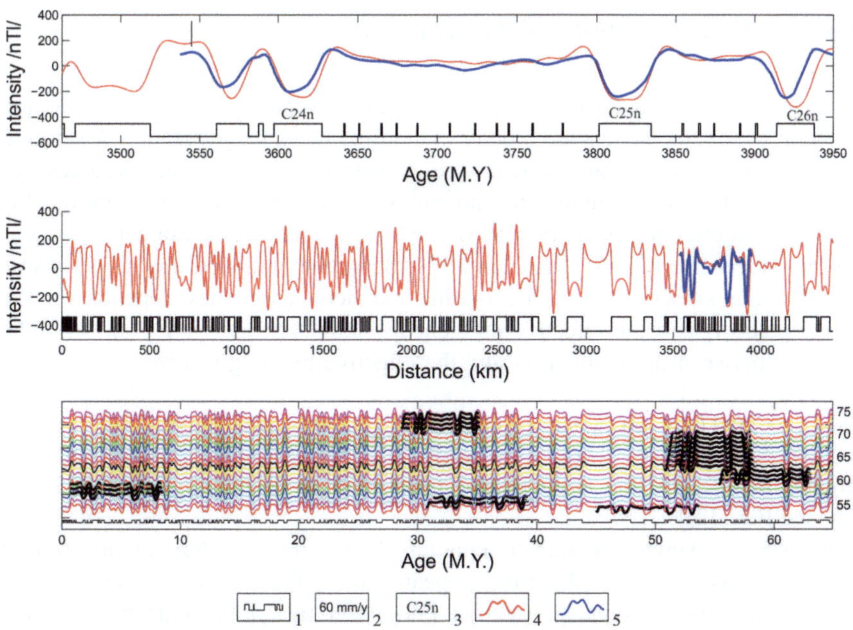

Fig. 32.2 Identification of the anomalies. Magnetization corresponding to the scale CK95 [Cande, S.C. and D.V. Kent], the spreading rate (2), polarity chron denotation (3), synthetic magnetic field (4), the observed profile from the Arabien basin (5). In the middle is the classical VM model whereas in the top and the bottom are smoothed ones

(red) illustrate the goodness of fit for the technique used for marine magnetic anomaly identification.

Determination of the Interval Spreading Rate

In the process of identification we obtain the map of the axes of the anomalies in the Arabian and Eastern Somalian basins and the locations of the reversals (edges of the chrons). Therefore we can estimate the interval spreading rate as a function of the age of the chrones. In the bottom of Fig. 33.3 is the magnetic scale and in the top there are the interval spreading rates for Arabian and Somali basins.

We see that the spreading rate is relatively stable for these ages. Comparing the spreading rates in both basins show that between 53.5 and 58 Ma the spreading rate in the Arabian basin is higher than in the Somalian one. This asymmetry was connected with the subduction zones (Jagoutz et al. 2015) or with the Deccan-Reunion plume (Cande and Stegman 2011). It is possible that the mechanisms give an equal contribution in this asymmetry. The variation of the interval spreading rate near the chron C24 is more complicated what probably connected with the rift propagating.

Skewness Determination of the Anomalies

The shape of marine magnetic anomalies depends on both the palaeolatitude and orientation of the magnetic source body at the time of its magnetization. Changing the palaeo-position of the magnetic source leads to the change of the skewness, or shape asymmetry, of the magnetic anomalies. In the Fourier representation this corresponds to the phase shift by $-\theta$, where θ is the skewness parameter.

In the papers (Bott 1967; Schouten 1971; Schouten and McCamy 1972) it was found that the skewness θ of the marine magnetic anomalies depends on the effective paleo I'_r and the effective ambient I'_0 inclination as $\theta = I'_r + I'_0 - \pi$. The connection between the inclination and the effective one is given by

$$\tan I'_0 = \tan I_0 / \sin(A - D_0), \tan I'_r = \tan I_r / \sin(A - D_r)$$

where D_0 and D_r are the ambient and the remanent declinations, and A is the strike, see Fig. 32.4.

Therefore the skewness parameter contains the information about the location of the paleomagnetic pole in the time appearance of the anomaly. The effective inclination is a useful concept since it eliminates the ambient declination, the remanent declination, and the strike from the equation for θ. It is important that the skewness is independent of depth, thickness of the layer, and the intensity of magnetization. Evidently, that the amplitude is dependent on the depth and the thickness of the layer.

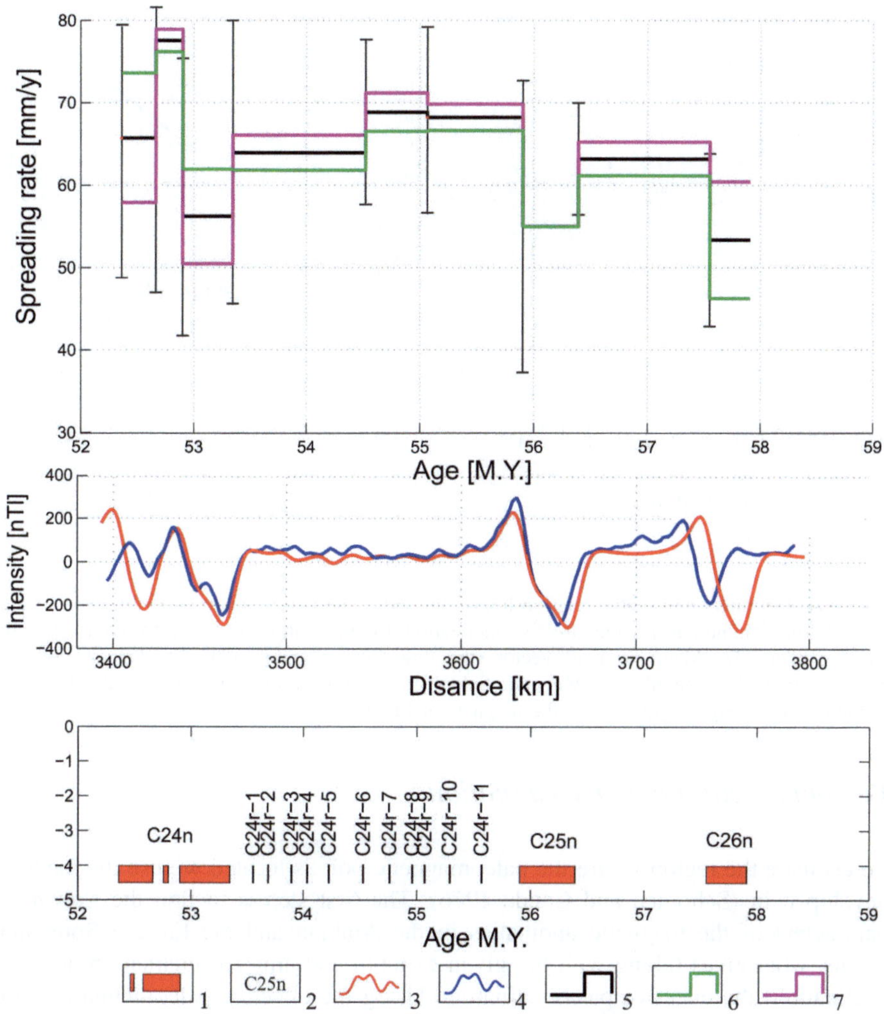

Fig. 32.3 Interval half-spreading rate variation versus time in the conjugate Arabian and Eastern Somali bassins magnetic layer in the standard spreading model with blocks of direct polarity colored red (1), polarity chron denotation (2), model magnetic field (3), and observed profile from the Somali basin (4). The average interval half-spreading rate (5), the interval half-spreading rate in the Somalian basin (6), the interval half-spreading rate in the Arabien basin (7)

To find the skewness of observed anomalies we apply the inverse phase filter and use several methods to find the most reliable estimate (Yoshida 1985; Arkani-Hamed 1991; Roest et al. 1992) (Fig. 32.5). The soft testing on the simulated profiles shows that the best method is to find the maximum of the correlation of the given profile and the profiles simulated with different values of the skewness parameter. Note that the visual evaluation of the similarity of the profiles usually gives satisfactory results.

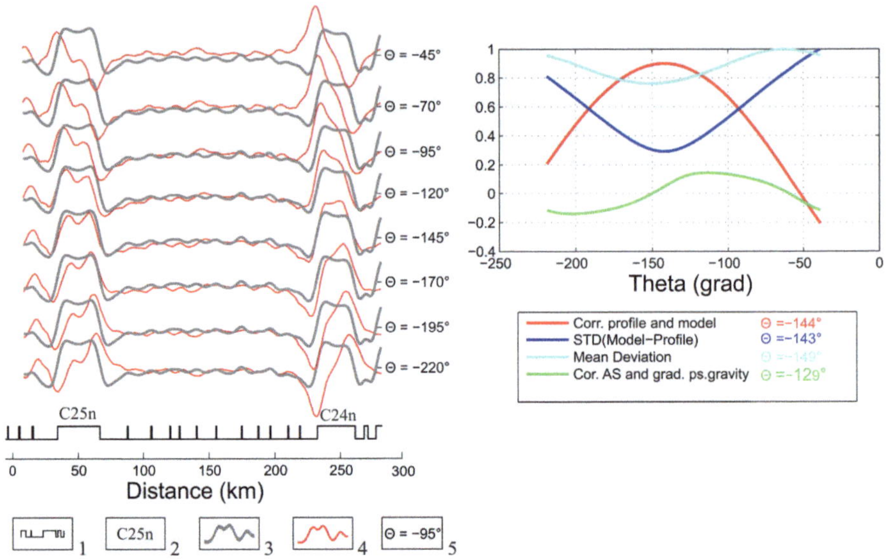

Fig. 32.4 Geometry and spatial relationship of the body. The x-axis is normal to the strike of the body. I—the inclination, I′—the effective inclination, the inclination of the vector component in the x-y plane, B—Magnetic field vector (the direction coincides with the direction of the magnetization), B′—the effective Magnetic field vector (the direction coincides with the direction of the effective magnetization), A—the magnetic azimuth

Paleomagnetic Pole Determination

To estimate the region where the paleomagnetic pole is located we use the method developed in (Schouten and Cande 1976). The first step is to find the skewness parameters of the magnetic anomalies in the Arabian and the Eastern Somalian basins, what gives (along with the given azimuth and ambient inclination and the declination) the effective paleoinclination. Using the effective paleoinclinations of pairs of the conjugate anomalies from these basins we find the locations of the paleomagnetic poles. The point is that a single anomaly gives semi-great circle as the locus of the magnetic pole. To find a unique pole it is necessary to determine the skewness for two conjugate anomalies. The two resulting semi-great circles should intersect. Because the skewness can be found with an error, the locus of the magnetic pole for a single anomalies is the region between two intersecting semi-great circles. For a pair of anomalies we have a quadrangle (Figs. 32.5 and 32.6).

We have obtained the location of the magnetic paleopole for the anomalies A24–A26 what give the average effective paleolatitude of the Carlsberg ridge. Our new estimates is 23°S, what differ from the published results. One of the first estimates

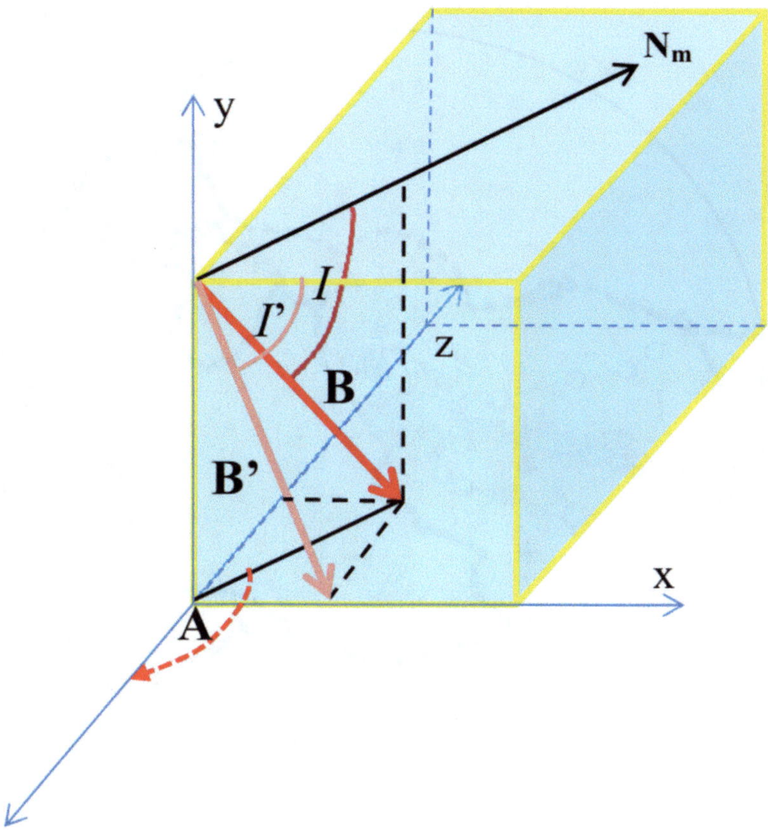

Fig. 32.5 The example of reduction to the pole. The magnetic anomalies after the phase filtering with the step 25°. The magnetization corresponding to the scale CK95 [Cande, S.C. and D.V. Kent] (1), the polarity chron denotation (3), synthetic anomalies from a standard block model computedat the pole are black (4), the observed profile for the given shift is red (4), the shift is on the right (5). On the right the application of several methods is shown: correlation coefficient between the model and the observed profiles (Arkani-Hamed 1991) (red), the root mean square error of the difference (blue), (Yoshida 1985) (aquamarine), and cross-correlation between absolute values of the analytic signal and the horizontal derivative of the pseudo-gravity (green) (Roest et al. 1992)

of the paleolatitude of the Carlsberg ridge is 10°S (McKenzie and Sclater 1971). In our earlier investigation of the mountains in the Eastern Somalian basin we have obtained 5°S (Brusilovsky 1991) and the paleographic reconstruction in (Besse and Courtillot 1988) is 17°S.

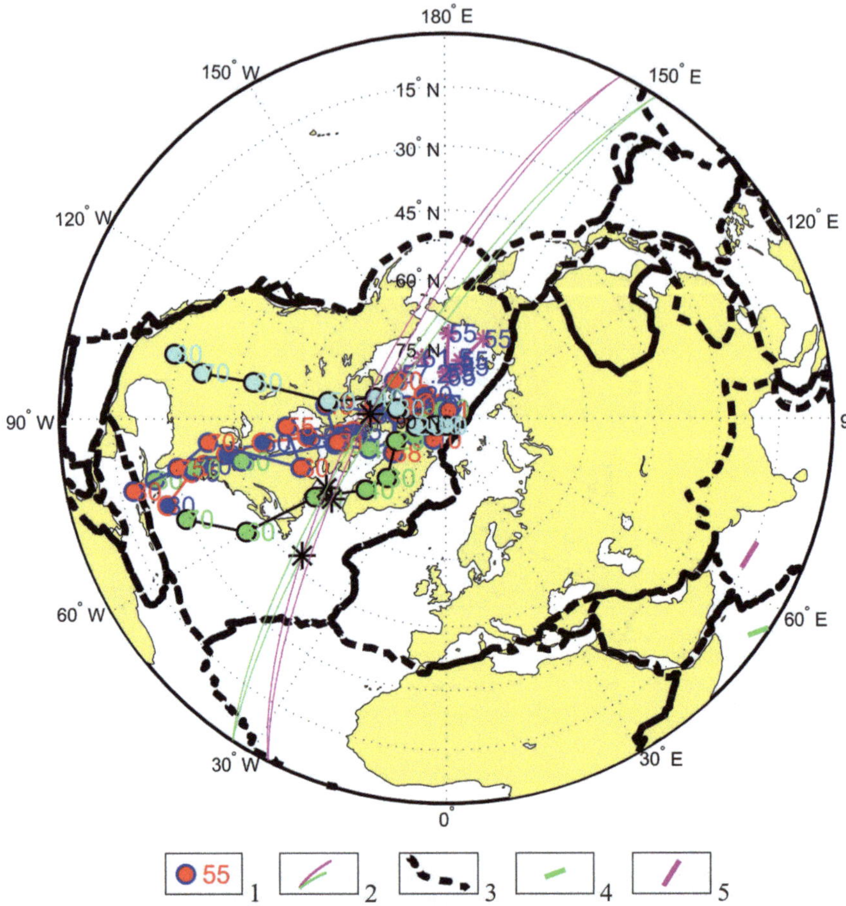

Fig. 32.6 Comparison between several apparent polar wander paths for India. (each of which has its own color of marker and line: green/black (Klootwijk and Peirce 1979), green/blue (Besse and Courtillot 1991), blue/red (Besse and Courtillot 2002), red/blue (Schettino and Scotese 2005), cyan/black (Torsvik et al. 2012)) and the location of the pole by spherical two angles for the anomalies A24–A26 (stars). The path and the location of the pole at the given age in Ma (1), the sides of the spherical two angles (2), the boundaries of the plates (3),the ancient anomaly in the Somalian basin (4), the ancient anomaly in the Arabian basin (5)

Conclusion

The geohistorical and paleomagnetic study of the marine magnetic anomalies in the northwestern part of the Indian Ocean give the following preliminary results about the ancient evolution of this region. We identify the anomalies A24–A26 using the reliable algorithm. We find the edges of the chrones and some cryptochrones (tiny wigles). This allow us to find the interval spreading rate, what is relatively stable in this period.

We show that spreading rates between 53.5 and 58 Ma is higher in the Arabian basin than in the Somalian one what is observed also in (Chaubey et al. 2002).

Also we find the skewness of the anomalies and obtain the estimate of the paleomagnetic poles.

Acknowledgements This work was supported by the Russia Foundation of Basic Research grants 15-05-06292. Figures were drafted using GMT software (Wessel and Smith 1991).

References

Aitchison, J. C., J. R. Ali, and A. M. Davis When and where did India and Asia collide?, J. Geophys. Res., 2007, 112, B05423, https://doi.org/10.1029/2006jb004706.

Arkani-Hamed, J., Thermoremanent magnetization of the oceanic lithosphere inferred from a thermal evolution model: Implications for the source of marine magnetic anomalies, Tectonophysics, v. 192, 1991, pp.81–96.

Besse, J., and V. Courtillot, Paleogeographic maps of the continents bordering the Indian Ocean since the Early Jurassic. Journal of Geophysical Research, 1988, v. 93, p. 11791–11808.

Besse, J., and V. Courtillot, Revised and Synthetic apparent polar wander path of the African, Eurasian, North American and Indian plates, and True polar wander since 200 Ma, J. Geophys. Res., 96, 4029–4050, 1991.

Besse, J., Courtillot, V.,.Apparent and true polar wander and the geometry of the geomagnetic field over the last 200 Myr, J. geophys. Res., 2002, 107(B11), 2300, https://doi.org/10.1029/2000jb000050.

Bott M.H.P. Solution of the linear inverse problem in magnetic interpretation with application to oceanic magnetic anomalies // Geophys, J. R. Astr. Soc. V. 13 P. 313. 1967.

Brusilovskii Yu. V., Gorodnitskii A.M., Merkur'ev S.A., Sochevanova N.A., Study of the see mountains of the Somalian basin by magnetic modeling, Doklady AN SSSR, 1991,v.316, n.4, pp. 868–871.

Cande, S. C. Stegman, D. R. Indian and African plate motions driven by the push force of the Reunion plume head. Nature, 475, 47–52 (2011).

Cande, S.C. and D.V. Kent, Revised calibration of the geomagnetic polarity timescale for the late Cretaceous and Cenozoic J. Geophys. Res., 100, 6,093–6,095, 1995.

Chaubey, A.K., Bhattacharya, G.C., Murty, G.P.S., Srinivas, K., Ramprasad, T. and Gopala Rao, D., 1998, Early Tertiary seafloor spreading magnetic anomalies and paleo-propagators in the northern Arabian Sea, Earth Planet. Sci. Lett. 154: 41–52.

Chaubey, A.K., Dyment, J., Bhattacharya, G.C., Royer, J.-Y., Srinivas, K., and Yatheesh, V.,. Paleogene magnetic isochrons and palaeo-propagators in the Arabian and Eastern Somali basins, NW Indian Ocean. In Clift, P.D., Kroon, D., Gaedicke, C., and Craig, J. (Eds.), The Tectonic and Climatic Evolution of the Arabian Sea Region. 2002, Geological Society Special Publication, 195(1):71–85.

Dyment, J., Evolution of the Carlsberg Ridge between 60 and 45 Ma: ridge propagation, spreading asymmetry, and the Deccan-Reunioon hotspot, J. Geophys. Res., 1998, 103: 24067–24084.

Jagoutz O., Royden L., Holt A.F., Becker T. W. Anomalously fast convergence of India and Eurasia caused by double subduction, Nature Geoscience, 2015, 8 (6), 475–478.

Klootwijk, C.T., and Peirce, J.W. India's and Australia's pole path since the late Mesozoic and the India-Asia collision. Nature, 1979, v. 282, pp. 605–607.

M. Seton, Whittaker J., P. Wessel, R. Müller, C. DeMets, S. Merkuryev, S. Cande, C. Gaina, G. Eagles, R. Granot, J. Stock, N. Wright, S. Williams, Community infrastructure and repository for marine magnetic identifications, Geochem. Geophys.Geosyst., 2014, Vol., 15, Issue 4, pp. 1629–1641, https://doi.org/10.1002/2013gc005176.

McKenzie, D. and Sclater, J. G. The Evolution of the Indian Ocean since the Late Cretaceous, Geophysical Journal International, 1971,v. 24, no. 5, pp. 437–528, https://doi.org/10.1111/j.1365-246x.1971.tb02190.x.

Merkouriev S.A., Sotchevanova N.A. The Carlsberg Ridge at the old and modern stage of opening, Bulletin of Kamchatka Regional Association "Educational-Scientific Center". Earth Sciences. ISSUE №11.№1. 2008, pp. 113–124.

Merkouriev S.A. and N.A.Sotchevanova Structure and evolution of the Carlsberg Ridge: Evidence for non-stationary spreading on old and modern spreading centers. Current Science, 2003b, Vol. 85, NO. 3, pp. 334–338.

Merkouriev, S., DeMets, C., 2006. Constraints on Indian plate motion since 20 Ma from dense Russian magnetic data: implications for Indian plate dynamics, Geochem. Geophys. Geosyst., 7, Q02002, https://doi.org/10.1029/2005gc001079.

Miles P.R, Munschy M, Segoufin J., Structure and early evolution of the Arabian Sea and East Somali Basin. Geophysical Journal International, 1998, 134 (3), 876–888.

Miles, P.R., and W.R. Roest, Earliest seafloor spreading magnetic anomalies in the North Arabian Sea and the ocean-continent transition, Geophysical Journal International, 1993, 115, pp. 1025–1031.

Roest, W.R., Arkani-Hamed, J., Verhoef, J. The seafloor spreading rate dependence of the anomalous skewness of marine magnetic anomalies, 1992, Geophysical Journal, 1992, vol. 109, issue 3, pp. 653–669.

S.A. Ivanov and S.A. Merkuryev. Some remarks on resolving and interpretation of short-wavelength marine magnetic anomalies. Proceedings of the 11th International School and Conference "Problems of Geocosmos",SPb.: VVM Publishing, 2016, pp. 37–44.

Schettino, A., and C.R. Scotese, 2005. Apparent polar wander paths for the major continents (200 Ma - Present Day): A paleomagnetic reference frame for global plate tectonic reconstructions, Geophysical Journal International, 163(2), 1–33.

Schouten H, Cande S.C. Palaeomagaetic Poles From Marine Magnetic Anomalies // Geophysical Journal International // V. 44. Issue 3. P. 567–575. 1976.

Schouten H., McCamy K. Filtering marine magnetic anomalies // J. Geophys. Res.V. 77. P. 7089–7099. 1972.

Schouten J.A.A. fundamental analysis of magnetic anomalies over oceanic ridges // Marine Geophysical Researches. V. 1. Issue 2. P. 111–144.1971.

Torsvik, T.H., et al. Phanerozoic polar wander, palaeogeography and dynamics. Earth-Science Reviews, (2012). 114, 325–368.

Wessel, P., and Smith W.H.F. Free software helps map and display data, Eos Trans. 1991, AGU, 72, 441–44.

White L.T., Lister G.S. The collision of India with Asia. Journal of Geodynamics 56–57 (2012) 7–17.

Yang T., et al., New insights into the India – Asia collision process from Cretaceous paleomagnetic and geochronologic results in the Lhasa terrane. Gondwana Research 28 (2015), 625–641.

Yoshida, K.K., Analysis of the parameter θ using mean deviation of anomaly patterns of marine magnetic lineations. Journal of Geomagnetism and Geoelectricity, 1985, v37, pp. 443–454.

Chapter 33
Geomagnetic Variations for Axial Dipole Hypothesis and Dynamo Parameters

S. V. Starchenko

Abstract A coarse observational/paleomagnetic confirmation is presented for the central hypothesis of paleomagnetism stating the predominance of axial dipole under averaging. The confirmation is based on the averaging time smaller than the full periods for axial dipole and larger than for equatorial one. The dipole-related average geodynamo scales are re-estimated as 18 km from the IGRF's dipole 1905–2020 variations. That leads to too small convection and geodynamo power estimations based on the known geodynamo-like scaling laws, while obtained here smallest magnetic and average hydrodynamic 7 km scale gives realistic convection power 0.4 TW.

Keywords Geomagnetic dipole · Geomagnetic and paleomagnetic variations
Geodynamo scale

Averaged geomagnetic dipole variations are about 10 μT/ka from the observations and archeo/paleo magnetic estimations. Corresponding decay/grow time is a few ka for the axial component of the dipole and a few hundreds years for the equatorial components. Assuming harmonically periodic behavior here, a coarse observational/paleomagnetic confirmation is presented for the central hypothesis of paleomagnetism stating the predominance of axial dipole under averaging. The confirmation is based on the averaging time smaller than the full periods for axial dipole and larger than for equatorial one. Thus, the equatorial or non-axisymmetric contribution to dipole is averaged to zero, while only the axisymmetric or central dipole remains. This averaging time could be up to a few ka for the inversion/excursion period and up to a few hundreds ka during stable polarity period.

The dipole-related geodynamo scales are re-estimated from the IGRF's dipole 1905–2020 variations following (Starchenko 2015). The average scale is corrected

S. V. Starchenko (✉)
Pushkov Institute of Terrestrial Magnetism, Ionosphere and Radio Wave Propagation, Russian Academy of Sciences (IZMIRAN), Kaluzhskoe Hwy 4, Moscow 108840, Troitsk, Russia
e-mail: sstarchenko@mail.ru

D. Nurgaliev et al. (eds.), *Recent Advances in Rock Magnetism, Environmental Magnetism and Paleomagnetism*, Springer Geophysics,
https://doi.org/10.1007/978-3-319-90437-5_33

to 3-times smaller value 18 km. That leads to too small convection and geodynamo power estimations based on the known after Christensen (2010a) and Starchenko (2014) geodynamo-like scaling laws. Thus, those laws are not supported by presented here observational estimations that are also in agreement with the recent detailed dynamo models (Cheng 2016; Aubert et al. 2017).

The obtained here 1905–2020 evolutions of the inverse geodynamo scales have minima and maxima corresponding to the smallest 7 km scale. This scale could be related to the average pure hydrodynamic scale that gives us the observational estimate of the realistic convection power 0.4 TW via pure hydrodynamic scaling laws well-known and well-confirmed after Rhines (1975). Corresponding average velocity ~ 1 mm/s coincides with the westward drift velocity of the magnetic heterogeneities in the liquid core of the Earth. Those estimated values are also compatible with the local turbulent and global time intervals characteristics of jerks, secular and paleomagnetic variations including reversals.

Axial Dipole Hypothesis Justifications and Averaging Times

A following (see text in italic) formulation of the central paleomagnetic axial dipole hypothesis is from (Khramov et al. 1982). *The geomagnetic field averaged over a relatively small interval in the geological time scale is the field of a central axial magnetic dipole whose axis coincides with the axis of rotation of the Earth. The geometric configuration of the magnetic field of such a dipole has a special feature for tectonic applications—the inclination of the geomagnetic field is determined by the latitude of the site.* Thus, the first rather practical goal of this paper is to give some simple justification to this hypothesis from the known variations and to determine a bit more precisely "*small interval in the geological time scale*". The directly observed (Jackson et al. 2000) averaged variation of the magnetic dipole is about 10 μT/ka, which agrees with the archaeomagnetic and paleomagnetic estimations (Macouin et al. 2004; Korte and Constable 2011; Ziegler et al. 2011; Johnson and McFadden 2015). Dividing the observed axial dipole value ~ 30 μT by ~ 10 μT/ka, we obtain a few ka, while an order of magnitude smaller equatorial dipole component gives a few hundred of years. Corresponding decay/grow time is a few ka for the axial component of the dipole and a few hundred of years for the equatorial components. Roughly assuming harmonic periodicity, a coarse observational/paleomagnetic confirmation is presented in Fig. 33.1 for this central hypothesis of paleomagnetism stating the predominance of axial dipole under averaging.

The confirmation is based on the averaging time smaller than the full periods for axial dipole and larger than for equatorial one. Thus, the equatorial or non-axisymmetric contribution to dipole is averaged to zero, while only the axisymmetric or central dipole remains. The averaging time could be up to a few ka for the inversion/excursion period and up to a few hundreds ka during stable polarity period. Correspondingly, the energy generation/dissipation processes for

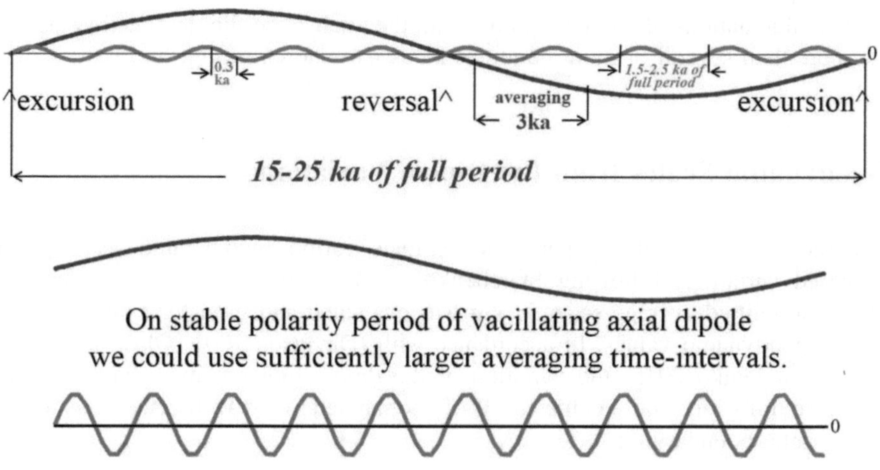

Fig. 33.1 Justifications to the axial dipole hypothesis are shown for 3 ka variation time of axial dipole and 0.3 ka for equatorial dipole. On the top—reversal and two nearest excursions for idealized harmonic oscillations. On the bottom—vacillating axial dipole and oscillating equatorial dipole

the non-symmetric components are by an order of magnitude faster than for the component that is symmetrical about the rotation axis of for the axial component. This conclusion also independently follows from the estimates of the westward geomagnetic drift and the change of the axial dipole over a few past hundred years, which have long since become classical (Jackson et al. 2000).

Let us coarsely assume that the considered generation/dissipation processes are strictly periodic. Then the harmonic period for the axial dipole component will be by an order of magnitude longer than the period for the components perpendicular to the rotation axis or equatorial components. Averaging of the fields over a time period that is significantly shorter than the axial period but longer than the equatorial period will yield practical zeroing of the equatorial components but retain the significant contribution of the axial component. Respectively, we obtain a coarse observational confirmation of the central hypothesis of paleomagnetism which states the predominance of the axial dipole under averaging (Khramov et al. 1982). Here, it should be noted that the cited estimates are not the harmonic periods; instead, they are the $\sim 1/e$ decay times (the decay to $\sim 1/3$ of the initial value). Hence, these exponential times should be increased by approximately a factor of 33.5–33.8 in order to obtain the duration of the full harmonic period as it is shown in Fig. 33.1.

Therefore, in the unstable polarity period the optimal averaging time is of order ka, whereas the considered time intervals should be longer in order to exceed the averaging time. These long time scales require the use of the paleomagnetic and/or geodynamo-like numerical and/or analytical models. Much longer averaging times appear in the case when the described coarse validation of the axial dipole hypothesis, instead of the full periodicity, relies on the so-called vacillations

(~10% fluctuations of the axial dipole with the unchanged dipole direction lasting for more than several or more hundreds ka as it is shown in Fig. 33.1).

Geodynamo Scales from the Dipole Variations

The second goal of this paper is to estimate various geodynamo scales from observed variations following Starchenko's (2015, 2016) papers where only a highly simplified average scale 57 km was evaluated and used for estimations of typical geodynamo values. In accordance with (Davidson 2001; Olson and Amit 2006; Starchenko 2015) the variation of the dipole moment **m** is linked to the magnetic diffusion (with coefficient λ) of geomagnetic field **B** directly beneath the core-mantle boundary G:

$$\frac{\partial \mathbf{m}}{\partial t} = -\frac{3\lambda}{2\mu_0}\int_G \hat{\mathbf{n}} \times \nabla \times \mathbf{B} d^2 r. \tag{33.1}$$

The Earth's dipole moment **m** is expressed through the standard Gaussian coefficients (g_1^1, h_1^1, g_1^0), Cartesian coordinates (x, y, z), and the radius of the planet R:

$$\mathbf{m} = \frac{4\pi}{\mu_0} R^3 (g_1^1 \hat{\mathbf{x}} + h_1^1 \hat{\mathbf{y}} + g_1^0 \hat{\mathbf{z}}). \tag{33.2}$$

I subdivide $\mathbf{B} = \mathbf{O} + \mathbf{V}$ into the observed potential part **O** ($\nabla \times \mathbf{O} = \mathbf{0}$) and the vortex part **V** ($\nabla \times \mathbf{V} \neq \mathbf{0}$) that is hidden in the core. Using (33.1, 33.2), I define the averaged radial derivatives (B'_X,B'_Y,B'_Z) of the vortex field through the variations of the dipole's components (R_c is the radius of the Earth's core):

$$B'_X \equiv \int_0^{\pi}\int_0^{2\pi} \sin\theta \left(\cos\theta\cos\phi \frac{\partial V_\theta}{\partial r} - \sin\phi \frac{\partial V_\phi}{\partial r} \right) \frac{d\theta d\phi}{4\pi} = \frac{2R^3}{3\lambda R_c^2}\frac{dg_1^1}{dt}, \tag{33.3}$$

$$B'_Y \equiv \int_0^{\pi}\int_0^{2\pi} \sin\theta \left(\cos\theta\sin\phi \frac{\partial V_\theta}{\partial r} + \cos\phi \frac{\partial V_\phi}{\partial r} \right) \frac{d\theta d\phi}{4\pi} = \frac{2R^3}{3\lambda R_c^2}\frac{dh_1^1}{dt}, \tag{33.4}$$

$$B'_Z \equiv \int_0^{\pi}\int_0^{2\pi} \sin^2\theta \frac{\partial V_\theta}{\partial r}\frac{d\theta d\phi}{4\pi} = -\frac{2R^3}{3\lambda R_c^2}\frac{dg_1^0}{dt}. \tag{33.5}$$

Estimations of the geodynamo scales require a value C of the magnetic field which should be divided by the radial derivatives from (33.3–33.5) in order to receive those scales. Previously in (Starchenko 2015) I used single C = 262 μT that is the known root-mean-square dipole value at core-mantle boundary. Here I prefer (the arguments are below) to use the averaged at $r = R_c$ as in (33.5) value of the axial dipole field:

$$C(t) = -\int_0^{\pi}\int_0^{2\pi} \sin^2\theta\,\hat{\theta}\cdot\nabla\left(\frac{R^3}{r^2}g_1^0\cos\theta\right)\frac{d\theta d\phi}{4\pi} = -\frac{2R^3}{3R_c^3}g_1^0(t). \quad (33.6)$$

The potential of the axial dipole field is here above in brackets and $C(1980) = 124$ μT for example.

The first argument for using C from (33.6) is that the corresponding axial dipole field is dominated not only at modern epoch but also in past due to the justified in the previous section axial dipole hypothesis. The second argument is that it is best of all to define this field value in the similar way as the derivatives were determined in (33.3–33.5). While in third—averaging of the axial dipole field in accordance with (33.3, 33.4) and (33.3–33.5)-type averaging of more complicated fields will preferably lead to zero or very small values. Therefore using (33.3–33.6) I express the inverse geodynamo scales $1/X$, $1/Y$ and $1/Z$ as

$$1/X = \frac{B'_X}{C} = -\frac{R_c}{\lambda g_1^0}\frac{dg_1^1}{dt} \quad (33.7)$$

$$1/Y = \frac{B'_Y}{C} = -\frac{R_c}{\lambda g_1^0}\frac{dh_1^1}{dt} \quad (33.8)$$

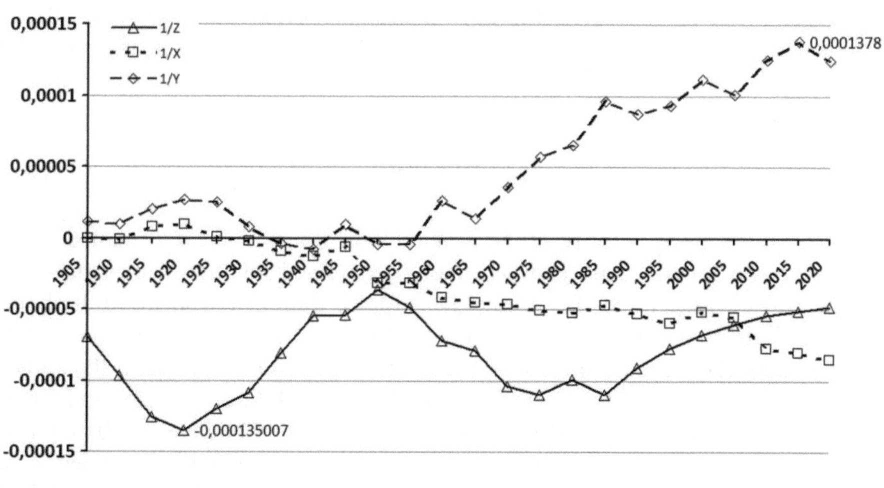

Fig. 33.2 The evolutions of the geodynamo inverse scales (33.7–33.9) directly under core-mantle boundary are shown (in 1/m) for the vortex geomagnetic field derived from IGRF 1905–2020 dipole variations

$$1/Z = \frac{B'_Z}{C} = \frac{R_c}{\lambda g_1^0}\frac{dg_1^0}{dt} \tag{33.9}$$

To specify the values in (33.7–33.9) I use the modern generally accepted International Geomagnetic Reference Field model (IGRF) for 1900–2020 which is described in detail on its official website http://www.ngdc.noaa.gov/IAGA/vmod/igrf.html. The value of magnetic diffusion λ = 0.8 m^2/s is taken from (Gomia et al. 2013) where it is determined from the experiments and first principles. It is worth noting that even before the pioneering works (Pozzo et al. 2012) on this determination; it was a common practice to use the severalfold higher λ values and the correspondingly strongly underestimated values for the adiabatic heat flows. The resulting evolutions of those geodynamo inverse scales are in Fig. 33.2.

Finally, hopefully correct average geodynamo scale S is estimated in the following obvious way:

$$S \equiv \left(\sum_{i=1}^{24} \frac{1/|X_i| + 1/|Y_i| + 1/|Z_i|}{72}\right)^{-1} = 18 \text{ km} \tag{33.10}$$

Here, index i enumerates each 5-year value of the inverse scales from (33.7–33.9) and Fig. 33.2.

Magnetic and Hydrodynamic Parameters of Geodynamo

The third goal of this paper is to evaluate geodynamo parameters from the observed variations. Let's try to use the average magnetic scale S from (33.10) in accordance with scaling laws from (Christensen 2010a) and (Starchenko and Pushkarev 2013b; Starchenko 2014). The last paper links the magnetic scale to the specific power of the convective or buoyancy transfer F that drives geodynamo and the F to S link in our notations could be written as

$$F = R_c^2 \Omega^3 (S/R_c)^{15/2} = 7.1 \cdot 10^{-17} \text{ W/kg} \tag{33.11}$$

Here, Ω is angular velocity of the rotation which, just as the core radius R_c, is exactly known for the Earth and is reliably determined for the other planets and moons. However the value of F from (33.11) for the Earth is too small. Multiplying it on the Earth's core mass we obtain the total convective geodynamo power P as small as 200 MW. That is three-four orders of magnitude smaller than the values from the known estimations (Braginsky and Roberts 1995; Starchenko and Jones 2002, 2013a; Christensen 2010b). Thus, the laws of (Christensen 2010a) and (Starchenko and Pushkarev 2013b, 2014) are not supported by presented here observational estimations that are also in agreement with the recent detailed dynamo models (Cheng and Aurnou 2016; Aubert et al. 2017).

The obtained here above 1905–2020 evolutions of the inverse geodynamo scales (see Fig. 33.2) have minima (−0.000135/m) and maxima (0.000138/m) corresponding to the smallest scale

$$H = 7\ \text{km}. \tag{33.12}$$

This scale could be related to the average hydrodynamic geodynamo scale (King and Buffett 2013). That gives the observational estimate of the realistic geodynamo convection power (M is mass of the planetary core)

$$P = FM = 0.4\,TW. \tag{33.13}$$

via pure hydrodynamic well-known and well-confirmed after Rhines (1975) scaling law

$$H = (FR_c^3/\Omega^3)^{1/5}. \tag{33.14}$$

Corresponding (to F = 0.2 pW/kg just obtained via H = 7 km) average velocity in the Earth's core

$$u = (F^2R_c/\Omega)^{1/5} = 1\ \text{mm/s}. \tag{33.15}$$

This coincides with the westward drift velocity of the magnetic heterogeneities in the liquid core of the Earth. Those estimated values are also compatible with the local turbulent and global time intervals characteristics of jerks, secular and paleomagnetic variations. The shortest time interval obtained from (33.12) and (33.15) is

H/u = 3 months,

which probably gives the minimal locally-turbulent hydrodynamic interval of the geodynamo. This interval corresponds to the turnover time of the convective cell. The next (by the length) time interval

S/u = 1 year

gives the minimal locally-turbulent interval of magnetic field which probably manifests itself in the correspondingly localized jerks. The global value

R_c/u = 100 years

coarsely specifies the order of magnitude of the secular geomagnetic variations.

The similar time intervals can also be determined on the side of the dissipation with the use of the coefficient of magnetic diffusion λ. The shortest interval

H^2/λ

is equal to the mentioned 1 year, and

S^2/λ

corresponds to decades of secular variation. The global diffusion time

$(R_c)^2/\lambda$

as long as half million years corresponds to the average time gaps between the reversals (Ogg et al. 2008). Between the reversal and secular time intervals there are intervals caused by the global nonlinear geodynamo mechanisms. These intervals can be roughly estimated as geometrical mean of the values estimated above. The time intervals starting from 1 Ma are associated with the viscous-diffusive transport ($\sim$ 1–10 Ma), effects of the mantle (10–100 Ma), and energy evolution of the whole core–mantle system (starting from 100 Ma).

Conclusions

Let us summarize the key results of this work.

1. Averaged geomagnetic dipole variations are about 10 μT/ka from the observations and archeo/paleo magnetic estimations. Corresponding decay/grow time is a few ka for the axial component of the dipole and a few hundreds years for the equatorial components. Assuming harmonically periodic behavior here, a coarse observational/paleomagnetic confirmation is presented for the central hypothesis of paleomagnetism stating the predominance of axial dipole under averaging. The confirmation is based on the averaging time smaller than the full periods for axial dipole and larger than for equatorial one. Thus, the equatorial or non-axisymmetric contribution to dipole is averaged to zero, while only the axisymmetric or central dipole remains. This averaging time could be up to a few ka for the inversion/excursion period and up to a few hundreds ka during stable polarity period, see Fig. 33.1 for details.
2. The dipole-related geodynamo scales are re-estimated (see Fig. 33.2) from the IGRF's dipole 1905–2020 variations following (Starchenko 2015). The average scale is corrected to 3-times smaller value 18 km. That leads to too small convection and geodynamo power estimations based on the known after Christensen (2010a) and Starchenko (2014) geodynamo-like scaling laws. Thus, those laws are not supported by presented here observational estimations that are also in agreement with the recent detailed dynamo models (Cheng and Aurnou 2016; Aubert et al. 2017).
3. The obtained here 1905–2020 evolutions of the inverse geodynamo scales have minima and maxima corresponding to the smallest 7 km scale, see Fig. 33.2. This scale could be related to the average pure hydrodynamic scale that gives us the observational estimate of the realistic convection power 0.4 TW via pure hydrodynamic scaling laws well-known and well-confirmed after Rhines (1975). Corresponding average velocity $\sim$ 1 mm/s coincides with the westward drift velocity of the magnetic heterogeneities in the liquid core of the Earth. Those estimated values are also compatible with the local turbulent and global time intervals characteristics of jerks, secular and paleomagnetic variations including reversals.

Acknowledgements The author is grateful to Valeriy Prohorovich Shcherbakov and Dmitry Dmitrievich Sokoloff for their reasonable criticism resulting in sufficient improving of this paper. The work on it was partly (estimations of magnetic geodynamo parameters) supported by the Russian Foundation for Basic Research with project No 16-05-00507a.

References

Aubert J., Gastine T., Fournier A., Spherical convective dynamos in the rapidly rotating asymptotic regime // J. Fluid Mech. V. 813. P. 558-593. 2017.

Braginsky S.I., Roberts P.H. Equations governing convection in the Earth's core and the geodynamo // Geoph. Astroph. Fluid Dyn. V. 79. P. 1–97. 1995.

Cheng J.S., Aurnou J.M., Tests of diffusion-free scaling behaviors in numerical dynamo datasets // Earth and Planetary Science Letters. V. 436. P. 121–129. 2016.

Christensen U.R. Dynamo Scaling Laws and Applications to the Planets // Space Sci Rev. V. 152, P. 565-590. 2010a.

Christensen U., Aubert J., Hulot G. Conditions for Earth-like geodynamo models // Earth Planet. Sci. Lett. V. 296. P. 487–496. 2010b.

Davidson P.A. An Introduction to Magnetohydrodynamics, Cambridge University Press, 2001.

Gomia H., Ohtaa K., Hirosea K. et al. The high conductivity of iron and thermal evolution of the Earth's core // Phys. Earth Planet. Int. V. 224. P. 88-103. 2013.

Jackson A., Jonkers A. R. T., Walker M. R. Four centuries of geomagnetic secular variation from historical records // Phil. Trans. R. Soc. Lond. V. A358. P. 957–990. 2000.

Johnson C.L., McFadden P. The Time-Averaged Field and Paleosecular Variation. In: Gerald Schubert (editor-in-chief) Treatise on Geophysics, 2nd edition, V. 5. Oxford: Elsevier; 2015. P. 385-417.

King E.M., Buffett B.A. Flow speeds and length scales in geodynamo models: The role of viscosity // Earth Planet. Sci. Lett. V. 371. P. 156–162. 2013.

Khramov A.N., Goncharov G.I., Komisarova R.A., et al. Paleomagnetologia. L.: Nedra. 1982. 312 pp.

Korte M., Constable C. Improving geomagnetic field reconstructions for 0-3 ka // Phys. Earth Planet. Int. V. 188. P. 247–259. 2011.

Macouin M., Valet J.-P., Besse J. Long-term evolution of the geomagnetic dipole moment // Phys. Earth Planet. Int. V. 147. P. 239–246. 2004.

Ogg J.G., Ogg G., Gradstein F.M. The Concise Geologic Time scale // Cambridge University Press. 2008. 150 pp.

Olson P., Amit H. Changes in earth's dipole // Naturwissenschaften. V. 93. P. 519-542, 2006.

Pozzo M., Davies C., Gubbins D., Alfe D. Thermal and electrical conductivity of iron at Earth's core conditions // Nature. V. 485. P. 355–358. 2012.

Rhines P.B. Waves and turbulence on a beta plane // J. Fluid Mech. V. 69. P. 417–433. 1975.

Starchenko S.V. Energetic Scaling of the Geodynamo // Geomagnetism and Aeronomy. V. 53. No 2. P. 243-247. 2013a.

Starchenko S.V. Analytic base of geodynamo-like scaling laws in the planets, geomagnetic periodicities and inversions // Geomagnetism and Aeronomy. V. 54. No 6. P. 694-701. 2014.

Starchenko, S.V., Observational estimate of magnetic field and geodynamo parameters under the surface of the Earth's core // Geomagnetism and Aeronomy. V. 55. No 5. P. 712-718. 2015.

Starchenko S.V. Hypothetical parameters of planetary dynamos deduced from direct observations, scaling laws, paleomagnetic and isotope reconstructions // Proceedings of the 11th International School and Conference "Problems of Geocosmos" (Oct. 03-07, 2016, St. Petersburg, Russia), edited by V.S. Semenov et al. P. 64-72. 2016.

Starchenko S.V., Jones C.A. Typical velocities and magnetic field strengths in planetary interiors // Icarus. V. 157. P. 426–435. 2002.
Starchenko S.V., Pushkarev Y.D. Magnetohydrodynamic scaling of geodynamo and a planetary protocore concept // Magnetohydrodynamics. V. 49. No. 1. P. 35-42. 2013b.
Ziegler L., Constable C., Johnson C., Tauxe L. PADM2 M: a penalized maximum likelihood model of the 0-2 Ma palaeomagnetic axial dipole moment. Geophys. J. Int. V. 184. P. 1069–1089. 2011.

Chapter 34
Simple Estimations for Planetary Convection Turbulence and Dynamo Magnetism from Optimized Scaling and Observations

S. V. Starchenko

Abstract My optimal similarity factors allow arguing the planetary turbulent convection to be close to its critical level with turbulent transport coefficients higher than the molecular values. These strongly simplify numerical models for the primordial and recent convection and correspondent magnetic field of the planets and moons. Mean arithmetic magnetic field in the dynamo region is well defined by the observable magnetic dipole. The Earth, Jupiter and Saturn have that mean fields of order 1 mT and small inclination of their dipoles to the axis of rotation. Thus, similar dynamo models are successful for those planets. The Uranus and Neptune also have ~ 1 mT, but their dipoles are strongly inclined requiring some modifications in dynamo modeling. The major problems are posed by too small and asymmetric magnetic field in Mercury/Ganymede and absence of an active dynamo in Venus/Mars.

Keywords Planetary interiors · Convection · Turbulence · MHD dynamo
Magnetism

Introduction

The system for combined (compositional and thermal) planetary-type convection was optimally scaled in plain layer to receive the moderate values of similarity factors and the simplest representation for turbulent processes. Such optimal similarity factors first allow arguing that the planetary turbulent convection should be close to its critical level. At second—correspondent turbulent transport coefficients are by several orders of magnitude higher than the molecular values. These two

S. V. Starchenko (✉)
Pushkov Institute of Terrestrial Magnetism, Ionosphere and Radio Wave Propagation, Russian Academy of Sciences (IZMIRAN), Kaluzhskoe hwy 4, Troitsk, Moscow 108840, Russia
e-mail: sstarchenko@mail.ru

D. Nurgaliev et al. (eds.), *Recent Advances in Rock Magnetism, Environmental Magnetism and Paleomagnetism*, Springer Geophysics,
https://doi.org/10.1007/978-3-319-90437-5_34

reasons strongly simplify numerical models used for the successful approximation of the planetary convective and MHD dynamo systems. Such dynamos are only possible at sufficiently powerful thermal or/and compositional convection. Consequently, the generation and its' absence are considered for the primordial and recent convection and correspondent magnetic field of the terrestrial planets and moons.

The previous researchers perhaps overlooked simple facts that mean magnetic field vector in the dynamo region is already known once magnetic dipole of a planet/moon is known. This mean field module in 2.7 times exceeds the root-mean-square (rms) dipole field at the outer dynamo boundary, while it is typically lower than the total rms field in the dynamo region. The Earth, Jupiter and Saturn have that mean fields of order 1 mT and small inclination of their dipoles to the axis of rotation. Thus, similar dynamo models are successful for those planets. The Uranus and Neptune also have ~ 1 mT, but their dipoles are strongly inclined requiring some modifications in dynamo modeling. The major problems for modeling are posed by too small and asymmetric magnetic field in Mercury/Ganymede and absence of an active dynamo in Venus/Mars. Some possible solutions to those problems are outlined.

The satellite geomagnetic and planetary observations only cover a few decades; the observatory measurements span 185 years, whereas the reliable direct historical observations lasted less than 500 years (Jackson et al. 2000). On the natural geomagnetic time scales exceeding one thousand years only the archaeomagnetic (Korte and Constable 2005, 2011) and paleomagnetic methods (Khramov et al. 1982; Macouin et al. 2004; Tarduno et al. 2015) work. Unfortunately, all these methods provide far smaller amounts of information (kilobits against terabits of the satellite and observatory data). Moreover, their spatial and, especially, time resolution is not better than tolerable. Here, the solution has unexpectedly come from the numerical geodynamo models that appeared to be unrealistic almost up to the end of the 20th century. The situation has radically changed with the publication of (Glatzmaier and Roberts 1995) where, with the key parameters far from planetary MHD dynamo, the author managed to create the first-principle numerical model much similar to the geodynamo. To date, hundreds of this type models have been developed (Christensen 2006a; Christensen and Aubert 2006b, 2010a; Christensen et al. 2010b; Olson and Christensen 2006; Yadav et al. 2013; Matsui et al. 2016; Aubert et al. 2017; etc.). These models are widely used in modeling/forecasting the dynamics and morphology of both the observed and reconstructed planetary magnetic field. The paradox that numerical dynamo simulations producing an Earth-like magnetic field though their parameters are not all realistic have already been documented (e.g. Christensen et al. 2010b; Aubert et al. 2017). The researchers found important to have a good ratio between a few chosen timescales to get Earth-like dynamos, while here part 2 contributes to the justification of using the parameters those are not so far from the critical values. The planetary-type plain layer convection equations are optimally scaled in part 2 similar to Starchenko (2017a). This allows receiving a new result of this paper that is simple estimations of turbulent Reynolds (1894) type transport coefficients; those are significantly higher than the commonly used molecular values. This partially removes the deviation of the cited numerical models from the key parameters related to viscosity and diffusion. The other—perhaps most important inconsistency is, as well, here resolved by the fact that turbulent convection under such a fast rotation is not as much above its critical level as it was commonly believed so far.

Another simple idea of this work (see Section "Observed Dipoles and Mean Dynamo Field") is to relate the observed magnetic planetary dipole to the mean magnetic field vector hidden in the region where MHD dynamo is working. This tremendously simple (and exact!) relation (34.5) via textbook (e.g., Davidson 2006) formula was seems overlooked so far. The Earth, Jupiter and Saturn have those mean fields of order 1 mT and small inclination of their dipoles to the axis of rotation. No wonder similar dynamo models are successful for those planets (Christensen 2010a, b; Schubert et al. 2011; Starchenko 2014, 2017b; Starchenko and Pushkarev 2013b). The Uranus and Neptune also have ~ 1 mT, but their dipoles are strongly inclined (Ness 2010) that require some modifications in the dynamo modeling (Ruzmaikin and Starchenko 1991; Starchenko 1995; Christensen and Aubert 2006b; Schubert et al. 2011; Yadav et al. 2013). The major problems for such modeling are posed by too small and asymmetric magnetic field in Mercury/Ganymede and absence of an active dynamo in Venus/Mars (Christensen 2006a; Schubert et al. 2011; Breuer et al. 2015).

Basing on Sections "Optimized Scaling, Criticality and Turbulence for Convection" and "Observed Dipoles and Mean Dynamo Field" estimates are presented in Section "Magneto-Convection Estimations for the Terrestrial Planets" for the typical and critical convective parameters, for the degree of turbulence and the corresponding transport coefficients in the Earth's core and in the cores of the known terrestrial planets. The known (Breuer et al. 2015; Starchenko 2016, 2017b) scenarios of the convective support and correspondent dynamos in the Earth and presumably the Moon, Venus and Mars are discussed. Meanwhile Mercury and Ganymede dynamos could have special type of 'contradicting' convection that could be supplemented or/and dominated by strong tidal effects from Sun or Jupiter respectively.

The results of this work are presented in the Conclusions Section "Conclusions".

Optimized Scaling, Criticality and Turbulence for Convection

Following (Cheng et al. 2015; Cheng and Aurnou 2016), for reaching the extreme planetary parameters, I consider the combined convection in a plane layer based on (Starchenko 2017a). This layer with thickness D rotates with angular velocity Ω in the gravitational field with acceleration g anti-parallel to z-axis. The layer's liquid has kinematic viscosity ν, diffusion λ, diffusivity κ, coefficients of thermal and compositional expansion α and β [for the Earth's core, about $\pm 10^{-5}$/K and +0.6, respectively (Starchenko and Jones 2002; King and Buffett 2013; Terasaki and Fischer 2016)]. The initial temperature and concentration are $(D - z)T'$ and $(D - z)\Xi'$. As in the Earth, the convection is determined by the compositional effects with its Prandtl number μ. Thermal impact leads to additional instability (at $\alpha > 0$) or stabilize convection (at $\alpha < 0$). Using relatively too small (Cheng and Aurnou 2016; Terasaki and Fischer 2016) transport coefficients and very fast rotation Starchenko (2017a) represented typical velocity U, new (δ, r, s) and classical (E, Ra) similarity factors as

$$U = \Omega D\delta^2, \delta = \frac{\sqrt{g\beta\Xi'}}{\Omega} = E\sqrt{\frac{Ra}{\mu}}, r = \frac{\alpha T'}{\beta\,\Xi'}, s = \frac{\nu\,\Omega^2/D^2}{(g\beta\Xi')^{3/2}} = E^{-2}\left(\frac{\mu}{Ra}\right)^{3/2} \tag{34.1}$$

Here: r is a new factor for thermal/composition effect ratio, the commonly used classical Ekman number $E = \nu/(\Omega D^2)$ is $\sim 10^{-15}$ for the molecular viscosity in the Earth' core determined from the first principles (Pozzo et al. 2013). Whereas the classical Rayleigh number, $Ra = g\beta\,\Xi' D^4/(\nu\lambda)$, is determined by the compositional convection at even more extreme level, about 10^{27} for the commonly accepted molecular diffusion of the light admixture in the Earth's core (Breuer et al. 2015; Terasaki and Fischer 2016). With such extreme values of those key geodynamo similarity factors, there is even not a gleam of hope that the real geodynamo will be successfully modeled in the near future. The use of the new similarity factors in front of higher order derivatives $\delta \sim 10^{-3}$ (instead of Ekman number $E \sim 10^{-15}$) and $s \sim 10^{-6}$ (instead of $1/Ra \sim 10^{-27}$) from (34.1) appears to be more promising. In this approach, the solution is likely to be possible even for the real geodynamo parameters with the use of the sufficiently powerful computers. This simplified system could be easily generalized on the planetary spherical layer based on the approach that was first suggested in (Starchenko and Kotelnikova 2013a).

Following new results of this paper are based on already somewhat standard (Smagorinsky 1963; Frick 2010) and inspired by Reynolds (1894) estimation of unified turbulent transport coefficient

$$kUD\delta \approx \nu \approx \kappa \approx \lambda. \tag{34.2}$$

Here, k is the numerical coefficient that should not to be too small. The new optimization here is that due to (1–2) this k is merely equal to s. So, the limits for those numbers can roughly be estimated as

$$0.01 \leq k = s < s_{cr}. \tag{34.3}$$

Here, s_{cr} is a critical number. Convection exists at $s < s_{cr}$ only. For pure compositional ($r = 0$) and turbulent ($\mu = 1$) convection one obtain $s_{cr} = 0.04$ using simple plain-layer formula for critical (index 'cr') values from equation (44) in (Starchenko 2017a):

$$s_{cr}^2 = \frac{4}{27}\frac{(\mu + r\sigma)^3}{\pi^4} = \frac{\mu^3}{E^4 Ra_{cr}^3}, \tag{34.4}$$

where σ is thermal Prandtl number. In the real spherical shell geometry, more parameters and complexities are added those are determined by the convective instability profile and the relative inner/outer shell radiuses. However, for the typical terrestrial-type turbulent parameters s_{cr} is still about a few hundredths, see correspondent $Ra_{cr} \sim E^{-4/3}$ for example in (Starchenko et al. 2006; Starchenko and Kotelnikova 2013a) where terrestrial type convection was investigated in spherical shells. In this case, the supercriticality of turbulent convection is only a slightly (higher $\sim s_{\text{cr}}/s$ times) grater than its critical level. The last is perhaps one of the important argument in favor of the successful 22-year (after Glatzmaier 1995) numerical modeling of the geodynamo-like systems. Another important fact is that the coefficients of turbulent transport (34.2) are by several orders of magnitude higher than the corresponding molecular quantities. For example, at $s = 0.01$, turbulent viscosity/diffusion is $0.02\ \text{m}^2/\text{s}$ and turbulent Ekman number is $E = 3 \times 10^{-11}$. The last can be achieved in the nearest future for the known geodynamo-like systems with the use of a sufficiently powerful supercomputer. Whereas the corresponding simplified convective system with the real turbulent $s \approx 0.01$ and even molecular $s \approx 10^{-6}$ parameters of the geodynamo-like systems and other planets can in principle be numerically solved even now.

Observed Dipoles and Mean Dynamo Field

It is natural to define the vector $\overline{\mathbf{B}}$ of mean magnetic field in planetary MHD dynamo region W as correspondent volume integral divided by the spherical region's volume of W with radius r_c. Due to the textbook's definition of magnetic dipole **M** via internal magnetic field **B** shown in Fig. 34.1 (for example see this formula on page 173 in Davidson's 2006 textbook) this mean field vector is exactly equal to the observed planetary dipole vector **M** divided on the volume of W times fixed constant $3/2\mu_0$:

$$\overline{\mathbf{B}} \equiv \frac{\int_W \mathbf{B} d^3 r}{4\pi r_c^3/3} = \frac{2\mu_0 \mathbf{M}}{4\pi r_c^3}. \tag{34.5}$$

Thus, perhaps surprisingly for the first time, the observed magnetic dipole of a planet is tremendously simply related to the mean magnetic field hidden in the region where dynamo is working. This mean field exceeds the dipole field extrapolated to the outer dynamo boundary. The last root-mean-squared (rms hereafter) field B_M is 0.26 mT for the well-known modern dipole field at the Earth's core-mantle boundary $r = r_c = 3400$ km. Correspondent to (34.5) module of the total mean geodynamo field B = 0.7 mT is 2.7 times larger than B_M as it is shown

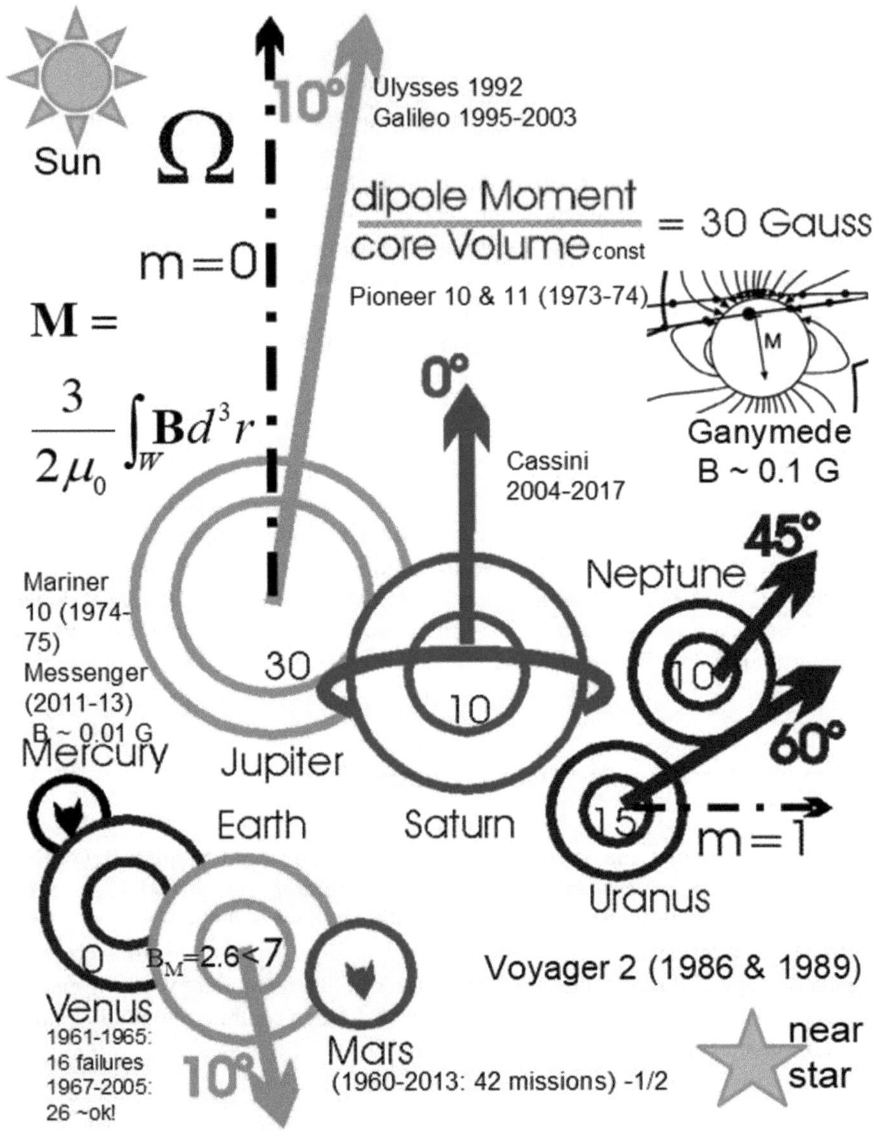

Fig. 34.1 The planetary/moon missions and observed by them own planetary magnetic dipoles are schematically shown together with the modules of the mean field B in MHD dynamo regions. Polar axis m = 0 is in direction of the rotation axis, while equatorial m = 1 axis is perpendicular. Dipole inclination angles are shown with degree marks

in Fig. 34.1. The same ratio will obviously be kept for any other planet or moon. This large mean field is typically yet lower than the total rms field in the dynamo region (Christensen 2006a; Christensen and Aubert 2006b, 2010a; Christensen et al. 2010b; Olson and Christensen 2006; Yadav et al. 2013; Matsui et al. 2016; Aubert et al. 2017; Starchenko 2013–2017; Starchenko and Pushkarev 2013b). That could be attributed to the cancelations of differently directed internal and somewhat chaotic vector fields **B** during the integration in (34.5).

The observed **M** and especially estimated r_c values in (34.5) are known not so well outside the Earth. For definition, I will further use those values from (Ness 2010; Schubert et al. 2011; Breuer et al. 2015). Then for Jupiter and Saturn we have from (34.5) semi-observed modules of mean fields B = 3 and 1 mT correspondingly with about a few dozen percents confident. The Earth, Jupiter and Saturn have very small inclination of their dipoles to the axis of rotation. No wonder similar dynamo models are successful for those planets (Glatzmaier 1995; Christensen and Aubert 2006b, 2010a; Christensen et al. 2010b; Olson and Christensen 2006; Schubert et al. 2011; King and Buffett 2013; Starchenko and Jones 2002, 2013–2017).

Almost as twin-like Uranus and Neptune also have B $\sim$1 mT, but their dipoles are strongly inclined to the axis of rotation (see Fig. 34.1). That requires some modifications in the dynamo modeling. Ruzmaikin and Starchenko (1989, 1991) and Starchenko (1995) supposed thin-layer kinematic mean-field dynamo in those ice giants. This allows them to make in 1989 successful prediction of the Neptune's dipole inclination. In our millennium Christensen and Aubert (2006b, 2010a), Schubert et al. (2011), Yadav et al. (2013) and others (e.g. see in Starchenko 2014, 2017b) develop more sophisticated and physically grounded models. While some difficulties remain to overcome nevertheless it is seems realistic successfully to model dynamo in all this five planets with unified mean field B $\sim$1 mT which with a few dozen percents confident is following directly from the known observation. Another thing unifying those five planets is their very fast rotation with almost negligible molecular transport effects that is utilized in the previous and next sections.

The sufficient problems for such modeling are posed by too small and asymmetric (with respect to equator) magnetic dipole in Mercury and Jovian moon Ganymede. Please see Fig. 34.1 where only an order of magnitude of the mean field module B is estimated due to large uncertainties in configurations of dynamo regions for those objects (e.g. Christensen 2006a; Schubert et al. 2011; Breuer et al. 2015). An absence of an active dynamo in the modern Venus and Mars is also problematic. Some possible ways to the solutions of those problems will be addressed below.

Magneto-Convection Estimations for the Terrestrial Planets

The Earth's thermal Prandtl number $\sigma \approx 10^{-2}$ and compositional $\mu \approx 10^3$ (Pozzo et al. 2013; King and Buffett 2013; Terasaki and Fischer 2016). Highly uncertain constraints are for r of (1, 4) which should roughly be $|r| < 1$ in the Earth and similar terrestrial planets. This uncertainty barely affects the critical value of (34.4) which is

$$s_{cr} = 1200 \tag{34.6}$$

This enormous value for the critical dissipation/generation ratio indicates that the convection is extremely easily generated in the Earth's core and is highly nonlinear when the real molecular $s \sim 10^{-6}$ of (34.1) is by nine (!) orders of magnitude lower than (34.6). This convection on the small scales can be naturally considered as extremely chaotic or turbulent with huge Reynolds numbers Re = $UD/\nu \approx 10^9$.

For the corresponding turbulent values σ and μ are about 1 on the large scales. If we neglect the thermal contribution assuming $r = 0$, the critical turbulent value from (34.4) will be:

$$s_{cr} = 0.04. \tag{34.7}$$

The similar moderate values are also obtained with all the other probable values of r from (34.1). They are likely to be significantly higher than -1, but rather smaller than 1 in the cores of the similar to Earth planets. However, as far as we approach -1, the non-standard weakly nonlinear or almost critical modes with the practically laminar flows become possible. This extremely strong stabilization (r is slightly above -1) is possible in the interiors of some terrestrial planets. For Mercury, which has an unusual magnetic field, this is partially supported by the numerical model (Christensen 2006a) and asymptotic model (Bassom et al. 2011). The similar situation is also probable on the Jovian satellite Ganymede. Thus, Mercury and Ganymede dynamos should have special type of those 'contradicting' convection that could be supplemented or dominated by strong tidal effects from Sun or Jupiter respectively. For the present Moon, Venus, Mars, and most of the large moons, the r value is probably ≤ -1, and the convection in this case could be not excited at all. On the deep past some non-Earth types of the composition convection could support rather strong MHD dynamo with small $|r|$ in Moon and Mars, while too slowly rotating Venus could never had convection valuable for the dynamo to exist (Breuer et al. 2015; Starchenko 2017b). Thus, for the present Moon, Venus, Mars, and most of the large moons, the mentioned contradiction could suppress the convection at all.

Conclusions

Let me summarize the key results of this work.

1. The observed magnetic dipole of a planet is exactly related to the mean magnetic field vector hidden in the region where MHD dynamo is working. This mean field in a few times exceeds the field extrapolated to the outer dynamo boundary.

The Earth, Jupiter and Saturn have those mean fields of order 1 mT and small inclination of their dipoles to the axis of rotation. No wonder similar dynamo models are successful for those planets. The Uranus and Neptune also have ~ 1 mT, but their dipoles are strongly inclined that require some modifications in the dynamo modeling.

2. Following (Starchenko 2017a) the convection system similar to that taking place in the core of the Earth and similar terrestrial planets is cardinally simplified with new similarity factor δ. This small parameter ($\delta \approx 2 \times 10^{-3}$ for the Earth) is equal to the ratio of the typical hydrodynamic size perpendicular to the rotation axis to the liquid core's thickness. The δ value that can be actually obtained on the present-day computers could successfully replace the Ekman number $E \approx 10^{-15}$, which has been used for one century and which can barely be achieved in the computations in the foreseeable future. Even more extreme compositional Rayleigh number ($\sim 10^{27}$ for the Earth's core) can be replaced in the similar way by the dissipation/generation ratio s suggested by me, which is $\sim 10^{-6}$ for the Earth's core.
3. For various terrestrial type convections, a very large critical value (34.6) of the dissipation/generation ratio s from (34.1) is obtained, while in the Earth's core $s \sim 10^{-6}$. This convection is very easily excited and is turbulent on the small scales. The correspondent turbulent s value on the large scales is limited by (34.3) and its definition (34.2). Thus, turbulent convection is not very much above its critical level, which is perhaps the most important substantiation of the successful 22-year numerical modeling of the geodynamo-type systems. Another important fact is that the turbulent transport coefficients (34.2) are by several orders of magnitude higher than the corresponding molecular values. These two features allow the strongly simplified numerical models to be used for the successful approximation of the planetary and geodynamo-like systems.
4. Domination of the composition convection over thermal one is outlined in the cores of the similar to Earth planets. However, with sufficient contradiction of the thermal effect to convection, the non-standard weakly nonlinear or almost critical modes with the practically laminar flows become possible. This could be in the interiors of Mercury/Ganymede with unusual magnetic fields that can additionally be supported by strong tidal effects from Sun/Jupiter. For the present Moon, Venus, Mars, and most of the large moons, the mentioned contradiction could suppress the convection at all. On the deep past, some non-Earth types of the composition convection could support rather strong MHD dynamo in Moon and Mars, while too slowly rotating Venus could never had convection valuable for the dynamo to exist.

Acknowledgements Author is deeply grateful to the anonymous reviewer and to Prof. Dmitry Sokoloff for their constructive criticism that resulted in sufficient improving of this work. This work was basically done under IZMIRAN budget. The work was also partly supported by the Russian Foundation for Basic Research (project no. 16-05-00507a) and 28th program of the presidium of Russian Academy of Sciences.

References

Aubert J., Gastine T., Fournier A. Spherical convective dynamos in the rapidly rotating asymptotic regime // J. Fluid Mech. V. 813. P. 558–593. 2017.

Bassom, A.P., Soward, A.M., Starchenko, S.V. The onset of strongly localized thermal convection in rotating spherical shells // J. Fluid Mech. V. 689. P. 376–416. 2011.

Breuer D., Rueckriemen T., Spohn T. Iron snow, crystal floats, and inner-core growth: modes of core solidification and implications for dynamos in terrestrial planets and moons // Progress in Earth and Planetary Science. V. 2. No. 39. 2015. https://doi.org/10.1186/s40645-015-0069-y.

Cheng J. S., Stellmach S., Ribeiro A., Grannan A., King E. M., Aurnou J. M. Laboratory-numerical models of rapidly rotating convection in planetary cores // Geophys. J. Int. V. 201. P. 1–17. 2015.

Cheng J.S., Aurnou J.M., Tests of diffusion-free scaling behaviors in numerical dynamo datasets // Earth and Planetary Science Letters. V. 436. P. 121–129. 2016.

Christensen U.R. A deep dynamo generating Mercury's magnetic field // Nature. V. 444. P. 1056–1058. 2006a.

Christensen U.R. Dynamo Scaling Laws and Applications to the Planets // Space Sci Rev. V. 152. P. 565–590. 2010a.

Christensen U.R., Aubert J. Scaling properties of convection-driven dynamos in rotating spherical shells and application to planetary magnetic fields // Geophys. J. Int. V. 166. P. 97–114. 2006b.

Christensen U., Aubert J., Hulot G. Conditions for Earth-like geodynamo models // Earth Planet. Sci. Lett. V. 296. P. 487–496. 2010b.

Davidson P.A. An Introduction to Magnetohydrodynamics, Cambridge Texts in Applied Mathematics, 2006.

Frick P.G. Turbulence: models and approaches. Lecture course (2nd edition in Russian). Perm State Technical University. Part I. Perm. 2010.

Glatzmaier G.A., Roberts P.H. A three-dimensional convective dynamo solution with rotating and finitely conducting inner core and mantle // Phys. Earth Planet. Int. V. 91 (1–3). P. 63–75. 1995.

Jackson A., Jonkers A. R. T., Walker M. R. Four centuries of geomagnetic secular variation from historical records // Phil. Trans. R. Soc. Lond. V. A358. P. 957–990. 2000.

Khramov A.N., Goncharov G.I., Komisarova R.A., et al. Paleomagnetologia. L.: Nedra. 1982. 312 pp.

King E.M., Buffett B.A. Flow speeds and length scales in geodynamo models: The role of viscosity // Earth Planet. Sci. Lett. V. 371. P. 156–162. 2013.

Korte, M., Constable, C.G. Continuous geomagnetic field models for the past 7 millennia: 2. CALS7 K // Geochem. Geophys. Geosys. 6 Q02H16. 2005.

Korte M., Constable C. Improving geomagnetic field reconstructions for 0-3 ka // Phys. Earth Planet. Int. V. 188. P. 247–259. 2011.

Macouin M., Valet J.-P., Besse J. Long-term evolution of the geomagnetic dipole moment // Phys. Earth Planet. Int. V. 147. P. 239–246. 2004.

Matsui H. et al. (37 authors), Performance benchmarks for a next generation numerical dynamo model // Geochem. Geophys. Geosyst. V. 17 (5). P. 1586–1607. 2016.

Ness N.F. Space Exploration of Planetary Magnetism // Space Sci Rev. V. 152. P. 5–22. 2010.

Olson P., Christensen U.R. Dipole moment scaling for convection-driven planetary dynamos // Earth Planet. Sci. Lett. V. 250. P. 561–571. 2006.

Pozzo M., Davies C., Gubbins D., Alfe D. Transport properties for liquid silicon-oxygen-iron mixtures at Earth's core conditions // Phys. Rev. B. V. 87. 014110. 2013.

Reynolds O. On the dynamical theory of incompressible viscous fluids and determination of the criterion // Roy. Soc. London. V. 186. P. 123–161. 1894.

Ruzmaikin A.A., Starchenko S.V. Generation of the large-scale magnetic fields of Uranus and Neptune by turbulent dynamo // Kosmicheskie issledovaniya. V. 27. P. 297–303. 1989.

Ruzmaikin A.A., Starchenko S.V. On the origin of Uranus and Neptune magnetic fields // Icarus. V. 93. P. 82–87. 1991.
Schubert G., Soderlund K.M. Planetary magnetic fields: Observations and models // Phys. Earth Planet. Inter. V. 187. P. 92–108. 2011.
Smagorinsky J. General circulation experiments with the primitive equations I. The basic experiments // Monthly Weather Review. V. 91. No. 3. P. 99–164. 1963.
Starchenko S.V. Generation of nonaxisymmetric magnetic structures in galaxies, the Sun, and planets // Astronomy Reports. V. 72. No 2. P. 257–262. 1995.
Starchenko S.V., Kotelnikova M.S., Maslov I.V. Marginal stability of almost adiabatic planetary convection // Geophys. Astrophys. Fluid Dyn. V. 100. P. 397–428. 2006.
Starchenko S.V. Analytic base of geodynamo-like scaling laws in the planets, geomagnetic periodicities and inversions // Geomagnetism and Aeronomy. V. 54. No 6. P. 694–701. 2014.
Starchenko S.V., Observational estimate of magnetic field and geodynamo parameters under the surface of the Earth's core // Geomagnetism and Aeronomy. V. 55. No 5. P. 712–718. 2015.
Starchenko S.V. Hypothetical parameters of planetary dynamos deduced from direct observations, scaling laws, paleomagnetic and isotope reconstructions // Proceedings of the 11th International School and Conference "Problems of Geocosmos" (Oct. 03–07, 2016, St. Petersburg, Russia), edited by V.S. Semenov et al. P. 64–72. 2016.
Starchenko S.V. Scaling and excitation of combined convection in a rapidly rotating plane layer // (JETP) Zh. Eksp. Teor. Fiz. V. 124. No. 2. P. 352–357. 2017a.
Starchenko S.V. Energy geodynamo parameters compatible with analytical, numerical, paleomagnetic models and observations // Fizika Zemli. No. 6. P. 110–124. 2017b.
Starchenko S.V., Jones C.A. Typical velocities and magnetic field strengths in planetary interiors // Icarus. V. 157. P. 426–435. 2002.
Starchenko S.V., Kotelnikova M.S. Critical stability of almost adiabatic convection in fast rotating and wide spherical shell. JETP (Zh. Eksp. Teor. Fiz.). V. 143. 2. P. 388–396. 2013a.
Starchenko S.V., Pushkarev Y.D. Magnetohydrodynamic scaling of geodynamo and a planetary protocore concept // Magnetohydrodynamics. V. 49. No. 1. P. 35–42. 2013b.
Tarduno J.A., Cottrell R.D., Davis W.J., Nimmo F., Bono R.K.. A Hadean to Paleoarchean geodynamo recorded by single zircon crystals // Science. V. 349. No. 6247. P. 521–524. 2015.
Terasaki H., Fischer R.A. Physics and Chemistry of the Lower Mantle and Core. John Wiley & Sons. pp. 312. 2016.
Yadav R. K., Gastine T., Christensen U. R. Scaling laws in spherical shell dynamos with free-slip boundaries. Icarus. V. 225(1). P. 185–193. 2013.

Chapter 35
MHD Sources, 1600-2005 Evolution and 1900-2005 Probabilistic Time Analysis for Logarithmic Time-Derivatives of Geomagnetic Spherical Harmonics

S. V. Starchenko and S. V. Yakovleva

Abstract MHD sources of geomagnetic spherical harmonics are estimated via logarithmic time derivatives (or exponential rates) of Gauss components basing on IGRF and gufm1 models. 1600–2005 rates' evolutions show long flat 'quiet' and short pick-shaped 'disturbed' fields. Our probabilistic time analysis identifies the probabilities to grow/decay, expected values for grow/decay and median rates together with periodicity/aperiodicity estimates. Total grow/decay variations are ~20 years and median variations are ~100 years with variation as an inverse rate. Since 1900, the quiet field has grow/decay variations ~30 years while disturbed fields' grow/decay variations ~10 years. The median variations are ~60 years for disturbed field and ~200 years for quiet one.

Keywords Time analysis · Geomagnetic variations · Spherical harmonics
Geodynamo

MHD sources of geomagnetic spherical harmonics are estimated via logarithmic time derivatives of Gauss components basing on IGRF and gufm1 models. Those derivatives (or equivalently exponential rates) are determined by averaged velocity gradient and angle between the geomagnetic field and the velocity of fluid in the Earth's core. 1600-2005 evolutions of those rates roughly show about two-century periodicities with somewhat flat 'quiet' and pick-shaped 'disturbed' fields. Rates related to axissymetric harmonics are typically two orders of magnitude smaller than the corresponding tesseral rates. The exclusion is axisymmetric quadrupole

---

S. V. Starchenko (✉) · S. V. Yakovleva
Pushkov Institute of Terrestrial Magnetism, Ionosphere and Radio Wave Propagation, Kaluzhskoe hwy 4, Troitsk, Moscow 108840, Russia
e-mail: sstarchenko@mail.ru

S. V. Yakovleva
e-mail: svyakov@izmiran.ru

D. Nurgaliev et al. (eds.), *Recent Advances in Rock Magnetism, Environmental Magnetism and Paleomagnetism*, Springer Geophysics,
https://doi.org/10.1007/978-3-319-90437-5_35

which's rate was extremely large ~4/year at 1650. This and similar clear non-artificial picks in the other harmonics could be related to a perhaps new jerk-type long-term strongly non-linear geodynamo behavior.

We propose a new probabilistic time analysis in order to identify the probabilities to grow/decay, expected values for grow/decay and median rates together with periodicity/aperiodicity estimates. The total field has relatively long intervals of quiet and short intervals of disturbed field. Those total grow/decay variations are ~20 years and median variations are ~100 years. Hereafter we define a variation as an inverse rate. Since 1900, the quiet field has grow/decay variations ~30 years on 1900–1930, 1945, 1960–1980 and 2000–2005. The typical disturbed fields' grow/decay variations decreases to ~10 years following by abrupt (1940, 1950–1955) or stepwise (1985–1995) return to quiet field. The median variations are ~60 years for disturbed field and ~200 years for quiet one. During quiet interval the field is usually slightly dominated with periodical (or more precisely—half-periodical) behavior, while during disturbed interval—periodicities almost absolutely dominate at the strongest variation ~7 years in 1950 and moderately dominate in the other moderately disturbed cases.

MHD Sources of Geomagnetic Spherical Harmonics

The main potential geomagnetic field $\mathbf{B}_U$ satisfies the well-known Gaussian expansion (Jackson et al. 2000), which is valid for interiors of the Earth from its radius a to the radius of the liquid core c:

$$\mathbf{B}_U = -\nabla U, U = a\sum_{n=1}^{N}\sum_{m=0}^{n}\left(\frac{a}{r}\right)^{n+1}\left[g_n^m(t)\cos m\phi + h_n^m(t)\sin m\phi\right]P_n^m(\cos\theta) \quad |r \geq c. \tag{35.1}$$

Inside the core, we use standard decomposition of the magnetic field on poloidal and toroidal part. The poloidal part $\mathbf{B}_P$ is one and only source for the observed potential field $\mathbf{B}_U$ hereafter. Those fields are equivalent to each other on core-mantle boundary where $\mathbf{B}_U = \mathbf{B}_P$. Similar to (35.1) we represent in standard way corresponding poloidal scalar P as series of spherical harmonics:

$$\mathbf{B}_P = \nabla \times \nabla \times (rP), P = \sum_{n=1}^{N}\sum_{m=0}^{n}\left[G_n^m(t,r)\cos m\phi + H_n^m(t,r)\sin m\phi\right]P_n^m(\cos\theta) \quad |r \leq c. \tag{35.2}$$

Here G_n^m and H_n^m are analogs of Gauss coefficients multiplied by corresponding radial functions.

The radial component of the standard magnetic induction equation equates the time derivative of the radial magnetic field (represented later through the poloidal scalar P and an operator $\hat{L}$ that is the angular part of the Stokes operator multiplied by r^2) to its advection-diffusion MHD source

$$\begin{aligned}\frac{\partial B_r}{\partial t} &= \frac{\partial}{\partial t}\left(-\frac{\hat{L}P}{r}\right) = \hat{\mathbf{r}}\cdot\nabla\times\left(\mathbf{V}\times\mathbf{B}-\frac{\nabla\times\mathbf{B}}{\mu\sigma}\right), \hat{L} \\ &= \frac{1}{\sin\theta}\frac{\partial}{\partial\theta}\left(\sin\theta\frac{\partial}{\partial\theta}\right)+\frac{1}{\sin^2\theta}\frac{\partial^2}{\partial\phi^2}.\end{aligned} \quad (35.3)$$

We substitute expansion of P from (35.2) to the first equation of (35.3). The operator $\hat{L}$ acts on each spherical harmonic from (35.2) exactly returning it timed by known number. Integrating over the sphere $r = const$ the result timed by each individual spherical harmonic, we obtain:

$$\begin{aligned}n\frac{n+1}{r}[\dot{G}_n^m, \dot{H}_n^m] &= \oint\limits_{r=const}[\cos m\phi, \sin m\phi]P_n^m\hat{\mathbf{r}}\cdot\nabla\times\left(\mathbf{V}\times\mathbf{B}-\frac{\nabla\times\mathbf{B}}{\mu\sigma}\right)d^2r \\ &\equiv [\bar{G}_n^m(t,r), \bar{H}_n^m(t,r)].\end{aligned} \quad (35.4)$$

Here in squared brackets are represented values corresponding to cos and sin respectively, while each direct MHD sources are defined in terms of the spherically averaged quantities $(\bar{G}_n^m, \bar{H}_n^m)$.

Then we determine 1/time rates (for more details see (35.7) below) (R_n^m, S_n^m) as logarithmic derivative of (G_n^m, H_n^m) from (35.2) or as the normalized $(\bar{G}_n^m, \bar{H}_n^m)$ from (35.4).

$$\begin{aligned}\left(\frac{\dot{g}_n^m}{g_n^m}, \frac{\dot{h}_n^m}{h_n^m}\right) \cong (R_n^m, S_n^m) \equiv \left(\frac{\dot{G}_n^m}{G_n^m}, \frac{\dot{H}_n^m}{H_n^m}\right) &= \left(\frac{\bar{G}_n^m}{n\frac{n+1}{r}G_n^m}, \frac{\bar{H}_n^m}{n\frac{n+1}{r}H_n^m}\right) \propto \frac{V}{d}s \\ &\approx \left(\frac{F}{c^2}\right)^{1/3} \propto \frac{1}{30\text{ yrs}}.\end{aligned} \quad (35.5)$$

Just below the core-mantle boundary, these rates are equal to the logarithmic derivatives of the corresponding Gauss coefficients. Lower, rough equality (35.5) becomes approximate assuming analogy with (35.1) or more generally—r-independent normalization from the 4th relation of (35.5).

Finally, in (35.5) we estimated the values of the rates through a typical velocity V, hydrodynamic scale d, and sinus s of the angle between $\mathbf{V}$ and $\mathbf{B}$. At the same time, we neglected the effects of magnetic diffusion in comparison with effects associated with the flow velocity. It is seems natural for the core of the Earth, where the magnetic Reynolds number is of the order of 10^3 or more. Rhines (1975) expressed the typical hydrodynamic quantities V and d in terms of the buoyancy flow or the relative convection power F. These expressions for a non-magnetic system rotating rapidly with angular velocity Ω coincided with the scaling laws for MHD systems like geodynamo (Christensen 2010; Starchenko and Pushkarev 2013), and Starchenko (2014, 2015) gives us an estimation of s:

$$V = F^{2/5}c^{1/5}\Omega^{-1/5}, d = F^{1/5}c^{3/5}\Omega^{-3/5}, s = F^{2/15}c^{-4/15}\Omega^{-2/5}. \tag{35.6}$$

These relations and $F = 10^{-14}$ W/kg from (Christensen 2010; Starchenko and Jones 2002; Starchenko and Pushkarev 2013; Starchenko 2014, 2015) give the last two estimations in (35.5). In addition, the following analysis of the observational data permit us to solve the inverse problem—to estimate F directly from the observations, and further to determine V, d and s from (35.6).

Initial Data and Their Exponential Representation

We used the data of the gufm1 model (Jackson et al. 2000) for the period from 1590 to 1990, and IGRF model for 1995–2015. Numbers of the epochs were chosen every 5 years to maintain the IGRF periodicity. The time derivatives of the Gauss coefficients were estimated for a certain epoch taking into account the two subsequent epochs. This was done considering the lower reliability of the old data compared to the new data. Thus, we obtained "the logarithmic derivatives" or ratios of the time derivatives to the corresponding Gauss coefficients from 1590 till 2005. To avoid dividing by a very small number, we approximated linearly adjacent coefficients of those rare too small Gauss coefficients. Following (35.5), we identify the logarithmic derivatives with rates (R_n^m, S_n^m) from (35.5).

We note that the obtained rates have not only relatively narrow physical meaning in accordance with (35.5), as the product of the averaged gradient of the flow velocity to the sinus of the angle between the magnetic field and the velocity. These logarithmic derivatives have much general mathematical meaning. Each positive rate specifies the instantaneous rate of exponential growth, and the negative—the rate of decrease (in absolute value) in accordance with the obvious expressions ($\Delta t_- = 5$ years and $\Delta t_+ = 0$ in our case):

$$g_n^m(t) = g_n^m(t_0)e^{R_n^m(t_0)(t-t_0)}, h_n^m(t) = h_n^m(t_0)e^{S_n^m(t_0)(t-t_0)}, t_0 - \Delta t_- \le t \le t_0 + \Delta t_+. \tag{35.7}$$

Here t_0 is the considered moment of time.

Evolution of Rates for Axisymmetric and Tesseral Harmonics

In order to compare at fixed n minimal spatial concentration with maximal the evolution of the axisymmetric R_n^0 (Starchenko and Yakovleva 2016) and tesseral

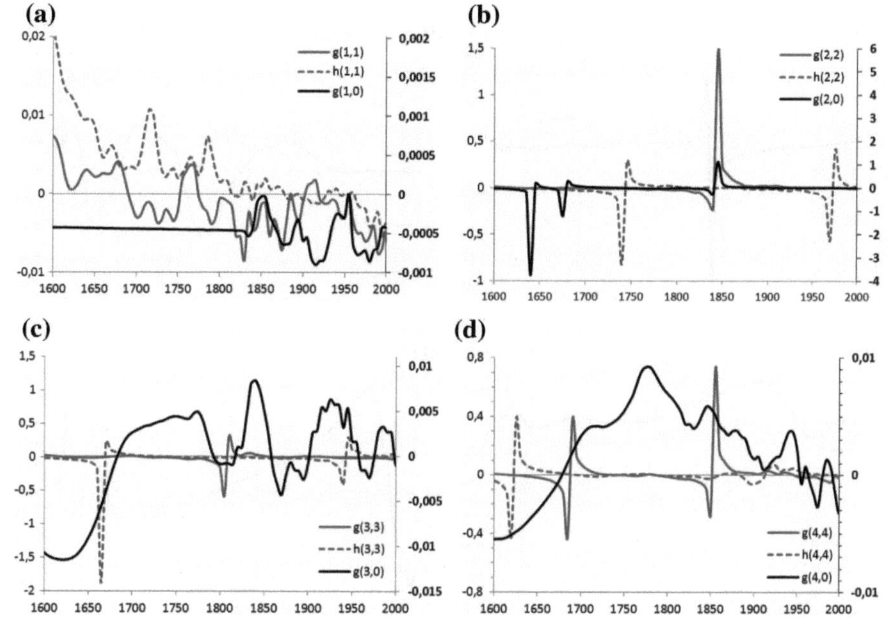

Fig. 35.1 Axisymmetric R_n^0 and tesseral $\left(R_n^n, S_n^n\right)$ rates for n = 1, 2, 3, 4 (a, b, c, d respectively) in terms of 1/year and correspondent Gauss harmonics g and h. Vertical left axis is for tesseral rates, while right one is for axisymmetric rates

$\left(R_n^n, S_n^n\right)$ rates from (5, 7) for n = 1–8 are shown in Figs. 35.1 and 35.2. It shows somewhat correlated peaks with roughly multiple to $\sim$50 and preferably $\sim$200 years periodicities (or more precisely to say—half-periodicities). Double picks with different signs and comparable amplitudes could have somehow artificial nature because they are mainly resulted from passing corresponding Gauss coefficient over the zero. There are also clearly single picks those are hardly could be well defined unfortunately. Significant geomagnetic disturbances could be near those single peaks. Follows (35.5), we can conclude that the physical nature of such disturbances is caused by a significant imbalance and, correspondingly, by increase of s, which is very small in ordinary (Starchenko and Pushkarev 2013; Starchenko 2014, 2015) or 'quiet' intervals, while it could grow up to its maximum ≤ 1 at 'disturbed' intervals (Starchenko and Yakovleva 2016). However, such correlation between Gauss coefficients of different degrees may have a more simple reason. Indeed, the gufm1 model (Jackson et al. 2000) was reconstructed by the methodology that used step-by-step inversions. Each inversion in turn is based on the regularized least square search in the parameter space, which is why the specific spatial data characteristics may lead to the perturbations in all degrees simultaneously. Nevertheless, from Figs. 35.1 and 35.2 we could see the most pronounced peaks around 1650, 1750, 1850 and 1975. The most variable or 'disturbed' are tesseral harmonics with n = 5, 6, 7 and axisymmetric one with n = 2.

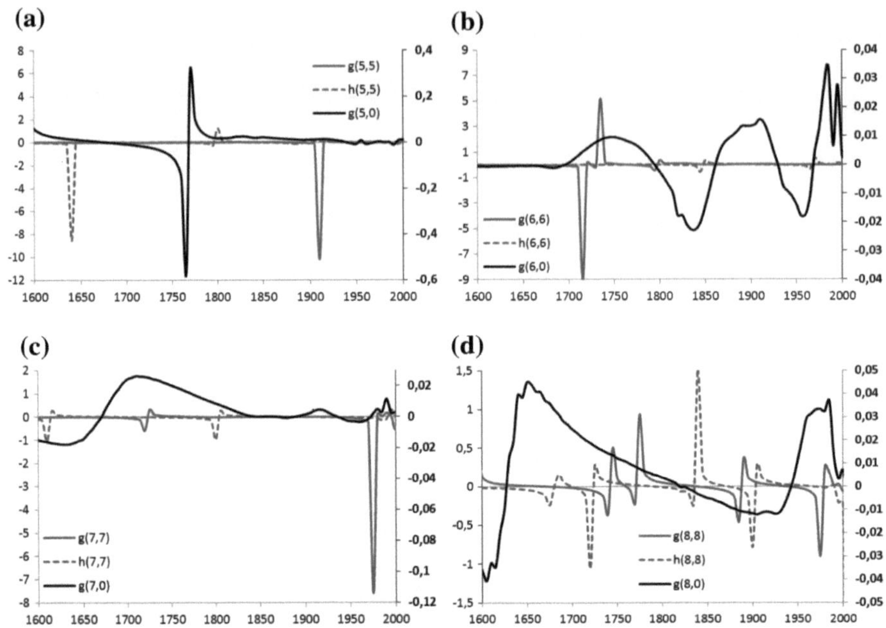

Fig. 35.2 The same as in Fig. 35.1, but for n = 5, 6, 7, 8 (a, b, c, d respectively)

Thus, 1600-2005 evolutions of those rates very roughly show about two-century periodicities with flat ‘quiet’ and pick-shaped ‘disturbed’ fields. Rates related to axisymmetric harmonics are typically two orders of magnitude smaller than the tesseral rates. The exclusion is axisymmetric quadrupole which’s pick rate module was extremely large—4/year at 1650, while it is small $\sim$1/century and with positive rate nowadays. Similar clear non-artificial pick in any harmonic could be related to a new jerk-type long-term strongly non-linear geodynamo behavior.

For more detailed and precise analysis, further we propose our new probabilistic exponential grow/decay time analysis.

Probabilistic Exponential Grow/Decay Time Analysis

Proposed exponential grow/decay time analysis is based on logarithmic time derivative or ratio of time-derivative at time-moment to the data value at this moment as in (35.5) and (35.7). This exponential rate R could take positive or negative value ranging from zero to infinity. To manage with so vast diapason we project R to non-dimension interval 0–1 via semi-obvious exponential transformation:

$$\rho = e^{-Ro/|R|}. \tag{35.8}$$

Here *Ro* is chosen to have $\rho \approx 0.5$ for the average data. For this, we define *Ro* from

$$0.5 = \left(\prod_{k=1}^{K} e^{-Ro/|R_k|}\right)^{1/K} \Rightarrow Ro = K \ln 2 / \sum_{k=1}^{K} 1/|R_k|. \qquad (35.9)$$

Here R_k values are taken from all or just representative part of the given data set.

The next step is to range all *L* available *R*–s from the smallest to largest in two: positive and negative data sets. Now we divide our non-dimension 0–1 interval on *l* parts (we use $l \approx \sqrt{L}$) and define the ***probability density*** (Frick 1999; Tijms 2007) for positive *Dp* and negative *Dn* values of *R* as

$$Dp(i) = D_i^+ \, l/L, Dn(i) = D_i^- \, l/L. \qquad (35.10)$$

Here D_i^+ is the number of positive *R*–s those after (35.8) fell into diapason $(i-1)/l < \rho \le i/l$ with *i* from 1 to *l*, while D_i^- is similar but for negative *R*–s.

Formulation (35.10) allows us to define aperiodicity probability *De* and probability *Pp*/*Pn* to execute positive/negative rate *R* or otherworld—grow/decay probability as

$$De = l^{-1}\sum_{i=1}^{l} |Dp(i) - Dn(i)|, Pp = l^{-1}\sum_{i=1}^{l} Dp(i), Pn = l^{-1}\sum_{i=1}^{l} Dn(i), Pp + Pn = 1. \qquad (35.11)$$

Correspondent ***expected values*** (Frick 1999; Tijms 2007) of ***median*** *Rm* and ***grow/decay*** *Pp*/*Pn* ***rates*** are defined as

$$Rm = \frac{Ro}{l}\sum_{i=1}^{l} \frac{Dn(i) - Dp(i)}{\ln\frac{i-0.5}{l}}, Rp = \frac{-Ro}{lPp}\sum_{i=1}^{l} \frac{Dp(i)}{\ln\frac{i-0.5}{l}}, Rn = \frac{-Ro}{lPn}\sum_{i=1}^{l} \frac{Dn(i)}{\ln\frac{i-0.5}{l}} \qquad (35.12)$$

Thus, this new probabilistic time analysis is proposed in order to identify the probabilities to grow/decay with some rate, expected values for grow/decay and median rates together with periodicity/aperiodicity estimates. In the frame of this new analysis, we could use different statistical hypothesis those are compatible with the analysis formulation (35.8)–(35.12) shown above.

The simplest (while rather rough) statistical hypothesis is to suppose that each individual rate for a fixed moment of time (epoch hereafter) appears in geomagnetic field with equal probability. Thus, using the most reliable data from 1900 we plot in Fig. 35.3 $Pp - Pn$ and *De* 0.5 for each epoch and for all rates supposing that each (*n*, *m*) rate from (35.5) appears each 5 years' moment with equal probability. We chose to plot $Pp - Pn$ because its positive value correspond to prevailing grow

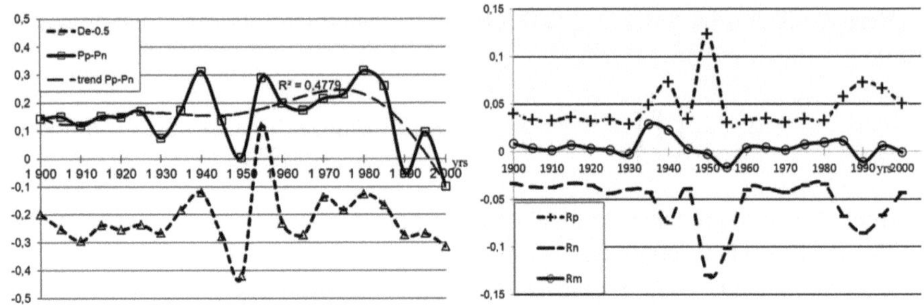

Fig. 35.3 On the left—non-dimensional aperiodicity/periodicity measure $De - 0.5$ and grow/decay measure $Pp - Pn$ from (35.11) are plotted for all the rates (35.5) at each 5-years' epoch. On the right—expected values of median Rm and grow/decay Pp/Pn rates formulated in (35.12) above

tendency, while negative—decay. Maximal value $Pp - Pn = 1$ corresponds to 100% probability for the harmonic module to grow, while minimal -1 corresponds to 100% probability to decay. Maximal value $De - 0.5 = 0.5$ corresponds to kind of absolute aperiodicity, minimal -0.5—absolute periodicity or/and half-periodicity, while $De - 0.5 = 0$ is sort of perfect balance between those two tendencies.

For clear interpretations and definition, let us define the inverse value of any above-mentioned rate as variation with value T in years. Than as in evolutions from Figs. 35.1 and 35.2, the averaged values for total field have relatively long intervals of quiet and short intervals of disturbed field in right panel at Fig. 35.3. Those total grow/decay variations are $\sim$20 years and median variations are $\sim$100 years. The quiet field has grow/decay variations $\sim$30 years on 1900–1930, 1945 and 1960–1980, 2000-2005-(?). The typical disturbed fields' grow/decay variations decreases to $\sim$10 years following by abrupt (1940, 1950–1955) or stepwise (1985–1995) return to quiet field. The median variations are $\sim$60 years for disturbed field and $\sim$200 years for quiet one. During quiet interval the field has a tendency (with some exclusions) to be in balance with slight domination of periodic behavior when $De \approx 0.25$, to see this please compare the right and left panels in Fig. 35.3. During disturbed interval—periodicity almost absolutely dominates for the largest rate pick (variation $T \sim 7$ years) at 1950, while almost perfect balance was in 1940 and moderate periodicity domination at 1990.

To have more detail picture of the difference between quiet and disturbed field let us compare probability density (35.10) for three different couples from disturbed and quiet epoch in Fig. 35.4. Comparing moderately disturbed (1940 and 1990) epoch with quiet preceding (1920) and quiet following (2005) epoch we see that in both cases the highest probability density for disturbed epoch is just moderately larger than for quiet one. Another rather natural difference is that this highest value appears at the lowest variation $T = 7$ years for disturbed epoch and at larger $T = 21$ years for quiet epoch. Besides there are some asymmetry in respect to horizontal $Dp = 0 = Dn$ axis that is especially clear for disturbed 1940-epoch. In contrary, the strongest disturbed 1950-epoch is almost perfectly symmetric at Fig. 35.4. This is in good agreement with the previous analysis based on Fig. 35.3

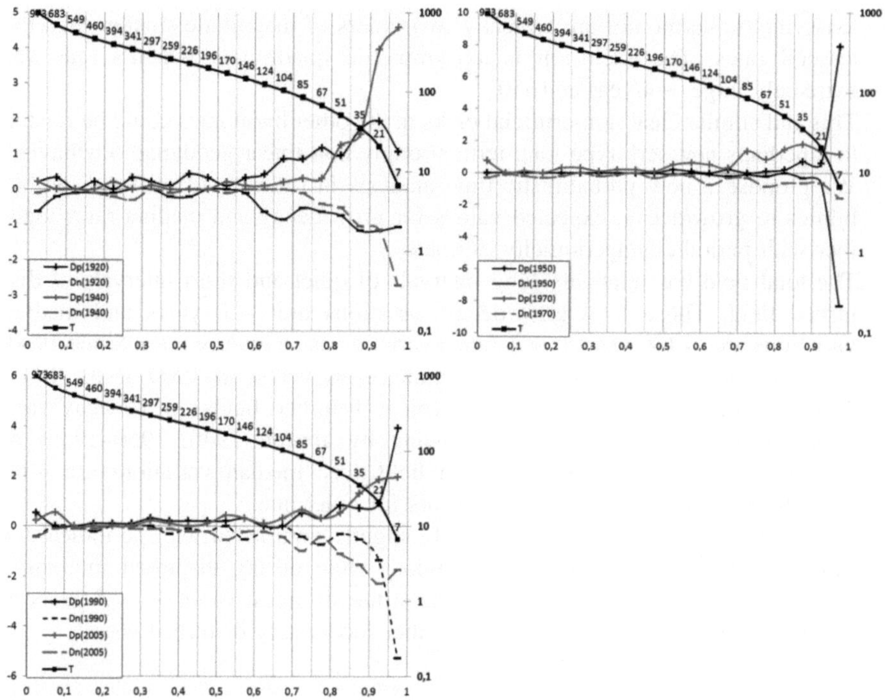

Fig. 35.4 On the upper left—positive part of the diagram contains probability densities *Dp* defined in (35.10) for moderately disturbed epoch at 1940 and preceding quiet epoch at 1920. Negative part contains *Dn*-s from (35.10) taken with negative sign in order to have better representation in terms of symmetry and comparisons. Variations' values *T* are shown on the right vertical axis and by numbers under dark squares for unified Ro = 1/(264 years) from (35.9) defined by 2005 epoch. On the upper right—similar diagram shows strongly disturbed epoch at 1950 and following quiet epoch at 1970. On the lower left—moderately disturbed epoch at 1990 and following quiet epoch at 2005 are shown

meaning almost absolute and short time periodicity for such a strong variation. Again, in contrast to the considered 1990–2005 moderate case we see that the quiet 1970-epoch is rather asymmetric and thus not so well periodical.

Conclusions

1. Natural MHD sources of geomagnetic spherical harmonics are estimated via logarithmic time derivatives of Gauss components basing on IGRF and gufm1 models. Those derivatives (or equivalently exponential rates) are determined by averaged velocity gradient and angle between the geomagnetic field and the velocity of fluid in the Earth's core.
2. Obtained 1600-2005 evolutions of those rates show about two-century periodicities with flat 'quiet' and pick-shaped 'disturbed' fields. Rates related to

axissymetric harmonics are typically two orders of magnitude smaller than the tesseral rates. The exclusion is axisymmetric quadrupole which's rate was extremely large ~4/year at 1650.

3. This and similar clear non-artificial picks in the other harmonics could be related to a perhaps new jerk-type long-term strongly non-linear geodynamo behavior.
4. We propose a new probabilistic time analysis in order to identify the probabilities to grow/decay, expected values for grow/decay and median rates together with periodicity/aperiodicity estimates.
5. The total field has relatively long intervals of quiet and short intervals of disturbed field. Those total grow/decay variations are ~20 years and median variations are ~100 years (we define a variation as an inverse rate). Since 1900, the quiet field has grow/decay variations ~30 years on 1900–1930, 1945, 1960–1980 and 2000–2005-(?). The typical disturbed fields' grow/decay variations decreases to ~10 years following by abrupt (1940, 1950–1955) or stepwise (1985–1995) return to quiet field. The median variations are ~60 years for disturbed field and ~200 years for quiet one.
6. During quiet interval the field is usually slightly dominated with periodical (or more precisely—half-periodical) behavior, while during disturbed interval—periodicities almost absolutely dominate at the strongest variation ~7 years in 1950 and moderately dominate in the other moderately disturbed cases.

Acknowledgements We are very grateful to anonymous reviewer who helped us to improve this work and we even inserted in our text three sentences of him/her. This work was partly supported by the Russian Foundation for Basic Research (project no. 16-05-00507a).`

References

Christensen U.R. Dynamo Scaling Laws and Applications to the Planets // Space Sci Rev. V.152. P. 565–590. 2010.

Frick P.G. Turbulence: models and approaches. Lecture course (in Russian). Perm State Technical University. Part II. Perm. 1999. ISBN 5-88151-193-X.

Jackson A., Jonkers A. R. T., Walker M. R. Four centuries of geomagnetic secular variation from historical records // Phil. Trans. R. Soc. Lond. V. A358. P. 957–990. 2000.

Rhines P.B. Waves and turbulence on a beta plane // J. Fluid Mech. V. 69. P. 417–433. 1975.

Starchenko S.V. Analytic base of geodynamo-like scaling laws in the planets, geomagnetic periodicities and inversions // Geomagnetism and Aeronomy. V. 54. No 6. P. 694–701. 2014.

Starchenko, S.V., Observational estimate of magnetic field and geodynamo parameters under the surface of the Earth's core // Geomagnetism and Aeronomy. V. 55. No 5. P. 712–718. 2015.

Starchenko S.V., Jones C.A. Typical velocities and magnetic field strengths in planetary interiors // Icarus. V. 157. P. 426–435. 2002.

Starchenko S.V., Pushkarev Y.D. Magnetohydrodynamic scaling of geodynamo and a planetary protocore concept // Magnetohydrodynamics. V. 49. No. 1. P. 35–42. 2013.

Starchenko S.V., Yakovleva S.V. MHD sources of multipoles in the Earth's core from 400 years of observations / Proceedings of the 11th International School and Conference "Problems of Geocosmos". P. 109–115. 2016.

Tijms H. Understanding Probability. 2nd edition. Cambridge University Press. 2007.

Chapter 36
Unmanned Airborne Magnetic Survey Technologies: Present and Future

Tsirel Vadim, Parshin Alexander, Ancev Vasily and Kapshtan Dmitry

Abstract The market of existing unmanned aerogeophysical systems is analyzed. The main types of unmanned aerial vehicles and their applicability for solving different geological problems are considered. The comparison of results obtained by different domestic magnetic survey systems at the reference site is presented. In addition, several results obtained by unmanned aeromagnetic survey systems under real-life conditions are also given. The possibility of replacing of the most part of both standard ground and aeromagnetic magnetic surveys by multirotor technologies at sites sized up to several hundred square kilometers are shown. Possible role of other types of Unmanned Aviation Vehicle designed for needs of magnetic survey is also analyzed.

Keywords Magnetic survey · Unmanned aerial systems · Geophysical prospecting

Now, most of Russian easily discovered deposits are greatly exhausted and the perspective to increase substantially our mineral resources base is linked with studies of difficult to access sites. Indeed, most of perspective sites are located in regions with heavy landscape conditions, such as strong landscape ruggedness, dense bushes, water obstacles etc. These factors make it difficult and expensive for the performance of standard ground surveys (geophysical ones in particular). The same is true for the standard aerogeophysical surveys which in addition do not

T. Vadim (✉)
Geologorazvedka JSC, Saint-Petersburg, Russia
e-mail: info@geolraz.com

P. Alexander
Institute of Geochemistry, SB RAS, SibGIS, LLC, Irkutsk National Research Technical University, Irkutsk, Russia

A. Vasily
Radar mms, JSC, Saint-Petersburg, Russia

K. Dmitry
Geoskan Ltd, Saint-Petersburg, Russia

D. Nurgaliev et al. (eds.), *Recent Advances in Rock Magnetism, Environmental Magnetism and Paleomagnetism*, Springer Geophysics,
https://doi.org/10.1007/978-3-319-90437-5_36

always provide necessary details in heavy landscape conditions. Besides, modern situation in the geological industry is characterized by a decrease of budget financial support and reorientation for investments from private companies by means of distribution of pioneer licenses to real economy enterprises to exploit insufficiently studied regions. The private capital encourages the introduction of new geological prospecting methods which allow to decrease significantly expenses and at the same time to increase geological survey operational efficiency. The magnetic survey is the most important one as a geologically universal prospecting method. It is important to note that modern geological survey requirements mean the development of technologies aiming at investigate remote areas not only quickly and cheaply but also supplying more exact data already in the process of the first prospecting survey stages. Moreover, the very term «stages of geological prospecting» undergoes now substantial changes.

Now, the use of Unmanned Aviation Vehicle (UAV) technologies is rapidly developing in the world. These technologies may perform survey at lower altitudes than those accepted by standard aerogeophysics without risk for members of vehicle team. At the same time UAV technologies increase the quality of the geophysical survey compared to the standard aerogeophysics data and reduces the workload of the ground survey. The use of unmanned systems creates other problems for aero geophysical system designers because traditional aerosensors complying with interference elimination methods are not applicable to lightweight UAV. Our exploring experience of such creation construction technologies and exploitation makes it possible for the authors to express their own view on how UAV-technologies would change the usual structure of geological prospecting works, explaining questionable conclusions by using their own obtained results.

Methods and Technologies

Three main UAV types are known: aircraft, helicopters and multicopters. Each type is characterized by its own advantages and disadvantages relating to geological survey problems.

UAV technologies may change the accepted way geological exploration and prospecting activities as demonstrated by the results of our created UAV complexes. The problem of each model type applicability has been considered in literature for a long time (Kroll 2013). Short information on foreign executed surveys is given in the paper (Wood et al. 2016).

Their opinion of the qualities of different varieties of UAV bearers for geophysical surveys, the authors already expressed (Parshin et al. 2017). We can see that the main advantage of helicopter and airplane systems, including UAV with Internal Combustion Engine (ICE), as electric ones do not significantly exceed copters on flight time and are significantly inferior in load mass, they have a longer flight time compared to multicopters. Besides, ICE planes require a take off strip availability that significantly limits its applicability under heavy landscape conditions. The main problem of combustion engine vehicles is the ignition system high together with electromagnetic noise level and significant vibrations. Though helicopters can theoretically fly following the landscape view, their maneuver level is lower than that of multirotor UAVs. The main drawback of multirotor UAV is the shortness of their flight time (15–40 min).

Fixed-Wing UAVs

Magnetic survey systems, most used in geology, appeared first in the world practice since 2011 (Semenova and Tsirel 2016). The first Canadian magnetic survey systems GeoSurv II and Venturer based on rather big unmanned planes are actually analogous to the standard aeromagnetic survey but they have substantially greater economic efficiency due to expense decrease on flight hours compared to big aviation. They also require the presence of significantly smaller takeoff strips and can be delivered to the work region by trucks. Considering foreign survey experience we can confirm that the efficiency of such systems in Russia may be rather high, for example, in the process of magnetic survey work in Yakutia regions which are favorable to diamond or raw hydrocarbon deposit surveys.

The main advantage of unmanned plane survey is a large flight time and accordingly possibility of investigation of significant areas from the takeoff point. The main disadvantages are impossibility of low speed flight and detailed following over the landscape, takeoff and landing areas preparation requirement. The important problem of further development of fixed-wing unmanned aerial systems is the complexity of realization of geophysical methods different from magnetic surveys. For example, the creation of effective gamma-ray spectrometry survey technology within UAV-plane model with the use of usually applicable sensors is hardly possible.

The fixed-wing aeromagnetic Unmanned Aviation System (UAS) development in Russia have received the State support since 2014–2016, within the Federal Program "Investigations and Surveys on priority areas of science and technology". The project works were performed by V.I. Vernadskiy State Geological Museum RAS and Geoskan Group of Companies (Cherkasov et al. 2016). This construction has not yet reached industry application, but many project innovation were used by Geoscan Company when performing UAV multirotor magnetic survey system which will be described further.

The creation of electric engine magnetometric airplanes with such motors and a light magnetometer arouses much interest because keeping rather long flight time combined with low vehicle cost and cheap exploitation, make it very attractive, but as the far as authors know, such constructions have not appeared at the market yet.

Helicopter Systems

Helicopter-based systems are characterized by a rather long flight time and in addition are able to perform rather complicated flight missions and are able to hover in flight. They can hover above appointed points. The authors have not found any information concerning such helicopter-based geological technologies examples in the foreign practice, that, appearangly is connected with sophisticated creation and exploitation of UAV helicopters, particularly in heavy conditions of geological expeditions. However, the specialists of JSC "Radar-mms" have created a magnetometric complex, including unmanned helicopter with on-board magnetometric system. The helicopter is equipped with the combustion engine and provides the achievement of the following main characteristics: carrying capacity up to 10 kg, flight time up to 2 h. A sufficiently large area can be surveyed at course speed from 40 to 60 km/h. The magnetometric system created by the authors consists of two magnetometers, to measure the total intensity of geomagnetic field with a four-barrel optically pumped quantum magnetometer and second fluxgate magnetometer used for getting data about magnetic interference from ICE, which further will be eliminated by software. The comparative results of survey by this system with pedestrian survey results in the reference site show that described magnetometric system provides correct measurements (Parshin et al. 2017).

Multirotor Systems

Multirotor geophysical UAV technologies are developed rather inadequately in the world practice, they appeared later than airplane models but now they are being developed rather rapidly. Apparently, such delay is associated with two main factors: relatively recent appearance of brushless electric motors and that up to a certain time some specialists believed that magnetometric systems manufacture based on multicopters is impossible because such magnetometric systems are characterized by noise character different from the standard aeromagnetic survey (and ICE unmanned airplanes) and so accordingly require new methods of their elimination. So far only two recent foreign constructions are known: SkyLance (Cunningham 2016) and UAV-MAG (Pioneer Exploration 2017). The first mentioned system, obviously, has not passed any significant approbation yet. The results discovered by the other system are notably inferior to pedestrian surveys in survey details and exactness (Pioneer Exploration 2017), because, first of all this survey was performed at an altitude of 45 m accessible for standard aerial surveys, and only standard UAV are used in that case. A good convergence with standard aerial survey is noted, for which reason this system can be considered as a technology effectively replacing the standard aerial survey when performing in small areas (up to 2000 linear kilometers), in fact, as it is positioned by the manufacturer. The survey cost is about 100$ per linear kilometer.

The Russian level of multirotor system development is in step with the world trend—anyway, at least two of three similar native developments appeared in Russia not later than abroad: magnetic survey technology of IPGG SB RAS, Novosibirsk (Firsov et al. 2015), SibGIS UAS technologies now, including magnetic survey technologies, gamma-spectrometry, multispectral photographic survey (Parshin 2015; Parshin and Morozov 2016) and a little later appeared a system for magnetic survey manufactured by Geoskan Company (the latest two models are shown in the Fig. 36.1). The authors of the first mentioned system have not yet represented any results of geological approbations in literature.

The second mentioned system, SibGIS UAS was initiatively developed for replacement of the traditional pedestrian surveys at the scale 1:10,000 and larger, in the most heavy landscape of natural and expedition conditions. That's why it includes not only UAV with geophysical sensors but technologies providing landscape modeling. The software for automated creation of flight tasks coordinated with the landscape of any complexity is an important part of the system. Now, two variants of magnetometric channels are used within the system—proton Overhauser magnetometer created on the basis of OEM-magnetometer POS-1 components (Sapunov et al. 2015) and fluxgate magnetometer, developed together with the research and development Centre "Specter" (Ufa, Russia). The second magnetometer model gives less exact results, but it is much lighter (400 g against 2 kg proton) and oriented first of all on the performance of modern magnetometric and gamma-spectrometric surveys on single UAV. In construction of the unmanned vehicle the whole complex of engineering solutions allowing minimization of the influence of magnetic and vibration interference to which this magnetometer is rather sensitive. Particularly, the suspension as visible in the Fig. 36.1b possessing a length which is more than 3 m. It's made of special material and designed for

(a) (b) (c)

Fig. 36.1 SibGIS UAS with fluxgate (**a**) and proton overhauser (**b**) magnetometers (Parshin et al. 2016a, b) and Geoscan UAS (**c**)

sensor remoteness from motor power lines and antennas of UAV equipment, at the same time it damps vibrations and prevents the sensor rotation. Standard deviation of obtained survey data of the scale 1:10,000 in real conditions are within 2 nT that allows to detect magnetic field anomalies of 3–4 nT (Parshin et al. 2016a, b). The accurate landscape modeling with the photogrammetry technology provides the possibility of significant decrease of measurement altitude (up to 5 m above vegetation). The surveying in landscape with flow mode, application of special construction UAV with reduced magnetic interference allow for achieving data quality compared to large-scale pedestrian survey in real conditions of field woks in rugged terrain (Fig. 36.2).

The example shown in Fig. 36.2 proves that data received by SibGIS UAS unmanned system in complex natural conditions exceeds the quality of pedestrian surveys, where some disadvantages appear. In long-time geological expeditions, when performing walk-mag works such disadvantages are rather typical. They can be caused by a human factor (for example, a metal spoon forgotten by the operator in the pack), malfunction of the magnetometer or accidental movement or malfunction of base station magnetometer system. Unfortunately, such significant disadvantages become visible rather often but only after field works are over, at the stage of final data processing when the additional survey is impossible. Such survey disadvantages may not influence directly the geological task showing, besides, that superiority of data quality obtained by SibGIS UAS system is free of doubt and objective—this automated system is at least free of the influence of human factor on measurement quality. Besides, due to high reliability and low costs of the airmobile part of this system and survey effectiveness increase compared to the pedestrian shows that one the work cost decrease in 1.5–2.5 compared to traditional pedestrian survey at the same scale. Now, the magnetic survey of one linear kilometer costs with the help of SibGIS UAS system approximately 35$ including base data processing (for reference, pedestrian survey cost in the same conditions is 85$, survey cost with MAG^{tm} UAV system—100$). The effective flight speed is 8–12 m/s under conditions of mid-mountain relief when performing magnetic survey.

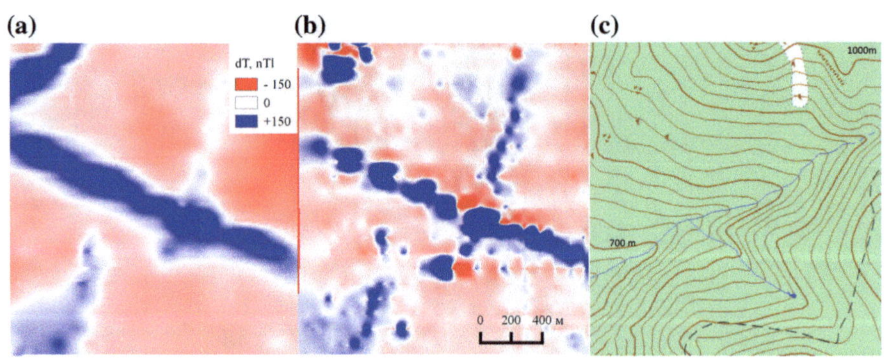

Fig. 36.2 Comparison of results of UAV magnetic survey by SibGIS UAS system (**a**) in comparison with walk-mag survey data (**b**), and information about landscape conditions (**c**)

The flight time under real conditions of mountainous area of Eastern Siberia is approximately half-hour (with proton magnetometer), for which reason it is more effectual to perform area survey from one takeoff point up to the first dozens square kilometers, the takeoff point is changed after survey of such site. During one flight mission, it's possible to perform up to 1 sq. km. survey at the scale of 1:10,000. That UAS is used in a wide range of environment conditions including negative temperatures and slight snowstorms, now, this system is successfully used when performing magnetic surveys for gold and uranium deposits in mountain areas of Irkutsk Region, Buryatia and Transbaikalia, some of which are extremely heavy to be effectively explored by the pedestrian method. The results received by SibGIS UAS system, in the authors' opinion, prove that the most part of pedestrian geophysical survey when performing prospecting and evaluation works at the scale 1:10,000–1:1000 can be replaced by multirotor technologies.

Geoskan Unmanned System (Fig. 36.1c) represents a conversion of original airplane magnetometric technology. A pseudoplane—a pole with airplane tail unit, where all the electronics of the magnetometric system and its satellite spatial binding system are located, is attached to the multirotor UAV with the help of long ropes. The tail unit is designed for stabilization of magnetometric sensor orientation in space and the decrease of orientation interference. However this system has no possibility to perform the survey considering the relief. Besides, its maneuverability necessary for the performance of survey at a fixed altitude above strongly dissected relief, may significantly decrease because of the long suspension which will allow the sensor swinging and deviate from the required survey profile when maneuvering or being under wind gusts. However, geospatial reference of obtained data will be correctly executed because GPS-antenna is located on pseudoplane as well as the magnetometer sensor.

Results of Demonstration Works at the Chernorud Site (Baikal Natural Area, Russia)

To a certain degree we can estimate the readiness of the completeness of described magnetic survey systems compared to industrial geophysical surveys on the base of the results of demonstration works at Chernorud site (Baikal Nature Territory, Russia) that was performed in 2017. But it is necessary to note that the aim of these performances was not to assess metrological and industrial possibilities of these systems and their comparison but geological specialist familiarization with new tendencies in the sphere of geological survey within the scientific and technical conference held there. So an easily accessible area directly adjacent to this base was used for demonstration flights, to note that it was fully suitable as a reference site because in connection with unique geological conditions of Baikal Region as well as significant local relief forms with a height of more than 20 m, are characterized by the presence of small relief structures with altitudes up to 10 m. The small morphological structures are not fixed on topographic maps and global digital relief models though they substantially influence magnetometric surveys at low altitudes. So, it was impossible to create flight tasks for UAV considering these landscape features without performing preliminary topographic works. That is why when performing a survey by SibGIS UAS system they were not considered, the flight was performed with constant altitude retention over the satellite digital relief model SRTM 1-Arcsecond which differs from actual relief in different parts of the reference site up to 5–15 m that additionally increased the error keeping the absolute altitude of sensor above the ground. As a result, it is impossible to say that the survey of this area performed by the SibGIS UAS system is methodologically correct and providing correct high quality data compared to the above-mentioned survey data fragment based on real industrial works, Fig. 36.3.

The surveys with the RadarMMS and Geoscan unmanned aerial systems were performed without any relief consideration, at the fixed altitude above takeoff point, because these systems did not include instruments for generation of flight missions considering digital terrain models. Besides, the profiles based on different measurements were performed with the help of various systems which also did not coincide and survey scales were different: the survey was performed by SibGIS UAS and RadarMMS systems at the scale 1:5000; the survey by the Geoscan UAS was performed considering a step of 30 m between the profiles. It is necessary to note that the above mentioned small relief structures affect less the measurement correctness by Geoskan and RadarMMS UAS than SibGIS UAS data because where they are localized the sensor height of the first two systems above ground was significantly greater than the third one. Taking into account the error of the digital model the error of keeping SibGIS UAS sensor altitude above the actual relief could be at the level of 50%. In absolute terms this sensor was at the altitude of 24–45 m above the ground. The survey results are specified in the Fig. 36.3. Due to the fact that measurements were performed with different spatial density ordinary Kriging interpolation was performed for perception improvement, parameters for each data set were chosen by automatic mode.

Despite described methodological disadvantages of the surveys, it follows (based on Fig. 36.3) that all magnetic survey systems detect any alterations of total intensity of earth's magnetic field (T). The general alteration trend of T including the increase of its values from the north-west to the south-east of the site is visible in data obtained by all systems. SibGIS UAS and Geoskan additionally fixed even more smaller field details, for example, positive anomaly in the northwestern part of the site. It's more vivid within Geoskan data because SibGIS UAS system survey covers only the edge of this anomaly and moreover, its altitude in this place was approximately 35 m (comparing to 20 m of Geoskan, Fig. 36.3d). At the same time in the southeastern part of the area SibGIS UAS system fixes local field anomalies in more details because its sensor nominal altitude above the relief is the same 35 m (against 40–60 in this place of the area with Geoskan and RadarMMS systems). In general, as relief difference in elevation is small at the reference site and located within the range of 40 m, the survey by the Geoskan and RadarMMS unmanned

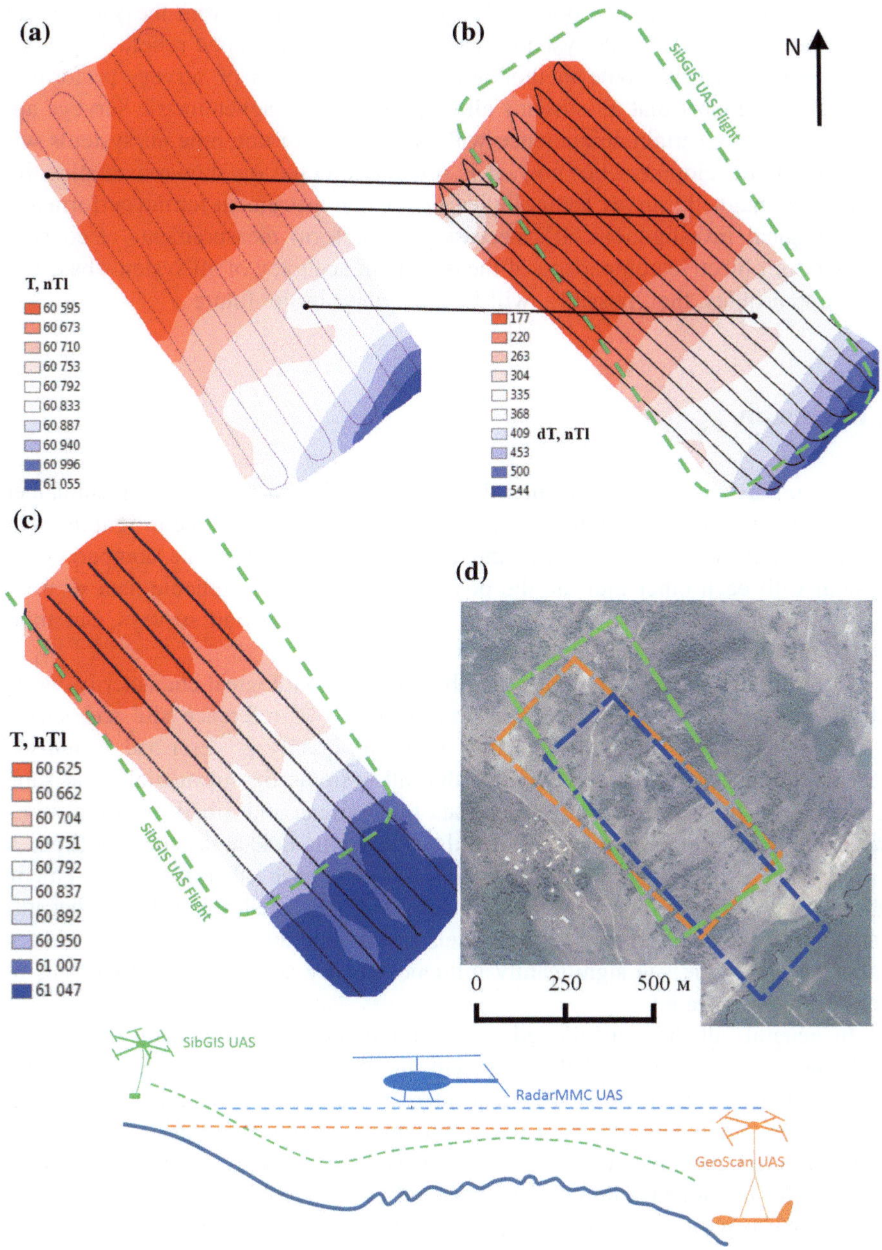

Fig. 36.3 Demonstration work results at Chernorud area: **a** SibGIS UAS system, sensor altitude above ground 35 ± 10 m (green). **b** Geoskan UAS (orange). **c** Radar MMS UAS (blue). **d** Flight Scheme in plan and in crossection and site topographic map

systems at constant altitude does not differ much in details from the survey data performed by SibGIS UAS system, made according to assumed relief.

In general, despite some methodological disadvantages of surveys performed, there are no doubts that all systems allow performing magnetometric surveys and characterized by certain coincidence with each other. In due time all systems will continue its development but now, it is obviously based on shown examples that unmanned magnetic survey technologies are the real instrument for the performance of geological investigations. It's necessary to perform measurements at constant altitude for objective comparison of measurement accuracy of all systems based on the same profiles, that, possibly, will be performed in the future.

Conclusion

In the authors' opinion, the results described in this article show that unmanned systems give new opportunities for geologists and it is impossible to ignore them. Shown results of methodological and industrial surveys and the comparison of system with each other indicate the possibility of obtaining reliable data with the help of unmanned technologies, equal to standard pedestrian surveys, but more economically effective and with a better productivity.

Unfortunately, due to demonstrative but not methodological character of works at Chernorud site, the not quite successful choice of reference site and survey methods, represented data do not allow judge on quality of described systems in real conditions, however it is obvious that all systems fix substantially the same geomagnetic field scheme and survey data by Geoscan and SibGIS UAS systems, which allows marking more precise details of magnetic field at the site. For accurate comparison of data quality of different systems with each other, the measurements should be performed on constant altitude and on equal flight tasks, because insufficiently accurate digital relief models in total including the presence of local relief excess at the site significantly influence survey accuracy, performed taking into account the relief.

In general, all three mentioned types of unmanned aeromagnetic prospecting technologies (fixed-wing, helicopter and multirotor) may be required by the industry that is confirmed by the great interest and even by the existence of company orders (including small business and small scientific groups) for which traditional aerial methods were previously inaccessible. Such groups of customers receive the instrument allowing quickly and cost-effectively investigate territory sites which it was impossible to explore by the standard pedestrian survey.

We think that above mentioned systems based on unmanned planes can mostly replace classic aeromagnetic survey including the increase of economical efficiency (according to foreign sources, the gain is 5 fold). This technology could enter the list of large service companies, performing state orders. In the process of performance of detailed surveys (scale of 1:1000 to 1:10,000) especially in poor relief conditions, the most effective variant is the multirotor one, which can be considered as a perspective replacement for walk-mag surveys and also aerial surveys in areas up to the first hundreds square kilometers, i.e. when performing works at small licensed sites when expenses for big aviation are not smoothed by its possible lower cost. The main advantage of multirotor systems is the ability of performance of high-detailed survey with accurate relief flow in comparison with UAV-airplanes and UAV-helicopters. Economic efficiency increase and performance efficiency are especially significant compared to the traditional pedestrian survey. Considering continuous parameter improvement of lithium-ion polymer battery we can expect a constant flight time improvement—for previous four years typical flight time of light hobby multicopters increased from 10–15 to 17–25 min, more professional solutions provide flight time up to 1 h.

Helicopter systems compared to copters have a larger prime-cost, that implies somewhat upper flight altitudes and accordingly somewhat smaller survey scale. Multirotor and helicopters applicability border, in authors opinion, is situated at border scale 1:10,000 and licensed sites of 100 km^2, and also depends on relief complexity, vegetation character and possibility of being closer to the survey place. Significantly longer flight time of combustion engine helicopters compared to multicopters allow to suppose that UAV helicopter model is a possible substitution of traditional helicopter and airplane airborne geophysics under conditions of simple relief.

We can state that in the sphere of multirotor and helicopter technologies at the Russian level and experience do not cede to the foreign ones. So, despite that unmanned technologies have not taken any significant market place yet in Russia, the authors are sure that discovered results indicate the possibility of most part pedestrian works and classical aerial surveys replacement by unmanned technologies. Unmanned geological and geophysical methods should be considered as a real instrument of geological investigations efficiency increase. A poorly developed normative base for UAV application in Russia is a significant obstacle to their general introduction in geological practice. Nevertheless, the authors are sure that unmanned technologies will significantly change the geological prospecting methodology in the nearest future.

References

Cunningham, M. (2016) Aeromagnetic surveying with unmanned aircraft systems // A thesis of Master of Science Diss., Carleton University, Canada, Ontario, 144 p.

Cherkasov, S., Sterligov, B., Semenov, A. (2016) Use of UAVs to geological prospecting. Aeromagnetic survey // Materials of the seminar "Geoscan technology application for remote sensing of land and monitoring of objects", https://files.geoscan.aero/public/seminar/Cherkasov.pdf (in Russian).

Firsov, A., Zlygostev, I., Dyadkov, P. et al. (2015) Application of a high-frequency magnetometer for lightweight UAVs at a geological and geophysical study of kimberlitic tubes // Mat. Conf. Interexpo-GeoSibir, Novosibirsk, 299–304 (in Russian).

Kroll, A. (2013) Evaluation of an unmanned aircraft for geophysical survey//23rd International Geophysical Conference and Expedition, Melbourne, Australia, 1–4.
Parshin, A. (2015) Prospects of using drones for geological exploration for on the ore deposits at Baikal mountainous area // Problems of Natural Sciences 6, 97–101 (in Russian).
Parshin, A., Bydyak, A., Blinov, A., Kosterev, A. Morozov, V., Mikhalev, A., Prosekin, S., Tarasova, Yu., Spiridonov, A. (2016a) Low-altitude unmanned aeromagnetic survey in management of large-scale structural-geological mapping and prospecting for ore deposits in composite topography. Part 1 // Geography and Natural Resources 6, 144–149 (in Russian).
Parshin, A., Bydyak, A., Blinov, A., Kosterev, A. Morozov, V., Mikhalev, A., Prosekin, S., Tarasova, Yu., Spiridonov, A. (2016b) Low-altitude unmanned aeromagnetic survey in management of large-scale structural-geological mapping and prospecting for ore deposits in composite topography. Part 2 // Geography and Natural Resources 6, 150–155 (in Russian).
Parshin, A., Morozov, V. (2016) SibGIS Flight Planner. Software patent for computers №2017615422.
Parshin, A., Tsirel, V.,Antsev, V. (2017) Present and future of unmanned aerogeophysics in Russia // Prospect and protection of mineral resources 8, 33–38 (in Russian).
Pioneer Explorations LTD: UAV-MAG - The leader in Unmanned Geophysics Surveying (2017) http://pioneeraerialsurveys.com/Pioneer_Aerial_Surveys_Promo_2017.pdf.
Sapunov, V., Narkhov, E., Fedorov, A., Sergeev, A., Denisov, A. (2015) Ground overhauser DNP geophysical devices // International Multidisciplinary Scientific GeoConference Surveying Geology and Mining Ecology Management. SGEM, 215–222.
Semenova, M.P., Tsirel, V.S., (2016) Development prospects unmanned air geophysics // Prospect and protection of mineral resources 8, 34–39 (in Russian).
Wood, A., Cook, I., Doyle, B., Cunningham, M., & Samson, C., (2016) Experimental aeromagnetic survey using an unmanned air system // The Leading Edge 35, 270–273.

MIX
Papier aus verantwortungsvollen Quellen
Paper from responsible sources
FSC® C105338

If you have any concerns about our products,
you can contact us on
ProductSafety@springernature.com

In case Publisher is established outside the EU,
the EU authorized representative is:
Springer Nature Customer Service Center GmbH
Europaplatz 3, 69115 Heidelberg, Germany

Printed by Libri Plureos GmbH
in Hamburg, Germany